Definitions, units and conversion factors

In the SI, units are given for both defining relations of the magnetization, $\boldsymbol{B}=\mu_0(\boldsymbol{H}+\boldsymbol{M})$ and $\boldsymbol{B}=\mu_0\boldsymbol{H}+\boldsymbol{M}$, respectively. $\mu_0=4\pi\cdot10^{-7}\,\mathrm{Vs\,A^{-1}\,m^{-1}}$, A: molar mass, ϱ: mass density, $\boldsymbol{P}$: magnetic moment, $\boldsymbol{M}$: magnetic moment per unit volume (magnetization, magnetic polarization).

Quantity	cgs/emu	SI	
$\boldsymbol{B}$	$\mathrm{G}=(\mathrm{erg\,cm^{-3}})^{1/2}$	$\mathrm{T}=\mathrm{Vs\,m^{-2}}$	
	$1\,\mathrm{G}\hat{=}$	$10^{-4}\,\mathrm{T}$	
$\boldsymbol{H}$	$\mathrm{Oe}=(\mathrm{erg\,cm^{-3}})^{1/2}$	$\mathrm{Am^{-1}}$	
	$1\,\mathrm{Oe}\hat{=}$	$10^3/4\pi\,\mathrm{Am^{-1}}$	
$\boldsymbol{M}$	$\boldsymbol{B}=\boldsymbol{H}+4\pi\boldsymbol{M}$	$\boldsymbol{B}=\mu_0(\boldsymbol{H}+\boldsymbol{M})$	$\boldsymbol{B}=\mu_0\boldsymbol{H}+\boldsymbol{M}$
	G	$\mathrm{Am^{-1}}$	T
	$1\,\mathrm{G}\hat{=}$	$10^3\,\mathrm{Am^{-1}}$	$4\pi\cdot10^{-4}\,\mathrm{T}$
$\boldsymbol{P}$	$\boldsymbol{P}=\boldsymbol{M}V$	$\boldsymbol{P}=\boldsymbol{M}V$	$\boldsymbol{P}=\boldsymbol{M}V$
	$\mathrm{G\,cm^3}$	$\mathrm{Am^2}$	$\mathrm{Vs\,m}$
	$1\,\mathrm{G\,cm^3}\hat{=}$	$10^{-3}\,\mathrm{Am^2}$	$4\pi\cdot10^{-10}\,\mathrm{Vs\,m}$
σ	$\sigma=M/\varrho$	$\sigma=M/\varrho$	$\sigma=M/\varrho$
	$\mathrm{G\,cm^3\,g^{-1}}$	$\mathrm{Am^2\,kg^{-1}}$	$\mathrm{Vs\,m\,kg^{-1}}$
	$1\,\mathrm{G\,cm^3\,g^{-1}}\hat{=}$	$1\,\mathrm{Am^2\,kg^{-1}}$	$4\pi\cdot10^{-7}\,\mathrm{Vs\,m\,kg^{-1}}$
σ_m	$\sigma_\mathrm{m}=\sigma A$	$\sigma_\mathrm{m}=\sigma A$	$\sigma_\mathrm{m}=\sigma A$
	$\mathrm{G\,cm^3\,mol^{-1}}$	$\mathrm{Am^2\,mol^{-1}}$	$\mathrm{Vs\,m\,mol^{-1}}$
	$1\,\mathrm{G\,cm^3\,mol^{-1}}\hat{=}$	$10^{-3}\,\mathrm{Am^2\,mol^{-1}}$	$4\pi\cdot10^{-10}\,\mathrm{Vs\,m\,mol^{-1}}$
χ	$\boldsymbol{P}=\chi\boldsymbol{H}$	$\boldsymbol{P}=\chi\boldsymbol{H}$	$\boldsymbol{P}=\chi\mu_0\boldsymbol{H}$
	$\mathrm{cm^3}$	$\mathrm{m^3}$	$\mathrm{m^3}$
	$1\,\mathrm{cm^3}\hat{=}$	$4\pi\cdot10^{-6}\,\mathrm{m^3}$	$4\pi\cdot10^{-6}\,\mathrm{m^3}$
χ_v	$\chi_v=\chi/V$	$\chi_v=\chi/V$	$\chi_v=\chi/V$
	$\mathrm{cm^3\,cm^{-3}}$	$\mathrm{m^3\,m^{-3}}$	$\mathrm{m^3\,m^{-3}}$
	$1\,\mathrm{cm^3\,cm^{-3}}\hat{=}$	$4\pi\,\mathrm{m^3\,m^{-3}}$	$4\pi\,\mathrm{m^3\,m^{-3}}$
χ_g	$\chi_\mathrm{g}=\chi_v/\varrho$	$\chi_\mathrm{g}=\chi_v/\varrho$	$\chi_\mathrm{g}=\chi_v/\varrho$
	$\mathrm{cm^3\,g^{-1}}$	$\mathrm{m^3\,kg^{-1}}$	$\mathrm{m^3\,kg^{-1}}$
	$1\,\mathrm{cm^3\,g^{-1}}\hat{=}$	$4\pi\cdot10^{-3}\,\mathrm{m^3\,kg^{-1}}$	$4\pi\cdot10^{-3}\,\mathrm{m^3\,kg^{-1}}$
χ_m	$\chi_\mathrm{m}=\chi_\mathrm{g}A$	$\chi_\mathrm{m}=\chi_\mathrm{g}A$	$\chi_\mathrm{m}=\chi_\mathrm{g}A$
	$\mathrm{cm^3\,mol^{-1}}$	$\mathrm{m^3\,mol^{-1}}$	$\mathrm{m^3\,mol^{-1}}$
	$1\,\mathrm{cm^3\,mol^{-1}}\hat{=}$	$4\pi\cdot10^{-6}\,\mathrm{m^3\,mol^{-1}}$	$4\pi\cdot10^{-6}\,\mathrm{m^3\,mol^{-1}}$

Experimental errors

In this volume, experimental errors are given in parentheses referring to the last decimal places. For example, 1.352 (12) stands for 1.352 ± 0.012, and 342.5 (21) stands for 342.5 ± 2.1.

Landolt-Börnstein / New Series

Springer

Berlin
Heidelberg
New York
Barcelona
Budapest
Hong Kong
London
Milan
Paris
Tokyo

Landolt-Börnstein

Numerical Data and Functional Relationships
in Science and Technology

New Series
Editor in Chief: W. Martienssen

Units and Fundamental Constants in Physics and Chemistry

Nuclear and Particle Physics (Group I)

Atomic and Molecular Physics (Group II)

Solid State Physics (Group III)

Macroscopic Properties of Matter (Group IV)

Geophysics (Group V)

Astronomy and Astrophysics (Group VI)

Biophysics (Group VII)

Landolt-Börnstein
Numerical Data and Functional Relationships in Science and Technology
New Series / Editor in Chief: W. Martienssen

Group III: Solid State Physics
Volume 27

Magnetic Properties of Non-Metallic Inorganic Compounds Based on Transition Elements

Subvolume J 2
Halides II

Editor: H.P.J. Wijn

Authors: A. Chelkowski and H.P.J. Wijn

Springer

ISSN 0942-7988 (Solid State Physics)

ISBN 3-540-56655-4 Springer-Verlag Berlin Heidelberg New York
ISBN 0-387-56655-4 Springer-Verlag New York Berlin Heidelberg

Library of Congress Cataloging in Publication Data
Zahlenwerte und Funktionen aus Naturwissenschaften und Technik, Neue Serie
Editor in Chief: W. Martienssen
Vol. III/27J2: Edited by H.P.J. Wijn
At head of title: Landolt-Börnstein. Added t.p. : Numerical data and functional relationships in science and technology.
Tables chiefly in English.
Intended to supersede the Physikalisch-chemische Tabellen by H. Landolt and R. Börnstein of which the 6th ed. began publication in 1950 under title: Zahlenwerte und Funktionen aus Physik, Chemie, Astronomie, Geophysik und Technik.
Vols. published after v. 1 of group I have imprint: Berlin, New York, Springer-Verlag
Includes bibliographies.
1. Physics--Tables. 2. Chemistry--Tables. 3. Engineering--Tables.
I. Börnstein, R. (Richard), 1852-1913. II. Landolt, H. (Hans), 1831-1910.
Physikalisch-chemische Tabellen. IV. Title: Numerical data and functional relationships in science and technology.
QC61.23 502'.12 62-53136

Product Liability: The data and other information in this handbook have been carefully extracted and evaluated by experts from the original literature. Furthermore they have been checked for correctness by the authors and the editoral staff before printing. Nevertheless the publisher can give no guarantee for the correctness of the data and information provided. In any individual case of application the respective user must check the correctness by consulting other relevant sources of information.

Production: PRODUserv Springer Produktions-Gesellschaft, Berlin
Typesetting: Macmillan India Ltd.
Printing: Druckhaus Langenscheidt KG, Berlin
Binding: Lüderitz & Bauer, Berlin

SPIN 10103363 63/3020 - 5 4 3 2 1 0 - Printed on acid-free paper

Editor

H. P. J. Wijn, Institut für Werkstoffe der Elektrotechnik der Rheinisch-Westfälischen Technischen Hochschule Aachen, Templergraben, 52056 Aachen, FRG

Authors

A. Chelkowski, Institute of Physics, Silesian University, 40-007 Katowice, Poland

H. P. J. Wijn, Institut für Werkstoffe der Elektrotechnik der Rheinisch-Westfälischen Technischen Hochschule Aachen, Templergraben, 52056 Aachen, FRG

Preface

In Volume III/27 of the New Series of Landolt-Börnstein, entitled "Magnetic Properties of Non-Metallic Inorganic Compounds Based on Transition Elements" the magnetic properties are given of inorganic compounds which contain, apart from other elements, at least one type of a transition element. In a way this volume can be considered as a supplement to the Volumes III/4 and III/12 "Magnetic and Other Properties of Magnetic Oxides and Related Compounds", published in 1970 and in the period 1978 - 1982, respectively. However it seemed appropriate to change the title of the volume, since not only magnetic properties of oxides, but also those of pnictides, chalcogenides and halides have been a major field for research in the last decades.

The magnetic properties of halides have been compiled in Subvolume III/27j, the second part of which is presented herewith. While the first part (Subvolume j1, published in 1994) deals with halides of the general formula MX_n with M being a 3d element and X a halogen element, the present Subvolume j2 covers the halides composed out of three types of elements, A, M, and X, where generally A represents an alkali element, M a 3d element and X a halogen element.

A survey of the contents of all subvolumes planned for LB III/27 is printed on the inside of the front cover.

The use of two systems of units illustrates the situation in magnetic research today, where conversion from CGS units to SI units and vice versa has to be frequently performed. The user of the tables and figures of this subvolume is helped to convert the units by the list of definitions, units and conversion factors for the magnetic quantities occurring most frequently. This list is reproduced in the front cover.

Many thanks are due to the authors for the agreeable co-operation, the Landolt-Börnstein editorial office in Darmstadt, especially Dr.W.Polzin, Frau R.Brangs and Frau I.Lenhart for the great help with the editorial work, and to Springer Verlag for their thoughtful help in the final preparation of this volume.

Aachen, February 1995 **The Editor**

Table of contents

Magnetic properties of non-metallic inorganic compounds based on transition elements

Subvolume j2: Halides II

---

*) The author's name (M. Hagiwara) has been unintentionally omitted in subvolume III/27j1.
Address: Dr. M. Hagiwara, Department of Electronics and Information Science, Kyoto Institute of Technology, Matsugasaki, Kyoto 606, Japan.

9 Magnetic properties of halides

(Chap. 9.1–9.5 see Subvolume III/27j1)

9.6 AMX_3 and $AMX_3 \cdot 2H_2O$ compounds (A, M = metal, X = halogen element)

9.6.0 Introduction

9.6.0.1 General remarks

The table and figures presented in this section contain data on magnetic and related properties of ionic ternary and pseudoternary perovskite-type compounds of the formula AMX_3 or $AMX_3 \cdot 2H_2O$ which contain besides the halogen elements F, Cl, Br and I at least two metallic elements, a monovalent element of the group 1A (Li, Na, K, Rb, Cs) and Tl or NH_4, represented by the symbol A and a bivalent element of the group 2A (Mg, Ca, Ba), the 3d transition elements (V, Cr, Mn, Fe, Co, Ni (Cu)) and Zn, Cd or Pb, represented by the symbol M.

The compounds listed in the table and figures are designated by their chemical formula. The compounds in the table are arranged in the order as their elements appear in the periodic system.

Thus the halogen elements (X) are given in the following order

F, Cl, Br, I.

Within one halide system the sequence of the group 1A and Tl elements or NH_4 group (A) is

Li, Na, K, Rb, Cs, NH_4, Tl.

The group 2A, 3d and further elements (M) are listed in the following order

Mg, Ca, Ba,V, Cr, Mn, Fe, Co, Ni, (Cu), Zn, Cd, Pb.

The table provides information on paramagnetic Curie temperature Θ, ferromagnetic Curie temperature T_C, Néel temperature T_N, magnetic susceptibility χ, effective paramagnetic moment p_{eff} and saturation or spontaneous magnetic moment p_s or p_0. Further data are given in "Remarks".

The observed effective paramagnetic moments as well as the saturation moments depend on the crystal electric field (CEF) splitting, which depends on the electron configuration and on the symmetry of the particular components. This splitting is characterized by the CEF Hamiltonian which can be expressed in different ways, in terms of the CEF parameters B_n [64H3], the CEF overall splitting or in the case of cubic symmetry (see Section 9.9) by the so-called Lea Leask and Wolf parameters W and x [62L1]. All these parameters are used in this compilation.

All symbols and abbreviations used frequently in this section are listed in Section 9.6.0.2.

For further crystallographic and magnetic data for AMX_3 and $AMX_3 \cdot 2H_2O$ compounds see also Landolt-Börnstein, Vol. III/4a (publ. 1970) and Vol. III/12a (publ. 1978).

9.6.0.2 List of frequently used symbols and abbreviations

(for Subchapters 9.6–9.11)

Symbols

Symbol	Unit	Property
a, b, c	Å	lattice parameters
A, a	cm^{-1}	spin Hamiltonian parameters
A		absorbance
B	T, G	magnetic induction
B_r		resonance field
$\Delta B_{(pp)}$	T	(peak-to-peak) resonance linewidth
B_n, B_n^m	cm^{-1}, K	CEF parameters
C_g	$cm^3\,K\,g^{-1}$	Curie constant per gram
C_m	$cm^3\,K\,mol^{-1}$	Curie constant per mole
C_p (C_v)	$cal\,mol^{-1}\,K^{-1}$, $J\,mol^{-1}\,K^{-1}$, $J\,g^{-1}\,K^{-1}$	heat capacity at constant pressure (volume)
C_L		lattice heat capacity
C_{mag}		magnetic heat capacity
c_{ij}	$dyn\,cm^{-2}$	elastic stiffness
D	cm^{-1}	single ion anisotropy energy (D/k in K)
E	$dyn\,cm^{-2}$	Young modulus
E	eV	energy (E/k in K)
E	$V\,cm^{-1}$	electric field
$(1/2)e^2qQ$	$mm\,s^{-1}$	quadrupole coupling constant
g		spectroscopic splitting factor
h	Oe	magnetic ac field
H	$A\,m^{-1}$, Oe	magnetic field; symbol mostly used for external (applied) field (also given as $\mu_0 H$ in tesla (T))
H_0		starting field (for demagnetization)
H_1		rf field
H_A		anisotropy field
H_b		field at the bicritical point
H_c		critical field for (phase) transitions
H_D		field due to Dzyaloshinsky–Moriya interaction
H_E		exchange field
H_{eff}		effective field
H_{hf}		hyperfine field
H_{int}		internal field
H_p		amplitude of oscillating magnetic field
H_r		resonance field
H_{SF}		spin flop transition field
$H^{a,b,c}$		magnetic field along crystallographic a, b, c axes
$\Delta H_{(pp)}$	Oe, $A\,m^{-1}$	(peak-to-peak) resonance linewidth

Symbols (continued)

Symbol	Unit	Property
I		intensity
I_{int}		integrated intensity
I_p		peak intensity
I_{tot}		total intensity
J	eV	exchange interaction energy (J/k in K, J/hc in cm^{-1})
$J_{AB}, J_{1,2,3}, J'$, $J_{a,b,c}$		exchange interaction energies; for special meaning see corresponding tables or figures
k		Boltzmann constant
k	$Å^{-1}$	wavevector
K_1, K_3	$erg\,cm^{-3}$, $erg\,mol^{-1}$	anisotropy constant
L	dyn cm, dyn cm mol^{-1}, dyn cm g^{-1}	torque
$\Delta l/l$		thermal expansion
m	g	mass
$M_{(0)}$	G	(spontaneous) magnetization
n		refractive index
Δn		birefringence
p	bar, dyn cm^{-2}	pressure
p	μ_B, A m^{-2}	magnetic moment
$p_{A(M)}$		magnetic moment per atom A (per ion M)
p_{eff}		effective paramagnetic moment
p_{FU}		magnetic moment per formula unit
p_s		saturation magnetic moment
p_0		spontaneous magnetic moment
q	$Å^{-1}$	wavevector
Q		reciprocal vector component for neutron scattering (in reciprocal lattice units)
R	J mol^{-1} K^{-1}	gas constant
S		spin quantum number
t	s	time
T	K	temperature
T_C		Curie temperature
T_N		Néel temperature
T_{SF}		spin flop transition temperature
T_0		decomposition temperature
$T_{0,1,2}, T_c, T_H$, T_t, T_X		magnetic or structural transition temperatures (for definition see respective figures)
T^*		"final" temperature (for special definition see chap. 9.9)
T_1	s	spin lattice relaxation time
v	cm s^{-1}	sound velocity
V	$Å^3$	volume of unit cell
V_{zz}		principal component of EFG tensor
W	K	CEF parameter

Symbols (continued)

Symbol	Unit	Property
x		CEF parameter
x, y, z		coordinate axes
Z		number of formula units per unit cell
α	cm^{-1}	optical absorption coefficient
α	$dB\,cm^{-1}$	ultrasonic attenuation coefficient
α	K^{-1}	linear thermal expansion coefficient
β	deg	angle of monoclinic unit cell
γ	$Hz\,Oe^{-1}$	gyromagnetic ratio
Γ	cm^{-1}	linewidth of Raman or Mössbauer line
δ	$mm\,s^{-1}$	isomer shift
Δ	$mm\,s^{-1}$	quadrupole splitting
ε	$mm\,s^{-1}$	quadrupole splitting
ζ		reduced wavevector coordinate
η		asymmetry parameter
θ	deg	angle
θ_F	$deg\,cm^{-1}$	Faraday rotation
Θ, Θ_p	K	paramagnetic Curie temperature
Θ_D	K	Debye temperature
κ		inverse correlation length in reciprocal lattice units (r.l.u)
κ	$W\,cm^{-1}\,K^{-1}$	thermal conductivity
λ	µm, nm	wavelength
$\lambda_{h,k,l}$		magnetostriction constant
μ_B	$J\,T^{-1}$	Bohr magneton
ν	Hz	frequency
$\bar{\nu}$	cm^{-1}	wavenumber
$\Delta\bar{\nu}$	cm^{-1}	Raman shift
ρ	$cm^3\,g^{-1}$	density
σ	$G\,cm^3\,g^{-1} = emu\,g^{-1}$	magnetic moment per gram = specific magnetization
σ_m	$G\,cm^3\,mol^{-1}$ $= emu\,mol^{-1}$	magnetic moment per mole = molar magnetization
σ_0		spontaneous magnetization
σ_r		remanent magnetization
σ_s		saturation magnetization
σ_{st}		staggered magnetization
$\sigma_{a,b,c}$		magnetization along crystallographic a, b, c axes
τ		fluorescence lifetime, relaxation time, pulse duration time
ϕ, φ	deg	angle, scattering angle
ϕ	$min\,Oe^{-1}\,cm^{-1}$	Verdet constant
χ_g	$cm^3\,g^{-1} = emu\,g^{-1}$	magnetic susceptibility per gram
χ_m	$cm^3\,mol^{-1}$ $= emu\,mol^{-1}$	magnetic susceptibility per mole

Symbols (continued)

Symbol	Unit	Property
χ_v	$cm^3\,cm^{-3} = emu\,cm^{-3}$	volume magnetic susceptibility
χ_{st}		staggered susceptibility
χ', χ''		real, imaginary part of susceptibility
$\chi_{a,b,c}$		susceptibility along crystallographic a, b, c axes
χ_T		isothermal susceptibility
ω	Hz	angular frequency

Abbreviations

ac	alternating current
AF	antiferromagnetic (phase)
AFMR	antiferromagnetic resonance
bcc	body centered cubic
BG	background
C	commensurate
calc	calculated
CEF(A)	correlated effective field (approximation)
EFG	electric field gradient
ENDOR	electron nuclear double resonance
EPR	electron paramagnetic resonance
ESR	electron spin resonance
exp	experimental
F	ferromagnetic (phase)
F, G	components of the magnetic mode
fcc	face centered cubic
FI	ferrimagnetic (phase)
FMR	ferromagnetic resonance
FU	formula unit
HTS	high temperature series (prediction)
IC	incommensurate
IM	intermediate (phase)
ind	induced
IR	infrared
M	metal
M	monoclinic
mag	magnetic
max, m	maximum
ME	Mössbauer effect
MF(A)	molecular field (approximation)
ND	neutron diffraction
NMR	nuclear magnetic resonance
NQR	nuclear quadrupole resonance
NS	neutron diffraction spectra
O	orthorhombic
OAF	oblique antiferromagnetic (phase)

P	paramagnetic (phase)
p	as subscript: perovskite
QMC	quantum Monte Carlo (calculation)
rel, r	relative
rf	radio frequency
RT	room temperature
sc	simple cubic
SF	spin-flop (phase)
SG	spin glass (phase)
SGS	single ground state
st	staggered
T	trigonal
TIP	temperature independent paramagnetism
tot	total
$\perp$, $\parallel$	perpendicular, parallel to a crystallographic axis
Γ_i	magnetic mode

9.6.1 Table

Compound	Θ [K]	T_C [K]	T_N [K]	χ	p_A [μ_B]	p_{eff} [μ_B]	Remarks	Ref.
$LiBaF_3$: Ti^{3+}							from EPR at 34 GHz and 77 K, Ti^{3+} center shows tetragonal symmetry, $g_\parallel = 1.9567$, $g_\perp = 1.9702$	79Y2
$LiBaF_3$: Cr^{3+}							from EPR at 34 GHz and 77 K, Cr^{3+} shows cubic symmetry, $g = 1.9702$	79Y2
$LiBaF_3$: Mn^{2+}							from EPR at 34 GHz and 77 K, Mn^{2+} shows cubic symmetry, spin Hamiltonian parameters $A = -88.3 \cdot 10^{-4}\ cm^{-1}$ $a = 5.5 \cdot 10^{-4}\ cm^{-1}$, $g = 2.0014$ $(\mathscr{H} = g\mu_B \boldsymbol{S} \cdot \boldsymbol{H} + A\boldsymbol{S} \cdot \boldsymbol{I} + \frac{a}{6}(S_x^4 + S_y^4 + S_z^4 - \frac{707}{16}))$	79Y2
$LiBaF_3$: Fe^{3+}							from EPR at 34 GHz and 77 K, Fe^{3+} shows cubic symmetry, spin Hamiltonian parameters $g = 2.0025$, $a = 120 \cdot 10^{-4}$ cm	79Y2
$LiBaF_3$: Co^{2+}							from EPR at 34 GHz and 4.2 K, Co^{2+} center shows cubic symmetry, spin Hamiltonian parameters $g = 4.3164$, $A = 94.7 \cdot 10^{-4}\ cm^{-1}$ Co^{2+} substitutes for the Li^+ ion which is surrounded by fluorine octahedron	79Y2
$NaMnF_3$			66.2				another magnetic transition at 145 K suggested	62T2
			60				no evidence for any higher temperature magnetic transition; short range order seems to persist above T_N	64P1
							AFMR in ordered single crystal	67S4
			Fig. 1 Fig. 2				lattice parameters vs. T: Fig. 1, optical birefringence vs. T: Fig. 2	88R1

Compound	Θ [K]	T_C [K]	T_N [K]	χ	p_A [μ_B]	p_{eff} [μ_B]	Remarks	Ref.
$NaCoF_3$							antiferromagnet, T_N below 77 K	67N1, 67N2
			74 Fig. 3				from neutron diffraction G type antiferromagnet, cubic space group Pbnm-D_{2h}^{16}	70F1
$NaNiF_3$			149				temperature dependence of the $^3A_{2g} \rightarrow {}^1E_g$ transition: Fig. 4	65P1, 66L1
							^{19}F NMR investigation of local magnetic field	66P1
			156	Figs. 5, 9			$\sigma(H, T)$: Figs. 6, 7 and $\sigma_0(T)$: Fig. 8	67J1
							maximum magnon frequency vs. T_N: Fig. 10	68B1
			149 Fig. 11				from neutron diffraction	68E1
							AFMR	68G1
			146				unit cell: Fig. 12; FM and AFM frequencies vs. T: Fig. 13	73P1
$KMgF_3$							fundamental lattice vibrations	67P2
							refraction index vs. T: Fig. 14; see also Fig. 117	84K4
$KMgF_3$: V^{2+}							exchange interaction between nearest-neighbor V^{2+} ions $J_{AB}/k = -7.0$ K	67H3
$KMg_{1-x}Mn_xF_3$							ESR, transition temperature vs. x: Fig. 15	66H1
				Figs. 16, 17				70B1
			Figs. 18, 19				EPR linewidth vs. x: Fig. 18	72G2
				Fig. 20			$\sigma_m(H)$: Fig. 21, Curie constant vs. x: Fig. 22	73B1
				Fig. 23				73Y2
			Fig. 24					79P1
			Figs. 25, 26					82A2

$KMgF_3$: Fe^{2+}					acoustic resonance	68W1
					EPR temperature dependence	78K1
$KMgF_3$: Co^{2+}					acoustic resonance	68W1
$KMg_{1-x}Co_xF_3$						
x = 0.025				Fig. 27		67S1
x = 0.05				Fig. 28		67S1
x = 0.11				Fig. 29		67S1
$KMg_{1-x}Ni_xF_3$						
x = 0.2					fundamental lattice vibrations	67P2
x = 0.5					fundamental lattice vibration	67P2
$KMgF_3$: Ni^{2+}					acoustic resonance	68W1
x = 0.035				Fig. 30		69Y1
x = 0.054				Fig. 31		69Y1
x = 0.081				Fig. 32		69Y1
x = 0.3···1			Fig. 33		magnon-pair mode frequency vs. x: Fig. 33	75F1
x = 0.2···1	Fig. 34		Fig. 35	Fig. 36		88S2
$KMgF_3$: Eu^{2+}					excited-state absorption by Eu^{2+}	90M1
$KCrF_3$				Fig. 37		66Y1
$KMnF_3$	− 238		~ 88.3			60S1
			Fig. 38			
		81.5	88		magnetic transitions at 81.5 and 88 K, distortion of regularity of octahedra at 65 K, transition from cubic to an orthorhombic phase Pbnm at 184 K, pseudocell dimensions: Fig. 39	61B1
			Fig. 39		crystal structure in the orthorhombic phase: Fig. 40, fluorine octahedron distortions: Fig. 41	

(continued)

Compound	Θ [K]	T_C [K]	T_N [K]	χ	p_A [μ_B]	p_{eff} [μ_B]	Remarks	Ref.
$KMnF_3$ (continued)		Fig. 42	Fig. 42	Fig. 43			below 88.3 K transition to uniaxial antiferromagnetism, below 81.5 K weak ferromagnetism, weak moment results from canting of magnetic sublattices because of differences in the sublattices anisotropy, between 81.5···88.3 K a moment appears only in strong magnetic fields, torque vs. T: Fig. 44, location of screw axes with respect to Mn^{2+} ions: Fig. 45, fluorine distortions: Fig. 46, torque vs. angle: Figs. 47···49	61H1
							double resonance and nuclear cooling	61H2
	4.5			Fig. 50			Θ from temperature range 350···600 K	63H1
							ultrasonic induced spin transition	64D1
							^{55}Mn NMR absorption, unusual intensity and strong temperature and field dependence associated with the strong coupling between nuclear and electronic modes in material with low anisotropy and large nuclear magnetization suggested	64N1
			88 Fig. 51				thermal conduction of single crystal: Fig. 51	64S1
							exchange effects in the electronic absorption spectrum of Mn^{2+}	65F1
			Fig. 52				critical magnetic scattering	66C1
							spin wave dispersion, nearest and next-nearest neighbor exchange interactions $J_1/k = 3.80$ K, $J_2/k = 0.11$ K, anisotropy field $H_A = 3.10$ kOe, J_1/k from neutron scattering 3.12 K; from T_N 3.1 K; Θ (molecular field) 4.5 K; AFMR and NMR 3.89 K	66P2

			magnon sideband in absorption spectrum	67A1
			fundamental lattice vibrations	67P2
			electron–magnon interaction	67Y1
			multiple magnons in absorption spectra of thin slabs	69S2
	Fig. 53		phonon sideband vs. T: Fig. 53	70B3
			structural transition temperature 184 K	70M1
	Fig. 54			71P1
	87.9	Fig. 55		72M2
Fig. 56	Fig. 57		T_C and T_N from ESR experiment	72S2
			AFMR	72Y2, 73S1
		Fig. 58	$\chi(H, T)$ at 10 MHz: Fig. 58	73M2
			AFMR: Fig. 59, second-order magnetic transition associated with spin reorientation at 50 K and $H = 0$: Fig. 60	74S1
			magnetic structure	75H1
81	88		from neutron diffraction, structural transition at 91 K accompanied with the R_{25} and M_3 soft-mode phonon condensation shows second-order character, T_N second order purely magnetic transition accompanied with development of the G-type antiferromagnetic structure, T_C first order structural and magnetic phase transition with temperature hysteresis: Fig. 61	75H2
			spin reorientation and AFMR in low temperature phase	75S1
	Fig. 62		birefringence vs. T: Fig. 62	76B2
Fig. 63	Figs. 63, 64		T_C and T_N from ESR investigation	76J2
Figs. 65, 66	Figs. 65, 66		T_C and T_N by AFMR investigation	78J2

(continued)

Compound	Θ [K]	T_C [K]	T_N [K]	χ	p_A [μ_B]	p_{eff} [μ_B]	Remarks	Ref.
$KMnF_3$ (continued)			88.5	Figs. 67, 70			thermal expansion: Figs. 68 and 69, first order structure transition at 186 K from cubic to tetragonal symmetry, second order structure transition at 91.5 K, third structure transition at 81 K associated with the onset of weak ferromagnetism	83B1
							heat capacity: Fig. 259	83B2
	Fig. 93	Fig. 92	Fig. 92	Fig. 92		Fig. 96		84C1
							magneto-optical activity at the structural phase transition in paramagnetic state: Fig. 71	85P2
		Figs. 72···76	Figs. 72···75	Fig. 72			uniaxial antiferromagnetic phase transition coincides with the crystallographic transition caused by the condensation of the M_3-type soft phonon at 88 K, transition at 83 K indicates weak ferromagnetic component by spin canting and a large rearrangement of crystallographic domains, the transitions at 88 K and 83 K are interpreted in terms of magneto-elastic coupling induced by super-exchange between Mn^{2+} spins and the tilting of MnF_6 octahedra related to the M_3 rotational mode; magnetic transition temperature; Figs. 73···76	86H3
							crystal structure near magnetic transition: Figs. 77, 78	88R1
			Fig. 79					90S2
		Fig. 80	Fig. 80				AFMR vs. T: Figs. 80···83	91S1

$KMnF_3$: Fe^{2+}		Figs. 84, 85	Figs. 84, 85			AFMR vs. *T*: Figs. 81···85	91S1
$KMnF_3$: Co^{2+}						magnetic excitation	69S4
$KMn_xCo_{1-x}F_3$			Fig. 86	Fig. 87		for temperature range 300 to 600 K:	63H1
x = 0.933	− 164		93			Curie constant per gram $3.00 \cdot 10^{-2}$ cm^3K/g	
x = 0.739	− 180		100			$2.88 \cdot 10^{-2}$ cm^3K/g	
x = 0.488	− 194		108			$2.69 \cdot 10^{-2}$ cm^3K/g	
x = 0.975						Co^{2+} ion absorption transition in magnetic host: Fig. 88	67H2
x = 0.99			Fig. 89			AFMR (*T*): Figs. 89, and 81···83	91S1
$KMn_xNi_{1-x}F_3$				Fig. 90		for $T > 450$ K:	63H1,
x = 0.905	− 164		100			Curie constant per gram $2.95 \cdot 10^{-2}$ cm^3K/g	69M3
x = 0.724	− 196		124			$2.48 \cdot 10^{-2}$ cm^3K/g	
x = 0.465	− 274		171			$2.04 \cdot 10^{-2}$ cm^3K/g	
x = 0.99						effect of magnetic interactions on the temperature dependence of the $^3A_2 \rightarrow {}^3T_1$ absorption-band peak of Ni^{2+}	65H1
						$J = 18$ cm^{-1} for MnNi pair in its electronic ground state, absorption of light by pairs of exchange-coupled Mn and Ni ions	66F2
				Fig. 91			69M3
				Figs. 92···95	Fig. 96	T_C-weak ferromagnetic transition temperature, α[%]: the ratio of Ni precipitation to all Ni content in sample	84C1
x = 1	− 142	81.5	88.5		6.07		
x = 0.98	− 131	82.0	90.5		5.97		
x = 0.94	− 137	90.2	95.5		5.86		
x = 0.74	− 135		122		5.40		
x = 0.72	− 152		124		5.29		
x = 0.66	− 150		134		5.17		

(continued)

Compound	Θ [K]	T_C [K]	T_N [K]	χ	p_A [μ_B]	p_{eff} [μ_B]	Remarks	Ref.
$KMn_xNi_{1-x}F_3$ (continued)								
x = 0.635	− 163		137			5.04	$\alpha = 0.035$	
x = 0.29	− 274		196			4.35	$\alpha = 0.22$	
x = 0.195	− 238		216			3.86		
x = 0.135	− 386		230			3.85	$\alpha = 0.275$	
x = 0.05	− 437		251.5			3.56	$\alpha = 0.32$	
x = 0	− 632		267			3.69	$\alpha = 0.35$, sample 1	
x = 0	− 612		267			3.32	$\alpha = 0.185$, sample 2	
			Figs. 97, 79				AFMR line width: Fig. 97, AFMR field: Fig. 98	90S2
x = 0.99			Fig. 99				AFMR linewidth: Fig. 83, AFMR field: Figs. 81, 82, 99, 100	91S1
$KMn_xZn_{1-x}F_3$				Fig. 101				73Y2
$KFeF_3$							anomalous thermal conduction of single crystal	64S2
							relaxation and exchange induced quadrupole effects, exchange integral for $T_N = 115$ K: $J/k = 5.38$ K	67F1
			115				exchange integral $J/k = 7$ K	68F2
		41	112				below T_C weak ferromagnetic, spin arrangement below T_C: Fig. 102	75T1
		Figs. 103, 104	Fig. 103					
$KCoF_3$			Fig. 86					63H1
			Fig. 105				hyperfine interactions investigations by NMR shifts of the ^{19}F	64H1
			114				T_N from thermal conductivity	64S1
			Fig. 106					
							^{19}F NMR	65S1
							fundamental lattice vibrations	67P2

				Fig. 107		77B3
			113	Fig. 108		77S6
			Fig. 108			
		Figs.109 ···111	Fig. 110			78T1
					domain wall motion	80C1
			119.25			82F2, 83F1
			118	Fig. 112		87M1
$KCo_{0.25}Ni_{0.75}F_3$					Co^{2+} ion absorption in magnetic host: Fig. 113	67H2
$KNiF_3$			253			60H1
	−851		275	Fig. 114	for $T > 650$ K, Curie constant $C_g = 1.61 \cdot 10^{-2}$ cm^{3}K/g	63H1
					covalency effect, average additional band peak shift of 320 cm^{-1} below T_N to higher energies as compared with magnetically disordered Ni^{2+} in $KMgF_3$	63K1
					^{19}F NMR in a single crystal	63S1
			275		thermal conduction of single crystal	64S1
					IR absorption related to the spin system	65T1
					phonon-induced spin-dependent IR absorption, exchange parameter determined from peak position, $J/k = 52.3$ K for $k_p = 0$ or $J/k = 55.3$ K for the phonons k_p near the edge of the Brillouin zone	66T2
					nearest neighbor exchange constant $J/k = 86$ K from χ ($T > T_N$); 90 K from $T(\chi_{max})$; 88 K from T_N; 93 K from χ (0)-zero temperature perpendicular susceptibility	67L1
			275		electron-magnon interaction, maximum magnon frequency vs. T_N: Fig. 10	68B1

(continued)

Compound	Θ [K]	T_C [K]	T_N [K]	χ	p_A [μ_B]	p_{eff} [μ_B]	Remarks	Ref.
$KNiF_3$ (continued)			273				T_N: Fig. 115	68P3
			Fig. 116				influence of magnetic ordering on the lattice vibration measured in far IR region, lattice constant vs. T: Fig. 116	68S2
				Fig. 91				69M3
							energy bands	72M1
			246				spin orientation parallel to [001]	74P1
			Fig. 117				$n(T)$: Figs. 117, 118	77S5
			Fig. 119				$T_N(H)$: Fig. 119, anomaly of thermal expansion: Fig. 120	78B2
			Fig. 121				NMR spin-lattice relaxation time vs. $1/T$: Fig. 121	79E1
							NMR observation of field-induced domain reorientation	79E2
			Fig. 122				elastic constant vs. T: Fig. 122	81G2
			246.55					82F2, 83F1
							magnetic structure: Fig. 123, magnon energies vs. T/T_N: Fig. 124	82G2
			247				$\Delta n(T,p)$: Fig. 125	83N2
				Fig. 95		Fig. 96		84C1
							T_N from Δn: Figs. 14 and 126, 127	84K4
			247				T_N from X-band AFMR: Fig. 128	85C2
			Fig. 79					90S2
$KCuF_3$							by neutron diffraction no magnetic ordering observed down to 4.2 K	61S1
							^{19}F NMR, one dimensional antiferromagnet, nearest neighbor exchange constant $J/k = -190$ K along c axis and 0.2 K along a axis, dull χ maximum at 243 K, no long range order down to 4.2 K	67H1

			Figs. 129, 130	$J/k = -190$ K along the chain	67K2
		Figs. 131, 133		magnetic structure by neutron diffraction: Fig. 132	69H1
				ESR	71I1
		38	Fig. 134	$g = 2.30$ along a axis and 2.16 along c axis, $J/k = -195$ K along c, T_N: Fig. 134	73H1
		Figs. 135, 136		from neutron diffraction	73I1
		39.51		J/k (c axis) $= -190$ K, J' (a axis)/J (c axis) $= 0.027$, quasi 1D, planar, spin-easy axis a or b	74I1
		Fig. 137	Fig. 137		75M1
				^{19}F NMR investigation	76K1
				magnetic hyperfine interactions	76K2
		39.5		T_N: Fig. 138	78I1
$KZnF_3$				thermal conduction of single crystal	64S1
$KZnF_3$: Mn^{2+}				nearest neighbor exchange constant for Mn^{2+} pairs $J/k = 9.9$ K	69K1
$KZnF_3$: MnNi				absorption of light by pairs of exchange-coupled Mn and Ni ions, $J = 18\ cm^{-1}$ for MnNi pair in its electronic ground state	66F2
$KCdF_3$: Cr^{3+}				EPR phase transition investigations	79A1
$RbMg_xNi_{1-x}F_3$	Fig. 139			$\sigma(x)$: Fig. 140	68M1
$RbCa_xNi_{1-x}F_3$	Fig. 139			$\sigma(x)$: Fig. 140	68M1
$RbMnF_3$				AFMR	62T1, 63F1
				^{55}Mn NMR, large enhancement of the signal from one of the NMR modes in a low anisotropy above the critical field for spin flopping	64H2

(continued)

Compound	Θ [K]	T_C [K]	T_N [K]	χ	p_A [μ_B]	p_{eff} [μ_B]	Remarks	Ref.
$RbMnF_3$ (continued)							relative intensity of fluorescence vs. T: Fig. 141, fluorescence profile maxima vs. T: Fig. 142	65P2
				82.6				66C2
							axial stress dependent AFMR	66E1
							AFMR below spin-flop transition	66I1
			83					66L1
			Fig. 143				thermal expansion; Fig. 143, magnetic heat capacity: Fig. 144	66T1
							second-order spin wave instability investigations at 4.2 K	67C1
	−136		83				AFMR, at 4.2 K: magnetostriction constant $\lambda_{100} = 3.0 \cdot 10^{-6}$, anisotropy field $H_A = 4.1$ Oe, exchange field $H_E = 860$ kOe; $\chi_v(T_N) = 3.7 \cdot 10^{-4}$ emu/cm^3	67E2
							magneto-elastic coupling by AFMR	67E3
							Faraday magneto-resonance effect: Fig. 145	67K3
							absorption spectrum below and above T_N	67M2
							fundamental lattice vibrations	67P2
							electron–magnon interaction	67Y1
			82.96				ultrasonic propagation near T_N: Fig. 146	68G3
							magnon–magnon interaction	68F1
							excitation of nuclear and electron spin waves	68H1
							AFMR, antiferromagnetic domains observation	68I1
							exciton dispersion	68M2
							^{19}F nuclear-acoustic resonance dispersion mode	68M3

				inelastic magnetic scattering near T_N	68N1
		82		anisotropy of the Zeeman effect	68Y1
				fluorescence life-time vs. T: Fig. 147	69G2
				ENDOR investigation of nuclear saturation	69I1
		Fig. 148		elastic properties	69M1
				magneto-elastic properties	69M2
				ESR	71T1
		83.0		$g = 1.975$, nearest neighbor exchange constant $J/k = -3.37$ K	72B2
		Fig. 149	Fig. 149		
		Fig. 150		excitation of $^4\Gamma_1$ line vs. T: Fig. 150	72S3
			Fig. 151		74J1
				NMR relexation above T_N	75T1
		Fig. 152		T_N vs. magnon frequencies: Fig. 152	76B3
				magnetic field dependence of thermal conductivity for pure and Fe, Ni or Co doped samples	77H1
				(T, H) phase diagram	77S4
		Figs. 153, 154		$T_N(H)$: Figs. 153, 154	78S1
		Fig. 155		$T_N(H)$: Fig. 155	78S2
			Fig. 156	$\sigma(H)$: Fig. 157	82C1
		82.6			82F2, 83F1
				$p_{Mn}(H)$: Fig. 158, heat capacity: Fig. 259	83B2
		Fig. 159		refraction index vs. T: Figs. 14, 159	84K4
$RbMn_xNi_{1-x}F_3$ $x = 0.99$				effect of magnetic interactions on the temperature dependence of $^3A_2 \rightarrow {^3T_1}$ absorption-band peak of Ni^{2+}	65H1
$x < 0.4$	Fig. 139			$\sigma(x)$: Fig. 139	68M1
$RbMnF_3$: Ni (5%)		Fig. 160		g vs. temperature from EPR: Fig. 160	72G1

(continued)

Compound	Θ [K]	T_C [K]	T_N [K]	χ	p_A [μ_B]	p_{eff} [μ_B]	Remarks	Ref.
$RbFeF_3$	−88		75	Fig. 161	0.27		p_A: weak ferromagnetic at 0 K, $C_m = 3.53$ cm^{3}K/mol, $g = 2.17$, $\sigma_m(T)$: Fig. 162	66W1
							magnetically induced quadrupole interaction	67G1
							$\sigma(T)$: Fig. 163, shear modulus vs. T: Fig. 163	67T1
	200					5.82	$\chi_0 = \chi^{TIP} = 20 \cdot 10^{-6}$ cm^3/g, spontaneous $p_{FU} = 0.515$ μ_B at 1.46 K, $\sigma(T)$: Fig. 164	67W1
							$\sigma(T)$: Fig. 165	68G2
		87	102				magnetic and optical behavior, torque vs. angle between H_0 and [001] axis: Fig. 166, at T_C undergoes first-order transition from a tetragonal antiferromagnetic phase to an orthorhombic ferromagnetic phase where $c/a > 1.003$, above T_N paramagnetic, between 87 and 40 K orthorhombic, at 40 K first order transition, below 40 K probably monoclinic	69G1
							local magnetic fields on ^{19}F, ^{87}Rb and ^{85}Rb measured by NMR	69P1
			100.5 Fig. 167				$p_{Fe} = 4.6$ μ_B from neutron diffraction: Figs. 167 and 168, critical exponent $\beta = 0.329$	71W1
		87	101				at 39 and 87 K first order phase transition, T_N from Mössbauer experiment: Figs. 169, 171, spin arrangement below 39 K: Fig. 170	81S1
			101				at 39 and 87 K successive phase transitions, $p_{II} - p_I = 0.68$ μ_B, spin arrangement: Fig. 172	81S2

$RbCoF_3$					Co^{2+} ion absorption transition in magnetic host: Fig. 173	67H2
			32			67P3
					^{19}F NMR investigation of spin density distribution	67S3
					spin-density space oscillations and hyperfine interaction	68P2
				Fig. 107		77B3
$RbCo_xNi_{1-x}F_3$		Fig. 139			$\sigma(x)$: Fig. 140	68M1
x = 0.23					effect of magnetic interactions on the temperature dependence of the $^3A_2 \rightarrow {}^3T_1$ absorption-band peak of Ni^{2+}	65H1
					Faraday rotation: Fig. 174, 175	68P1
$RbNiF_3$	− 130	145		Fig. 176	$\sigma(H)$: Fig. 177, 178 and 179; $\sigma(T)$: Fig. 180, $g = 2.2$	66P1, 66S1
					Faraday effect: Fig. 181	67P1
				Fig. 182		67S2, 80K1
		145			electron-magnon interaction, maximum magnon frequency dependence on magnetic transition temperature: Fig. 10	68B1
		Fig. 139			σ: Fig. 140	68M1
		Fig. 183			second-order Raman shift: Fig. 183	69C1
		Fig. 184			light scattering from phonons and magnons	69F1
				Fig. 185	magnetic local field vs. T from NMR: Fig. 186	69S5
					NMR investigations of magnetization: Fig. 187	70P1
		Fig. 188			$(n_{\parallel} - n_{\perp})(H, T)$: Fig. 188	70S3
		Fig. 189		Figs. 190, 192	ESR line width vs. T: Figs. 189, 191	71T1
		139			two inequivalent sites of Ni^{2+}	74P1
		139			$H_S(T)$: Fig. 193, hexagonal ferromagnet	80B2

(continued)

Compound	Θ [K]	T_C [K]	T_N [K]	χ	p_A [μ_B]	p_{eff} [μ_B]	Remarks	Ref.
$RbNi_{1-x}Cd_xF_3$		Fig. 139					$\sigma(x)$: Fig. 139	68M1
$CsMnF_3$			53.5 Fig. 194				T_N from magnetic anisotropy, $\chi_m[10^{-3}\ cm^3/mol] = 10.6$ at 298 K; 27.9 at 77 K and 39.7 at 4.2 K, from ESR $g = 1.9989$, anisotropy energy vs. T: Fig. 195	63L1
							^{55}Mn NMR modes	67W2
							nuclear and electron spin-wave relaxation	69S3
			Fig. 196				T_N from AFMR: Fig. 196	70B2
							exciton, magnon and phonon structures of the absorption spectra in the 0.3900 μm region	70S1
			52.328 Fig. 197				AFMR and EPR, critical exponent $p = 0.511$	72W1
	−91		53.5	Fig. 198			$\sigma_m(H)$: Fig. 199, see also Fig. 210; Curie constant 4.49 cm^3 K/mol	75Y2
			Fig. 152				T_N vs. magnon frequencies: Fig. 152	76B3
			53.06 Fig. 200				$C_p(T)$: Fig. 200	77I1
							spin-waves, magnon dispersion: Fig. 201	77K4
							magnetic field and temperature dependence of light absorption	78Y1
							$T_N(H)$	79O1
							$T_N(H^2)$: Fig. 202	79S1
			Figs. 203···206				T_N from thermal expansion: Figs. 203, 204	80S2
							six-sublattice antiferromagnet suggested	84S1

$CsMn_{1-x}Co_xF_3$							with $0.4 \le x \le 0.05$ ferrimagnetic with a Curie temperature of 50 K and $p_{FU} = 1.77\ \mu_B$ at 4.2 K; $p_{FU}(H, T)$: Fig. 207, $p_{FU}(x)$: Fig. 208,	77Y1
x = 0.5	− 110						from $\chi_\perp$	
	− 10						from $\chi_\parallel$, Curie constant 5.9 $cm^3\,K\,mol^{-1}$	
$CsMn_{1-x}Ni_xF_3$			Fig. 209	Fig. 209			$\sigma_m(H)$: Fig. 210	75Y2
x = 0.09	− 90		51.5					
							with $0.3 \le x \le 0.5$ ferrimagnetic with Curie temperature of 50 K, $p_{FU}(x)$: Fig. 208, $p_{FU}(H, T)$: Fig. 211	77Y1
x = 0.35	− 50.0						Curie constant 3.3 $cm^3\,K\,mol^{-1}$, $p_{FU} = 1.42\ \mu_B$ at 4.2 K and $\boldsymbol{H} \perp c$	
x = 0.4					1.21		for $p_A = p_{FU}$.	
x = 0.75	− 26		5	Fig. 212			structure type 9H	79D2
x = 0.5	− 129	63		Fig. 213			$p^S_{FU}(T = 0\ K) = 0.44\ \mu_B$, high temperature 6H structure type, $p^s_{FU}(T)$: Fig. 213	79D2
$CsFeF_3$							$\sigma(H)$: Fig. 214	70M3
$CsCoF_3$				Fig. 215			structural forms: Fig. 216, $\sigma_m(H)$: Fig. 217	70L1
	62		8				9L phase, anisotropy constant K $= 5 \cdot 10^{-4}$ emu/mol, transition at $H_c \approx 11$ kOe, $C_m = 3.7\ cm^3\,K/mol$	
	65	50					6L phase, ferrimagnetic below 50 K, magnetization not saturated at 17.20 kOe and 4.2 K where $p_A = 0.8\ \mu_B/Co^{2+}$	
$CsNiF_3$	− 75	111 Fig. 218		Fig. 218	0.6	3.2	$p_A = p_s/Ni$ at 4.2 K and 17.2 kOe, $g = 2.3$, $\sigma(T)$: Fig. 218	69L1
	8			Fig. 219		3.6	from neutron diffraction one-dimensional magnetic order of moments along c axis, space group $P6_3/mmc$, hexagonal $BaNiO_3$ type $\chi_m = 0.00168\ cm^3/mol$ at 100 K; 0.79 at 10 K; 2.48 at 4.2 K and 3.06 at 3.5 K	71S2

(continued)

Compound	Θ [K]	T_C [K]	T_N [K]	χ	p_A [μ_B]	p_{eff} [μ_B]	Remarks	Ref.
$CsNiF_3$ (continued)			2.67 Fig. 220					72S1
	22		2.613	Fig. 221			$J/k = 8.3$ K, heat capacity vs. T: Figs. 222 ⋯ 224	73L1
			2.7 Fig. 225					74S2
			2.61	Fig. 226			from susceptibility $J/k = 11.8$ K, $A/k = 4.5$ K $\mathscr{H} = -J\Sigma S_i S_{i+1} + A\Sigma(S_i^z)^2$, from neutron diffraction $T_N = 2.7$ K: Fig. 227	76S1
			2.32				from susceptibility for sample with 2% of Mg: Fig. 226	
			2.37				from elastic neutron scattering: Fig. 227	
	20			Figs. 228, 229				76S2
			2.7 Fig. 230				Young modulus: Figs. 230, 231	77B2
				Fig. 232			log-log plot of χ_m vs. T: Fig. 233	77D1
							FMR: Figs. 234 and 235	77G1
			2.61				FMR at 4.2 K and $\nu = 35$ GHz: Figs. 236, 237; $\sigma(H)$: Fig. 238	77R1
							magnetization vs. H: Fig. 239	77S2
			2.65	Fig. 240			magnetic ordering: Fig. 241	77S3
			Figs. 242, 243				AFMR: Figs. 242, 243	79Y1
			2.65					80G1

		Fig. 244 2.664		from neutron scattering, critical indices; $\beta = 0.34$ for $3 \cdot 10^{-3} \leq \varepsilon \leq 3.55 \cdot 10^{-2}$ ($\varepsilon = T/T_N - 1$); ν, γ for $T > T_N$ were found to be different in different ε-ranges, $\nu = 0.68$ and $\gamma = 1.45$ for $1 \cdot 10^{-3} \leq \varepsilon \leq 1.2 \cdot 10^{-2}$ and $\nu = 0.54$, $\gamma = 1.1$ for $1.4 \cdot 10^{-2} \leq \varepsilon \leq 8.6 \cdot 10^{-2}$, where ν is defined in Fig. 245 and γ in eq. $\chi(q, T)_{q=0} \propto \left(\frac{T - T_N}{T_N}\right)^{-\gamma}$ with q being the distance from the Brillouin zone center; T_N: Fig. 245	80S3
				magnetic solitons by NMR studies	83G1
			Fig. 246		84I1
				from refractive index and linear magnetic birefringence measurements the magnetic short-range order was found to extend beyond 120 K	85L1
				heat capacity	85R1
		2.61		3-dimensional antiferromagnet	85T3
		2.7	Fig. 247		86C1
		Fig. 247			
		2.61 Fig. 249		magnetic field dependence of spin structure: Fig. 248	86L1
			Figs. 250···252	anomaly at ≈ 190 K observed in structural expansion coefficients; it is suggested to be a structural phase transition induced by pressure: Fig. 251	87S1
				$p(H)$: Figs. 253, 254. $p(T)$: Fig. 255	91C1
$CsCuF_3$	−5			no long-range ordering observed	66B1
$CsEuF_3$	−5			^{19}F NMR, no long-range ordering observed	66B1

(continued)

Compound	Θ [K]	T_C [K]	T_N [K]	χ	p_A [μ_B]	p_{eff} [μ_B]	Remarks	Ref.
NH_4MnF_3			75.1 Fig. 257	Figs. 260⋯264			heat capacity: Figs. 256 ⋯ 259, $p_{Mn}(H)$: Fig. 265 and 158, $J/k = -3.11$ K, simple cubic perovskite structure at room temperature; at 182.1 K structural phase transition to pseudo-tetragonal symmetry occurs	83B2
NH_4CoF_3		124.5		Fig. 266			$\sigma_m(H)$: Figs. 267 and 268, $\chi(T)$: Figs. 107, 269	77B3
$TlMnF_3$			83				^{19}F, ^{203}Tl and ^{205}Tl NMR	65P3
			83					66L1
	− 125						NMR	66Z1
	− 138		76				AFMR at 4.2 K: anisotropy field $H_A = 6.8$ Oe, exchange field $H_E = 630$ kOe, magnetostriction constant $\lambda_{100} = -3.2 \cdot 10^{-6}$; $\chi_v(T_N) = 4.2 \cdot 10^{-4}$ emu/cm^3	67E2
$TlMnF_3$: Co^{2+}							AFMR for $0 \leq x \leq 0.001$	67E1
$TlFeF_3$		~ 77					generation of coherent sound by large linear magnetostrictive coupling excited by an r.f. electromagnetic field, the inverse (sound r.f.) effect observed as well	69T1
$TlCoF_3$			94 Fig. 270				magnon peak frequency vs. T: Fig. 270	73R1
			94				$J/k = -0.8$ K	73T1
				Fig. 107				77B3
$TlNiF_3$				Fig. 271			$\sigma(T)$: Fig. 271	67K1
							Figs. 139, 140	68M1
$LiCuCl_3 \cdot 2H_2O$							monoclinic, $a = 6.078$ Å, $b = 11.145$ Å, $c = 9.145$ Å, $\beta = 108°50'$; magnetic phase diagram: Figs. 272, 273	71F1

	4.440			from AFMR, $g = 2.224$, spin-ordering direction lies at most 1° out of the *ac* plane, 16° from *z*; $g_x = 2.050$, $g_b (\approx g_y) = 2.136$, $g_z = 2.24$, along principal magnetic axes	72Z1
	4.4	Fig. 274		$\sigma(H)$: Fig. 275, magnetic phase diagram: Figs. 276, 277	73M3
$NaMnCl_3$	Fig. 280	Fig. 278		$\sigma_m(H, T)$: Fig. 279	78B1
	Figs. 281, 282			$\sigma_m(H)$: Fig. 281, spin-flip field vs. *T*: Fig. 282	80F1
	Figs. 283, 284				85P1
$KMnCl_3$	100 Fig. 286			EPR: Fig. 285, AFMR: Fig. 286	65K1
(β-phase)	Fig. 287 2.1			can be obtained in two orthorhombic (space group Pb nm), but distinctly different crystallographic phases, (α, β), one of these phases (β) undergoes a transition to magnetically ordered (antiferromagnetic) phase below 2.1 K; the magnetic structure below T_N consists of a spiral propagating along *b* axis with a period of 3.17 *b*, the magnetic moment per Mn^{2+} is 3.78 μ_B and no component was observed along *b* axis	79G1
(α-phase)	Fig. 288 102		4.6	p_A magnetic moment per Mn^{2+} ion below T_N, magnetic structure below T_N is antiferromagnetic G-type, $\sigma(T)$: Fig. 289, induced rotation on the Cl^- octahedra: Figs. 290, 291	82G1

(continued)

Compound	Θ [K]	T_C [K]	T_N [K]	χ	p_A [μ_B]	p_{eff} [μ_B]	Remarks	Ref.
$KMnCl_3$ (continued)								
(β-phase)	1.5 − 1.9		2.1 98	Figs. 294, 295, 297	Fig. 296	5.91	$p_{Mn}(H)$: Fig. 296, below 98 K two sublattice antiferromagnetic, with increasing applied field magnetic structure changes from helical into fan structure and then into induced ferromagnetic state, $\Theta = 1.5$ and 1.9 K for $T > 24.3$ and $T < 23$ K respectively, $J_1/k = 0.06$ K, $J_2/k = -0.07$ K, $J_3/k = -0.11$ K ($J_{1,2}$: intrachain exchange interactions, J_3: interchain exchange interaction; see [84F1])	83V3, 84F1
(α-phase)			98	Fig. 292			$J_1/k = -4.1$ K, $J_2/k = -1.0$ K, $\sigma_m(H)$: Fig. 293, hysteresis loops: Fig. 293	89F1, 84F1
$KMnCl_3 \cdot 2H_2O$			Fig. 298				$\sigma/\sigma_0(T)$: Fig. 299, $C_p(T)$: Fig. 298	68F3
$KFeCl_3$	10		16 Fig. 301	Fig. 300	4.1	5.5	T_N from neutron diffraction, space group Pnma, antiferromagnetically coupled ferromagnetic chains parallel to the b axis, antiferromagnetic axis along b, magnetic moment $p_A = p_{Fe^{2+}}$ calculated from 4.2 K pattern	74G1
			18.5 Figs. 303, 304				from Mössbauer experiment, $H_{eff}(T)$: Fig. 302	74P2, 75P1, 77K2
$KCuCl_3$				Fig. 305			at 17.5 K ordering takes place: Fig. 305	67M1

$RbMg_xMn_{1-x}Cl_3$			Fig. 306 Fig. 24	Figs. 308, 309			EPR peak halfwidth vs. T: Figs. 306, 307, values for $\chi_g(0\,K)$ in $10^{-6}\,cm^3/g$ are given in χ-column.	79P1
x = 0.45	− 190		41	250			in paramagnetic region $\Delta\chi_g^{-1}/\Delta T = 90\,g/cm^3\,K$	
x = 0.35	− 120		55	103			in paramagnetic region $\Delta\chi_g^{-1}/\Delta T = 75\,g/cm^3\,K$	
x = 0.18	− 140		78	56			in paramagnetic region $\Delta\chi_g^{-1}/\Delta T = 61\,g/cm^3\,K$	
x = 0	− 210		91	40			in paramagnetic region $\Delta\chi_g^{-1}/\Delta T = 50\,g/cm^3\,K$	
$RbMg_xFe_{1-x}Cl_3$							magnetic transition temperature: Fig. 310, magnetic correlations: Fig. 311, from neutron diffraction	90H1
$RbVCl_3$				Fig. 312			$g = 1.98$	77N2
		19 Fig. 313			2.3		p_A magnetic moment per V^{2+}, extrapolated to 0 K.	83H1, 85H1
$RbCrCl_3$							possible scheme for the $\alpha(P6_3/mmc)(470\,K)$ −β(monoclinic) (201 K) − γ(monoclinic) structural phase transitions	80P1
				Figs. 314···316			$g = 1.988$, two crystallographic phase transitions: Fig. 315	84T1
$RbMnCl_3$			86				from EPR: Fig. 285	65K1
			Figs. 317, 318				AFMR, resonance field vs. T: Fig. 317, resonance amplitude vs. T: Fig. 318	66S2
			Fig. 319				$\sigma(T)$: Fig. 320, $p_{Mn} = 3.9\,\mu_B$ and $4.2\mu_B$ calculated from the relation of the (100) + (003) + (101) and (102) magnetic lines intensities at 4.2 K to nuclear intensities, respectively	71M2
			94		3.9		from neutron diffraction	
			95			5.3	from susceptibility	

(continued)

Compound	Θ [K]	T_C [K]	T_N [K]	χ	p_A [μ_B]	p_{eff} [μ_B]	Remarks	Ref.
$RbMnCl_3$ (continued)			94.6	Figs. 321, 323			$\sigma(H)$: Fig. 322, antiferromagnet with weak ferromagnetism suggested, $\sigma_s = 0.41$ Gcm^3/g at 78 K	75F3
			Fig. 152				T_N vs. magnon frequencies: Fig. 152	76B3
			94				σ_m: Fig. 324	78B5
	− 204		94.6	17.9			in χ column χ_m [10^{-3} cm^3/mol] at T_N for the hexagonal phase	79F1
	− 263		112.5	Fig. 325			for cubic (high-pressure) phase	
			Fig. 306					79P1
					Fig. 326		$p_{Mn}(H)$: Figs. 326, 327	81M1
			94.6				ferroelastic phase transition at 272 K, below 272 K a metastable twin structure is formed which consists of all possible orientation states and twin boundaries of one W-type wall; domain walls parallel to the basal plane of crystal; symmetry of the ferroelastic phase not higher than monoclinic with the diad axis perpendicular to the C_6 axis of the paraelastic phase suggested	84B1
	− 193			10		6.26	for T from 130 to 300 K, in χ column χ_m [10^{-3} cm^3/mol] at 293 K, at 4.2 K	88P1
				Fig. 329			$\chi_m = 17.0 \cdot 10^{-3}$ cm^3/mol, $\chi_m(H)$: Fig. 330	
α-$RbMnCl_3 \cdot 2H_2O$							heat capacity vs. T: Fig. 328	77K3
$RbMnCl_3(:Sm^{3+})$	− 135		96	10.1		5.85	for T from 130 to 300 K, in χ column χ_m [10^{-3} cm^3/mol] at 293 K, at 4.2 K	88P1
				Fig. 329			$\chi_m = 36.6 \cdot 10^{-3}$ cm^3/mol, for $Sm^{3+} < 1\%$, anomaly of $\chi_m(T)$ at 30 K is suggested to be connected with the onset of short-range ordering	

$RbFeCl_3$	− 50			Fig. 331	6.70	for $T < 105$ K, max. $\chi_m^{\parallel} = 97.7 \cdot 10^{-3}$ cm^3/mol, $\chi_m^{\parallel}(0) = 39 \cdot 10^{-3}$ cm^3/mol, from susceptibility $g_{\parallel} = 4.73$ and $J_{\parallel}/k = -11.5$ K, maximum of $\chi_m^{\parallel}$ at 12.6 K	69A1
	− 3			Fig. 332	6.35	for $T < 85$ K, max. $\chi_m^{\perp} = 0.492$ cm^3/mol, $\chi_m^{\perp}(0) = 0.492$ cm^3/mol, from susceptibility $g_{\perp} = 4.49$ and $J_{\perp}/k = -3.8$ K, maximum of $\chi_m^{\perp}$ at 0 K ($J_{\parallel}, J_{\perp}$: nearest neighbor exchange constant between magnetic ions and along c and in c plane, respectively).	
	− 18				6.09	for $T > 105$ K	
	10				5.8	for $T > 85$ K, linear chain antiferromagnet	
	− 134	109		Fig. 334	5.58	polytypes suggested: Fig. 333, $\sigma(T)$: Fig. 334, high-pressure phase suggested	71L1
			2.55				73M1
	− 3			Fig. 335	5.95	for $T < 80$ K, $J_{\perp}/k = -1.6$ K, $J_{\parallel}/k = 6$ K, $g_{\perp} = 4.25$, $g_{\parallel} = 4.6$ ($J_{\parallel}$, $J_{\perp}$: see above [69A1])	73W1
	8			Fig. 337	5.52	for $T > 80$ K, $\sigma_m(H)$: Fig. 336; orders at 2.45 K, from Mössbauer effect	74M1
			2.55	Figs. 338, 339			75E1
						heat capacity: Fig. 340, phase diagram: Fig. 341	81H1
			Fig. 342			softening of magnetic excitations in small region around $q = 0$ observed, hexagonal $P6_3/mmc$	81Y1

(continued)

Compound	Θ [K]	T_C [K]	T_N [K]	χ	p_A [μ_B]	p_{eff} [μ_B]	Remarks	Ref.
$RbFeCl_3$ (continued)				Fig. 344	Fig. 343		three transitions at $T_{N1} = 2.5$ K, $T_{N2} =$ 2.35 K, $T_{N3} = 1.95$ K; one-dimensional chains in which spins interact feromagnetically with XY-like anisotropy, these chains are coupled antiferromagnetically in the hexagonal basal plane; for $T < T_{N3}$ 120°-structure with first order phase transition; two different incommensurate structures in the basal plane at $T_{N3} < T < T_{N2}$ and $T_{N2} < T < T_{N1}$ of the second order; $g_{\perp} = 3.89$, $g_{\parallel} = 2.35$, $2J_{\perp}/k = 6.3$ K, $2J_{\parallel}/k = 2.9$ K, $D/k = 23.9$ K	82W1
							magnetic ordering observed at 2.55 K, at 4.2 K magnetic ordering occurred along c axis of single crystal with applied fields $B > 1T$,	83B3
							phase diagram	83S3
							possibility of a triple-Q incommensurate phase suggested	83S2
				Figs. 345···348			triangular XY-like antiferromagnet, shows successive incommensurate – incommensurate – commensurate phase transitions with decreasing temperature in zero field: Fig. 349, new ordered states found in the magnetic field between 5 and 22 kOe: Fig. 349, magnetic after-effects observed in the commensurate phase at the lowest temperatures in zero field: Figs. 345···348	83W1
			1.95				NMR echo dependence on r.f. field H_1: Fig. 350, NMR echo dependence on H: Fig. 351	83W2

$RbFeCl_3 \cdot 2H_2O$		remanent magnetization: Figs. 352···355	85K2
		$p_{Fe}(H)$: Figs. 356···358, $p_s = 2.6\ \mu_B$ per Fe, $g_{\parallel} = 2.6$	88A2
		magnetic field induced phase transition	90S3
		$\sigma(T)$: Fig. 359, magnetic heat capacity $C_{mag}(T)$: Fig. 360	77K1
		$\sigma_m(H)$: Fig. 361, magnetic structure: Figs. 362, 363	78B4
	11.96	spin-cluster resonance shows characteristics of ferromagnetic chain and an antiferromagnetic chain for external magnetic field parallel to *c*- and *a* axes respectively, experimental data for $\boldsymbol{H} \parallel c$ could be described by pure Ising model with $J_b/k = 0.76$ K, $J_c/k = -0.21$ K, (J_{bc}/k (diagonally in the *bc* plane) = -0.13 K, $p\boldsymbol{H}/\lvert\boldsymbol{H}\rvert$ (magnetic moment per Fe along *c*) = $1.5\ \mu_B$, correction for demagnetizing field, magnetic structure in ordered state: Fig. 362, phase diagram: Fig. 364	79V1
	11.96	phase diagram: Fig. 364, magnetic structure: Fig. 362, from spin-cluster excitations $p_{Fe} = 4.2\ \mu_B$ and for $\boldsymbol{H} \parallel c$, $p_{Fe} = 1.6\ \mu_B$, from spin-cluster resonance $p_{Fe} = 4.5\ \mu_B$ and for $\boldsymbol{H} \parallel c$, $p_{Fe} = 4.5\ \mu_B$, from NMR $p_{Fe} = 4.6\ \mu_B$, from neutron diffraction $p_{Fe} = 3.9\ \mu_B$, from magnetization for $\boldsymbol{H} \parallel c$, $p_{Fe} = 1.6\ \mu_B$	80V2
	Fig. 365	Mössbauer relaxation study of domain walls in the quasi 1D antiferromagnet, proposed magnetic structure: Fig. 362	85S1

(continued)

Compound	Θ [K]	T_C [K]	T_N [K]	χ	p_A [μ_B]	p_{eff} [μ_B]	Remarks	Ref.
$RbFeCl_3 \cdot 2H_2O$ (continued)							$\sigma_m(H)$: Fig. 366, magnetic structure; Fig. 363, magnetic moment orientation: Fig. 367, 3D long range order magnetic peak vs. H: Fig. 368	85S2
							magnetic chain and soliton: Fig. 369, magneto-light absorption: Fig. 370	90T2
$RbFeCl_3 \cdot 2D_2O$							magnetic structure: Fig. 363	83S3
			11.85				magnetic phase diagram: Figs. 371···374	85S2
$RbFeCl_{3-x}Br_x$							optical Raman and magnetic circular dichroism spectra of single crystal, temperature dependence of absorption intensities vs. T; Fig. 375	82B1
x = 0			2.5					
x = 3			5.5					
			Figs. 376, 377		Fig. 378		mixed pseudo-one-dimensional ferromagnetic-antiferromagnetic system; on the $RbFeCl_3$ side of the phase diagram the magnetic correlations appear finite for x as small as 0.03, the periodicities of these correlations are similar to those of the incommensurate and commensurate magnetic phase of pure $RbFeCl_3$, on the $RbFeBr_3$ side of the phase diagram the ordering vector remains at the H point ($Q = (1/3, 1/3, 1)_N$), at an intermediate composition (x = 1.0) there are no detectable one-dimensional ferro- or antiferromagnetic correlations down to 1.3 K, $p_{Fe}(H)$: Fig. 378, ordering temperature: Figs. 376, 377, phase diagram: Figs. 379, 380	89H1

x = 0.03					magnetic correlations: Fig. 381 from neutron diffraction, magnetic transition temperature: Fig. 310	90H1
$RbCoCl_3$		28			three magnetic phases; a quasi 1D phase for $T > T_N^1 \approx 28$ K, a partly ordered 3D phase for $T_N^1 > T > T_N \approx 11$ K, and a differently ordered 3D phase for $T < T_N$, band parameters vs. T: Fig. 382	83L1
		11				
			Fig. 383			89P1
					$p_{Co}(H)$: Fig. 384	90H2
$RbCoCl_3{:}^{57}Fe$		28			T_N from Mössbauer experiment: Fig. 385	88B1
		Fig. 385				
		11				
$RbCoCl_3 \cdot 2H_2O$		2.975	Fig. 386		$T_N(H)$: Figs. 387, 388	75F2
$RbNiCl_3$	− 100	11	Figs. 389, 525	3.34	p_{eff} and Θ-values for polycrystal, $g = 2.36$ from susceptibility, for single crystal; $g_{\parallel} = 2.21$ and $g_{\perp}$ 2.23 from ESR at 77 K for $\boldsymbol{H} \parallel c$ and $\boldsymbol{H} \perp c$, respectively; nearest neighbor exchange constants $J/k = -11.0$ K and -14.1 K from T_{max} and -12.1 K and 16.1 K from χ_{max} for $\boldsymbol{H} \parallel c$ and $\boldsymbol{H} \perp c$, respectively	69A1
		11			magnetic structure consists of an antiferromagnetic sequence of the spin along c axis chains, along with a screw spiral which is commensurate with the lattice; p_s for Ni^{2+} extrapolated to 0 K 1.6 μ_B, magnetic structure: Fig. 390	70M2, 71M1

(continued)

Compound	Θ [k]	T_C [k]	T_N [k]	χ	p_A μ_B	p_{eff} μ_B	Remarks	Ref.
$RbNiCl_3$ (continued)	− 130					3.3	Curie constant 1.398 cm^3/g	71E1
				Fig. 391			$J = 13.2\ cm^{-1}$, $D = 1.1\ cm^{-1}$ (J: nearest neighbor exchange constant, D: single ion anisotropy constant)	72M3
			11.15		1.3		$p_A = p_{Ni^{2+}}$ extrapolated to 0 K, magnetization vs. T: Fig. 392	72Y1
	− 101		11.1				$g = 2.23$ from ESR, $J/k = 18.5$ K, $J'/J = 2 \cdot 10^{-2}$ (J, J': intra and interchain exchange constants of nearest neighbors)	74W1
							phase diagram: Fig. 393	79J1
			11.0 Fig. 394					83I1
							no light scattering from magnons observed	83L1
							magnetic field dependence of magnetostriction: Figs. 395···397	84R1
			11.2				ESR, at 19.3 kOe and 1.65 K a transition takes place from plane triangular structure phase to the spin flop phase, angle between c axis and spin moment $> 57°$, $g_{\parallel c} = g_{\perp c} = 2.23$, spin = 1	88T1
			Fig. 398				Ising-like Heisenberg antiferromagnet on triangular lattices for $p \parallel c$ axis $T_N = 11.25$ K, for $p \perp c$-axis $T_N = 11.11$ K, p_{Ni} (T): Fig. 398, $\beta_{\parallel} = 0.27$, $\beta_{\perp} = 0.28$	91O1

$RbCuCl_3$			Figs. 399… 401		for γ-phase $C_\perp = 0.396$ and $C_\parallel = 0.398$ cm³ K/mol, $\Theta_\perp = -9$ and $\Theta_\parallel = -11$ K; for β-phase $C_\perp = 0.396$ and $C_\parallel = 0.398$ cm³ K/mol, $\Theta_\perp = 0$ and $\Theta_\parallel = 2$ K; (perpendicular and along *c*), structural phase transitions at 339 K from α-phase ($P6_3/mmc$) to β-phase (Pbcn) and at 260 K to γ-phase (C_2)	86T1
$RbCdCl_3$					structural phase transitions by NMR and NQR, at 387.5 K first order, at 363 K second order and at 340.5 K first order phase transition	78P1
$Rb_{1-x}Cs_xFeCl_3$		Figs. 402, 403 2.5			magnetization vs. *T*: Fig. 403, magnetic correlations observed below 2.5 K with *Q* vector similar to that of the IC_1 phase of $RbFeCl_3$, changing to that of the commensurate phase on application of a magnetic field, the correlation of the moments in an applied magnetic field depends on whether the sample is cooled in the magnetic field or cooled in zero field, as found in other random systems	86H1
$CsMg_{1-x}Co_xCl_3$						
x = 0.01	0		Fig. 404	6.45	for $T < 160$ K, *H* parallel to *c* axis, $g = 7.4$ and 7.32 obtained from susceptibility and ESR at 77 K respectively	69A1
	0			3.5	for $T < 25$ K, *H* perpendicular to *c* axis, $g = 4.3$ from susceptibility	
	− 32			5.7	for $T > 160$ K, *H* parallel to *c* axis	
	37			5.8	for $T > 25$ K, *H* perpendicular to *c* axis	

(continued)

Compound	Θ [K]	T_C [K]	T_N [K]	χ	p_A [μ_B]	p_{eff} [μ_B]	Remarks	Ref.
$CsMg_{1-x}Co_xCl_3$ (continued)							disordered one dimensional antiferromagnet for $0 < x < 0.2$ and $0.6 < x \leq 1$ from Raman effect	82J1
$0.9 < x \leq 1$			Figs. 405, 498	Figs. 406, 407			$T_{N1} \approx 21$ K is suggested to be the temperature where 2/3 of short-range ordered magnetic chains align antiferromagnetically, $T_{N2} \approx 9$ K is the temperature where the uncorrelated magnetic chains are ordered; see also $CsCoCl_3$: Mg [90M4]	87M3
$CsVCl_3$				Fig. 408			$CsNiCl_3$ structure type, space group $P6_3/mmc$, $g = 1.98$, $J/k = -115K$	77N1
						2.84	$g = 1.98$	77N2
			Fig. 409 13.3				spin structure: Fig. 410	82H1
			Fig. 313		1.9		single crystal, p_A magnetic moment per V^{2+} extrapolated to 0 K	83H1
			13.3	Fig. 411			one dimensional Heisenberg antiferromagnet, 1D chains with $J/k = -169$ K along c axis, chains are coupled by the antiferromagnetic interaction $J'/k = -0.046$ K	83H2
					1.97		$J'/J = 2 \cdot 10^{-4}$	85H1
			Fig. 412				from neutron diffraction	85T1
$CsCrCl_3$			Fig. 413 16		3.2		p_A from neutron diffraction; crystallographic first order phase transition at 171 K, spins lie in the basal plane, antiferromagnetic chains along c axis, at 4.5 K strong dispersion of the spin-wave energy along c-axis, no measurable dispersion perpendicular to c axis	79D1

		Fig. 414	Figs. 414, 415, 316			structural phase transition at 171 K caused by cooperative Jahn-Teller effect from a hexagonal α-phase $P6_3/mmc - D_{6h}^4$ to a monoclinic γ-phase $C2–C_2^3$, $g = 1.932$	84T1
$CsMnCl_3$		69				EPR: Fig. 285, AFMR: Fig. 416	65K1
	−145	Fig. 417		5.5	6.13	$p_A = p_{Mn}$ from neutron diffraction	71M3
		67					
		Figs. 418 419	Fig. 418			torque vs. T: Fig. 419	78Y2
		67					80A2
$CsMnCl_3 \cdot 2H_2O$			Figs. 420, 421				68S1
						$\sigma(\phi)$ at 1.07 K: Fig. 422	69B1
		4.89				from ^{35}Cl NMR vs. T: Fig. 423, proton frequencies vs. T: Fig. 424	69S1
		4.89				from neutron diffraction, critical exponent $\beta = 0.30$	70S2
		Figs. 425, 426					
						short-range order effect on ESR	72N1
					4.88	magnetic phase diagram: Fig. 427, $\Delta l/l$ vs. T: Fig. 428	73B2
		Figs. 427, 428					
						magnon sideband	74E1
						spin wave analysis	75I2
						magnetic heat capacity: Fig. 429	75K1
		4.89				$\sigma(T)$ from NMR: Fig. 430	76J1
						$\chi_m(H)$: Figs. 431 ··· 433	76K4
						heat capacity: Fig. 434	77K3
						magnetic phase diagram near the bicritical point	80V1
		4.90		4		p_A at 0 K from neutron diffraction	82A1
		4.89					82N1
		4.886	Figs. 435 ··· 439			susceptibility dispersion: Fig. 435...439, bicritical point at 4.354 K	85C1

(continued)

Compound	Θ [K]	T_C [K]	T_N [K]	χ	p_A [μ_B]	p_{eff} [μ_B]	Remarks	Ref.
$CsMnCl_3 \cdot 2D_2O$			5.06				triple point 4.34 K, 21.1 kOe: Fig. 440 C_p vs. T: Fig. 441	72F1
$CsMn_{1-x}Co_xCl_3 \cdot 2H_2O$							EPR investigations	78N1
				Fig. 442			for $0.04 < x < 0.07$ oblique antiferromagnetic phase exists, χ anisotropy: Fig. 442, anisotropy constant K_1 (x) at 4.2 K: Fig. 443	80Y1
x = 0			4.90		4		p_A at 0 K for all samples from neutron diffraction, neutron magnetic reflection vs. x: Fig. 444	82A1
x = 0.035			4.80					
x = 0.05			4.74					
x = 0.075			4.66					
x = 0.0034			4.875					83V1
x = 0.04			4.71				bicritical filed 1.05 and 0.70T	
							phase diagram: Fig. 445, structure model: Fig. 446	84K1
							(H, x) phase diagram: Fig. 447, proposed spin structure: Fig. 448	84K2, 84K3
							for x = 0.032 $C_p(T)$: Fig. 452	85Y1
x = 0			4.89				phase diagram: Fig. 449, zero field	86N1,
x = 0.3			3.38				resonance frequencies vs. x: Fig. 450, $\omega/\gamma - H$ diagrams: Fig. 451, magnetism of randomly mixed antiferromagnetism	90I1
							phase diagram: Fig. 447	86K1
							NMR vs. H: Fig. 455, transition temperatures: Figs. 453, 454	89K1

$CsMn_{0.96}Co_{0.04}Cl_3 \cdot 2D_2O$		4.80 Fig 456			T_N from neutron diffraction, easy axis tilts substantially (~ 40° of the easy axis of Mn containing salt to that of pure Co salt at 3.8K) even for such a dilute mixture	81A1
$CsMn_{1-x}Cu_xCl_3 \cdot 2H_2O$		Fig. 457				80S1
					H_b (bicritical field) is in good agreement with a theoretical calculation for a quasi 1D antiferromagnet with diamagnetic impurities	83V1
x = 0		4.882			$\mu_0 H_b = 2.014$ T	
x = 0.0042		4.800			$\mu_0 H_b = 1.955$ T	
x = 0.007		4.76			$\mu_0 H_b = 1.905$ T, phase diagram: Fig. 458	
x = 0.017		4.61			$\mu_0 H_b = 1.83$ T	
x = 0.0265		4.40			$\mu_0 H_b = 1.66$ T	
x = 0.034		4.32			$\mu_0 H_b = 1.585$ T	
$CsFeCl_3$	− 1		Figs. 459,	5.76	Θ and p_{eff} for $T \lesssim 90$ K,	73W1
	2		460	5.66	Θ and p_{eff} for $T \gtrsim 90$ K, perpendicular and parallel intrachain exchange constants $J_\perp/k = -0.5$K, $J_\parallel/k = -7$ K, $g_\perp = 4.20$, $g_\parallel = 4.6$	
					Mössbauer experiment down to 1.4 K shows no transition to three-dimensional ordered state	74M1
			Fig. 461		heat capacity: Fig. 462, phase diagram: Fig. 463	81H1
					from Mössbauer experiment no magnetic ordering down to 80 mK, magnetic ordering occurred in single crystals with field along *a* axis by $B \geq$ 3T at 2.2 K, and $B \geq$ 4T at 3.0 K, at 4.2 K no ordering could be induced with field up to 10T	83B3

(continued)

Compound	Θ [K]	T_C [K]	T_N [K]	χ	p_A [μ_B]	p_{eff} [μ_B]	Remarks	Ref.
$CsFeCl_3$ (continued)							light transmittance vs. H: Fig. 464	85T2
					Figs. 465···467		single ground state magnet, p_{Fe} (B): Figs. 465 ··· 467	87C1
$CsFeCl_3 \cdot 2H_2O$							$\sigma_m(H)$: Fig. 361, magnetic structure: Figs. 362, 363	78B4
			12.6				$\|J/k\| = 42K$, $\|J'/k\| = 0.32$ K (J, J': intra and interchain exchange constants), heat capacity vs. T: Figs. 468, 469	85K1
			12.75	Fig. 471			3D long range order, $\sigma_m(H)$: Figs. 471, 366, $\sigma(t)$: Fig. 470, magnetic moment orientation: Figs. 367, 363, magnetic peak vs. H: Fig. 368	85S2
							p_{Fe} (H): Fig. 472 orthorhombic crystal structure Pcca, exchange parameters along three principal axes $J_a/k = -40$ K; $J_b/k = -0.45$ K; $J_c/k = -0.15$ K, absorption spectra at various magnetic fields along a-axis at 4.2 K: Fig. 473, energy diagram: Fig. 474	89T1
							energy shift vs. H: Fig. 475	90O1
$CsFeCl_3 \cdot 2D_2O$							magnetic structure: Fig. 363	83S3
$CsCoCl_3$	− 90		Fig. 456			6.14	max $\chi_m^{\parallel} = 21.5 \cdot 10^{-3}$ cm^3/mol, $J_{\parallel}/k = -86$ K, $g_{\parallel} = 7.32$, $\chi_m^{\parallel}(0) = 5.8 \cdot 10^{-3}$cm^3/mol	69A1
	− 176					5.54	max $\chi_m^{\perp} = 18.5 \cdot 10^{-3}$ cm^3/mol, $J_{\perp}/k = -17$ K, $\chi_m^{\perp}(0) = 18.1 \cdot 10^{-3}$ cm^3/mol (along c axis and in c plane), linear chain antiferromagnet, single crystal $\chi(T)$: Fig. 476	69A1

21.5		two three-dimensional magnetic phase transitions: Fig. 477; from neutron diffraction at 4.2 K: collinear magnetic structure consists of ferromagnetic *ab* planes stacked antiferromagnetically along *c* axis and spins parallel to the *c* axis, $p_{Co} = 3.0$ and $2.3\ \mu_B$ according to powder and single crystal data, powder and single crystal data indicate the magnetic structure near 21.5 K to be different from that at 4.2 K, magnetic unit cell: Fig. 478	74M2
Fig. 481		sublattice magnetization vs. *T*: Figs. 479, 480; for sublattice 1, $R = 0.17$, $S_1 = 2.66$, $S_2 = 2.66$, $S_3 = -2.88$; for sublattice 2, $R = 0.07$, $S_1 = 1.70$, $S_2 = 2.34$, $S_3 = -3.21$; for sublattice 3, $R = 0.11$, $S_1 = 0.00$, $S_2 = 3.61$, $S_3 = -3.6\ \mu_B$ (*R*: magnetic structure factor)	78M1
20.82 Fig. 482		at 8 K second magnetic phase transition observed: Fig. 483	79B2
Figs. 484, 485			79Y3
Fig. 486		T_N from EPR: Figs. 486 ··· 488	81A2
		ESR due to propagating domain walls	83A1
21.5		phase transitions: Fig. 489	86I1
		spin frustration, by neutron diffraction, NMR and ESR measurements	86M1
Fig. 405	Fig. 407		87M3
		optical investigation, magnetic field dependence of absorption spectra	87M4
			78M1

(continued)

Compound	Θ [K]	T_C [K]	T_N [K]	χ	p_A [μ_B]	p_{eff} [μ_B]	Remarks	Ref.
$CsCoCl_3$ (continued)			Fig. 490 21.2 Fig. 491				trigonal and second order axial CEF parameters $B_0^2 = 2400\,cm^{-1}$, $B_0^4 = 2650\,cm^{-1}$, cubic CEF parameter $B_c^4 = -10150\,cm^{-1}$ from Mössbauer study, pseudo-one-dimensional Ising like antiferromagnet, Co^{2+} ions arranged in chains parallel to hexagonal c axis, $T_{N1} = 21.2$ K, $T_{N2} = 8.5$ K, spatial ionic Co^{2+} arrangement: Fig. 492	87W1
							NMR study, time averaged Co^{2+} moment $\langle p_{Co} \rangle = 2.3\ \mu_B$ suggested	89K2
							energy shift vs. H: Fig. 493	89M1
							p_{Co} vs. H: Fig. 384	90H2
							magnetic transition, magneto-optics	90H3
							^{133}Cs nuclear relaxation vs. T^{-1}: Fig. 494 μ^+ relaxation rate vs. T: Fig. 495	90M3
$CsCoCl_3$: Mg							0, 0.58, and 1.7% of Mg, diffuse neutron scattering vs. T: Fig. 496 and 497	90M4
$CsCoCl_3$: ^{57}Fe			Fig. 385				T_N from Mössbauer experiment: Fig. 385	88B1
$CsCo_{1-x}Zn_xCl_3$			Fig. 498					87M3
$CsCoCl_3 \cdot 2H_2O$							NMR frequency vs. T	71J1
			Figs. 499…501				magnetic moments arrangement: Fig. 519, $\sigma(T)/\sigma(0)$: Fig. 499, 502	72B1
			3.38	Figs. 503, 504			canted Ising linear chain lying in ac plane with moments 10° from c axis and a net moment in the a direction; heat capacity: Fig. 505, NMR frequency vs. T: Fig. 506, $\sigma(H)$: Fig. 507, $\sigma(T)$: Fig. 508, magnetization orientation dependence: Fig. 509, magnetic structure; Fig. 519, critical field vs. T: Fig. 510	72H1, 86M2

					magnetic phase diagram: Fig. 520	77B1
					heat capacity: Fig. 511	77K1
			Fig. 512		canted antiferromagnetic linear chain, $J/k = -52$ K, $g_b = 4.18$	77M1
		3.3 Figs. 513, 514			NMR ^{133}Cs relaxation time T_1: Figs. 513, 514	81G1
					$\sigma(H)$ at 1.3 K: Fig. 515, magnetic structure: Fig. 519, energy shift of lines vs. H: Fig. 516 and 517	86M2
					exciton-magnon transition	86M3
$CsCoCl_3 \cdot 2D_2O$		3.2			metamagnet with tricritical point at 2 K, $\sigma(H)$ and phase diagram: Fig. 518	74B2
		3.3			magnetic structure: Fig. 519, magnetic phase diagram: Fig. 520	76B1
					neutron diffraction study, $\sigma(T)$: Figs. 521 ⋯ 524, magnetic structure: Fig. 519, phase transition contains a tricritical point	77B1
$CsNiCl_3$	− 69	4.5	Figs. 525 ⋯ 527	3.41	linear chain antiferromagnet, from χ $g = 2.41$, T_m (maximum of χ_m maximum) $= 35$ K, maximum of $\chi_m \cdot 10^3 = 12.1$ and 12.2 cm^3/mol, maximum of $\chi_m(0) \cdot 10^3 = 3.5$ and 7.3 cm^3/mol, from χ_m(max) nearest neighbor interaction constant $J/k = -12.1$ and -16.1 K, from T_m $J/k = -11.0$ and -14.1 K, from ESR $g = 2.24$ and 2.28 for H parallel and perpendicular to c axis, respectively.	69A1
		4.5			magnetic structure consists of an antiferromagnetic sequence of chains with the spin along c axis, along with a screw being commensurate with the lattice; Ni^{2+} moment extrapolated to 0 K 1.5 μ_B, magnetic structure: Fig. 390	70M2

(continued)

Compound	Θ [K]	T_C [K]	T_N [K]	χ	p_A [μ_B]	p_{eff} [μ_B]	Remarks	Ref.
$CsNiCl_3$ (continued)			4.3				$p_{Ni}(T)$: Figs. 528, 529	71C1
			Figs. 530, 531				ultrasonic attenuation: Figs. 530, 531	72A1, 77A1
			Fig. 532				internal magnetic field vs. T: Fig. 532	72C1
				Fig. 391			$J = 11.1\ cm^{-1}$, $D = 0.25\ cm^{-1}$ (J: nearest neighbor exchange constant, D: single ion anisotropy constant)	72M3
			Fig. 533				two transition temperatures 4.27 and 4.86	73A1
			Fig. 534				neutron diffraction	73Y1
	− 56		4.5				$J/k = -13$ K, $J'/J = 6 \cdot 10^{-3}$, $g = 2.25$ from ESR (J,J': intra and interchain exchange constants of nearest neighbors)	74W1
							phase diagram: Fig. 535	75A1, 77A1
							ultrasonic attenuation	77A1
							nuclear spin lattice relaxation	78E1
			4.88	Figs. 536, 538			$\sigma(H)$: Fig. 537, phase diagram: Figs. 539, 540, spin rearrangement temperature $T_R(= T_{N2}) = 4.40$ K	79J1
							magnetostriction: Figs. 541 ···543	83R1, 84R1
			Fig. 544	Figs. 545, 546, 547			neutron diffraction, two transition temperatures	87K1
							p_{Ni} from NMR 3.6 μ_B, ESR 2.2 μ_B and neutron scattering 1.04 μ_B	88M1
							phase diagram: Figs. 539, 540	88P2

		4.84 (T_{N1}) 4.40 (T_{N2})			ESR investigations; at 19.2 kOe and 1.65 K a transition takes place from *ac* plane triangular structure phase to the spin-flop phase, $S = 1$, $\theta = 50.3°$ (angle between *c* axis and the spin moment), $g_{\parallel} = 2.18$ and $g_{\perp} = 2.20$, interchain exchange constant $J_1/k = -2.8 \cdot 10^{-2}$ K; T_{N1}, T_{N2} are transition temperatures corresponding to AF alignment of Ni spins along *c* and perpendicular part of moment	88T1
$CsCuCl_3$	3		Figs. 548 ⋯ 550	1.89	single crystal, linear chain antiferromagnet, T_m (temperature of χ_m maximum) = 7.7 K and 9.0 K, maximum of $\chi_m \cdot 10^3 = 18$ and 21 cm^3/mol, maximum of $\chi_m(0) \cdot 10^3 = 17.2$ and 20.1 cm^3/mol, from χ_m (max) nearest neighbor exchange constant $J/k = -7.3$ and -6.3 K, from T_m $J/k = -7.0$ and -6.0 K, from ESR $g = 2.19$ and 2.19 for *H* parallel and perpendicular to *c* axis, respectively	69A1
		10.4 Fig. 552	Figs. 551, 553		heat capacity vs. *T*: Fig. 552	69R1
			Figs. 554 ⋯ 556		linear antiferromagnet, $J/k = 4$ K, molar Curie constants (in cm^3 K/mol) for powder is 0.458, averages for single crystal 0.458, along *a* axis 0.464, along *c* axis 0.445, *g* (along *a* axis) = 2.22, *g* (along *c* axis) = 2.18	70R1
					magnetic hyperfine interactions	76 K2

(continued)

Compound	Θ [K]	T_C [K]	T_N [K]	χ	p_A [μ_B]	p_{eff} [μ_B]	Remarks	Ref.
$CsCuCl_3$ (continued)							NMR for ^{63}Cu at 169.2 MHz and 1.77 K, the analyses indicate that the spin direction of Cu^{2+} ions is parallel to z axis	76K5
							^{63}Cu NMR investigations	77D2
			10.70 Fig. 557				neutron diffraction, magnetic moments lie in c-plane and rotate by 5.1° along the chain or the c axis, the moments rotate by 120° in the c plane, magnitude of the magnetic moment is 0.58 μ_B at 4.2 K, critical exponent of magnetization is found to be $\beta = 0.358$ around T_N	80A1
			10.3 Fig. 558				$\Delta n(T)$: Fig. 558, dn/dT: Fig. 559	81H2
				Figs. 560 ···563			Jahn-Teller cooperative phase transition at $T_t = 418$ K, temperature interval of hysteresis $\Delta T_t = 3$ K, intrachain and interchain (J_1) exchange constants: $J/k = 24$ K, $J_1/J = -0.16$; for $T < T_t$ $C_\perp = 0.42$ cm^3 K/mol, $\Theta_\perp = 10$ K and $C_\parallel = 0.43$ cm^3 K/mol, $\Theta_\parallel = 8$ K; for $T > T_t$ $C_\perp = 0.41$ cm^3 K/mol, $\Theta_\perp = 32$ K and $C_\parallel = 0.42$ cm^3 K/mol, $\Theta_\parallel = 33$ K parallel and perpendicular to c axis, respectively, from Curie-Weiss law after correction of core diamagnetism and Van Vleck paramagnetism	81T1

			structural phase transition by ESR: Fig. 564	83T1
	10.7 Fig. 565		$T_t = 423$ K, C_{mag} vs. T: Fig. 565, $\sigma(T)$: Fig. 566 $\sigma(H)$: Fig. 567, intrachain and interchain exchange constants: $J/k = 25$ K, $J'/k = -3.9$ K	85F1
		Figs. 568, 569	for $T < T_t$ $C_{\parallel a} = 0.415$ cm³ K/mol, $\Theta_{\parallel a} = 30.5$ K, $g_{\parallel a} = 2.10$, $C_{\parallel c} = 0.419$ cm³ K/mol, $\Theta_{\parallel c} = 21.5$ K, $g_{\parallel c} = 2.11$; for $T > T_t$ $C_{\parallel a} = 0.385$ cm³ K/mol, $\Theta_{\parallel a} = 62.6$ K, $g_{\parallel a} = 2.03$, $C_{\parallel c} = 0.389$ cm³ K/mol, $\Theta_{\parallel c} = 50.8$ K, $g_{\parallel c} = 2.06$, for large single crystals a Jahn-Teller transition at 420 K takes place, a cooperative dynamic Jahn-Teller effect is occurring in domains	86H2
			phase diagram: Fig. 570, above critical field H_c sinusoidal phase with linear polarization of spins, new intermediate phase is due to a conical-point instability caused by weak dipole interactions, therefore in different temperature ranges two modulated structures are suggested	89G2
	Fig. 571		magnetic resonances: Fig. 571	90P2, 90P3
$CsPbCl_3$			structural phase transition by NMR-NQR, transition at 320 K first order, at 315 K second order and at 310 K first order	78P1
$CsPbCl_3{:}Mn^{2+}$			single crystal EPR at 35 GHz can be assigned to two and possibly four crystallographically equivalent sets of magnetically inequivalent sites	69C2

(continued)

Compound	Θ [K]	T_C [K]	T_N [K]	χ	p_A [μ_B]	p_{eff} [μ_B]	Remarks	Ref.
NH_4MnCl_3			100				from neutron diffraction	71S3
			105				EPR vs. T: Fig. 572, $g = 2.0099$	75T2
NH_4FeCl_3	– 2.5			Figs. 573, 574		5.79	Θ and p_{eff} for $T < 80$ K, perpendicular and parallel intrachain constants $J_\perp/k = -1.5$ K, $J_\parallel/k = -8$ K,	73W1
	3.0					5.65	Θ and p_{eff} for $T > 80$ K, $g_\perp = 4.15$, $g_\parallel = 4.5$	
NH_4NiCl_3	– 78		9	Fig. 575			hex. structure, $a = 6.927$, $c = 5.919$ Å, $g = 2.23$, intra- and interchain exchange constansts: $J/k = 16.9$ K, $\lvert J'/J \rvert = 2 \cdot 10^{-2}$	74W1
$TlMnCl_3$			118		4.7		from neutron diffraction, p_A per Mn atom at 4.2 K, transition to weak ferromagnetism observed at 108 K	71M4
	– 231			Fig. 576		6.25	molar Curie constant 4.89 cm^3 K mol^{-1}	73V1
	– 259	110.5	120 Fig. 579	Figs. 577, 580			weak ferromagnetic with moment parallel to [0 0 1], σ_s: Fig. 578, from neutron diffraction $T_N = 113$ K	77F1
α-$TlMnCl_3$			112	Fig. 581			$\sigma_m(T)$: Fig. 582	87V1
$TlFeCl_3$	– 5.5			Figs. 583, 337		5.90	Θ and p_{eff} for $T \lesssim 80$ K, $J_\perp/k = -2.7$ K, $J_\parallel/k = -8$ K,	73W1
	2					5.61	Θ and p_{eff} for $T \gtrsim 80$ K, $g_\perp = 4.20$, $g_\parallel = 4.6$	
$TlNiCl_3$			13				$J/k = -22$ K, $\lvert J'/J \rvert = 2 \cdot 10^{-2}$, $g = 2.24$ (J,J' intra and interchain exchange constants of nearest neighbors)	74W1
$KFeBr_3$			9.5 Fig. 584		3.7		from neutron diffraction, space group Pnma	82G3
$RbVBr_3$		30 Fig. 313			1.5		ordered magnetic moment per V^{2+} ion, extrapolated to 0 K, structure is hexagonal ($P6_3/mmc$)	83H1

		28			1.53		from neutron diffraction, $J'/J = 3 \cdot 10^{-4}$ (J,J': intra and interchain exchange constants)	85H1
$RbFeBr_3$				Fig. 585				75L1
					2.3		p_s/Fe, $p_{Fe}(H)$: Fig. 586; p_A magnetic moment per Fe^{2+}	88A2
$RbCoBr_3$			36		3.4		spins aligned along c axis with a $(+\ +\ -)$ configuration for the three signs in the basal plane; p_A per Co ion.	71M1
$RbNiBr_3$	− 155		23	Fig. 587			$J/k = -25.5$ K, $\lvert J'/J \rvert = 4 \cdot 10^{-2}$, $g = 2.22$ (J,J': intra and interchain exchange constants)	74W1
							resonance field vs. T: Fig. 588 $T_{N1} = 23.50$ K, $T_{N2} = 21.47$ K	90T1
$CsMg_xMn_{1-x}Br_3$ x = 0.92	− 17			Fig. 589		6.2		74M3
							magnetic interactions by neutron diffraction	87F2
$CsVBr_3$				Fig. 590			$CsNiCl_3$ structure type, space group $P6_3/mmc$, $g = 2.00$, $J/k = -90$ K to -80K, increases with temperature	77N1, 77N2
		22			1.8		magnetic neutron diffraction intensity vs. T: Fig. 313; p_A per V^{2+} ion, extrapolated to 0 K	83H1
		20.4 Fig. 313			1.87		from neutron diffraction, $J'/J = 1 \cdot 10^{-3}$ (J,J': intra and interchain exchange constants)	85H1
$CsMnBr_3$	− 167			Fig. 589		6.4		74M3
							magnon investigation by neutron scattering	77B5
			8.3	Fig. 591			$J'/k = -0.021$ K	82F1
			8.3				magnetic interaction by inelastic neutron scattering, spin-wave excitations of the triangular antiferromagnet, two spin wave branches of transverse and longitudinal symmetry described on the basis of a spin wave model	87F1

(continued)

Compound	Θ [K]	T_C [K]	T_N [K]	χ	p_A [μ_B]	p_{eff} [μ_B]	Remarks	Ref.
$CsMnBr_3$ (continued)			8.6				at $H = 37$ kOe along [100] the variation of the staggered magnetization measured at (1/3, 1/3, 1/3) vs. T shows two critical phase transitions at 8.9 and 7.2 K	87G1
			Fig. 592				from neutron scattering critical exponent $\beta = 0.22$	87M2
			8.42				antiferromagnet on stacked triangular lattice, critical exponent $\beta = 0.25$	88A1
							critical exponents from neutron diffraction $\gamma = 1.10$, $\nu = 0.57$ in the reduced temperature range $0.015 < \varepsilon < 0.1$ above T_N ($\varepsilon = T/T_N - 1$)	88K1
			8.32 Fig. 593				phase diagram: Fig. 621, $\beta = 0.29$, space group $P6_3/mmc$	89G1
							from neutron diffraction, $\gamma = 1.01$, $\nu = 0.54$, staced, XY-line antiferromagnet on a triangular lattice	89G3
			8.32				σ vs. T: Fig. 594	90A1
							(H, T) phase diagram: Fig. 595	90C1
				Figs. 596, 597				90M2
							magnetic phase diagram: Fig. 598	90P1
							phase diagram	90P3
				Fig. 599			σ vs. H: Fig. 600, 601, H_c vs. T: Fig. 602	91K1
$CsMnBr_3 \cdot 2H_2O$			5.75 Fig. 603				NMR, $\sigma(T)$: Fig. 604	74S3
							heat capacity: Fig. 605	77K3
				Figs. 606 ⋯612			dispersion of χ: Figs. 606 ⋯ 612	83C1

$CsMnBr_3 \cdot 2D_2O$					phase diagram: Fig. 613	78B3
					spin-flop system, $(\sigma_{st})^2(H)$: Fig. 614, (H^2,T) phase diagram: Fig. 615	79B1
					neutron scattering, phase diagram	80B1
$CsFeBr_3$	− 45	17	Fig. 616	6.02		74T1
$CsCoBr_3$		28 (14)	Fig. 618		heat capacity vs. T: Fig. 620, T_N: Figs. 617 ··· 620, in single crystal between T_N and 14 K magnetic structure is consistent with orthorhombic space group $Cmc2_1$ with one-third of antiferromagnetic cobalt chains being disordered and other two-thirds antiferromagnetically coupled in the basal plane, between 4 and 14 K the structure may be described by the space group $Cm'c2'_1$ and is similar to the collinear arrangement reported for $CsCoCl_3$ and $RbCoCl_3$ but with a small canting of about 10 deg, orthorhombic model: Fig. 621	75Y1
					magnons and excitons in 3D ordered phases	82L1
					domain walls propagation	82N2
					p_{Co} vs. H: Figs. 384, 619	90H2
$CsCoBr_3$:^{57}Fe		12			complete three dimensional magnetic order temperature T_{N2}	88B1
$CsNiBr_3$	− 95		Fig. 587		$g = 2.27$, $J/k = -18.5$ K, $\lvert J'/J \rvert = 3 \cdot 10^{-2}$ (J,J': intra and interchain exchange constants)	74W1
	− 85	Figs. 622 ···625	Figs. 625 ···628		heat capacity vs. T: Figs. 622 ··· 624, $T_{N2} = 11.75$ K, $T_{N1} = 14.25$ K, $J/k = -17$ K, $D/k = -1.5$ K	75B1, 77B4

(continued)

Compound	Θ [K]	T_C [K]	T_N [K]	χ	p_A [μ_B]	p_{eff} [μ_B]	Remarks	Ref.
$CsNiBr_3$ (continued)			11.75 14.25				isotropic Heisenberg system with $S = 1$, $J/k = -17.0$ K, $D/k = -1.5$ K, above 14.25 K 1D antiferromagnet	85T4
			Fig. 629				phase transition: Fig. 629	86I1
			14.2 11.5	Figs. 630, 631			$H - T$ phase diagram: Fig. 632	89P1
							resonance field vs. T: Fig. 633	90T1
$RbVI_3$		25			1.4		ordered magnetic moment per V ion, extrapolated to 0 K, crystallographic structure is hexagonal $P6_3/mmc$	83H1
		25			1.44			85H1
$CsVI_3$				Fig. 634			$g = 2.06$, $J/k = -67$ to -54 K increase with temperature, $CsNiCl_3$ structure type, space group $P6_3/mmc$	77N1, 77N2
		34.8 Fig. 313			1.6		ordered magnetic moment per V^{2+}, extrapolated to 0 K	83H1, 85H1
					1.64		neutron diffraction, $J'/J = 9 \cdot 10^{-3}$ (J,J': intra and interchain exchange constants)	
							interchain exchange constant $J_1/k = -7.5 \cdot 10^{-2}$ K	88T1
$CsCrI_3$							neutron diffraction, magnetic structure of β-$CsCrI_3$, low temperature space group Pbcn at 1.2 K is found to consist of an antiparallel sequence of ferromagnetic (0 0 1) planes with magnetic moment in the [0 0 1] direction of 3.16 μ_B	80Z1

$CsMnI_3$	11.1	Figs. 635, 636, 639	3.7	p_A per Mn atom at 0 K, $p_{Mn}(T)$: Fig. 637, $p_{Mn}(H)$: Fig. 638, $J/k = 9.1$ K, $J'/J = 1 \cdot 10^{-3}$, from neutron diffraction (J, J': intra and interchain exchange constants)	80Z2
	11.42 8.20			ESR investigations, at $H_c = 54$ kOe (at $T = 2$ K) transition from *ac*-plane triangular structure phase to the spin-flop phase, angle $\Theta = 55°$ between *c* axis and the spin moment, $g_\parallel = 1.99$, $g_\perp = 2.00$	88T1
				birefringence vs. T: Fig. 640 and 641	88I1
				NMR relaxation vs. T: Fig. 642	90M5
				ESR field vs. T: Fig. 643	90T1
	Fig. 644 Fig. 645			neutron diffraction: Figs. 644, 645 $T_{N1} = 11.41$ K, $T_{N2} = 8.21$ K	91H2

9.6.2 Figures

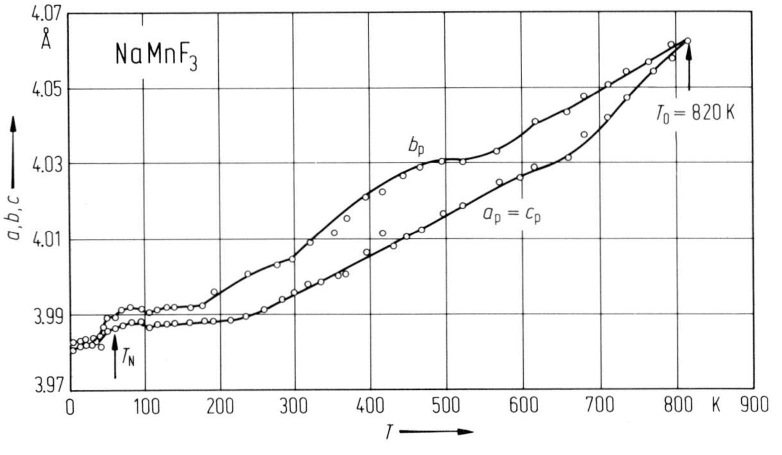

Fig. 1. $NaMnF_3$. Temperature dependence of lattice parameters, T_0: decomposition temperature [88R1]. See also Figs. 40 and 72.

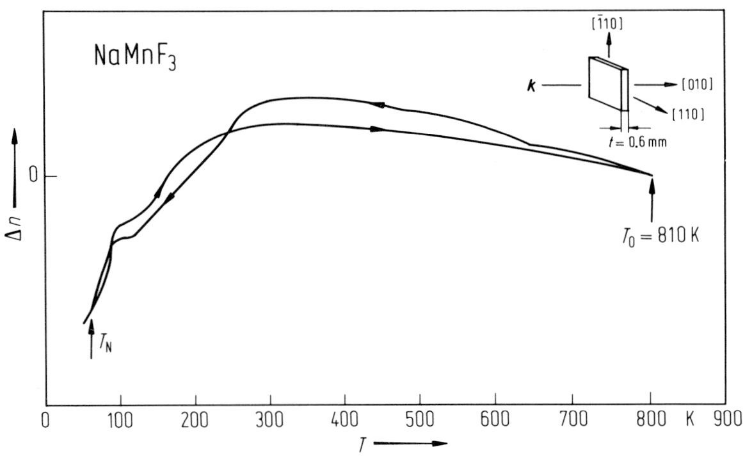

Fig. 2. $NaMnF_3$. Temperature dependence of optical retardation [88R1]. (Δn for $\lambda = 632.8$ nm.)

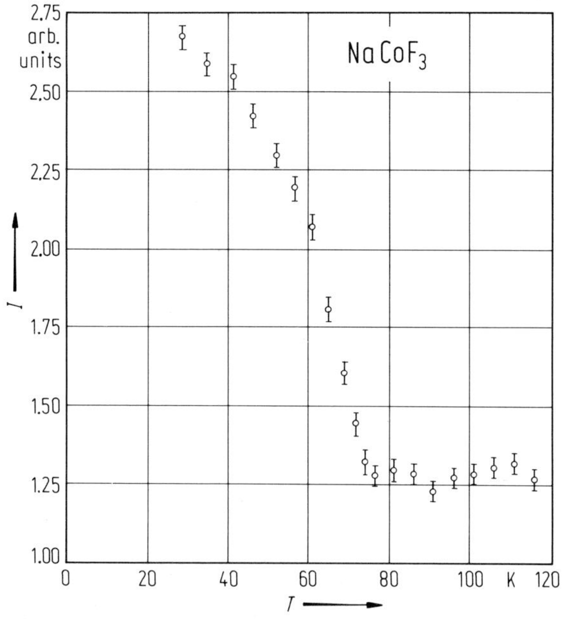

Fig. 3. $NaCoF_3$. Temperature dependence of the (0, 1, 1) and (1, 0, 1) magnetic neutron reflection intensity [70F1].

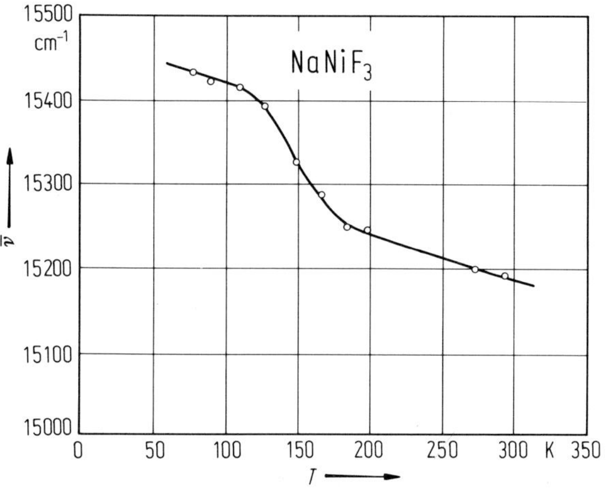

Fig. 4. $NaNiF_3$. Temperature dependence of the $^3A_{2g} \rightarrow {}^1E_g$ transition [65P1].

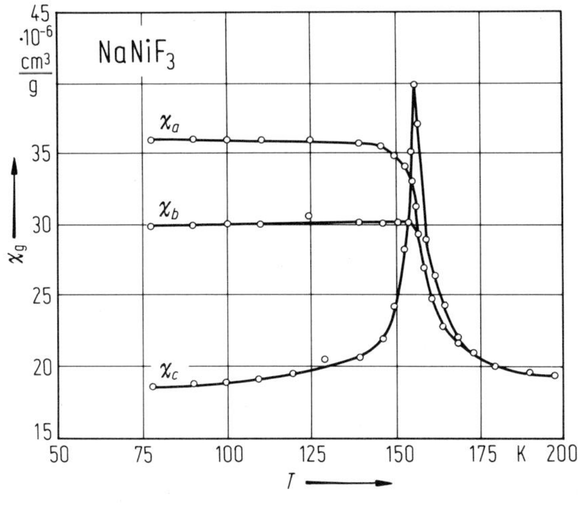

Fig. 5. $NaNiF_3$. Temperature dependence of χ_g for principal crystallographic axes [67J1].

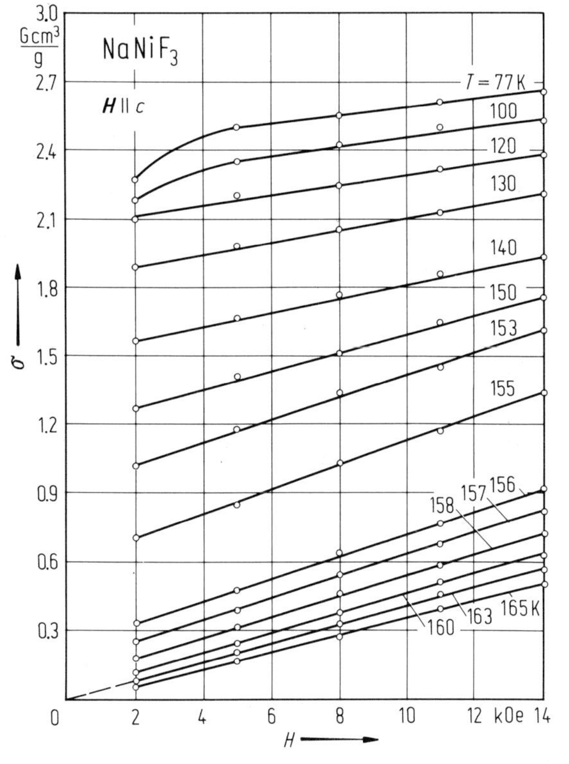

Fig. 6. $NaNiF_3$. Magnetic field dependence of σ along the c axis at various temperatures [67J1].

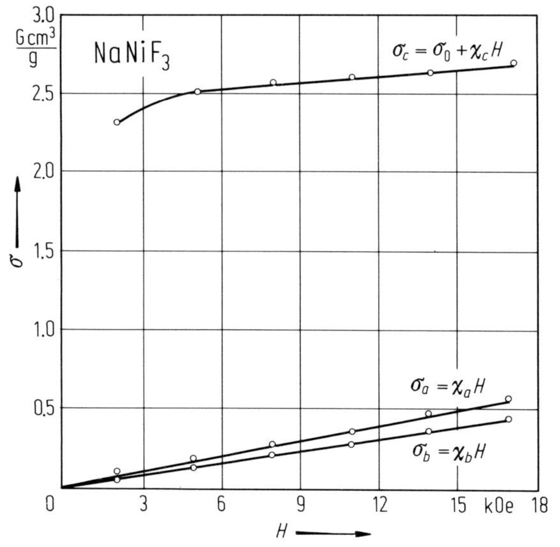

Fig. 7. $NaNiF_3$. Magnetic field dependence of σ along the principal crystallographic axes at 77K [67J1].

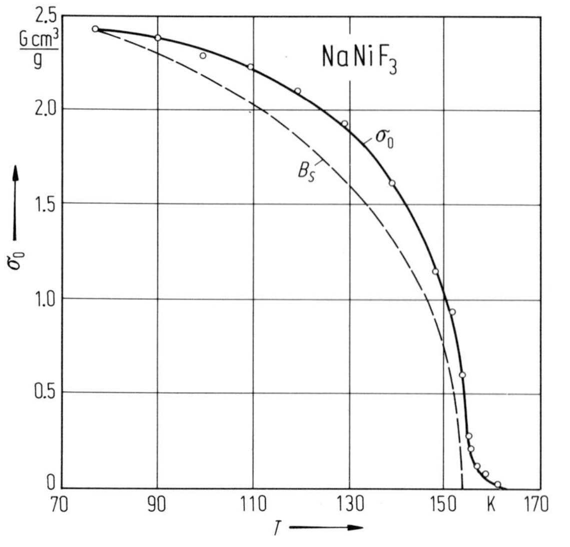

Fig. 8. $NaNiF_3$. Temperature dependence of the spontaneous moment σ_o along the c axis. B_s: Brillouin function for spin $S = 1$ normalized for $T = 77$K and $T = T_N = 156$K [67J1].

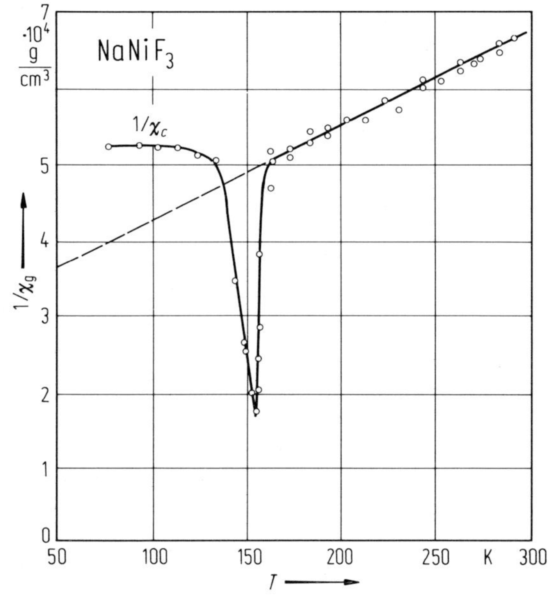

Fig. 9. $NaNiF_3$. Temperature dependence of χ_g^{-1} [67J1].

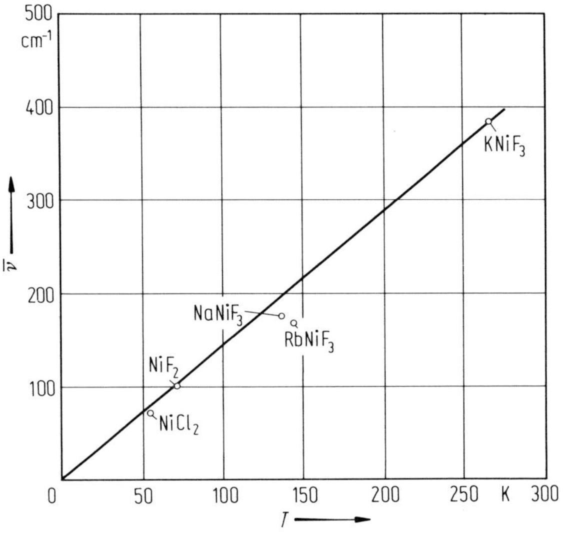

Fig. 10. $NaNiF_3$, $KNiF_3$, $RbNiF_3$. Dependence of maximum magnon frequency on magnetic transition temperatures [68B1].

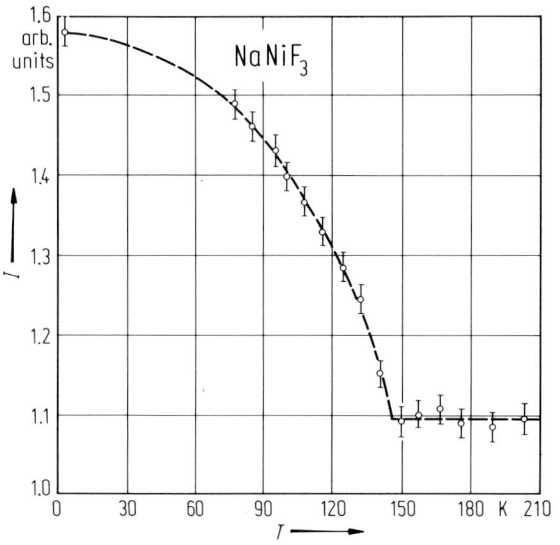

Fig. 11. $NaNiF_3$. Temperature dependence of the (0,1,1) and (1,0,1) magnetic neutron reflection intensity [68E1].

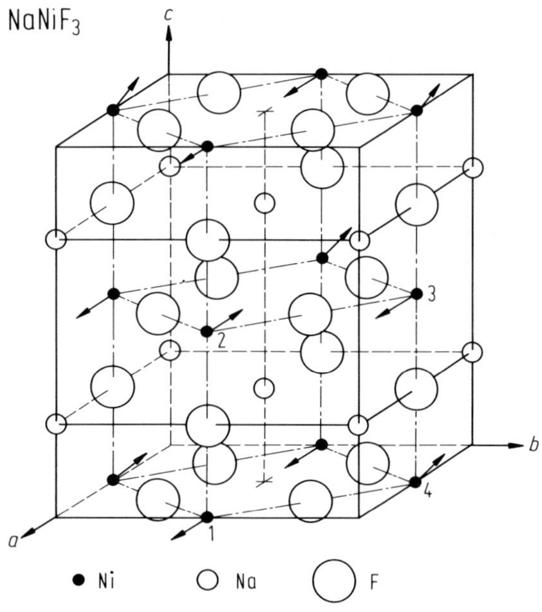

Fig. 12. $NaNiF_3$. Unit cell. There are four magnetic ions per unit cell [73P1].

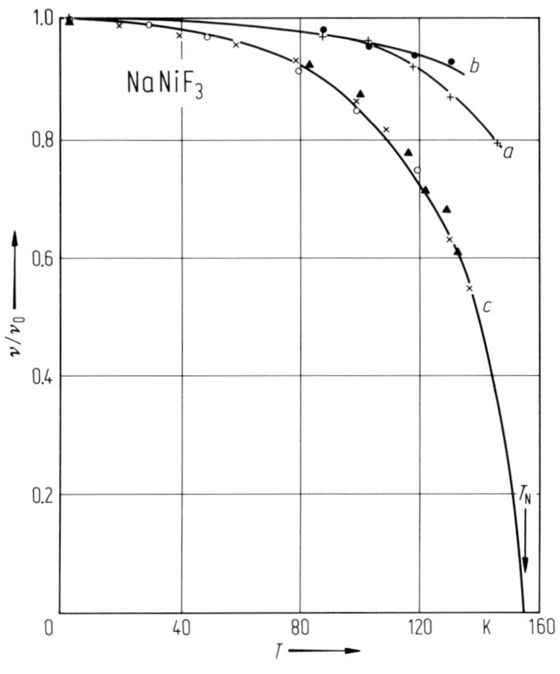

Fig. 13. $NaNiF_3$. Temperature dependence of FM (crosses) and (triangles) from [69G3] and AFM (open circles), two-magnon, + and (curve *a*), and (tentatively identified) four-magnon (full circles) and (curve *b*), normalized Raman frequencies at $H = 0$, (curve *c*) Brillouin law [73P1].

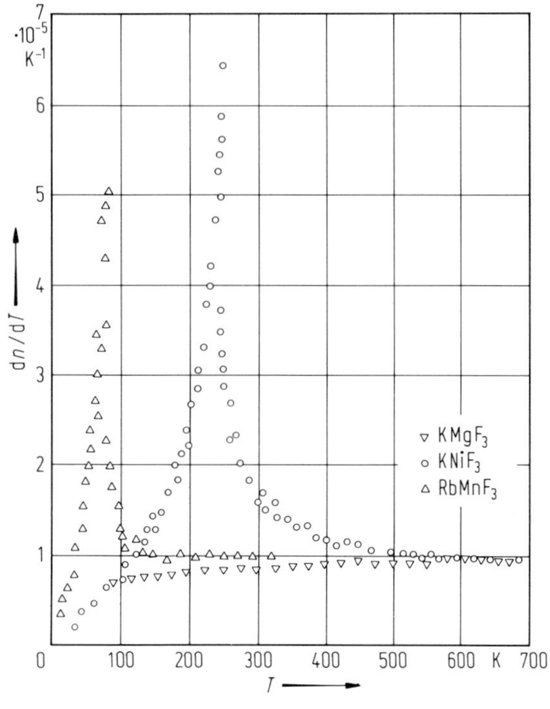

Fig. 14. $KMgF_3$, $KNiF_3$, $RbMnF_3$. Temperature dependence of dn/dT for $\lambda = 0.6328$ µm [84K4].

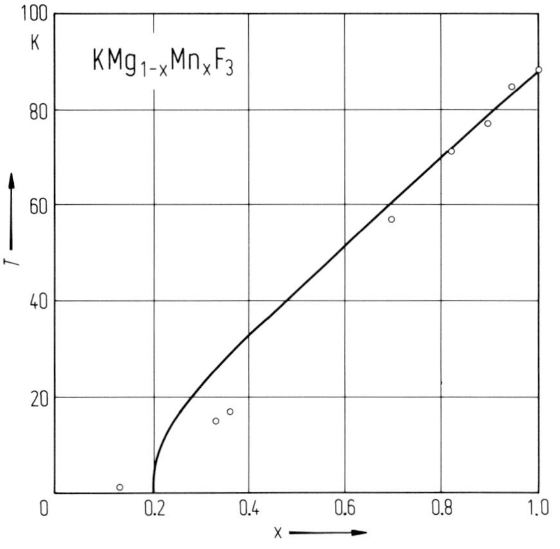

Fig. 15. $KMg_{1-x}Mn_xF_3$. Concentration dependence of the transition temperature at which the ESR field started to shift from the room temperature value [66H1].

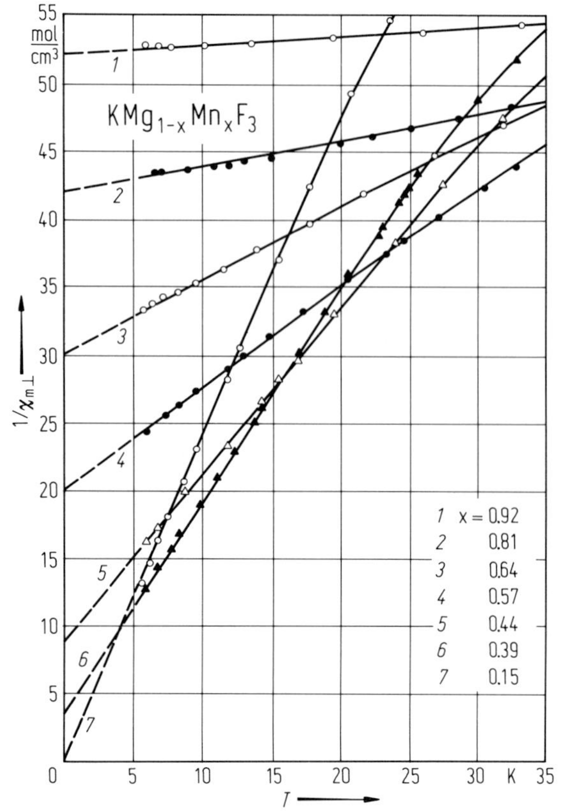

Fig. 16. $KMg_{1-x}Mn_xF_3$. Temperature dependence of $\chi_{m\perp}^{-1}$ at $H = 5 \cdots 10$ kOe for different x [70B1].

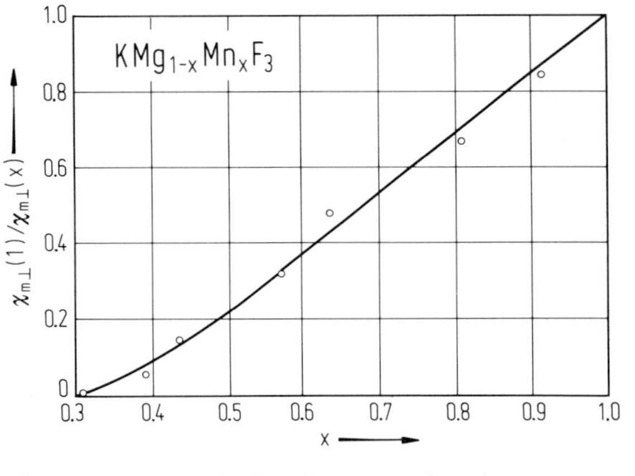

Fig. 17. $KMg_{1-x}Mn_xF_3$. Concentration dependence of zero temperature χ_m^{-1}. The zero temperature values are obtained by extrapolation of the experimental values measured at 9640Oe. For pure $KMnF_3$ $\chi_{m\perp}$ is corrected for the influence of canted ferromagnetism [70B1].

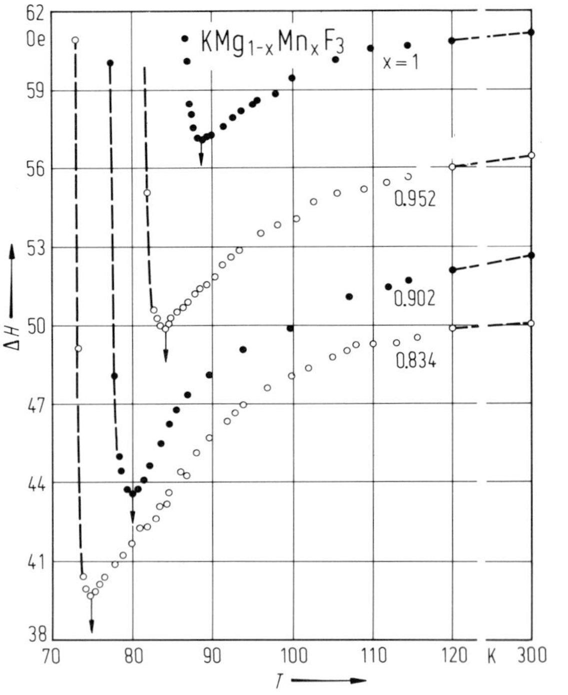

Fig. 18. $KMg_{1-x}Mn_xF_3$. Temperature dependence of EPR linewidth for single crystals with different x. Position of arrows indicate T_N [72G2].

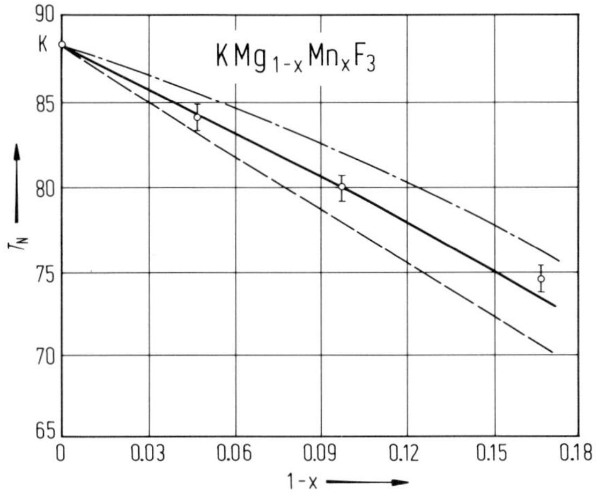

Fig. 19. $KMg_{1-x}Mn_xF_3$. Concentration dependence of T_N from EPR (circles). Three curves are calculated with different parameters [72G2].

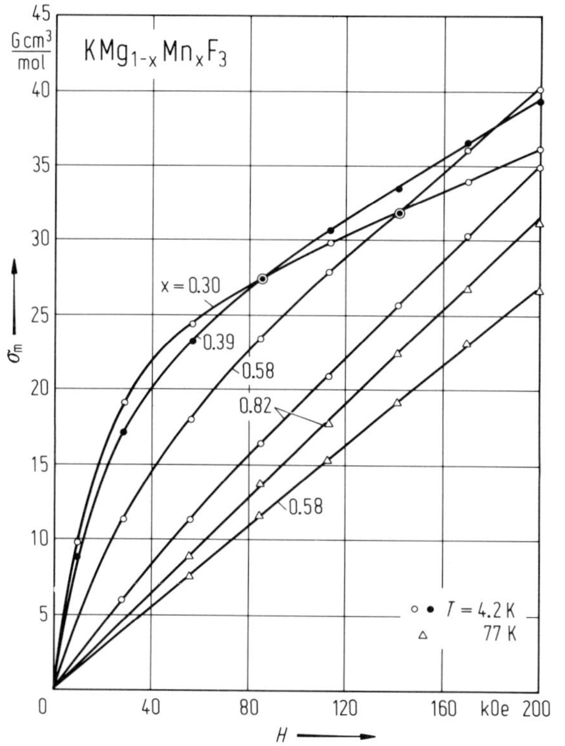

Fig. 21. $KMg_{1-x}Mn_xF_3$. Magnetic field dependence of σ_m for single crystals. This magnetization is independent of the crystal orientation [73B1].

Fig. 22. $KMg_{1-x}Mn_xF_3$. Concentration dependence of the effective Curie constant measured at low temperatures for the perpendicular susceptibility. $C(1)$ is the high-temperature Curie constant of one mole manganese compound [73B1].

→

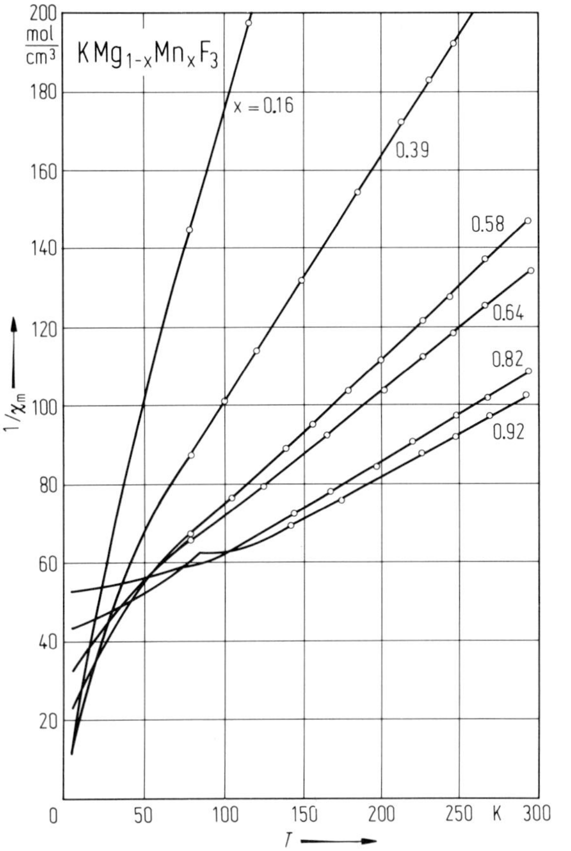

Fig. 20. $KMg_{1-x}Mn_xF_3$. Temperature dependence of χ_m^{-1} for single crystals at H from 7 to 10 kOe. At lower temperatures H = 9.640 kOe. χ_m can be considered as perpendicular susceptibility due to the lack of anisotropy below the ordering temperature [73B1].

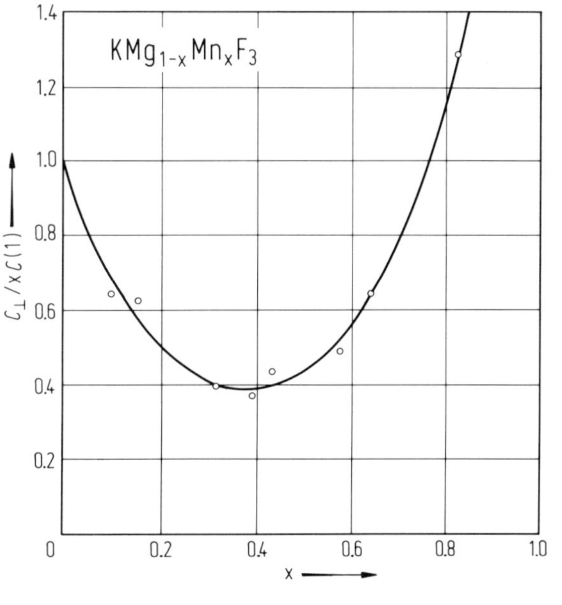

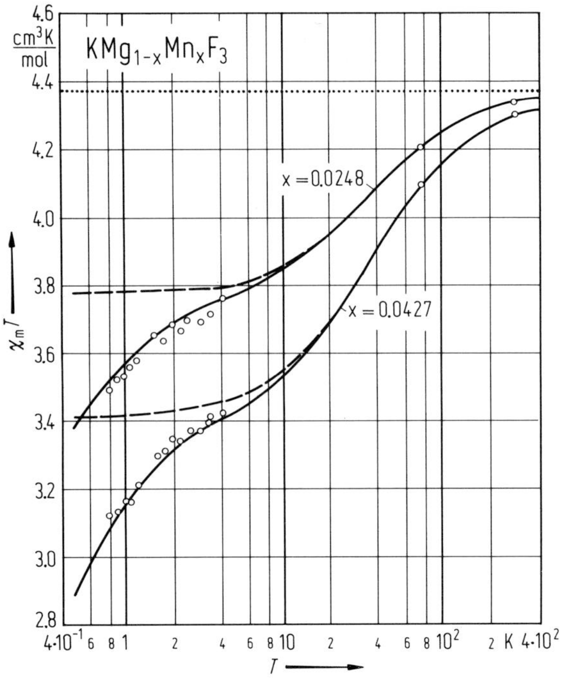

Fig. 23. $KMg_{1-x}Mn_xF_3$. Temperature dependence of $\chi_m T$. Lines are calculated with $J_1/k = J_2/k = 0$ (dotted), $J_1/k = 8.0$ K, $J_2/k = 0$ (dashed) and $J_1/k = 8.0$ K, $J_2/k = 0.08$ K (solid) [73Y2].

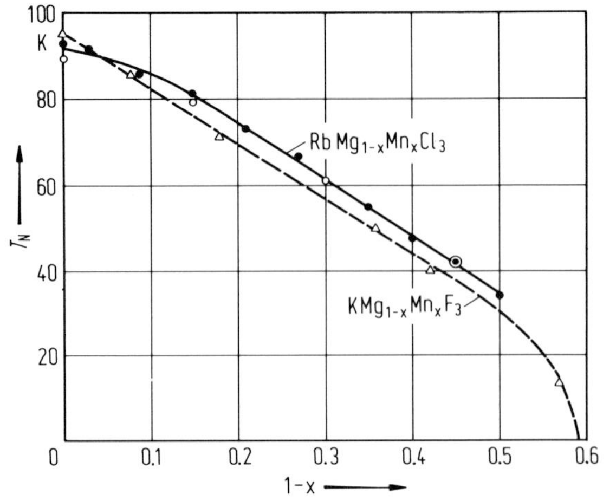

Fig. 24. $KMg_{1-x}Mn_xF_3$. Concentration dependence of T_N (dashed line) [70B1]. For $RbMg_{1-x}Mn_xCl_3$ the full circles are from χ measurements and the open circles from EPR [79P1].

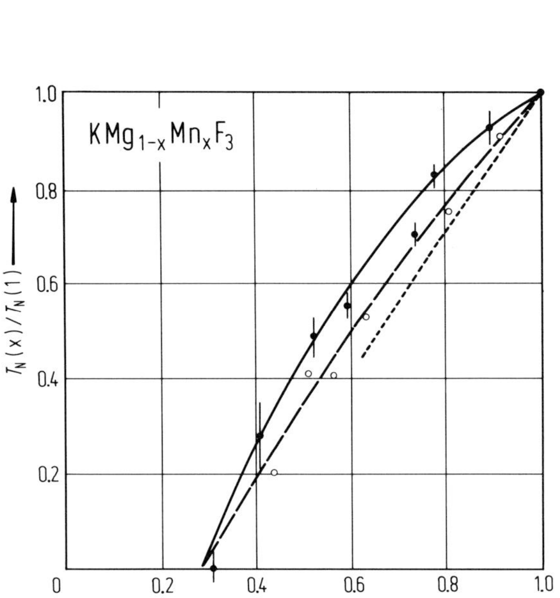

Fig. 25. $KMg_{1-x}Mn_xF_3$. Concentration dependence of normalized average transition temperature (open circles: data from [70B1, 73B1]). Curves are theoretical predictions (see original paper) [82A2].

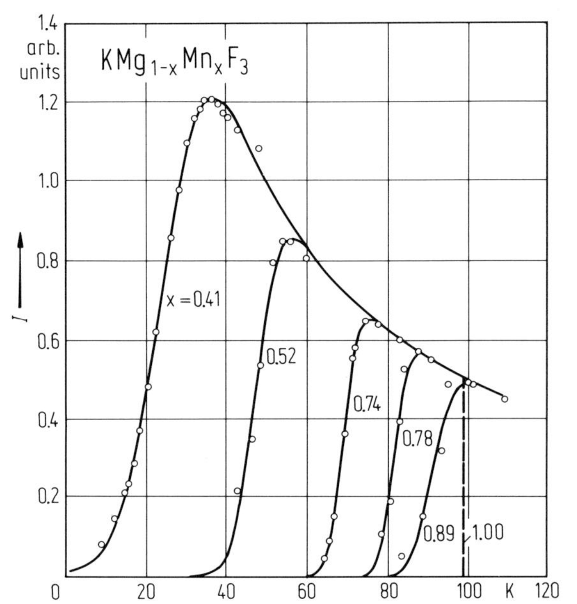

Fig. 26. $KMg_{1-x}Mn_xF_3$. Temperature dependence of the intensity of the modulated rf signal at 22 MHz in zero-external field for different x [82A2].

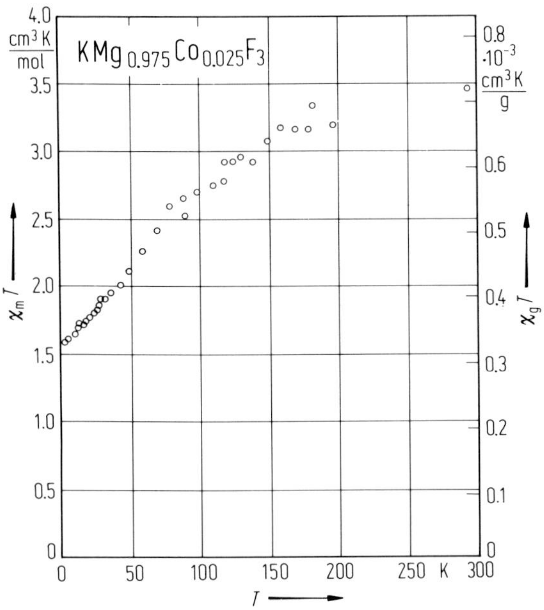

Fig. 27. $KMg_{0.975}Co_{0.025}F_3$. Temperature dependence of χT [67S1].

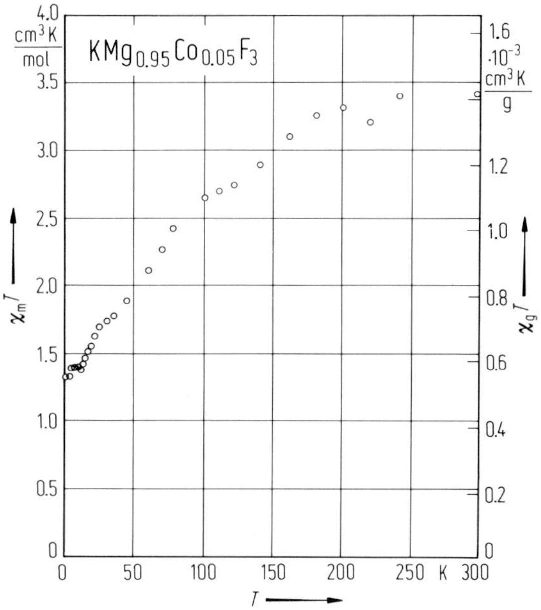

Fig. 28. $KMg_{0.95}Co_{0.05}F_3$. Temperature dependence of χT [67S1].

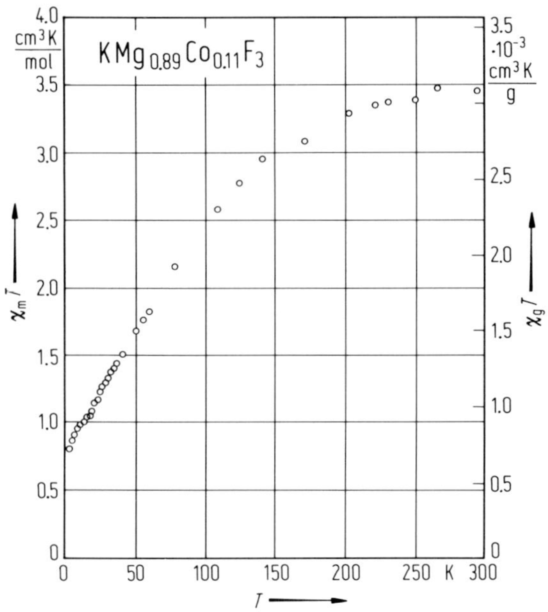

Fig. 29. $KMg_{0.89}Co_{0.11}F_3$. Temperature dependence of χT [67S1].

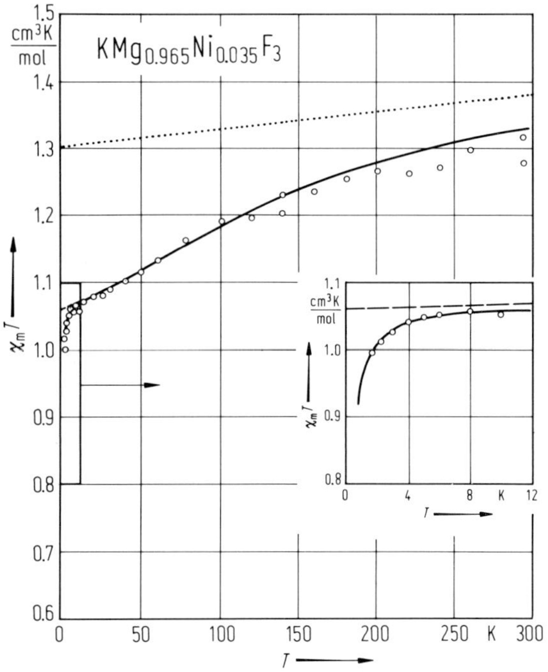

Fig. 30. $KMg_{0.965}Ni_{0.035}F_3$. Temperature dependence of $\chi_m T$. The curves are calculated with $J_1/k = J_2/k = 0$ (dotted), $J_1/k = 110$ K, $J_2/k = 0$ (dashed) and $J_1/k = 110$ K, $J_2/k = 0.4$ K (solid) [69Y1].

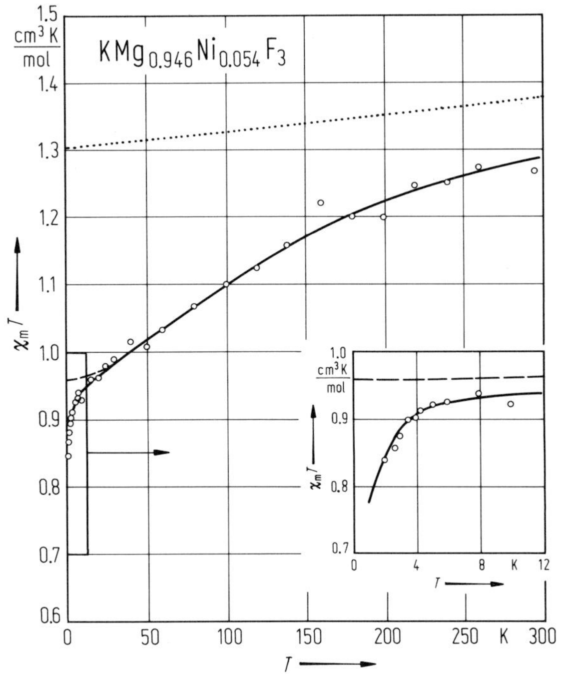

Fig. 31. $KMg_{0.946}Ni_{0.054}F_3$. Temperature dependence of $\chi_m T$. The curves are calculated with $J_1/k = J_2/k = 0$ (dotted), $J_1/k = 110$ K, $J_2/k = 0$ (dashed) and $J_1/k = 110$ K, $J_2/k = 0.6$ K (solid) [69Y1].

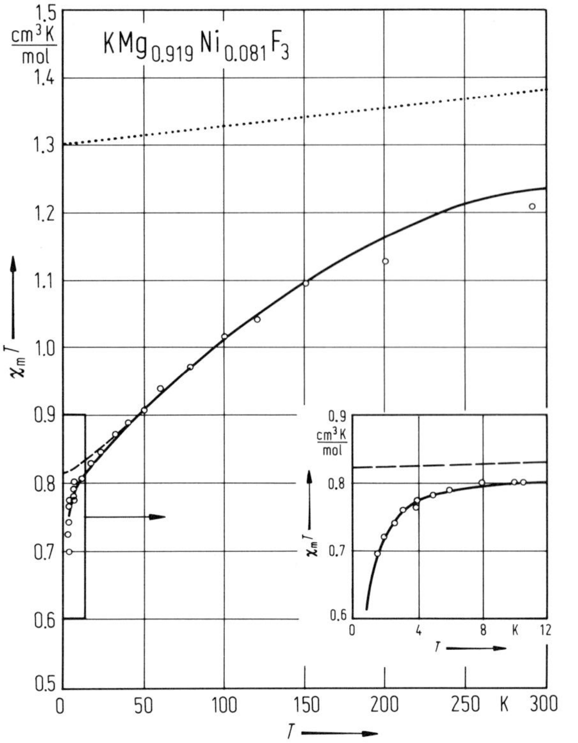

Fig. 32. $KMg_{0.919}Ni_{0.081}F_3$. Temperature dependence of $\chi_m T$. The curves are calculated with $J_1/k = J_2/k = 0$ (dotted), $J_1/k = 100$ K, $J_2/k = 0$ (dashed) and $J_1/k = 100$ K, $J_2/k = 0.5$ K (solid) [69Y1].

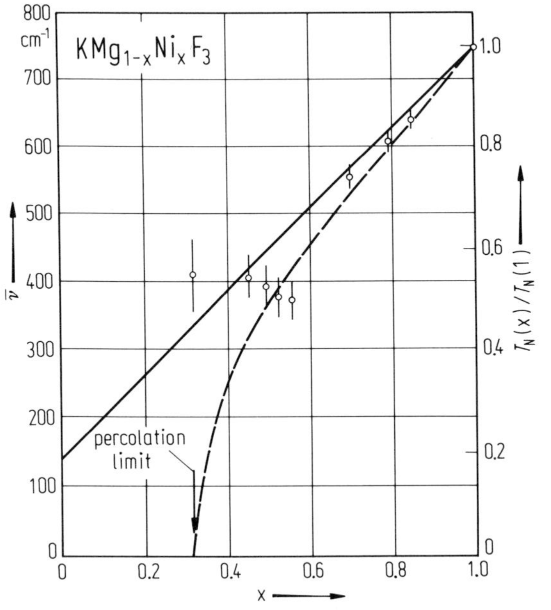

Fig. 33. $KMg_{1-x}Ni_xF_3$. Concentration dependence of T_N (dashed curve) and low-temperature magnon-pair mode frequency of non-magnetic ion (circles). Solid curve is theoretical (Ising model) [75F1].

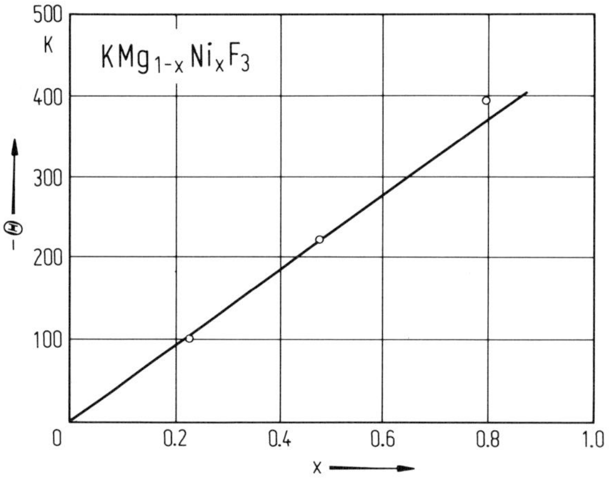

Fig. 34. $KMg_{1-x}Ni_xF_3$. Concentration dependence of Θ [88S2].

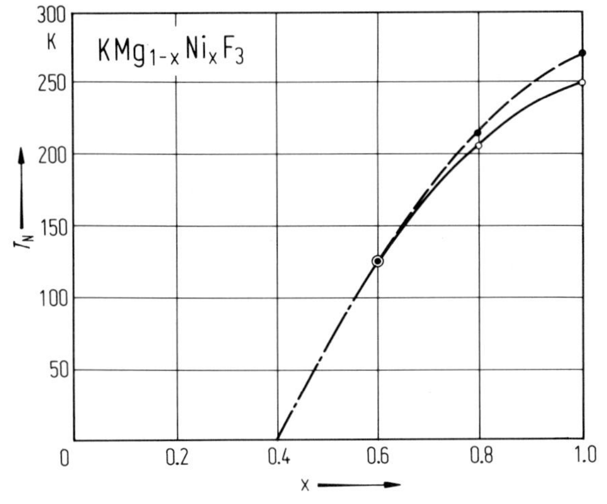

Fig. 35. $KMg_{1-x}Ni_xF_3$. Concentration dependence of T_N, from ——EPR; ---χ; ·-·-·-extrapolation [88S2].

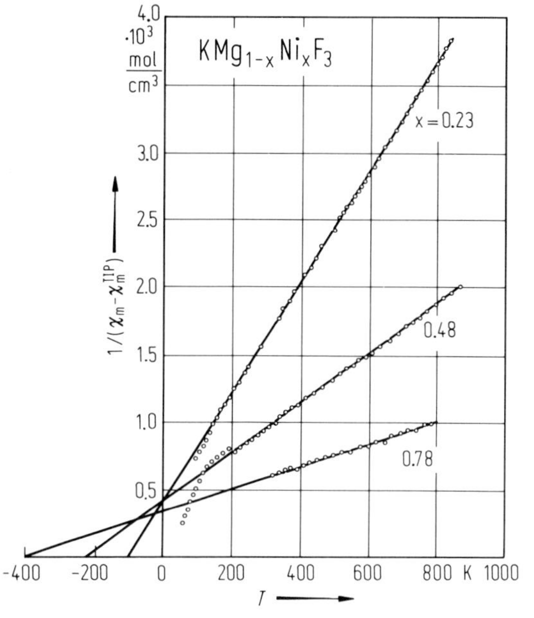

Fig. 36. $KMg_{1-x}Ni_xF_3$. Temperature dependence of χ_m^{-1} for different x. χ_m^{TIP}: temperature-independent susceptibility [88S2].

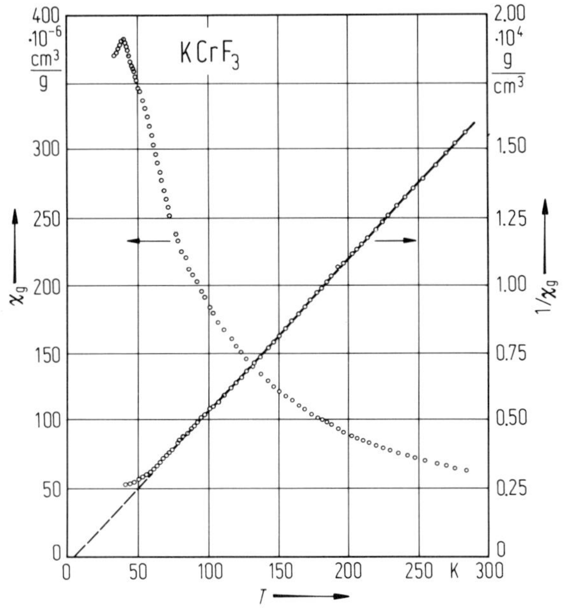

Fig. 37. $KCrF_3$. Temperature dependence of χ_g and χ_g^{-1} [66Y1].

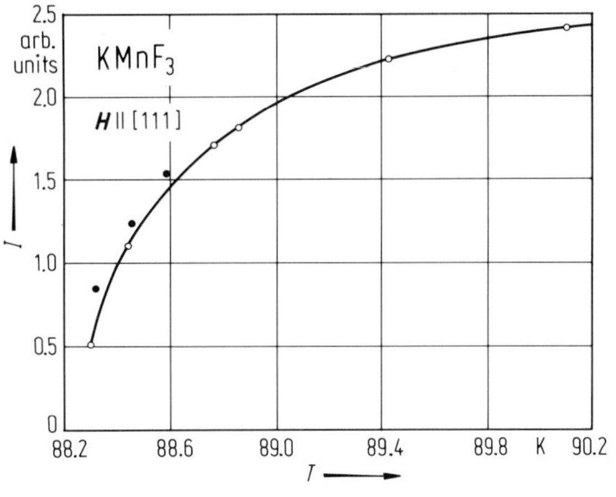

Fig. 38. $KMnF_3$. Temperature dependence of NMR ^{19}F line near T_N obtained by decreasing (open circles) and increasing (full circles) temperature [60S1].

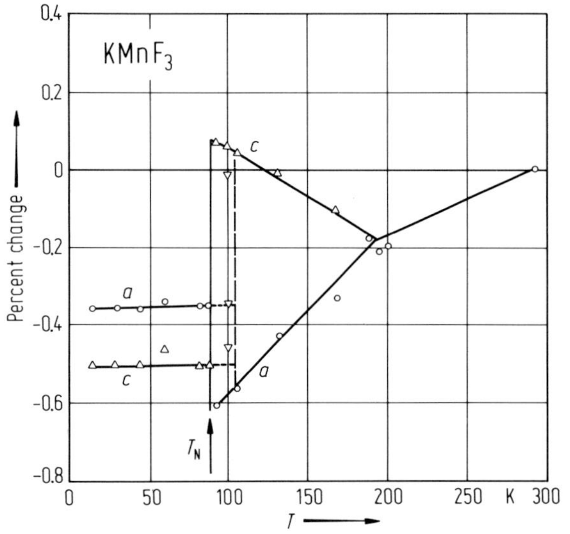

Fig. 39. $KMnF_3$. Temperature dependence of pseudocell dimensions [61B1].

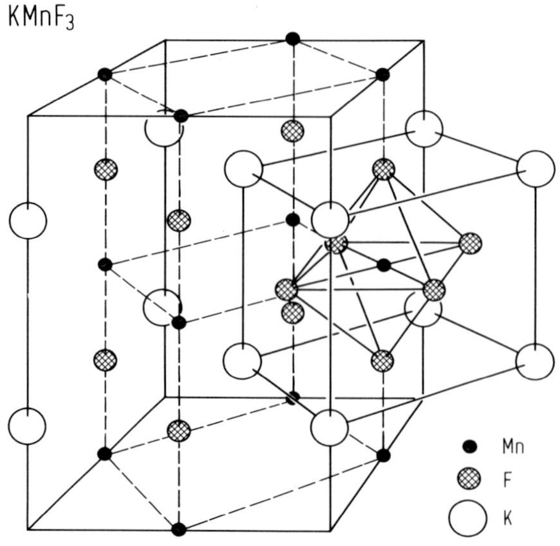

Fig. 40. $KMnF_3$. Low temperature unit cell in orthorhombic phase containing four molecules. The relation to the high temperature cubic unit cell containing one molecule is shown. This cell is at low temperatures no longer cubic and is called pseudocell. [61B1]. See also Fig. 12.

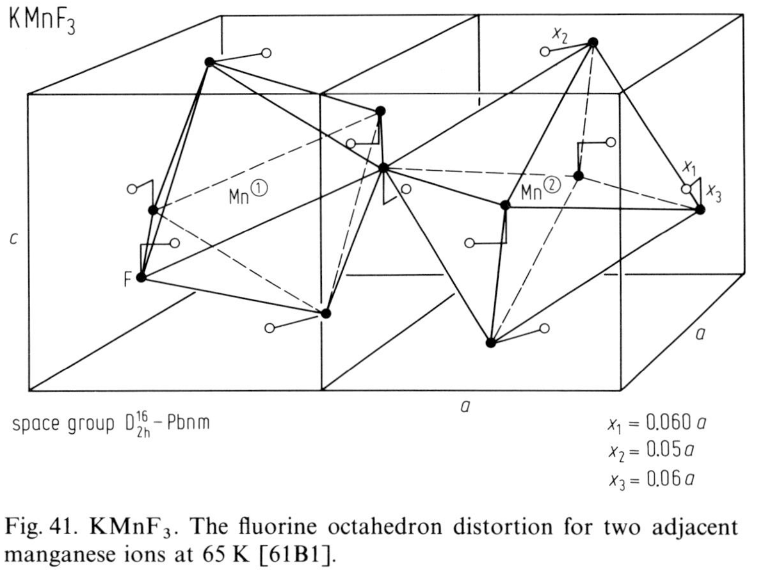

Fig. 41. $KMnF_3$. The fluorine octahedron distortion for two adjacent manganese ions at 65 K [61B1].

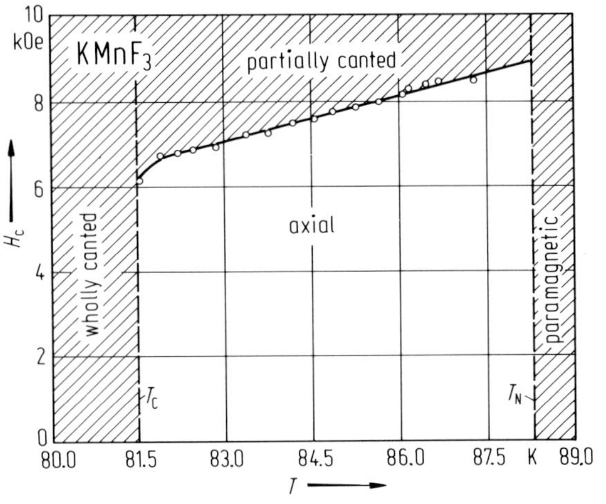

Fig. 42. $KMnF_3$. Temperature dependence of the critical field at which canting is produced [61H1].

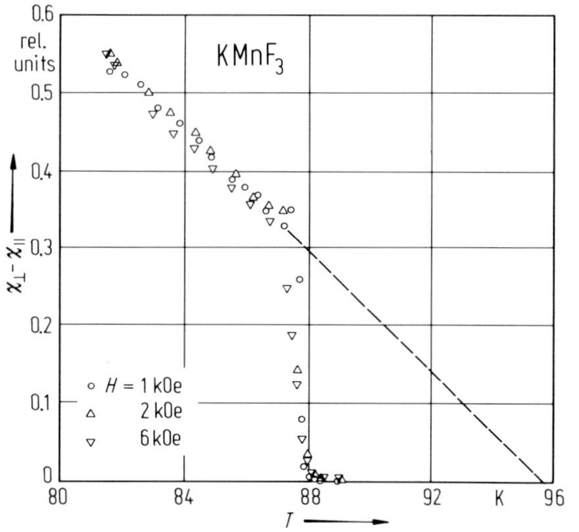

Fig. 43. $KMnF_3$. Temperature dependence of the susceptibility anisotropy for different H [61H1].

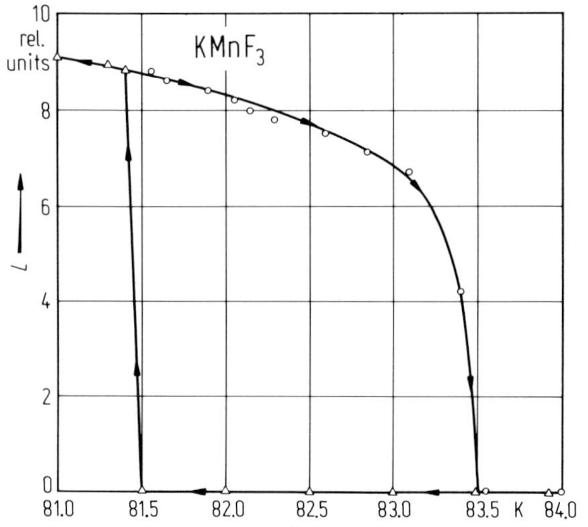

Fig. 44. $KMnF_3$. Temperature dependence of the torque at 2 kOe for the magnetic field along a [100] direction on cooling and heating [61H1].

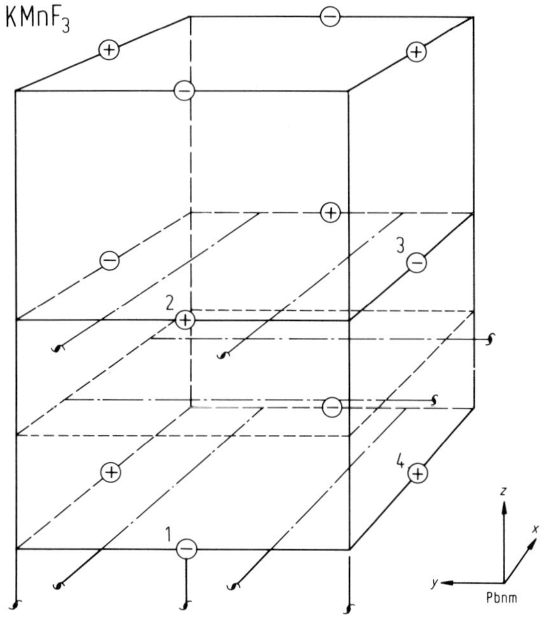

Fig. 45. $KMnF_3$. Screw axes with respect to Mn^{2+} ions; see also Fig. 40 [61H1]. ⊕ and ⊖ indicate different Mn^{2+} moment orientations.

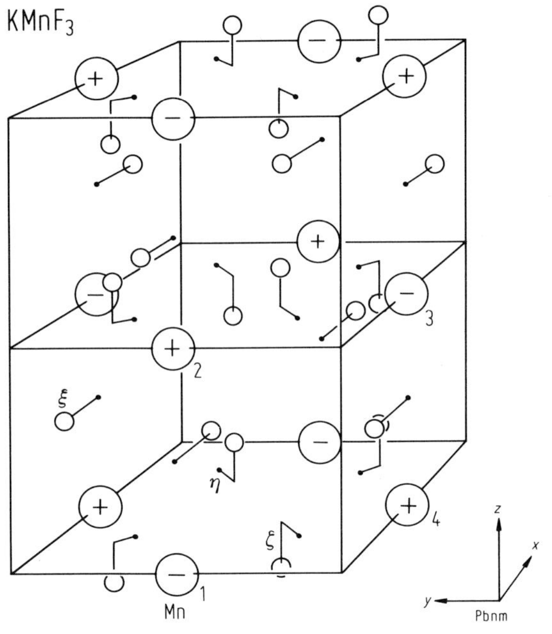

Fig. 46. $KMnF_3$. Crystal structure with the fluorine displacements ξ, η, ζ [61H1]. See also Figs. 40 and 45.

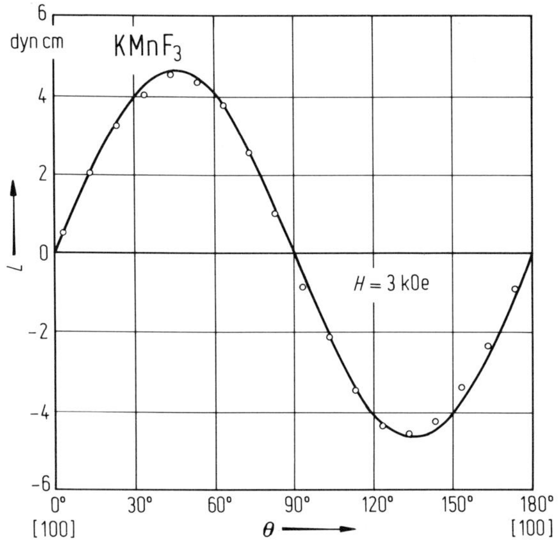

Fig. 47. $KMnF_3$. Magnetic torsion of a single crystal at 83.2 K in the uncanted state for a field in the (100) plane. Torque is observed to vary as the square of the field and with period of 180° [61H1]. Unit of L as given in the original paper.

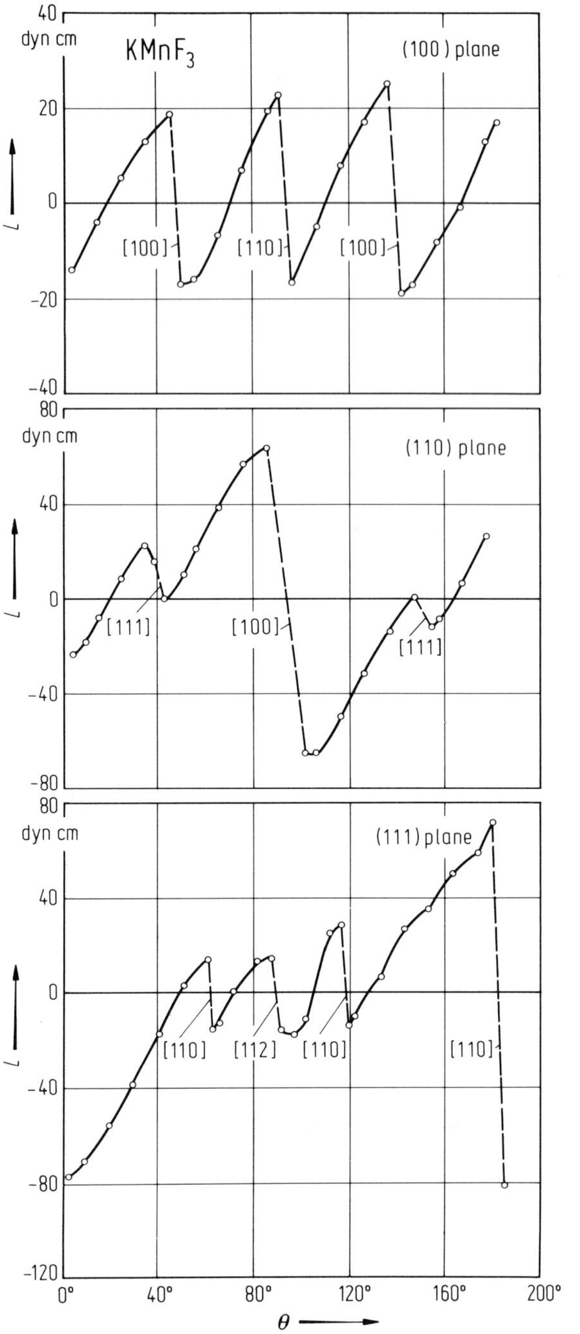

Fig. 48. $KMnF_3$. Magnetic torsion at 4.2 K and 2 kOe. The discontinuities observed along [100] direction for magnetic field in a (100) plane tend to disappear at high magnetic fields, at 77 K in a field above 1 kOe and at 4.2 K above 10 kOe [61H1].

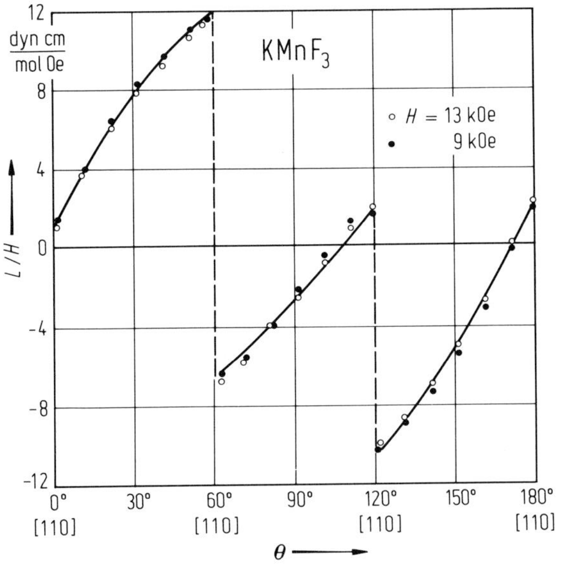

Fig. 49. $KMnF_3$. Torque per unit field vs. angle of field from a [110] direction for the field in a (111) plane below 81.5 K. Solid line is a theoretical fit for the sample suspended along a [111] direction for the Mn sublattice along the ⊕ and ⊖ axes (see Fig. 45 and 46) and canted so as to produce a weak moment along z axis [61H1].

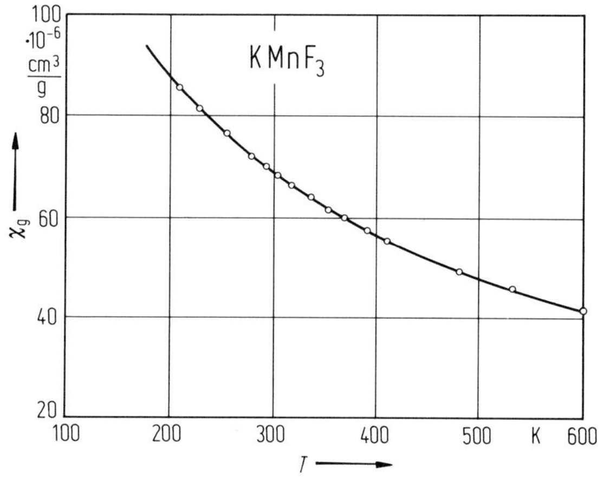

Fig. 50. $KMnF_3$. Temperature dependence of χ_g [63H1].

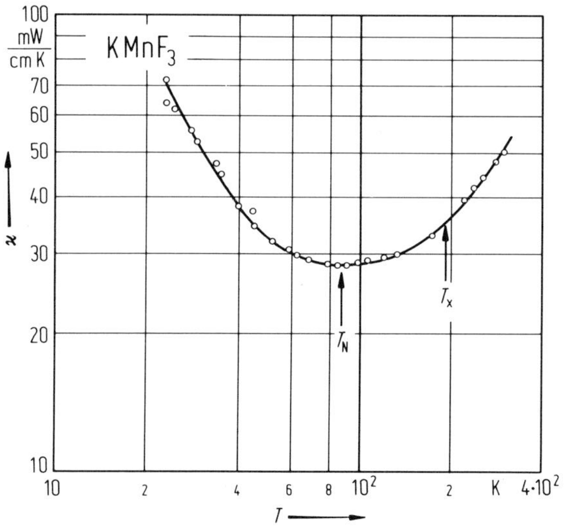

Fig. 51. $KMnF_3$. Temperature dependence of thermal conductivity (T_x: transition temperature for lattice distortion) [64S1].

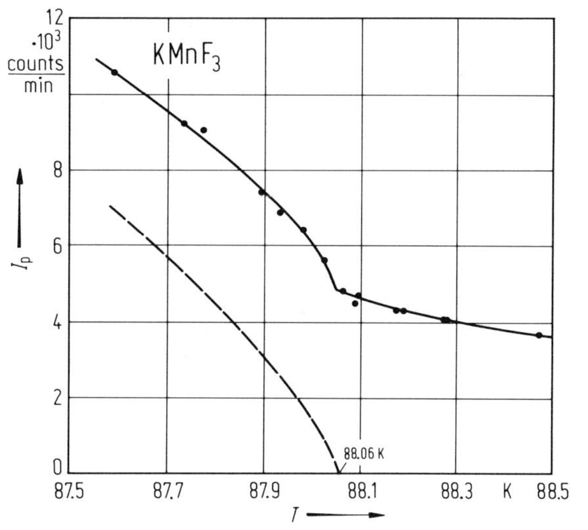

Fig. 52. $KMnF_3$. Temperature dependence of the $(\frac{1}{2}, \frac{1}{2}, \frac{1}{2})$ neutron peak intensity I_p close to T_N. Dashed line: antiferromagnetic Bragg peak intensity determined by assuming the critical magnetic scattering to be symmetrical about T_N [66C1].

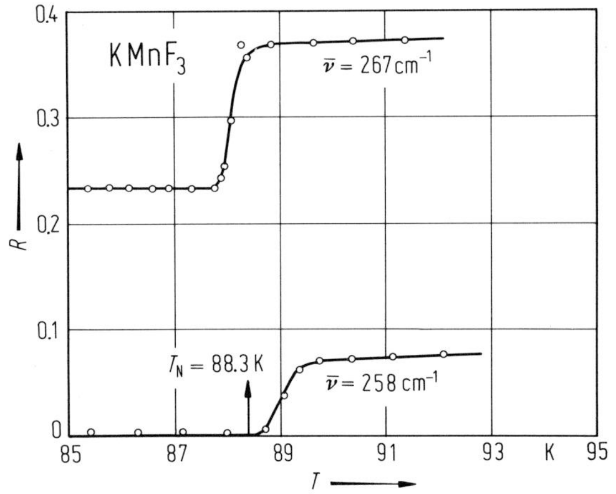

Fig. 53. $KMnF_3$. Temperature dependence of two-phonon sideband near T_N [70B3]. R: reflectivity.

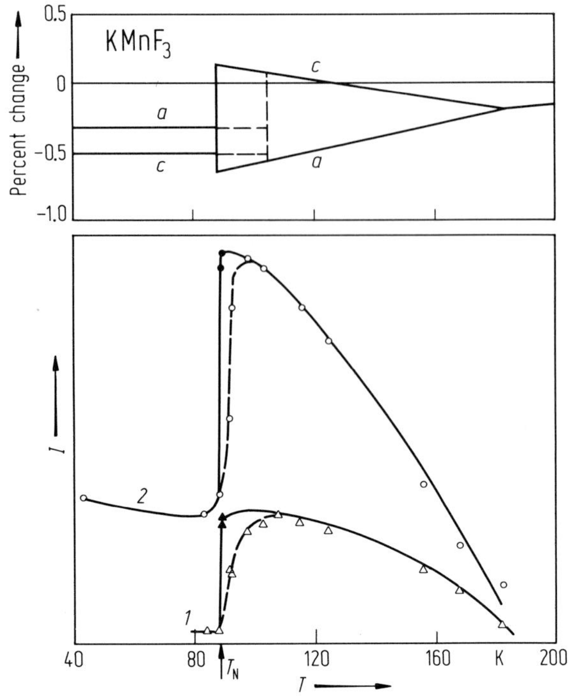

Fig. 54. $KMnF_3$. Temperature dependence of Raman lines intensity (curves *1*) 117 cm^{-1}, (*2*) 230 cm^{-1}. Solid line by cooling, dashed line by heating. Upper part: temperature dependence of lattice parameters (see Fig. 39) [71P1].

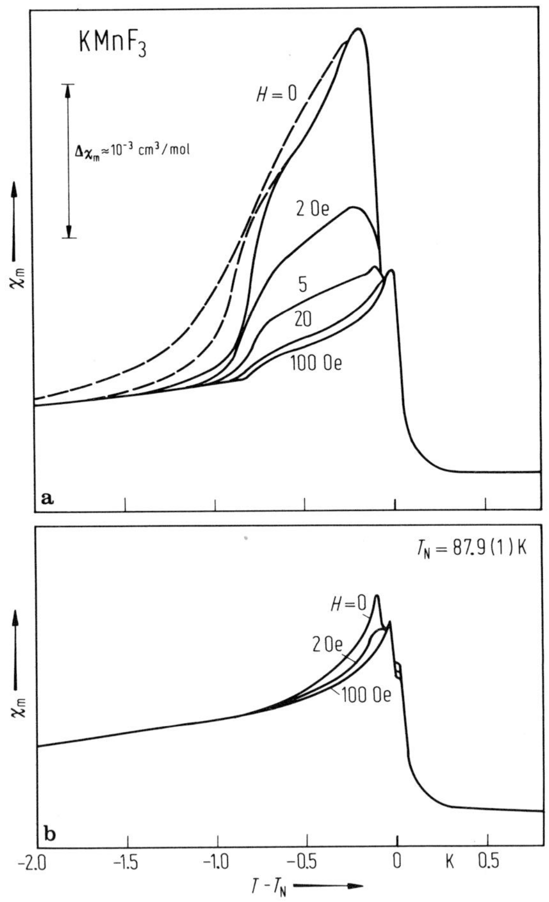

Fig. 55. $KMnF_3$. Temperature dependence of χ_m at 10 MHz with rf field and a biasing field $\boldsymbol{H}$ parallel to a [100] direction. (a) Temperature decreasing. Broken curves show the initial susceptibility for the first two coolings. (b) Temperature increasing. Minimum cooling temperature 4 K [72M2].

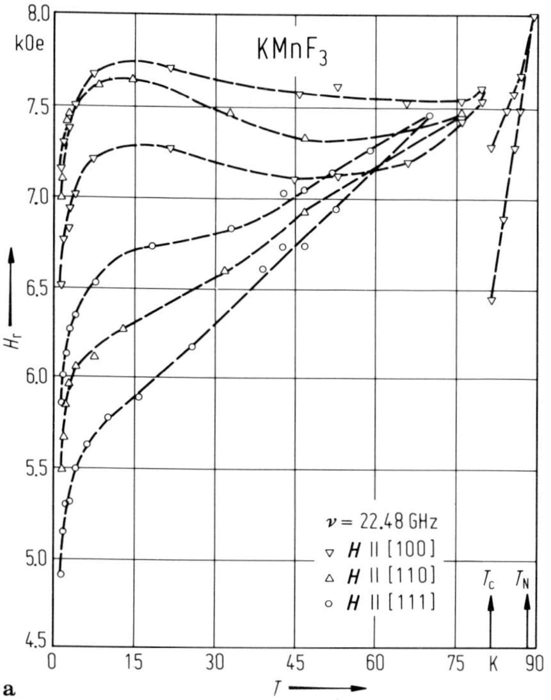

a

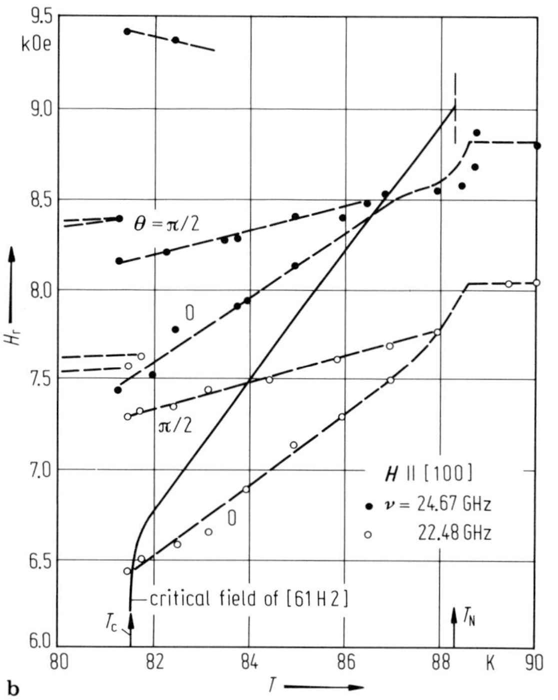

b

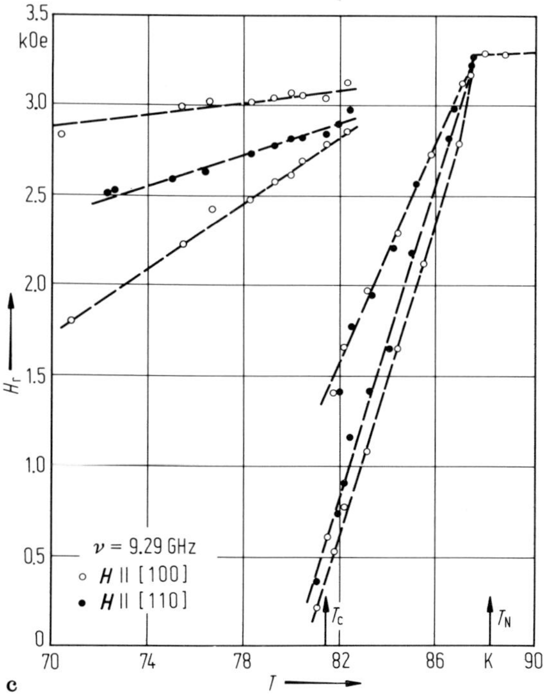

c

Fig. 56. $KMnF_3$. Temperature dependence of the magnetic resonance fields parallel to the [100], [110] and [111] directions of the tetragonal pseudocell at different frequencies [72S2]. Solid line in (b): critical field of [61H2]. θ: angle between applied field and easy axis.

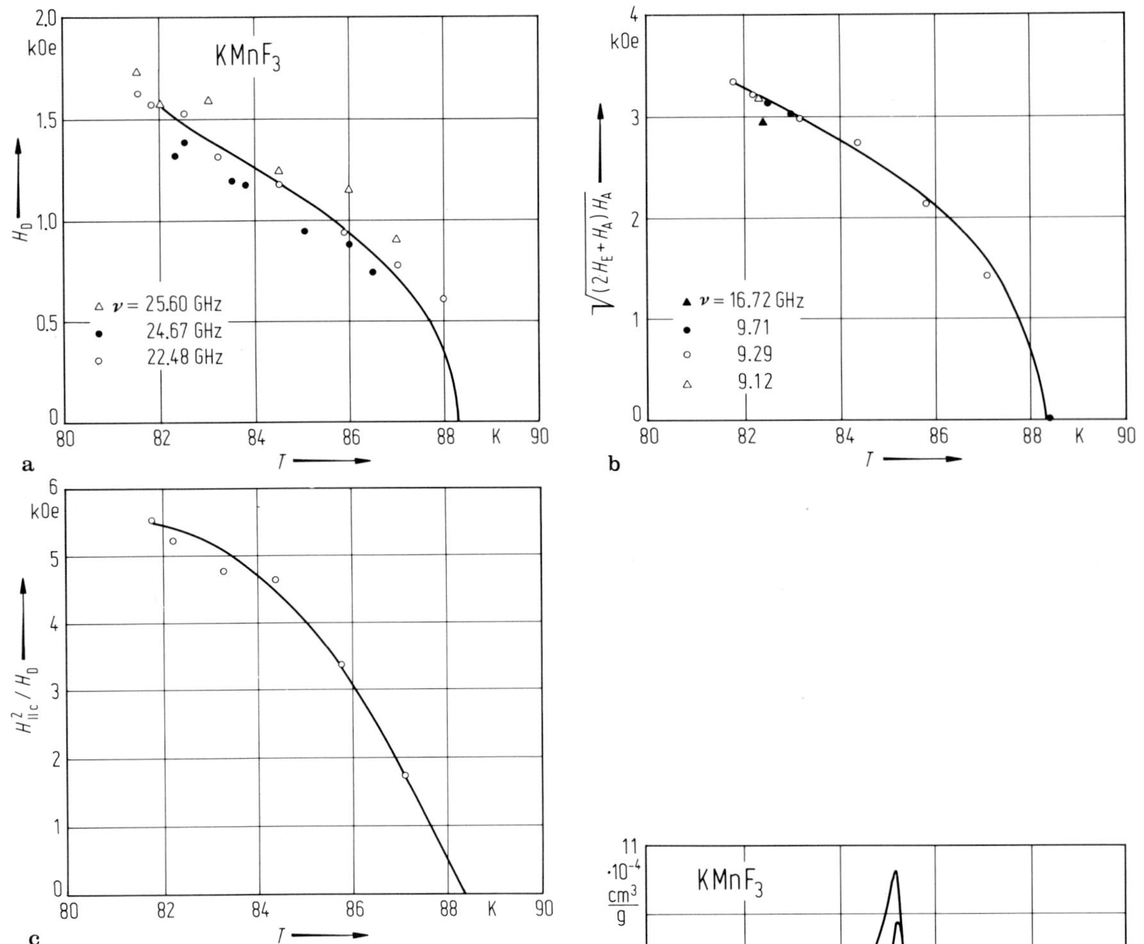

Fig. 57. $KMnF_3$. Temperature dependence of the effective fields obtained from EPR at different frequencies. H_D: effective field due to Dzyaloshinsky-Moriya interaction; H_E: exchange field; H_A: anisotropy field. $H_{\|c} = \{(2H_E + H_A)H_A - H_D^2\}^{1/2}$, $H_{\|c}^2/H_D$ is the field at which magnetization becomes perpendicular to the easy axis [72S2]. See also [76J2].

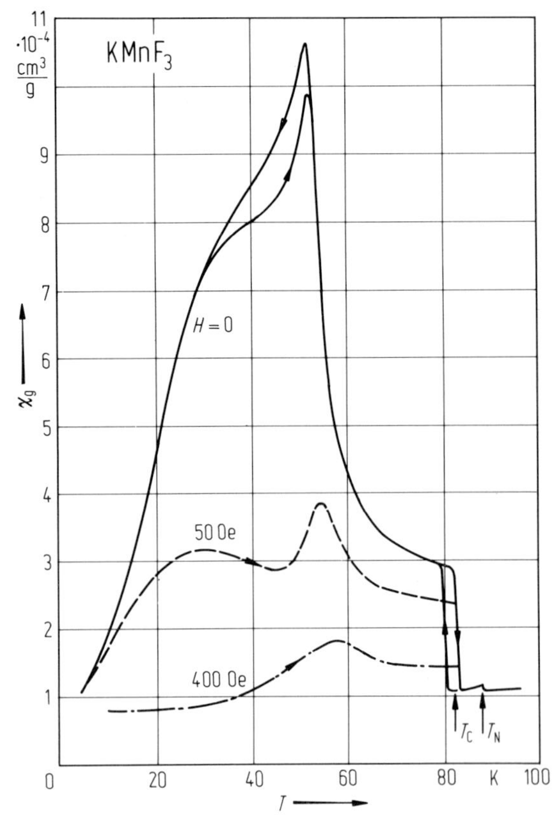

Fig. 58. $KMnF_3$. Temperature dependence of ac susceptibility at 10 MHz and different H [73M2].

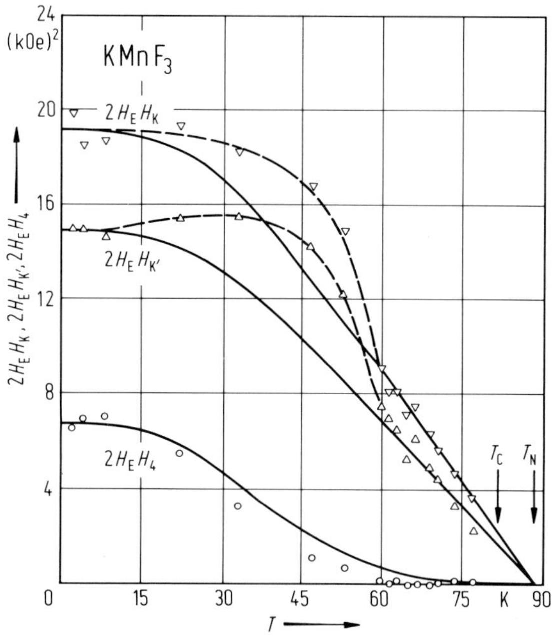

Fig. 59. $KMnF_3$. Temperature dependence of $2H_EH_K$, $2H_EH_{K'}$, and $2H_EH_4$. The solid curves represent the theoretical dependences of the anisotropy fields and are normalized at 0 K. $H_K = K/M_0$, $H_{K'} = K'/M_0$, $H_4 = K_4/M_0$, $H_D = DM_0$, $H_E = JM_0$ with K,K': orthorhombic anisotropy constants, K_4: cubic anisotropy constant, D: Dzyaloshinsky-Moriya vector [74S1].

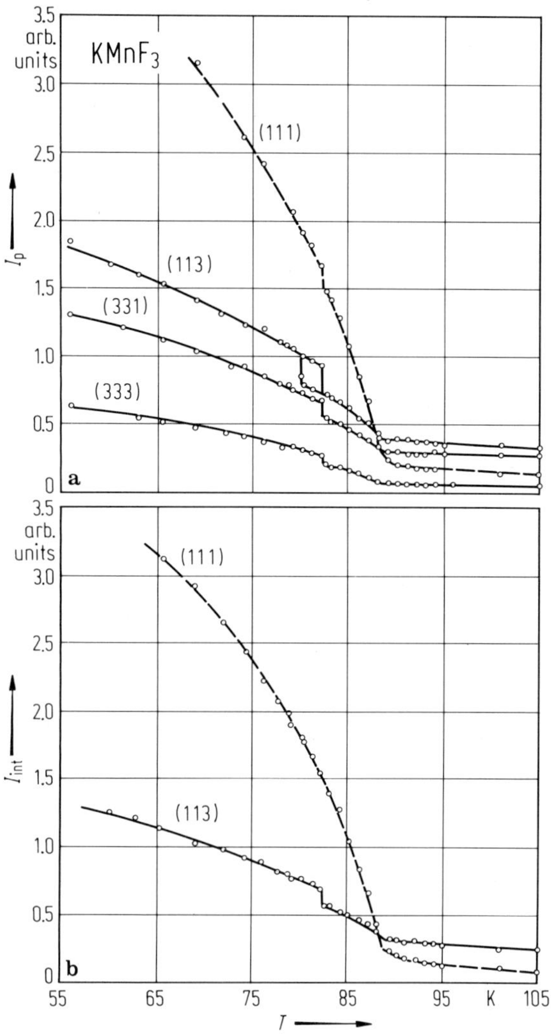

Fig. 61. $KMnF_3$. Temperature dependence of the intensity of Bragg reflections. The peak intensity I_p (a) and integrated intensity I_{int} (b) are in arbitrary units. The hysteresis loop at the 81 K transition is shown only for the case of (113) in (a) [75H2].

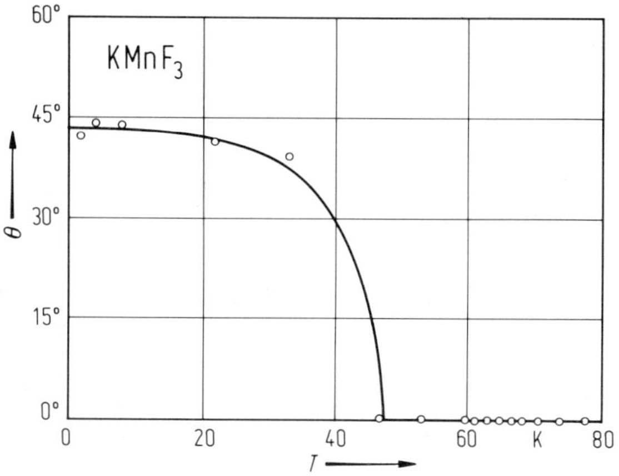

Fig. 60. $KMnF_3$. Temperature dependence of the angle θ, between the direction of the sublattice magnetization and the c axis [74S1].

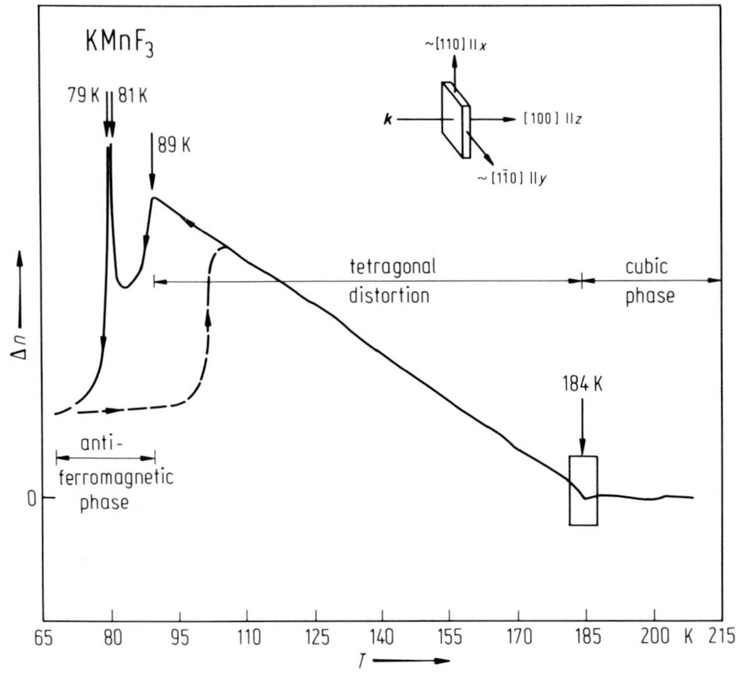

Fig. 62. $KMnF_3$. Temperature dependence of birefringence, measured with a He–Ne laser, vertical scale not calibrated [76B2].

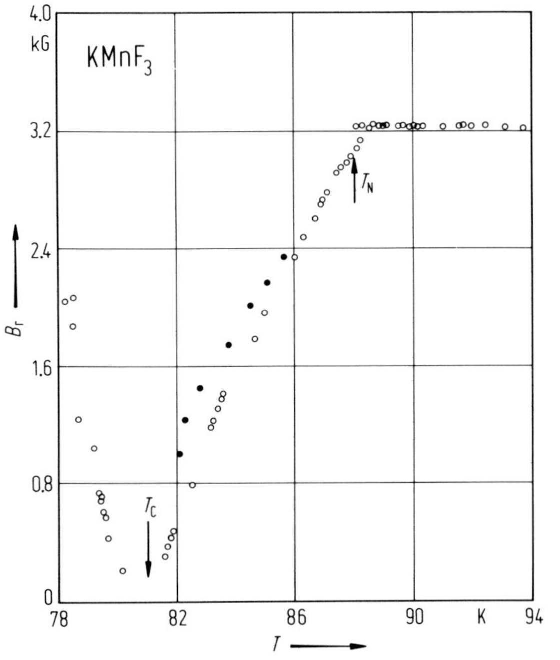

Fig. 63. $KMnF_3$. Temperature dependence of the ESR field at 9.25 GHz and $\boldsymbol{H}$ parallel to [110] for different lines [76J2].

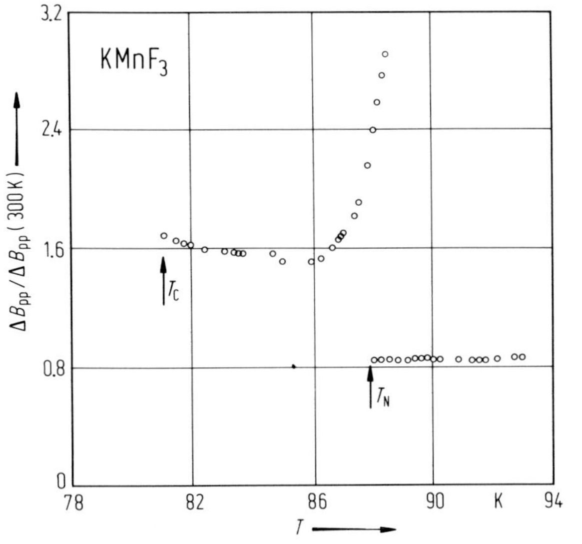

Fig. 64. $KMnF_3$. Temperature dependence of the relative resonance linewidth (peak-to-peak) [76J2].

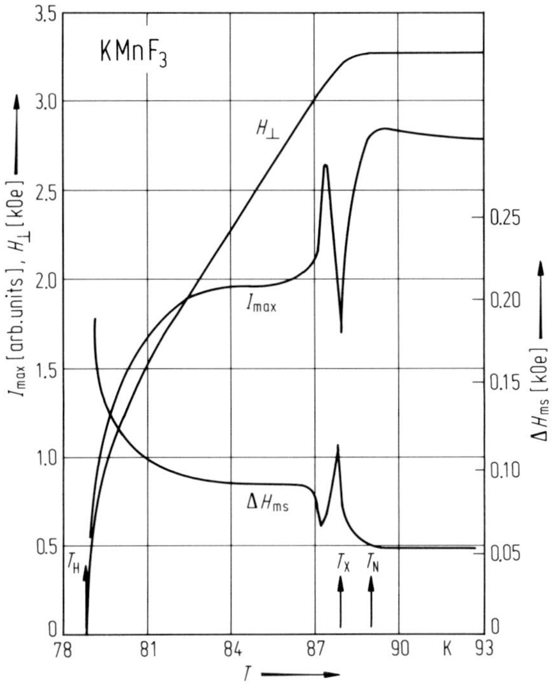

Fig. 65. $KMnF_3$. Temperature dependence of the resonance field $\boldsymbol{H}_\perp$ aligned along [100] direction and perpendicular to the easy axes for both kinds of domains in two-domain single crystal. ΔH_{ms}: absorption derivative linewidth, I_{max}: maximal intensity, T_x: temperature at which two lines $H_\perp$ and $H_\parallel$ are observed at the same H-value (for T_H see Fig. 66) [78J2].

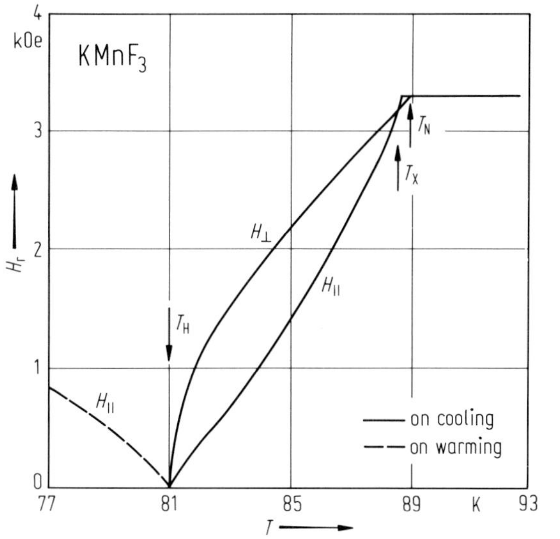

Fig. 66. $KMnF_3$. Temperature dependence of the resonance field $\boldsymbol{H}_r$ parallel to [100] and perpendicular to the easy axis in one tetragonal domain and parallel to the easy axis in the other. T_H is the temperature at which both the lines join at zero magnetic field. For other symbols see Fig. 65 [78J2].

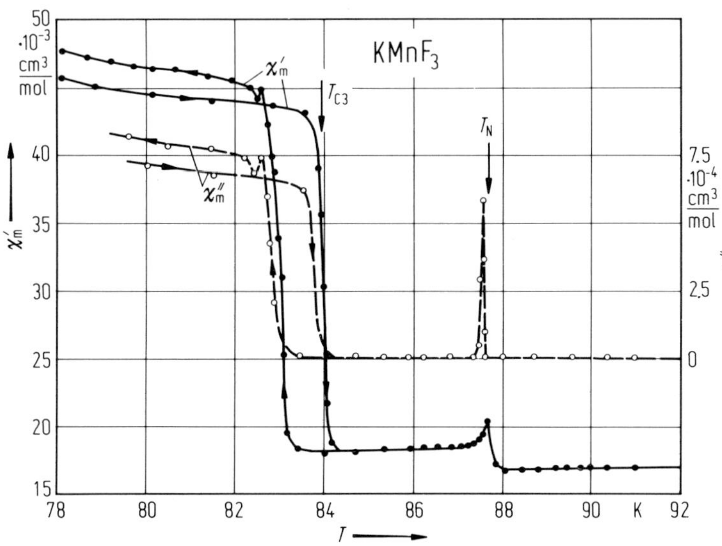

Fig. 67. $KMnF_3$. Temperature dependence of ac χ'_m and χ''_m for single crystals in the [100] direction at 332 Hz and 3 Oe. Arrows indicate the temperature scanning sense [83B1]. T_{C3}: see also Figs. 68, 69.

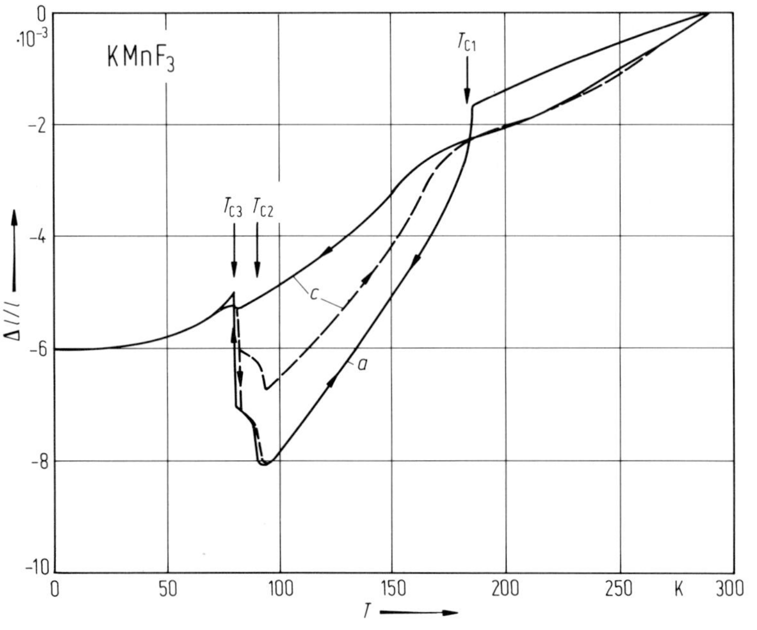

Fig. 68. $KMnF_3$. Temperature dependence of linear thermal expansion $\Delta l/l$ of a platelet single crystal. Curves *a* and *c* correspond to measurements parallel and perpendicular to the plane, respectively. Arrows indicate the temperature scanning sense [83B1].

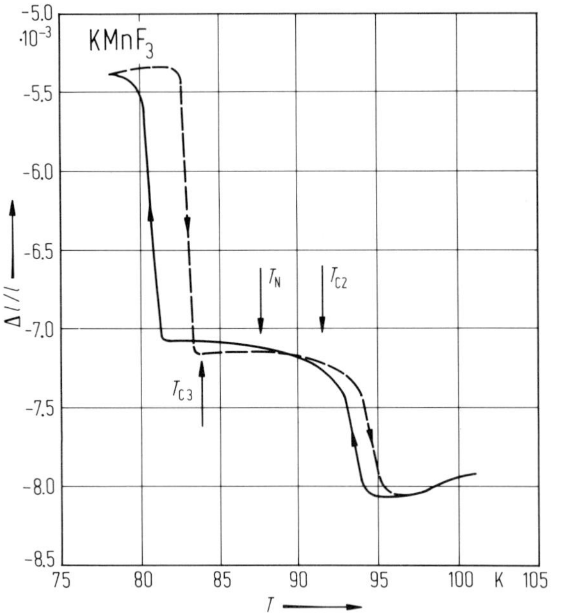

Fig. 69. $KMnF_3$. Temperature dependence of linear thermal expansion $\Delta l/l$ near T_N for a single crystal (see Fig. 68) [83B1].

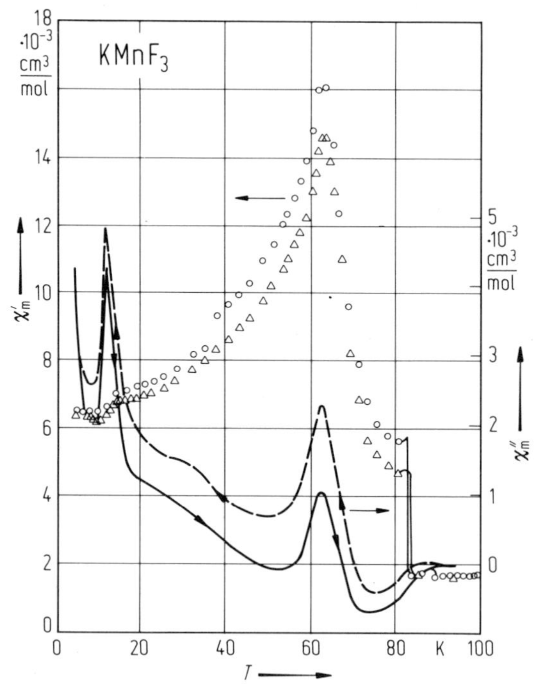

Fig. 70. $KMnF_3$. Temperature dependence of ac χ'_m and χ''_m for a platelet single crystal in the [100] direction at 332 Hz. Triangles denote heating and circles cooling runs of χ'_m, respectively [83B1].

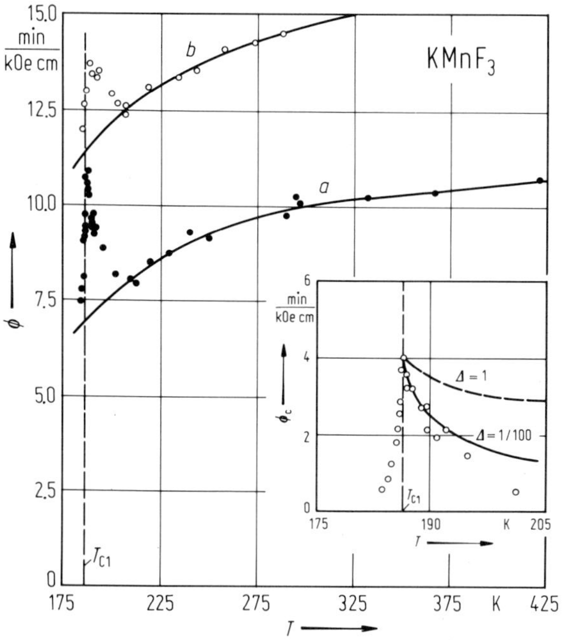

Fig. 71. $KMnF_3$. Curve *a*: Temperature dependence of the Verdet constants ϕ for $\lambda = 0.6328$ µm. The solid line is the best fit to the law $\phi = 16 - 3185/(T + 202)$ min kOe^{-1} cm^{-1}. In the insert, the critical contribution around T_{C1} after subtraction of the best-fit behavior for the ordinary part is shown. The dashed and solid lines are calculated with different anisotropy parameters Δ (see original paper). Curve *b*: ϕ for $\lambda \cong 0.4700$ (100) µm with the best-fit line $\phi = 24.7 - 4941/(T + 202)$ min kOe^{-1} cm^{-1} [85P2].

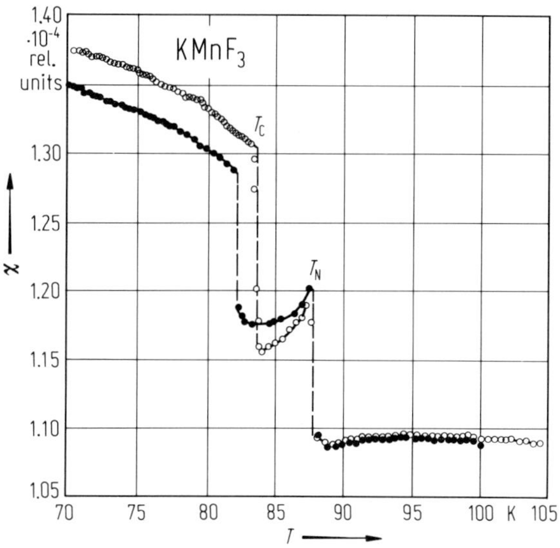

Fig. 72. $KMnF_3$. Temperature dependence of relative χ for a single crystal, measured by the Faraday method for ***H*** parallel to $[100]_p$ (solid circles) and parallel to $[110]_p$ (open circles). The suffix "p" refers to the pseudo perovskite cell ($a_p \times b_p \times c_p$); see Fig. 40 [86H3].

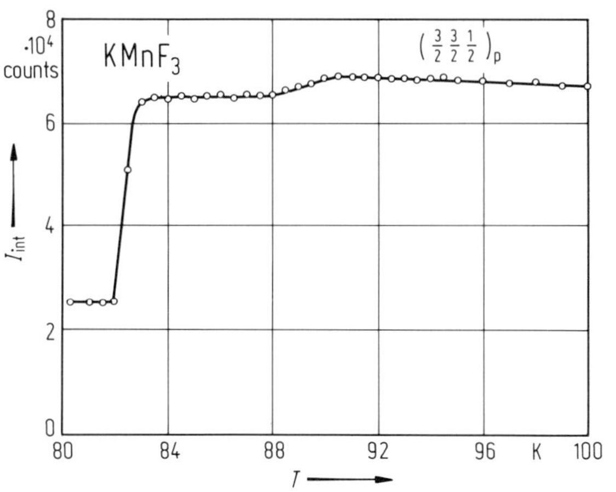

Fig. 73. $KMnF_3$. Temperature dependence of the integrated X-ray intensity of the $(\frac{3}{2},\frac{3}{2},\frac{1}{2})_p$ reflection for the transitions observed below 186 K [86H3]. Suffix p, see Fig. 72.

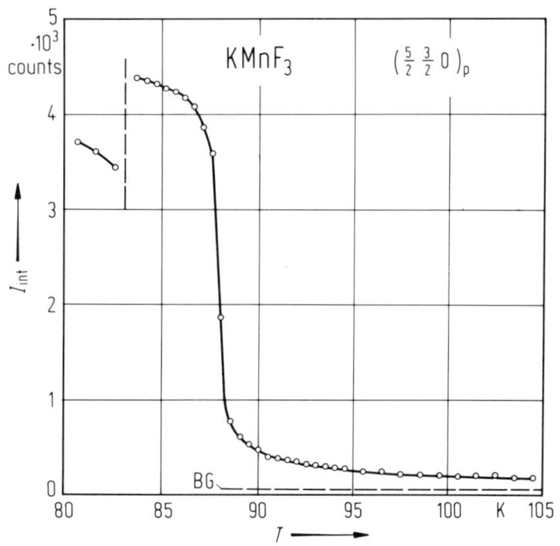

Fig. 74. $KMnF_3$. Temperature dependence of integrated intensity of the $(\frac{5}{2},\frac{3}{2},0)_p$ X-ray reflection [86H3]. Suffix p, see Fig. 72, BG: Background.

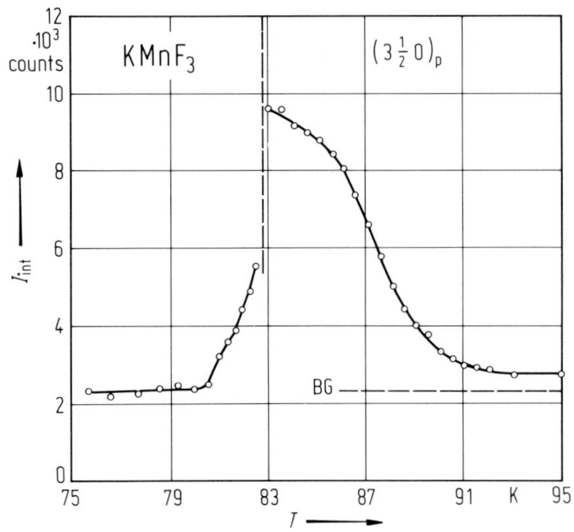

Fig. 75. $KMnF_3$. Temperature dependence of the X-ray integrated intensity of the superlattice reflection $(3,\frac{1}{2},0)_p$ [86H3]. Suffix p, see Fig. 72.

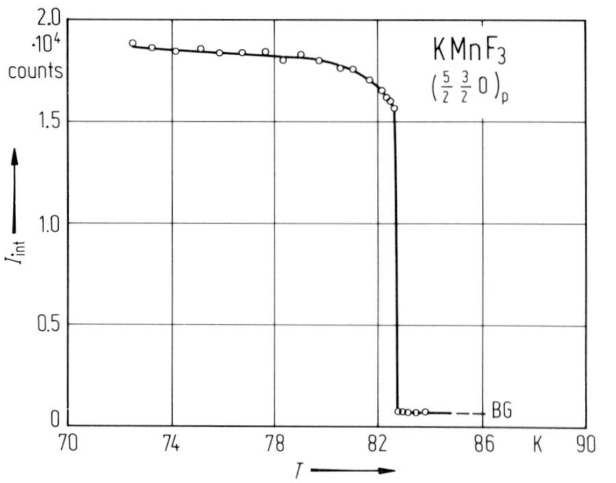

Fig. 76. $KMnF_3$. Temperature dependence of the X-ray integrated intensity of the $(\frac{5}{2},\frac{3}{2},0)_p$ reflection [86H3]. Suffix p, see Fig. 72. Note the difference to Fig. 74.

For Fig. 78, see next page.

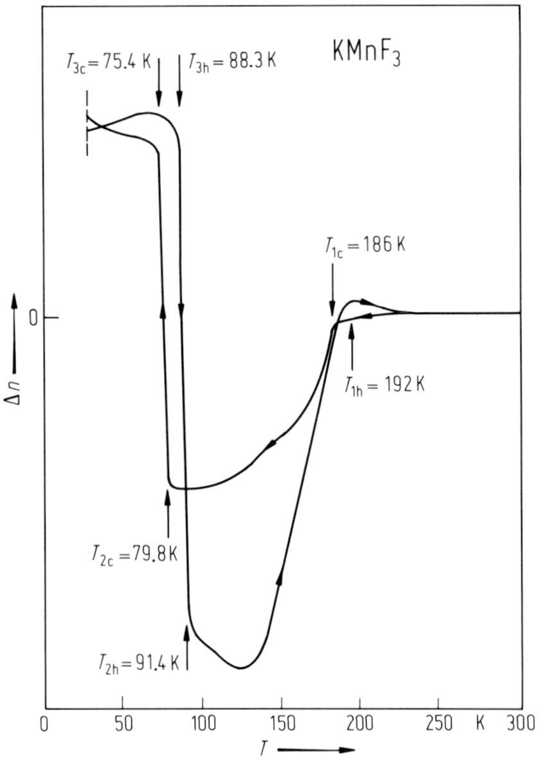

Fig. 77. $KMnF_3$. Temperature dependence of optical retardation on heating (T_{nh}) and cooling (T_{nc}) [88R1]. $\lambda = 632.8$ nm.

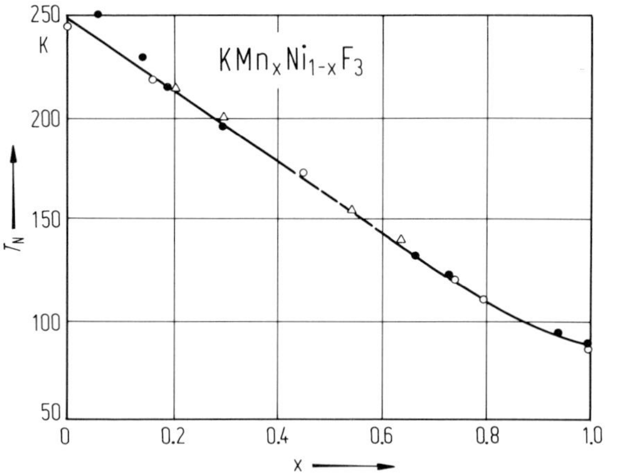

Fig. 79. $KMn_xNi_{1-x}F_3$. Concentration dependence of T_N (open circles) from optical data [79L1], (full circles) static susceptibility [84C1], and (triangles) ESR [90S2]. See also [63H1].

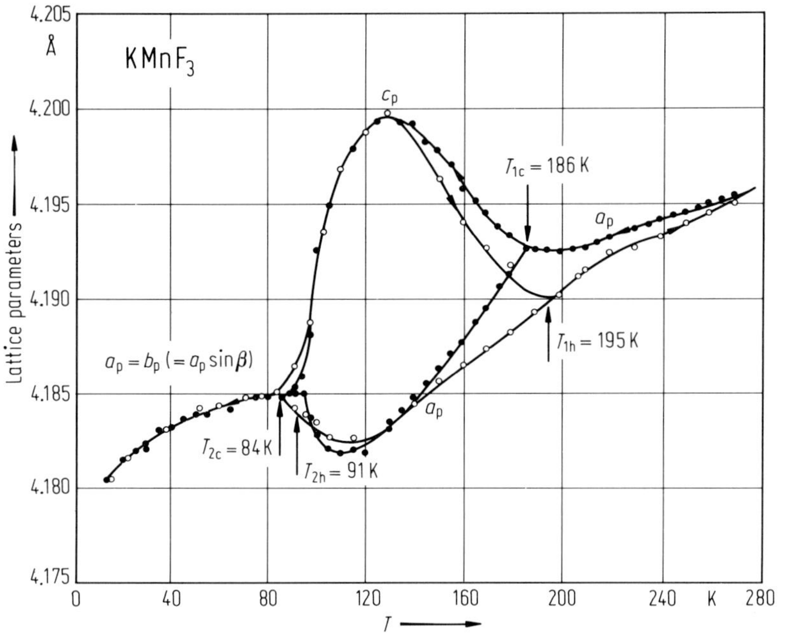

Fig. 78. $KMnF_3$. Temperature dependence of lattice parameters near magnetic and structural transitions (see also Fig. 77) [88R1]. Suffix p, see Fig. 72 and 40.

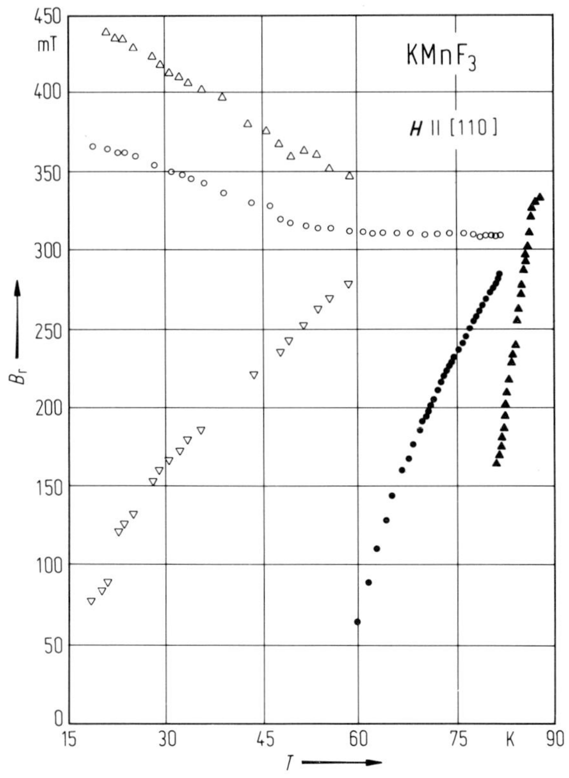

Fig. 80. $KMnF_3$. Temperature dependence of single crystal resonance fields B_r of a single crystal for $\boldsymbol{H} \parallel [110]$. Different symbols indicate different lines [91S1].

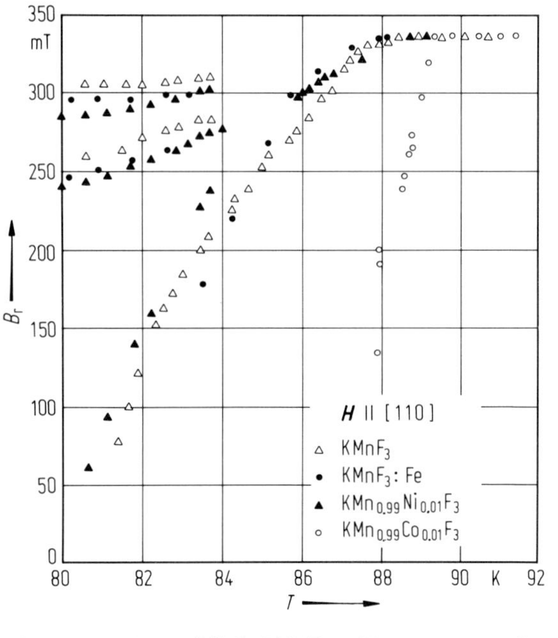

Fig. 81. $KMn_{1-x}(Ni,Co,Fe)_xF_3$. Temperature dependence of (AFMR) resonance fields B_r for $\boldsymbol{H} \parallel [110]$ and $T_C < T < T_N$ [91S1].

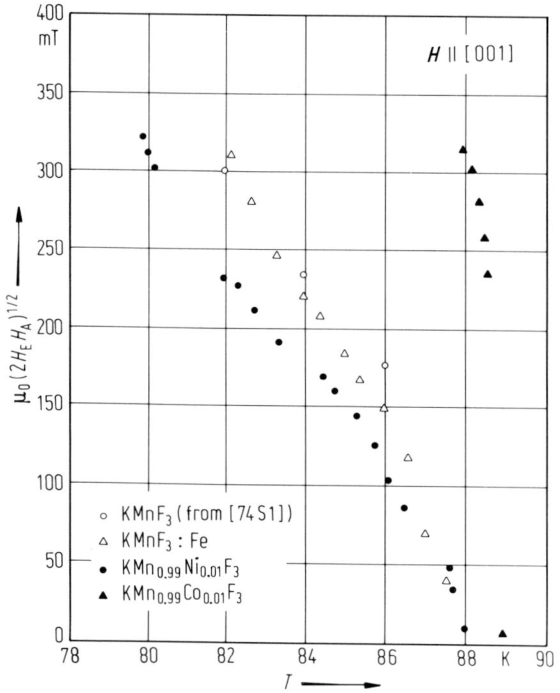

Fig. 82. $KMn_{1-x}(Ni,Co,Fe)_xF_3$. Temperature dependence of $(2H_EH_A)^{1/2}$ of single crystal for $\boldsymbol{H} \parallel [001]$ and $T_C < T < T_N$ [91S1].

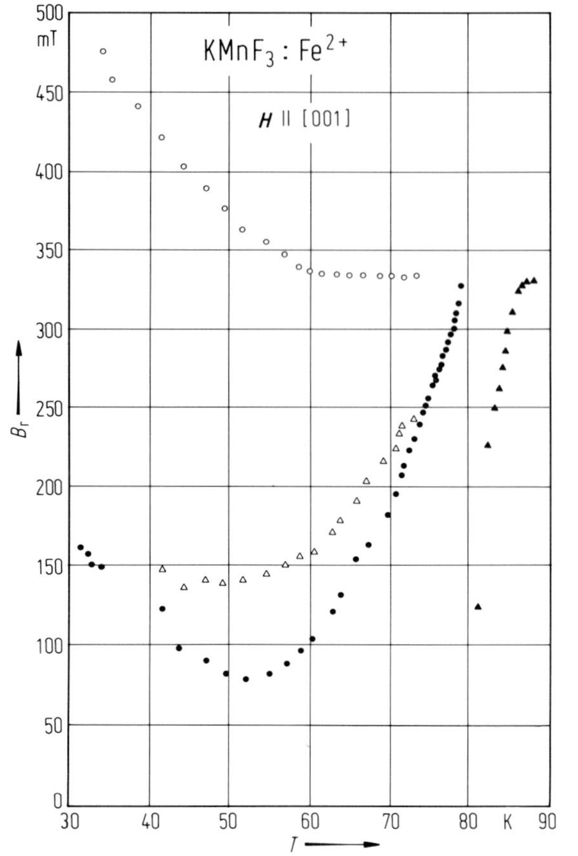

Fig. 84. $KMnF_3{:}Fe^{2+}$. Temperature dependence of single crystal (AFMR) resonance fields B_r of a single crystal for $\boldsymbol{H} \parallel [001]$. Different symbols indicate different lines [91S1].

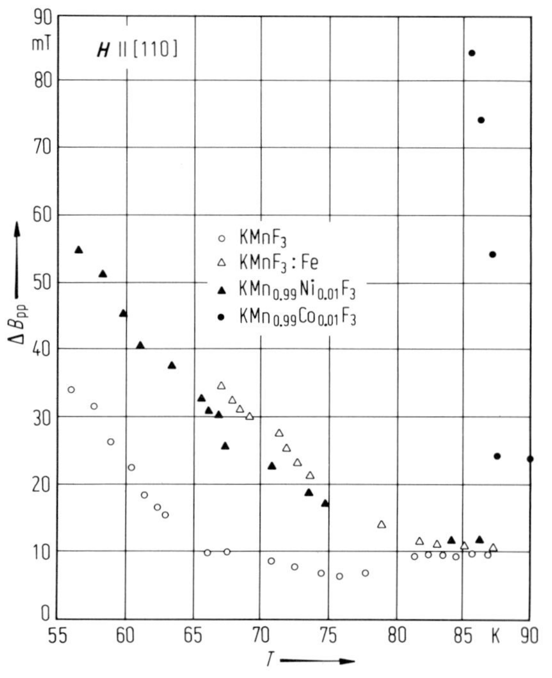

Fig. 83. $KMn_{1-x}(Ni,Co,Fe)_xF_3$. Temperature dependence of the (AFMR) resonance linewidth of a single crystal for $\boldsymbol{H} \parallel [110]$ [91S1].

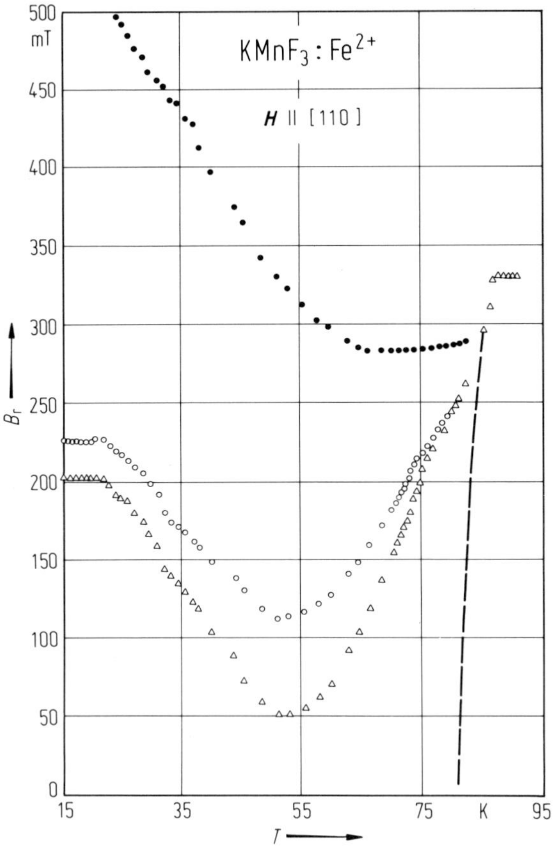

Fig. 85. $KMnF_3$:Fe^{2+}. Temperature dependence of single crystal (AFMR) resonance fields B_r of a single crystal for $\boldsymbol{H} \parallel [110]$. Different symbols indicate different lines [91S1].

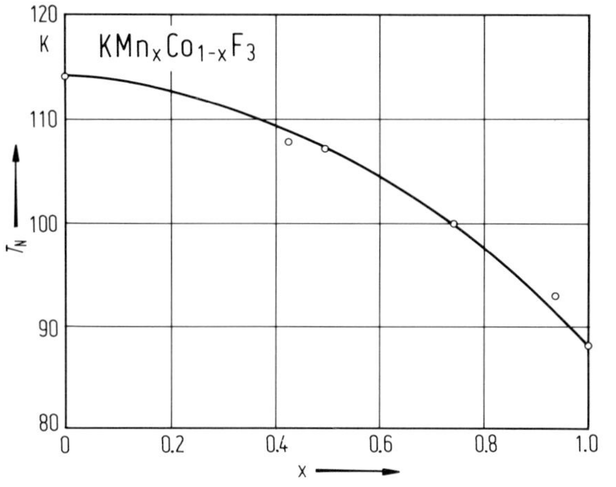

Fig. 86. $KMn_xCo_{1-x}F_3$. Concentration dependence of T_N [63H1].

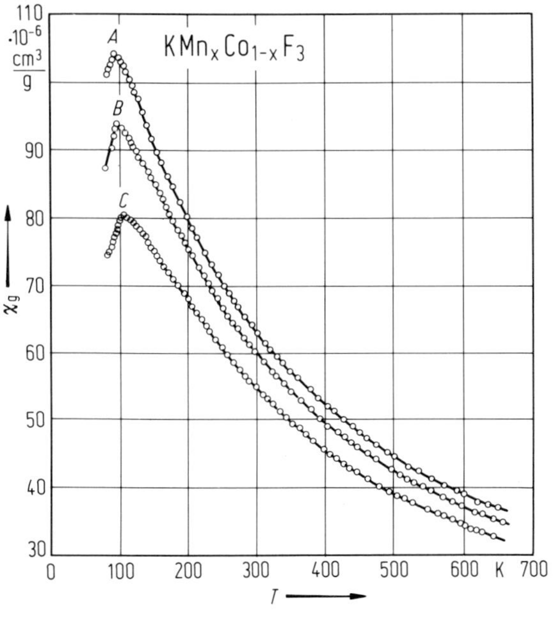

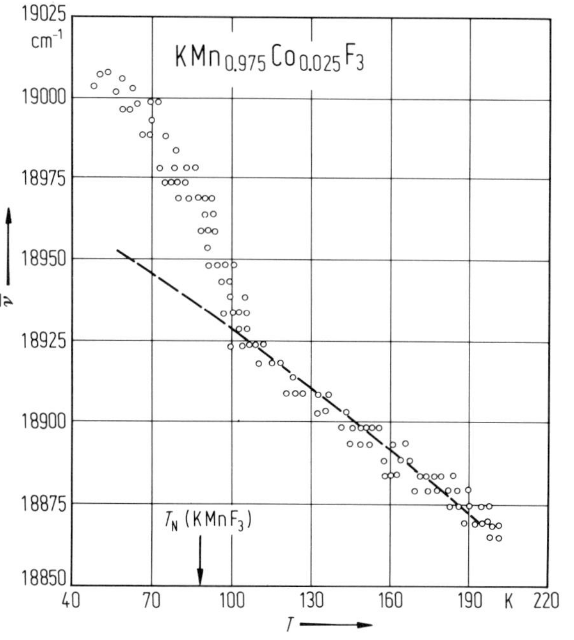

Fig. 88. $KMn_{0.975}Co_{0.025}F_3$. Temperature dependence of the $^4T_{1g} \rightarrow {}^4T_{1g}(P)$ ($\sim 19000\ cm^{-1}$) optical absorption band maximum of the Co^{2+} ion. The dashed line indicates purely thermal behavior to emphasize the magnetically induced shifts [67H2].

←

Fig. 87. $KMn_xCo_{1-x}F_3$. Temperature dependence of χ_g for x = 0.933 (curve *A*), 0.739 (*B*) and 0.488 (*C*) [63H1].

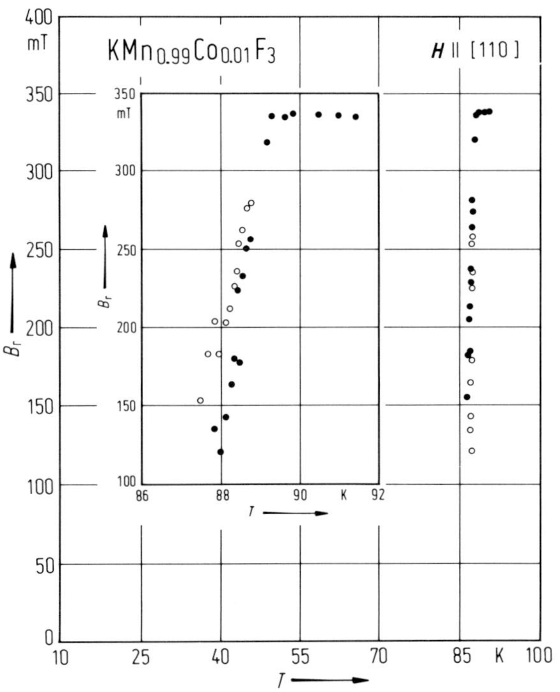

Fig. 89. $KMn_{0.99}Co_{0.01}F_3$. Temperature dependence of resonance (AFMR) fields B_r of a single crystal for $\boldsymbol{H} \parallel [110]$. Full and open circles indicate different lines [91S1].

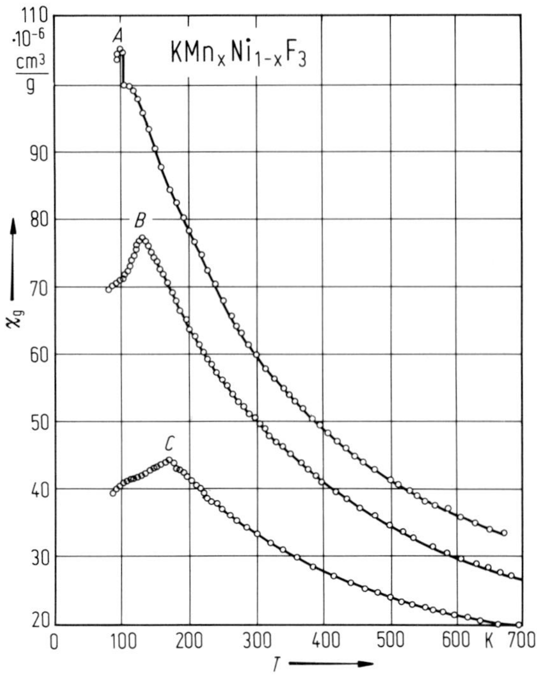

Fig. 90. $KMn_xNi_{1-x}F_3$. Temperature dependence of χ_g for x = 0.891 (curve *A*), 0.724 (*B*) and 0.465 (*C*) [63H1].

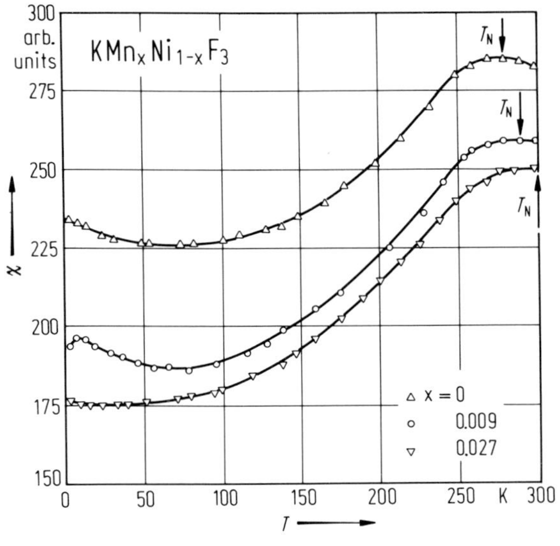

Fig. 91. $KMn_xNi_{1-x}F_3$. Temperature dependence of χ for different x [69M3].

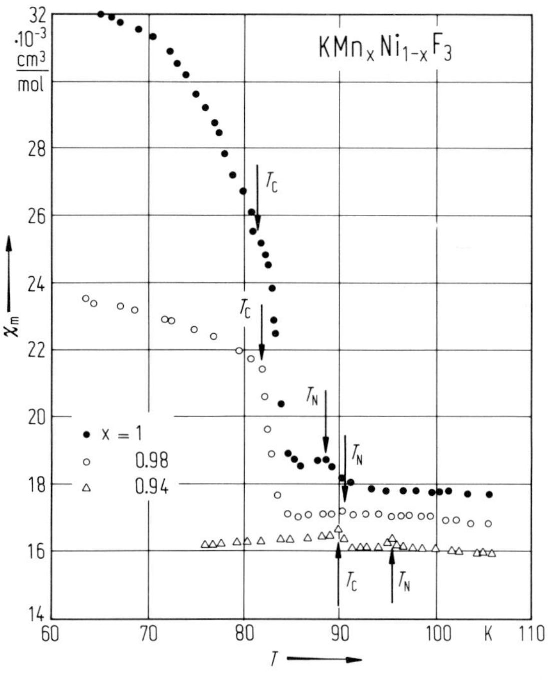

Fig. 92. $KMn_xNi_{1-x}F_3$. Temperature dependence of χ_m for different x. T_C transition temperature to the weak ferromagnetic phase. For x = 1 and 0.98 χ_m was measured along *c* axis (the other sample measured without orientations) [84C1].

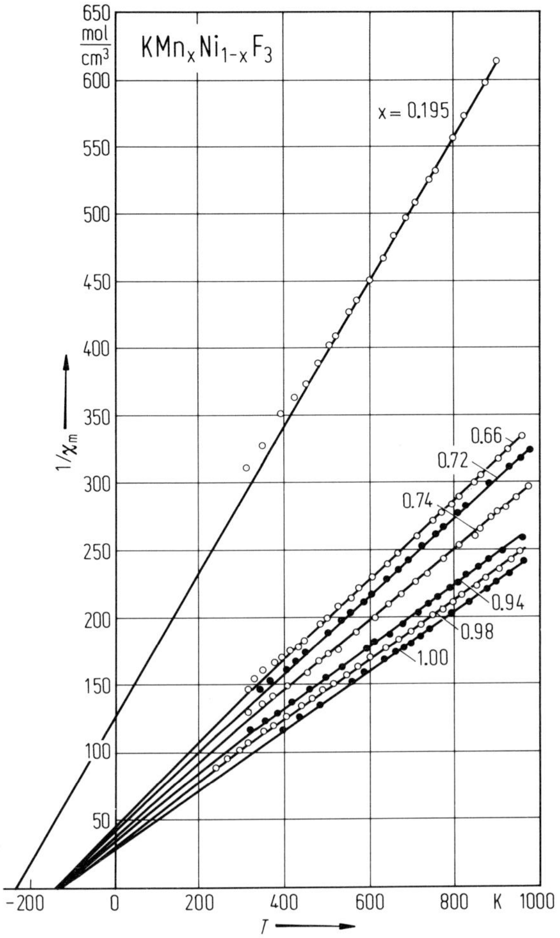

Fig. 93. $KMn_xNi_{1-x}F_3$. Temperature dependence of χ_m^{-1} for different x [84C1].

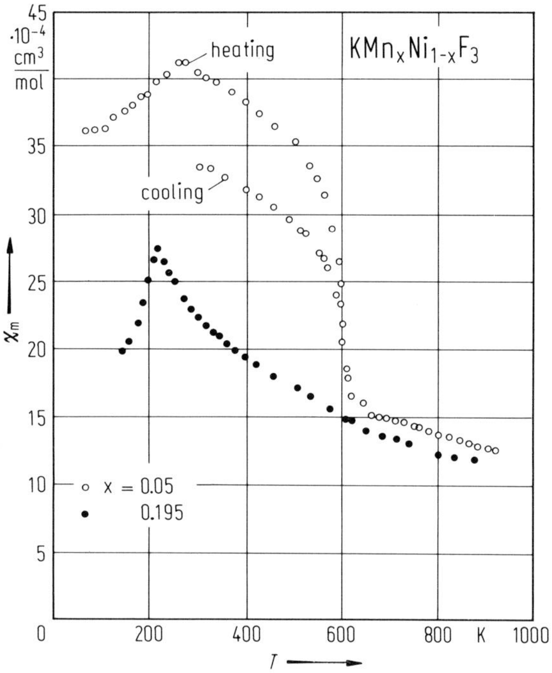

Fig. 94. $KMn_xNi_{1-x}F_3$. Temperature dependence of χ_m for different x (see also Fig. 95) [84C1].

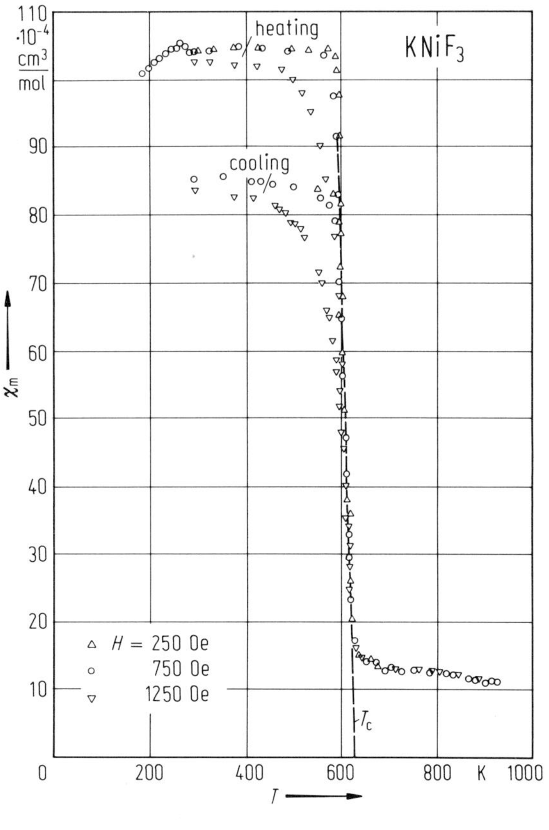

Fig. 95. $KNiF_3$. Temperature dependence of χ_m at different magnetic fields. Second phase is due to Ni precipitations (T_C(Ni) = 634 K). Authors suggest that single phase cannot be obtained in this compound. [84C1].

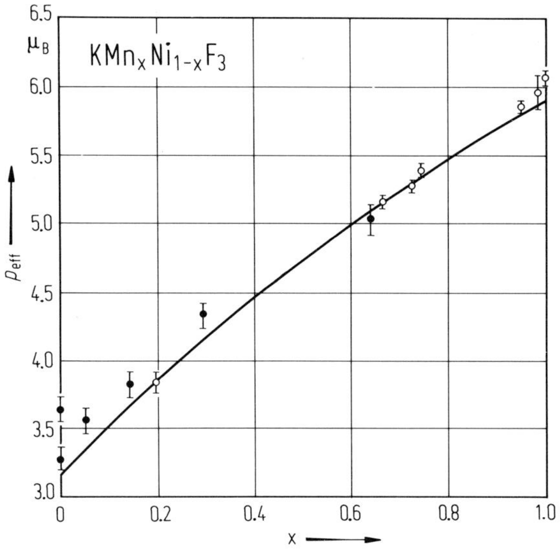

Fig. 96. $KMn_xNi_{1-x}F_3$. Concentration dependence of p_{eff}. Full circles: samples with nickel precipitation, solid line theoretical according to $p_{eff} = \sqrt{p_{eff}^2(Mn)x + p_{eff}^2(Ni)(1-x)}$ [84C1].

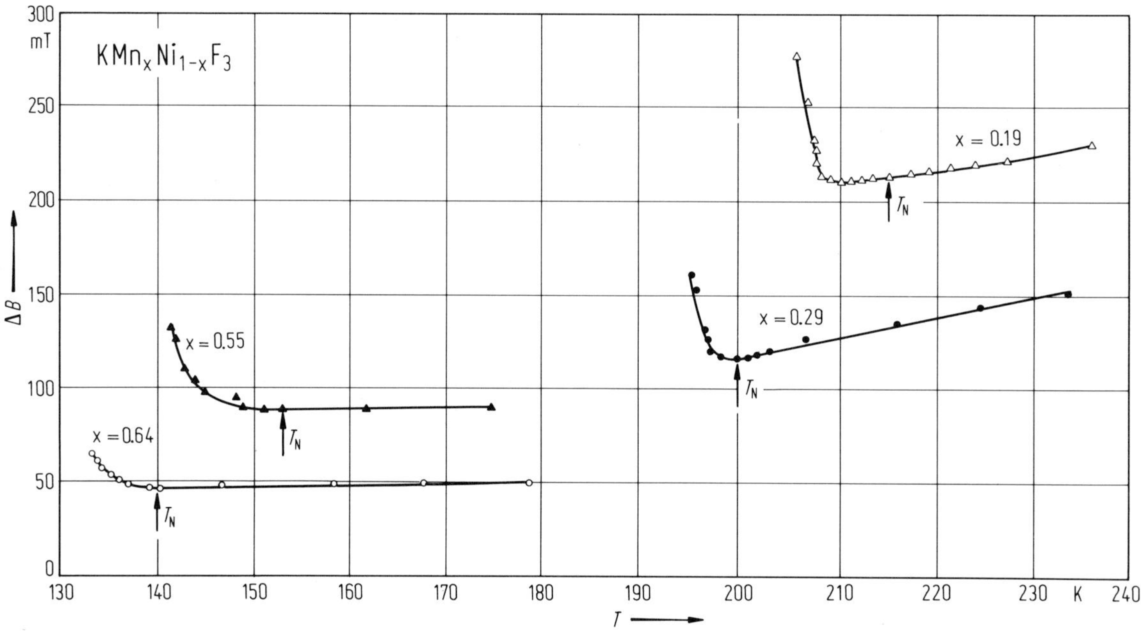

Fig. 97. $KMn_xNi_{1-x}F_3$. Temperature dependence of EPR resonance linewidth near T_N for single crystals (no anisotropy observed) [90S2].

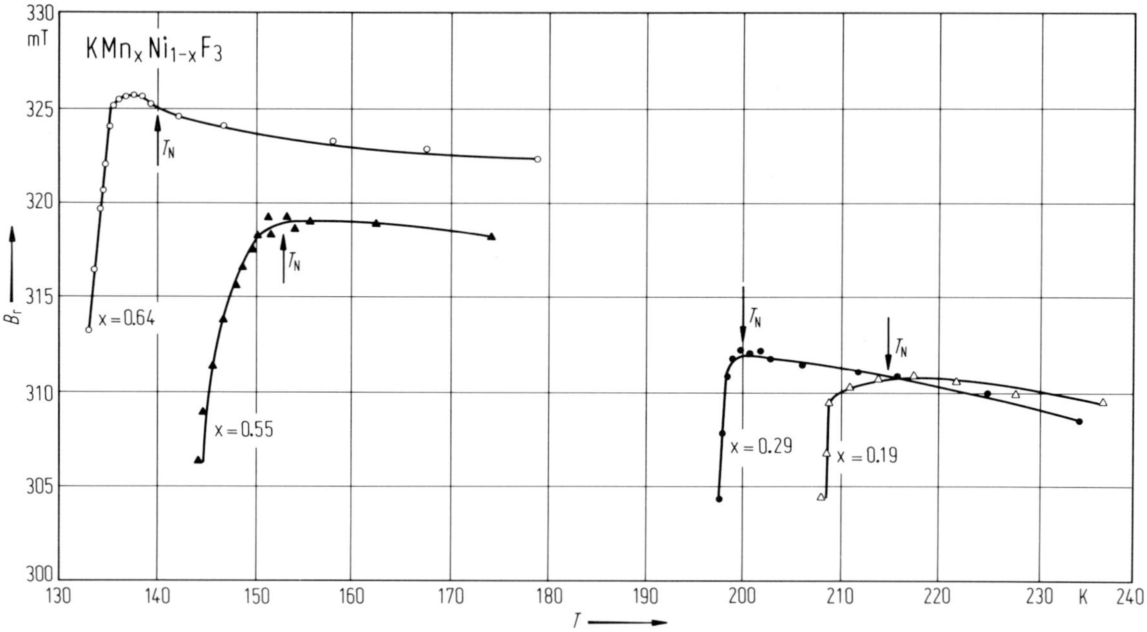

Fig. 98. $KMn_xNi_{1-x}F_3$. Temperature dependence of (AFMR) resonance fields B_r near T_N for single crystals (no anisotropy observed) [90S2].

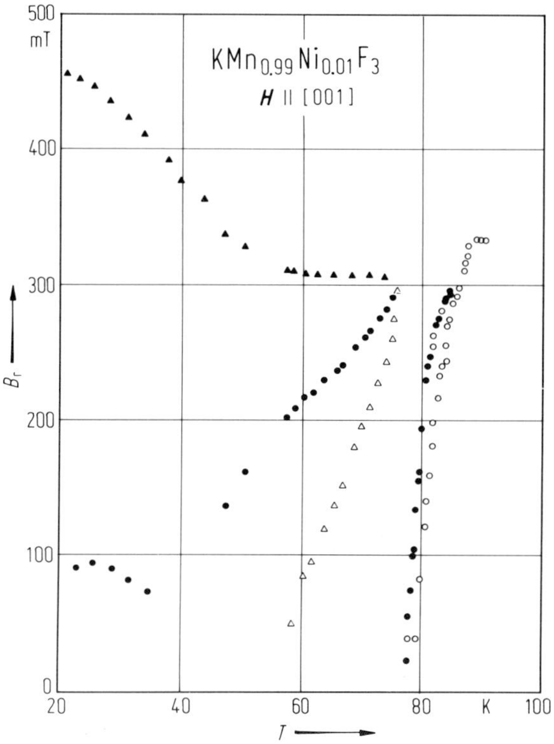

Fig. 99. $KMn_{0.99}Ni_{0.01}F_3$. Temperature dependence of resonance (AFMR) fields B_r of a single crystal for $\boldsymbol{H}\parallel$ [001]. Different symbols indicate different lines [91S1].

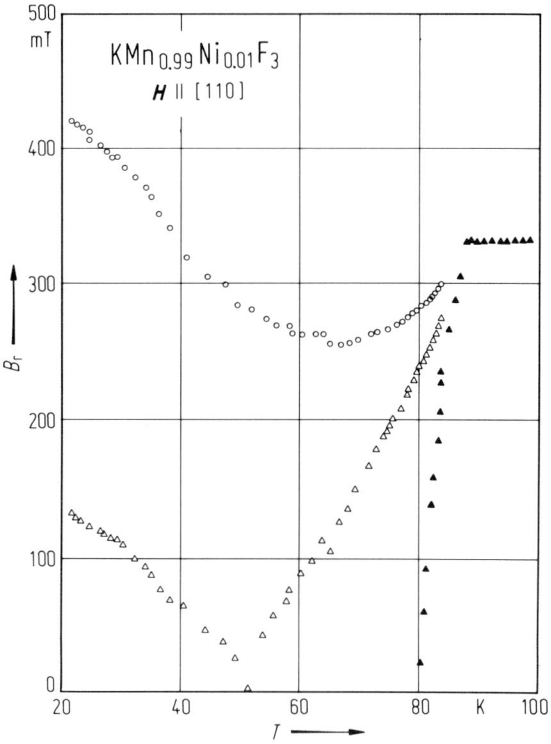

Fig. 100. $KMn_{0.99}Ni_{0.01}F_3$. Temperature dependence of resonance (AFMR) fields B_r of a single crystal for $\boldsymbol{H} \parallel$ [110]. Different symbols indicate different lines [91S1].

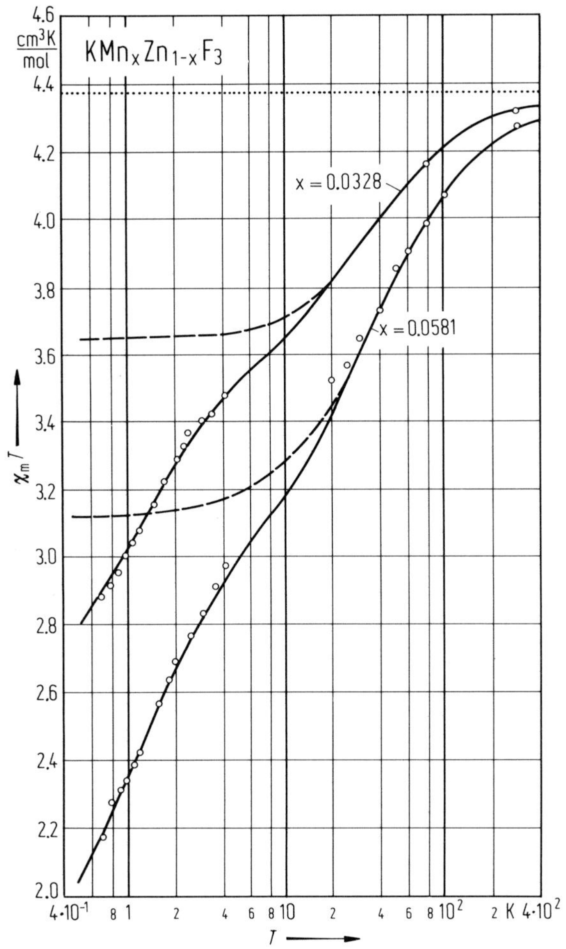

Fig. 101. KMn$_x$Zn$_{1-x}$F$_3$. Temperature dependence of $\chi_m T$. Lines are calculated with $J_1/k = J_2/k = 0$ (dotted), $J_1/k = 8.2$ K, $J_2/k = 0$ (dashed) and $J_1/k = 8.2$ K, $J_2/k = 0.22$ K (solid) [73Y2].

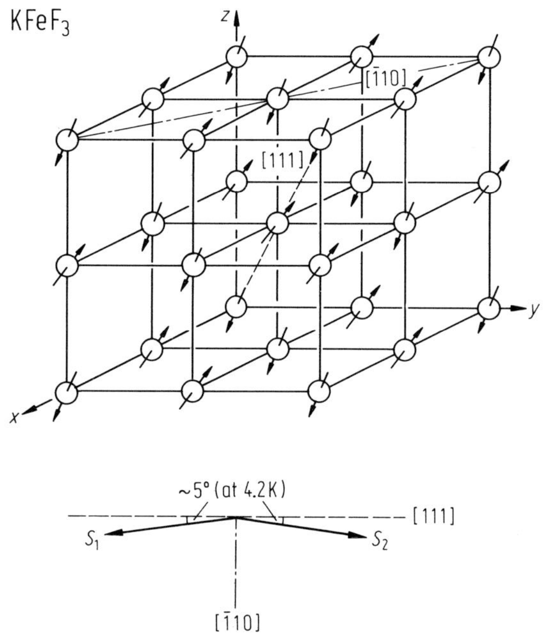

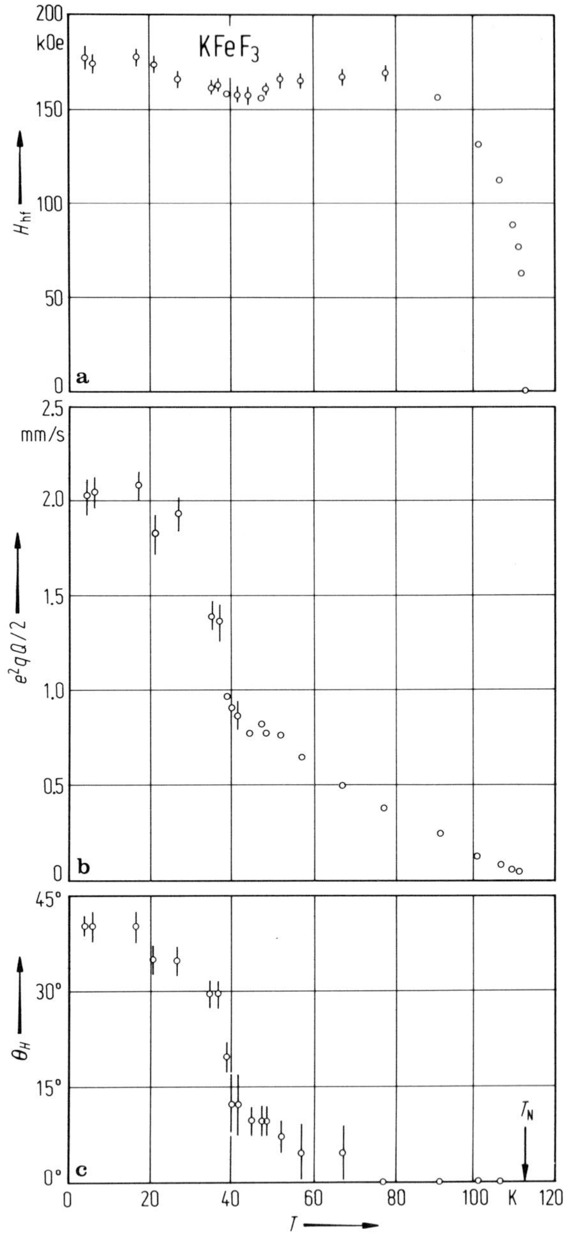

Fig. 103. KFeF$_3$. Temperature dependence of (a) hyperfine field H_{hf}, (b) $1/2e^2qQ$, (c) angle θ_H between H_{hf} and the principal axis z of the EFG tensor determined from Mössbauer experiment [75I1].

Fig. 102. KFeF$_3$. Spin arrangement below T_C [75I1].

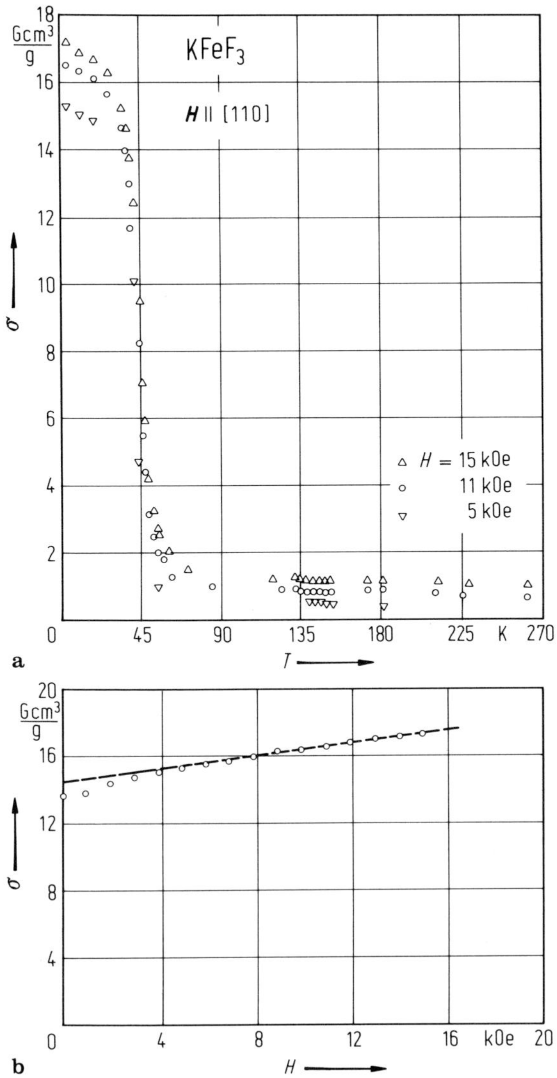

Fig. 104. $KFeF_3$. (a) Temperature dependence of σ at different H for $\boldsymbol{H} \parallel [110]$. (b) magnetic field dependence of magnetization ∥ [110] at 4.2 K [75I1].

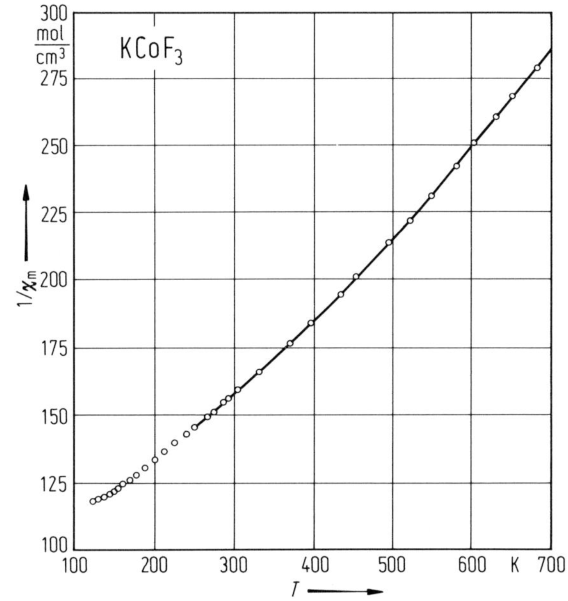

Fig. 105. $KCoF_3$. Temperature dependence of χ_m^{-1} [64H1].

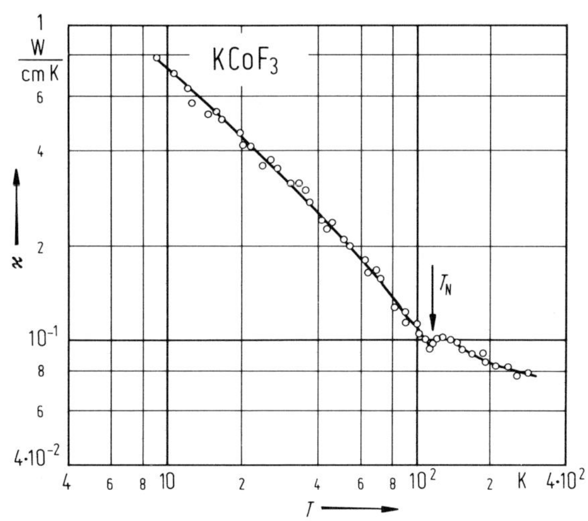

Fig. 106. $KCoF_3$. Temperature dependence of thermal conductivity [64S1].

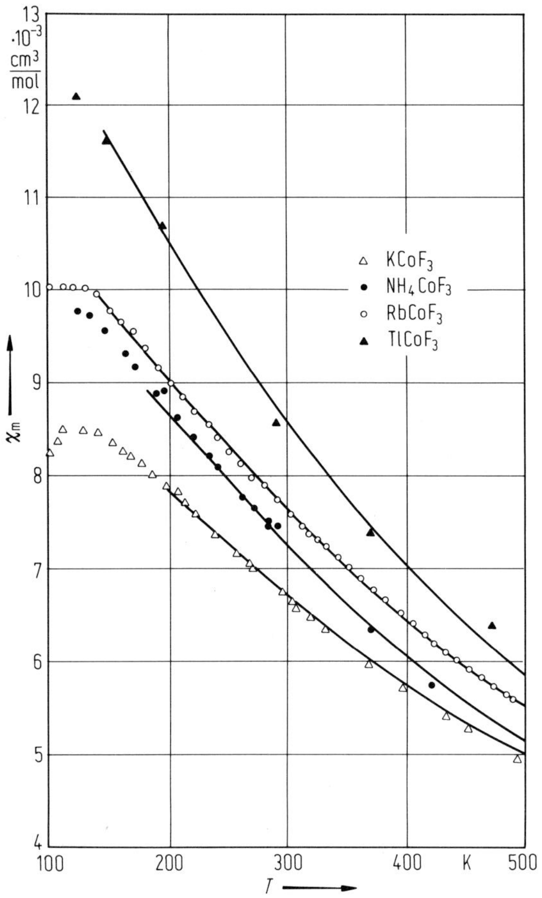

Fig. 107. $KCoF_3$, $RbCoF_3$, NH_4CoF_3, $TlCoF_3$. Temperature dependence of χ_m. Solid curves represent the calculations for χ_{MF} for each of the compounds [77B3]. Data for $KCoF_3$, $RbCoF_3$, $TlCoF_3$ are from other literature.

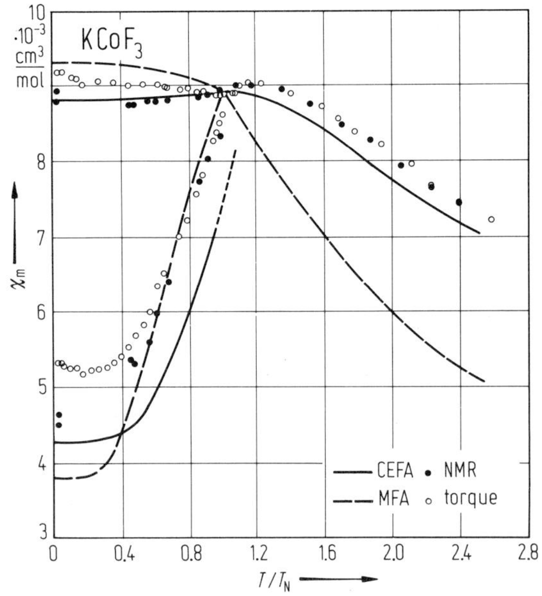

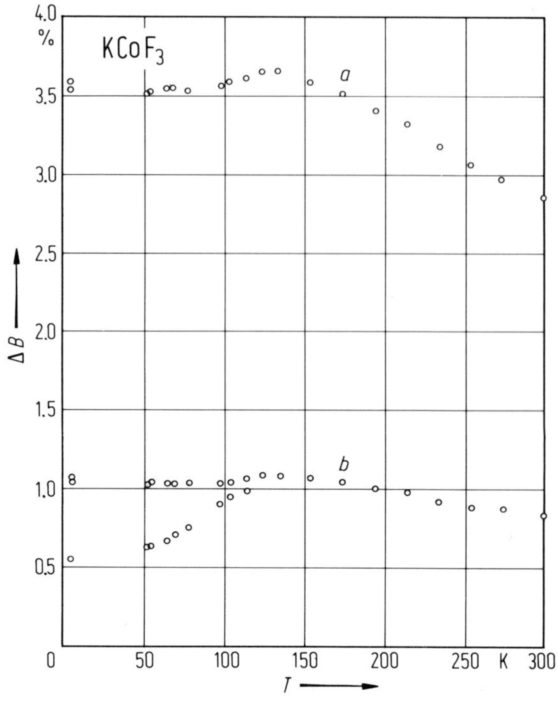

Fig. 109. $KCoF_3$. Temperature dependence of the fractional shift of the ^{19}F NMR field ΔB in $KCoF_3$. Splitting observed below T_N (curve b) corresponds to the difference between $\chi_\perp$ and $\chi_\parallel$ at the angle between H and [100] axis $\psi = 0°$ [78T1].

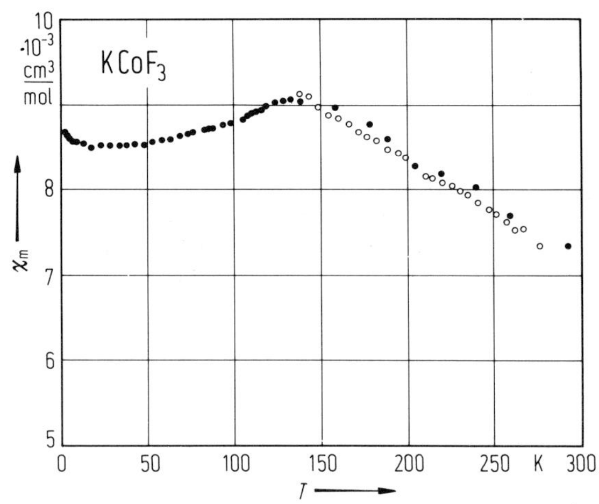

Fig. 110. $KCoF_3$. Temperature dependence of the bulk χ_m at 9.4 kOe along z axis. Full circles: data with the external stress of 30(5) kg/cm²; along y axis. Open circles: data without external stress [78T1].

←

Fig. 108. $KCoF_3$. Temperature dependence of χ_m. Solid and dashed curves are calculated by CEFA and MFA respectively [77S6].

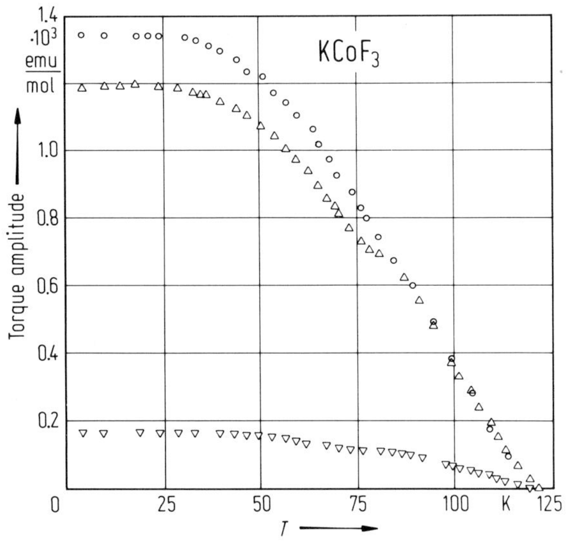

Fig. 111. $KCoF_3$. Temperature dependence of the torque amplitude in the x plane (Δ), the y plane (∇) and the z plane (○) at 9.4 kOe under stress along y axis. The single crystal was a rectangular prism of $1.3(x) \times 2.5(y) \times 0.6(z)$ mm [78T1].

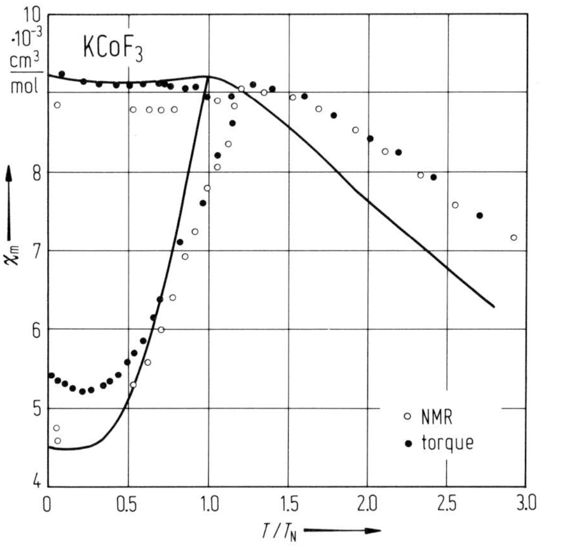

Fig. 112. $KCoF_3$. Temperature dependence of χ_m. Solid line is the best theoretical fit for $T_N = 118$ K; $J = -9.5\ \mathrm{cm}^{-1}$, orbital quenching factor $\kappa = 1.4$; spin-orbit coupling parameter $\lambda = 211\ \mathrm{cm}^{-1}$, tetragonal crystal field parameter $C = 0$ (see also [77S6, 78T1])[87M1].

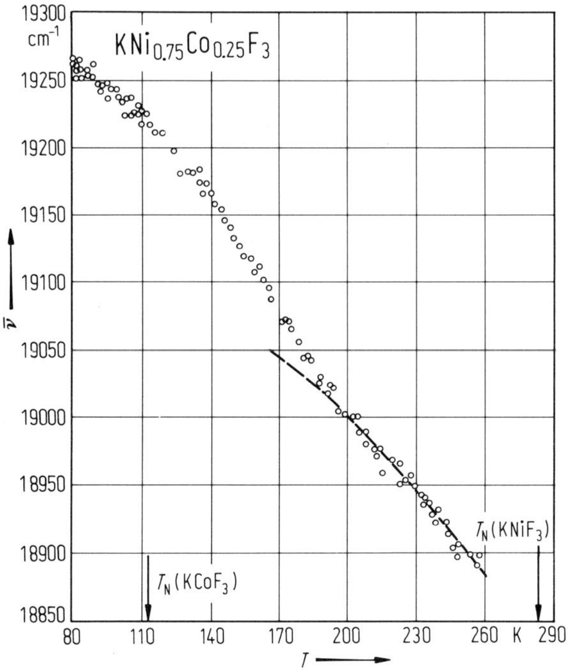

Fig. 113. $KNi_{0.75}Co_{0.25}F_3$. Temperature dependence of the ${}^4T_{1g} \rightarrow {}^4T_{1g}(P)$ ($\sim 19000\ \mathrm{cm}^{-1}$) absorption-band maximum of the Co^{2+} ion. The dashed line indicates purely thermal behaviour to emphasize the magnetically induced shifts [67H2].

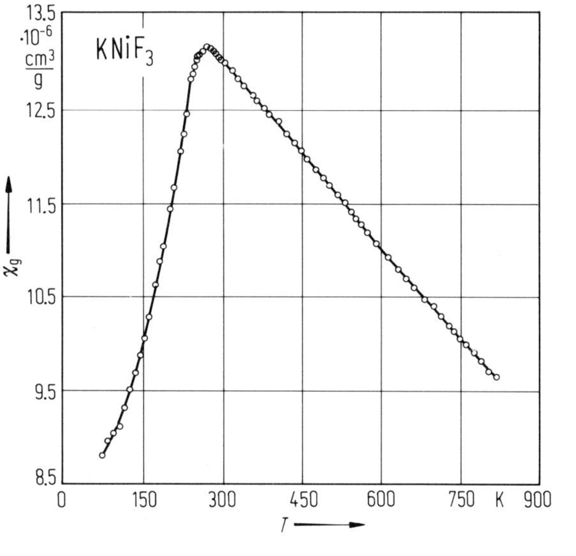

Fig. 114. $KNiF_3$. Temperature dependence of χ_g [63H1].

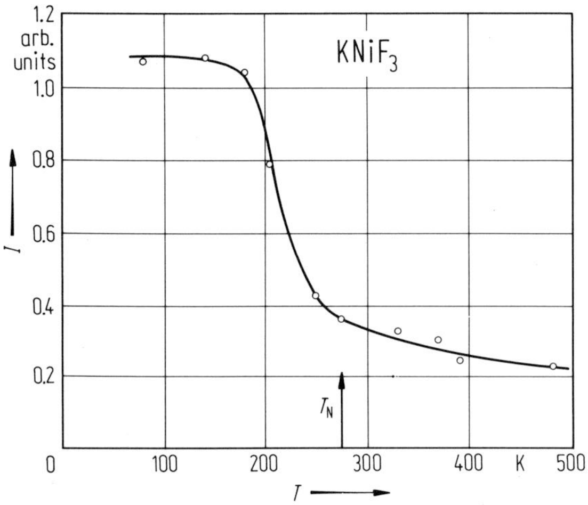

Fig. 115. $KNiF_3$. Temperature dependence of the intensity of the $\bar{\nu} = 36000\ cm^{-1}$ line of the transition $^3A_2 \rightarrow (^1E_a + {}^1T_{a/2})$ [68P3].

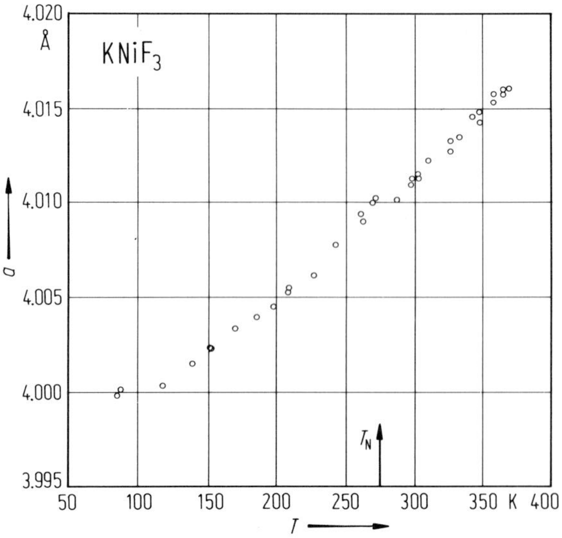

Fig. 116. $KNiF_3$. Temperature dependence of the lattice constant [68S2].

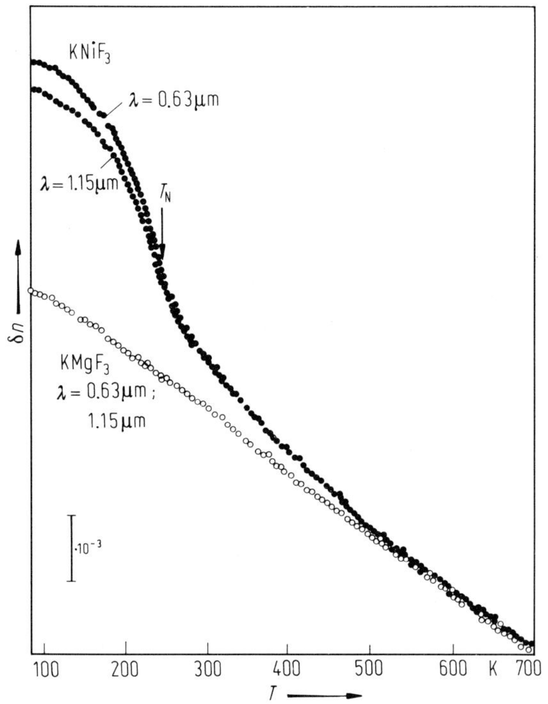

Fig. 117. $KNiF_3$, $KMgF_3$. Temperature dependence of the refractive index [77S5]. See also [84K4].

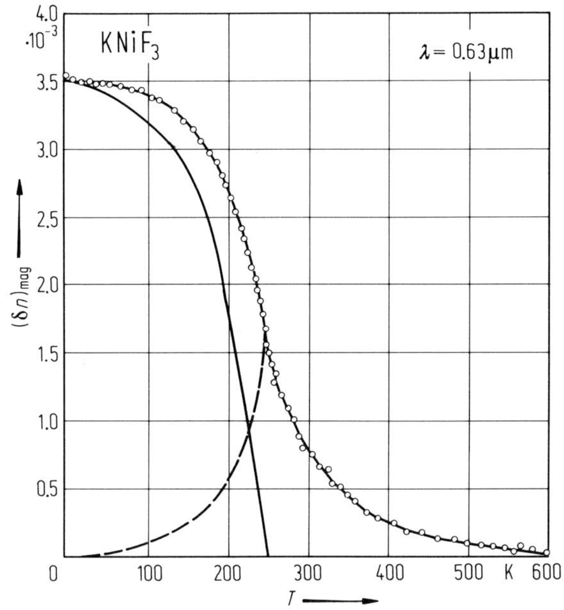

Fig. 118. $KNiF_3$. Isotropic magnetic contribution to the refractive index (circles), the square of the antiferromagnetic moment (solid line), the contribution of fluctuations of an antiferromagnetic moment to the refractive index below T_N (dashed curve) [77S5].

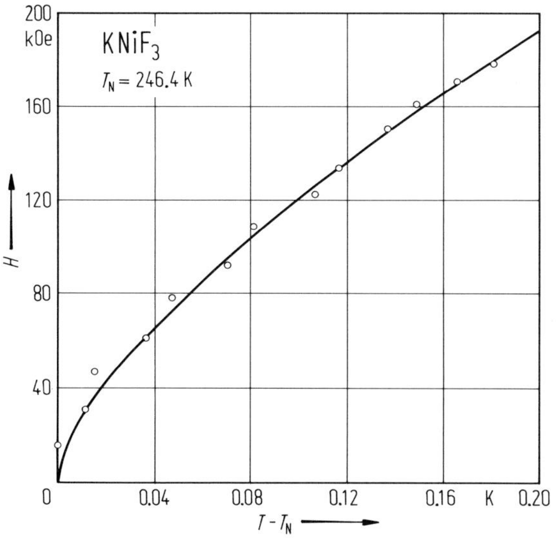

Fig. 119. $KNiF_3$. Magnetic field dependence of the ordering temperature T_c. $\boldsymbol{H}\|[001]$. The solid line is a fit to the data above 50 kOe according to Eq. $T_c - T_N = aH^{2/\phi} - bH^2$, where a and b are constants (adjustable parameters) and ϕ is the crossover exponent. For ϕ is taken its theoretical value 1.25 [78B2].

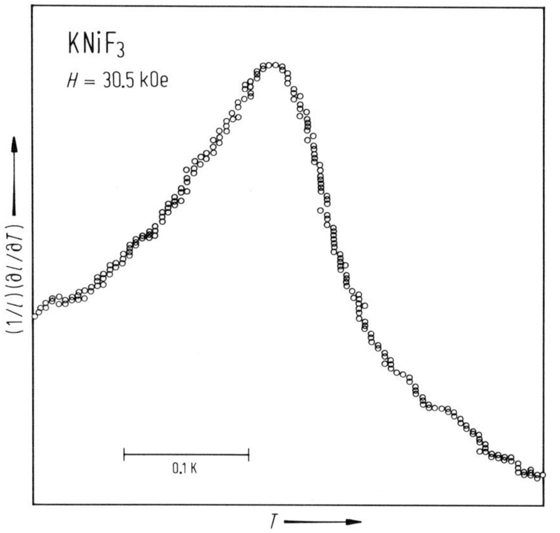

Fig. 120. $KNiF_3$. λ anomaly in thermal expansion coefficient at 30.5 kOe for $\boldsymbol{H}\|[001]$ [78B2].

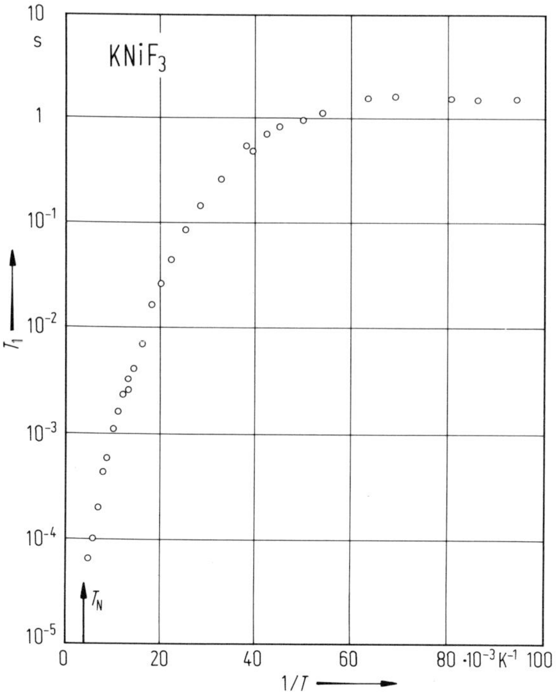

Fig. 121. $KNiF_3$. ^{19}F spin-lattice relaxation time vs. $1/T$. The external magnetic field $\boldsymbol{H}\|[100]$ at $\nu = 30$ MHz [79E1].

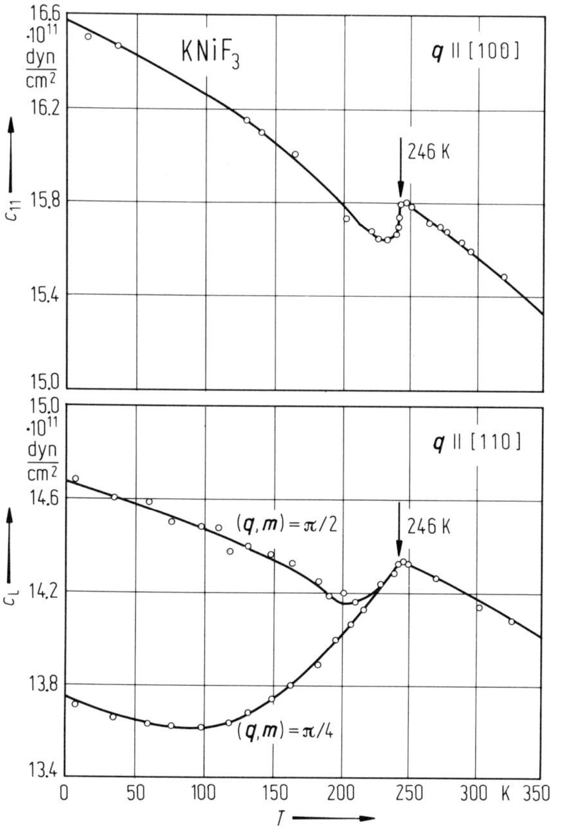

Fig. 122. $KNiF_3$. Temperature dependence of the elastic constants [81G2]. $m = S_1 - S_2$ at equilibrium ($S_1 S_2$: unit vectors related to up and down sublattice); q: wave vector; $c_L = \frac{1}{2}(c_{11} + c_{12} + 2c_{44})$.

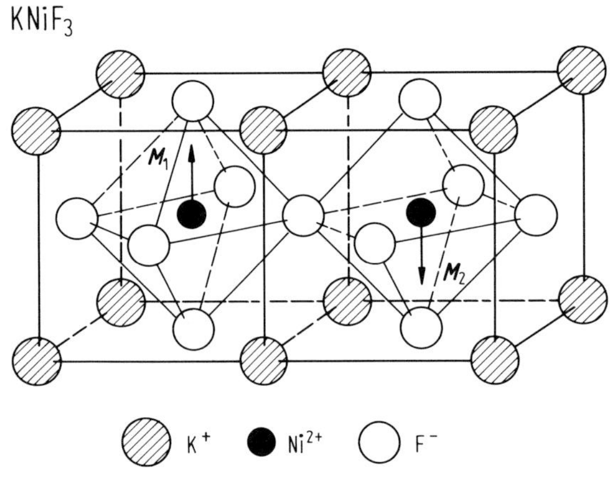

Fig. 123. $KNiF_3$. Magnetic crystal structure [82G2].

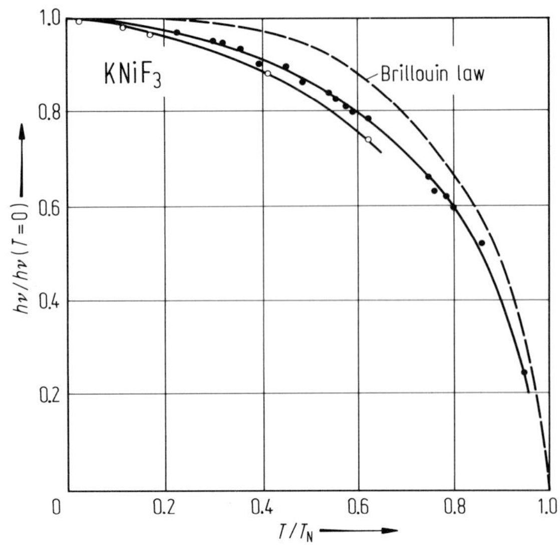

Fig. 124. $KNiF_3$. Relative temperature dependence of magnon energies for two scattering angles φ: (full circles) $h\nu(\varphi = \pi)$ (5 cm^{-1} at $T = 0$ K), (open circles) $h\nu(\varphi = \pi/2)$ (4.3 cm^{-1} at $T = 0$ K) [82G2].

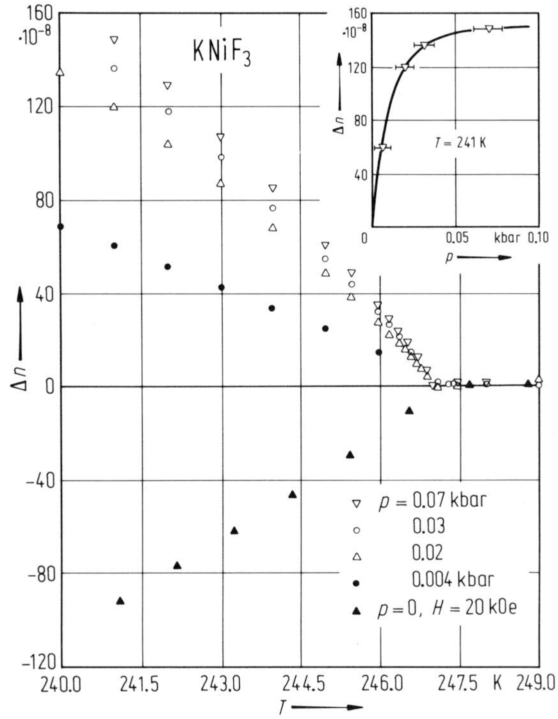

Fig. 125. $KNiF_3$. Temperature dependence of birefringence Δn near T_N for $\lambda = 632.8$ nm and different applied uniaxial pressures and for $H = 20$ kOe. In the insert uniaxial pressure dependence of Δn at 241 K is shown [83N2].

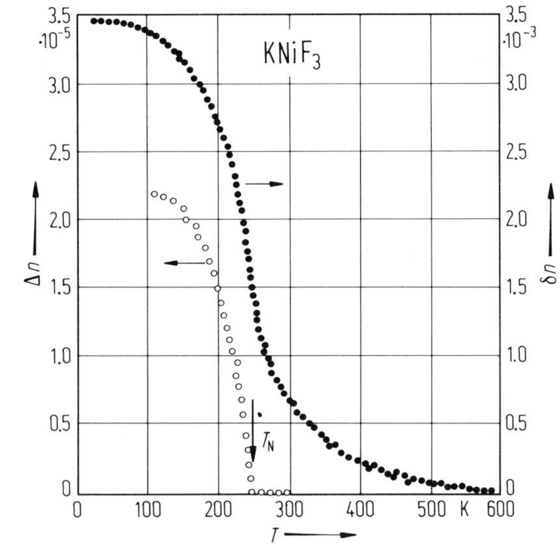

Fig. 126. $KNiF_3$. Temperature dependence of total (full circles) and anisotropic (open circles) magnetic contribution to refraction of light for $\lambda = 0.6328$ μm [84K4].

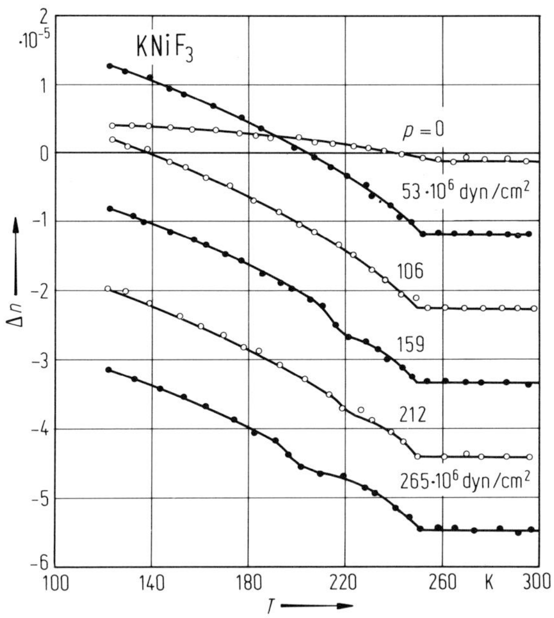

Fig. 127. $KNiF_3$. Temperature dependence of birefringence for $\lambda = 0.6328$ µm at different pressures along [0 0 1] axis [84K4]. 1 dyn/cm² = 10^{-6} bar.

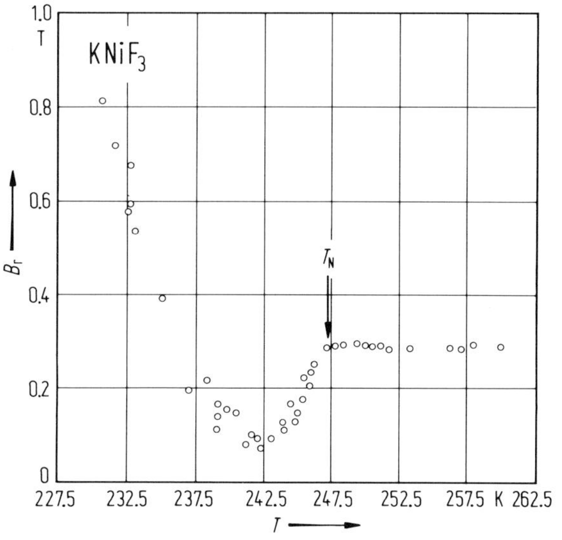

Fig. 128. $KNiF_3$. Temperature dependence of resonance fields from AFMR [85C2].

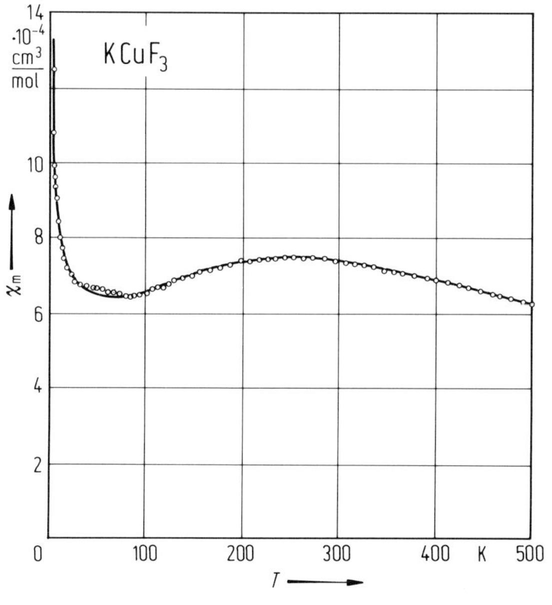

Fig. 129. $KCuF_3$. Temperature dependence of χ_m for powder [67K2].

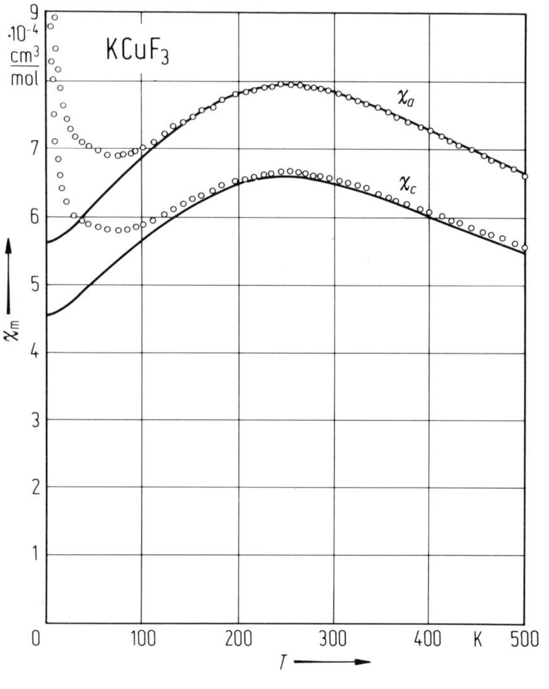

Fig. 130. $KCuF_3$. Temperature dependence of χ_m for a single crystal. Solid lines are theoretical [67K2].

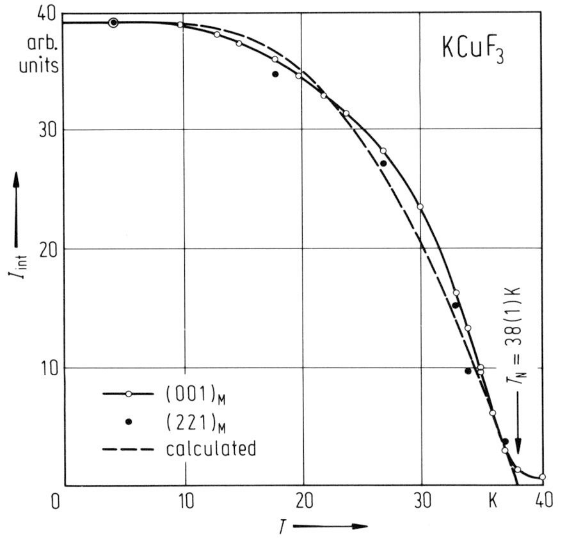

Fig. 131. $KCuF_3$. Temperature dependence of the (0 0 1) and (2 2 1) magnetic neutron integrated intensity I_{int} for single crystal I (compare Fig. 133). The sample was ~ 96% single crystal domain and showed multicrystalline growth with a mosaic spread of 1°, each neutron Bragg reflection consisting of two or three peaks [69H1].

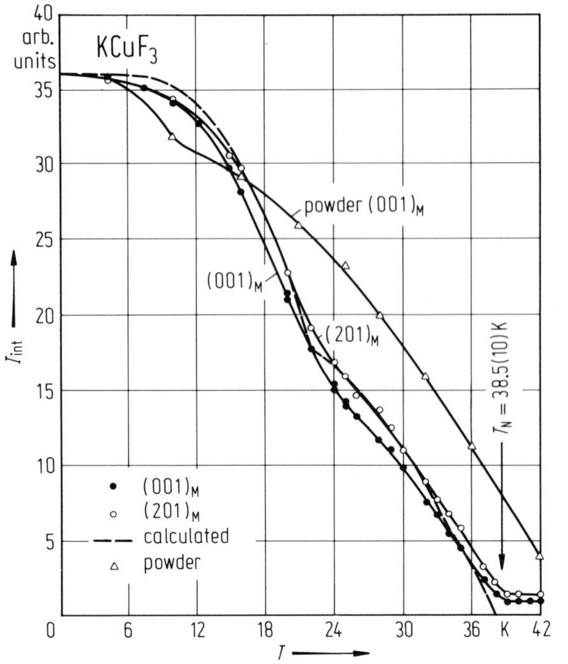

Fig. 133. $KCuF_3$. Temperature dependence of the (0 0 1) and (2 0 1) magnetic neutron integrated intensity I_{int} of single crystal II (compare Fig. 131). The sample had narrower mosaic of ~ 0.4° with some peak splitting and was ~ 93% single domain. This figure and Fig. 131 are for two different single crystals [69H1].

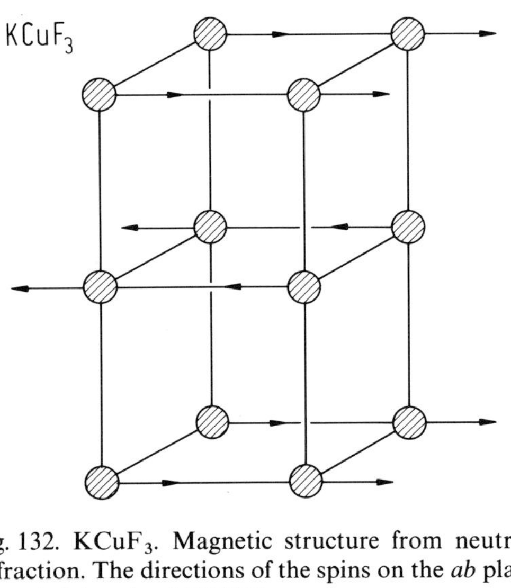

Fig. 132. $KCuF_3$. Magnetic structure from neutron diffraction. The directions of the spins on the *ab* plane are not determined [69H1].

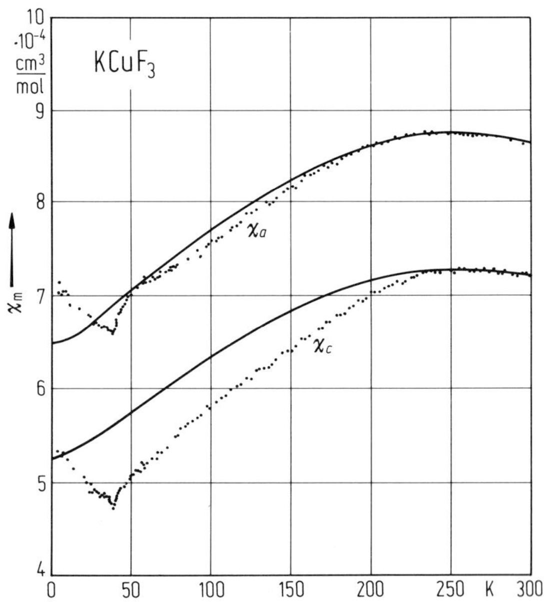

Fig. 134. $KCuF_3$. Temperature dependence of χ_m for a single crystal along *c* and *a* axes at 3.3 kOe [73H1]. The solid lines are calculated using the theory of Bonner and Fisher [64B1].

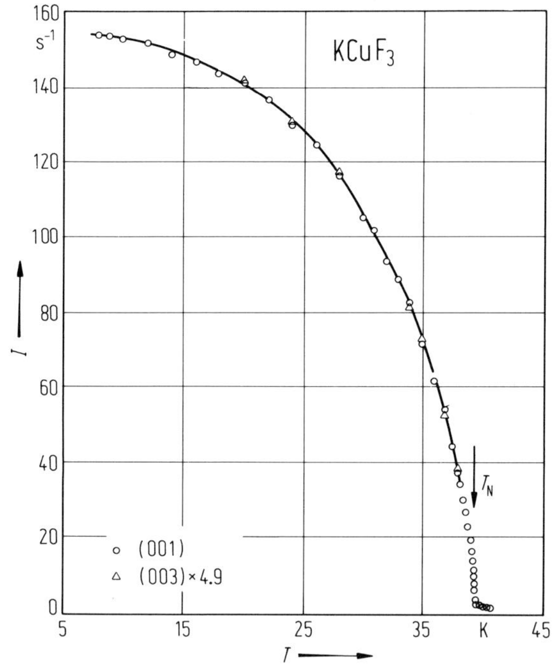

Fig. 135. $KCuF_3$. Temperature dependence of neutron-diffraction (0, 0, 1) and (0, 0, 3) intensities. The (0, 0, 3) intensities were normalized to the (0, 0, 1) ones at 20 K [73I1].

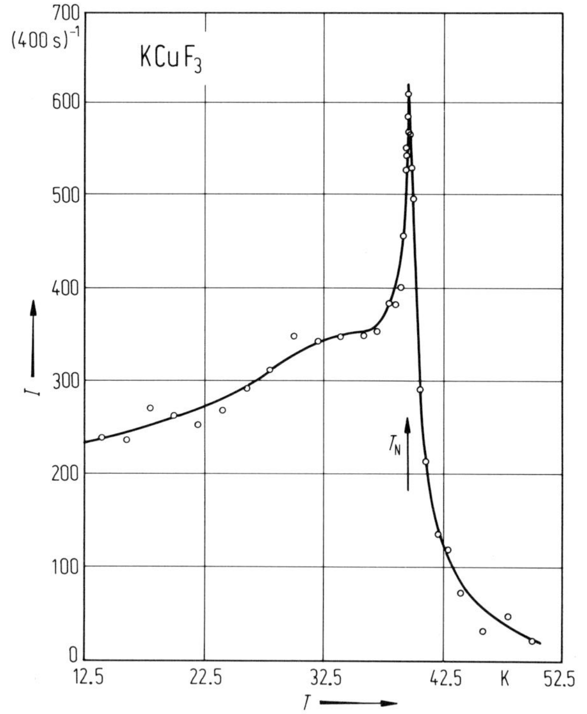

Fig. 136. $KCuF_3$. Temperature dependence of neutron-diffraction magnetic diffuse intensities at (− 0.02, 0, 1) position in reciprocal space [73I1].

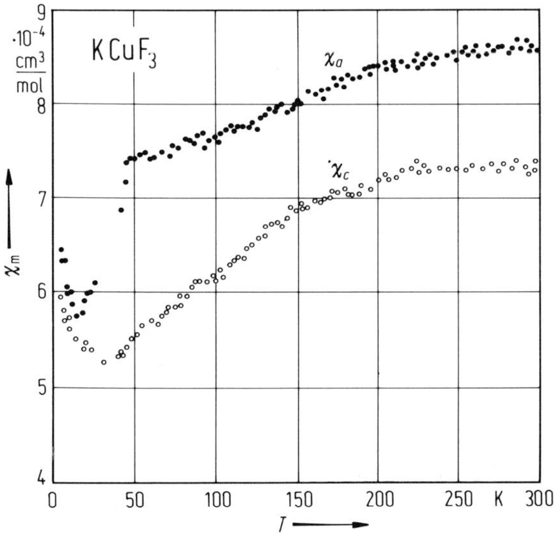

Fig. 137. $KCuF_3$. Temperature dependence of χ_m at 620 Oe for a single crystal along a axis (χ_a) and c axis (χ_c) (see also Fig. 130, 134) [75M1].

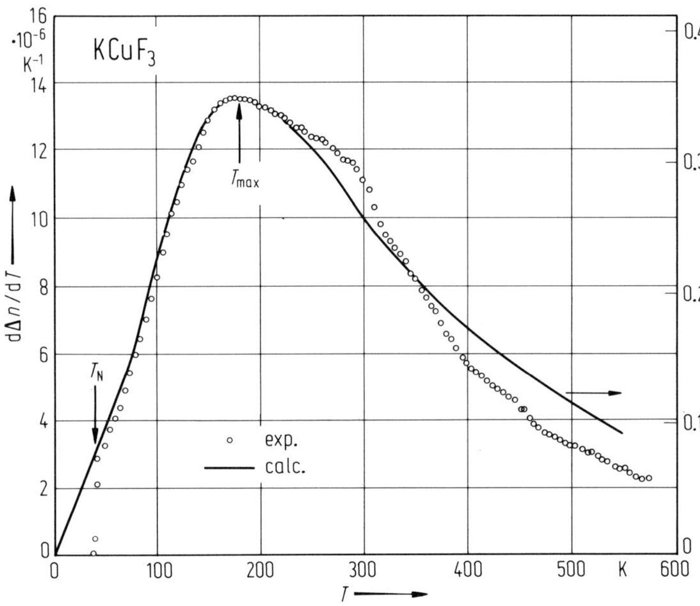

Fig. 138. $KCuF_3$. Temperature dependence of temperature derivative of Δn, $-\mathrm{d}\Delta n/\mathrm{d}T$, and calculated magnetic specific heat capacity C_{mag} (solid line, calculated according to Bonner-Fisher's theory) [78I1]. $\Delta n = n_a - n_c$; $\lambda = 0.6328$ µm.

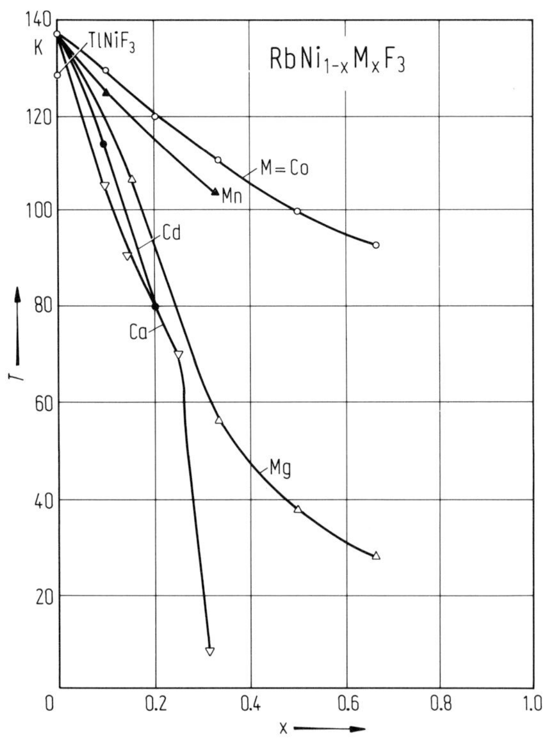

Fig. 139. $RbNi_{1-x}M_xF_3$, M = Co, Mn, Cd, Ca, Mg. Concentration dependence of magnetic transition temperature [68M1]. $TlNiF_3$ for comparison.

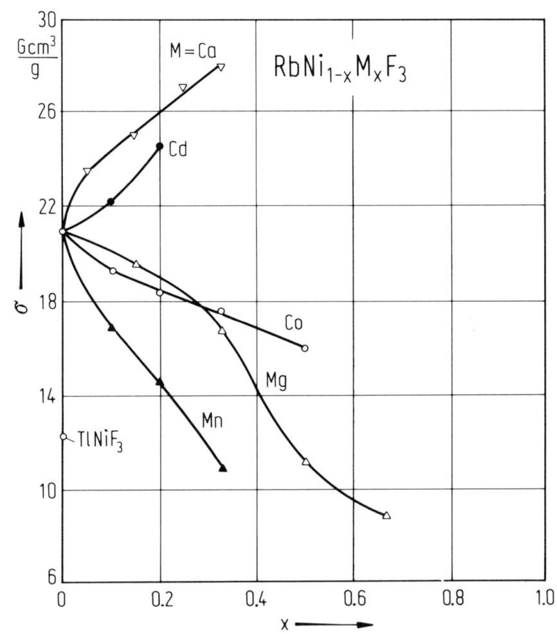

Fig. 140. $RbNi_{1-x}M_xF_3$, M = Ca, Cd, Co, Mg, Mn. Concentration dependence of the magnetization σ at 4.2 K and 50 kOe (Ca at 25 kOe) [68M1]. $TlNiF_3$ for comparison.

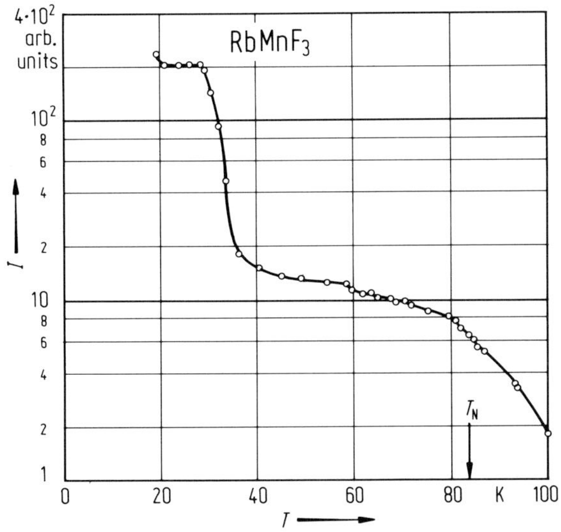

Fig. 141. $RbMnF_3$. Temperature dependence of the relative intensity of the fluorescence [65P2].

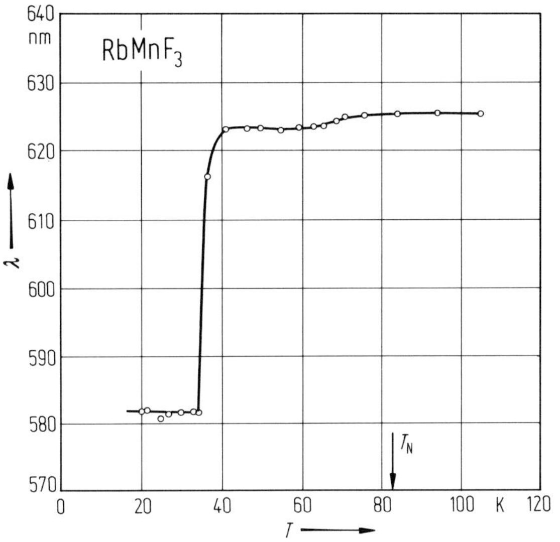

Fig. 142. $RbMnF_3$. Temperature dependence of the fluorescence profile maxima. The variation from 35 to 40 K is attributed to the alignment at the excited Mn^{2+} ion with respect to the ordered ground state [65P2].

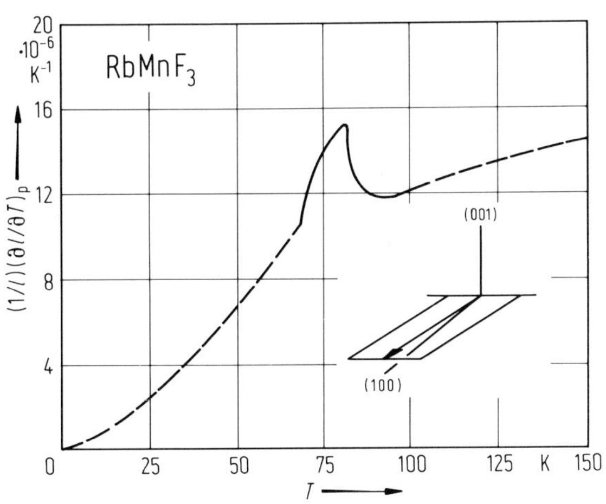

Fig. 143. $RbMnF_3$. Temperature dependence of the thermal expansion coefficient $\left(\frac{1}{l}\frac{\partial l}{\partial T}\right)_p$ near T_N [66T1].

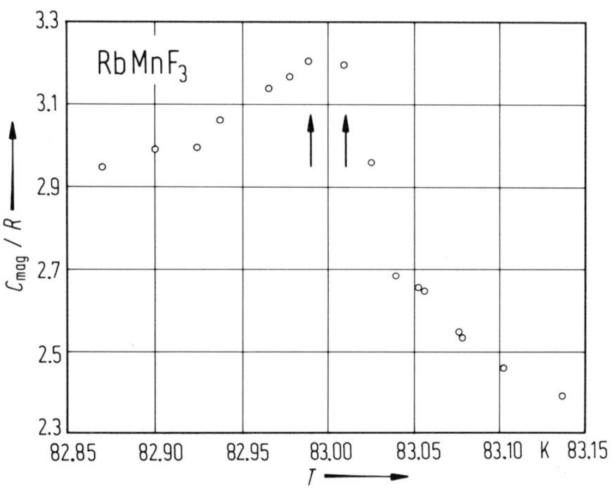

Fig. 144. $RbMnF_3$. Temperature dependence of the magnetic heat capacity near T_N [66T1]. Arrows indicate maxima.

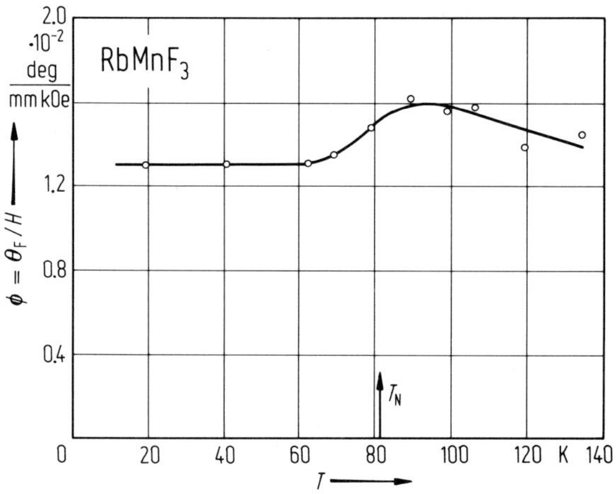

Fig. 145. $RbMnF_3$. Temperature dependence of magneto-resonance Faraday effect at $H > 20$ kOe [67K3]. ϕ: Verdet constant, $\lambda = 632.8$ nm.

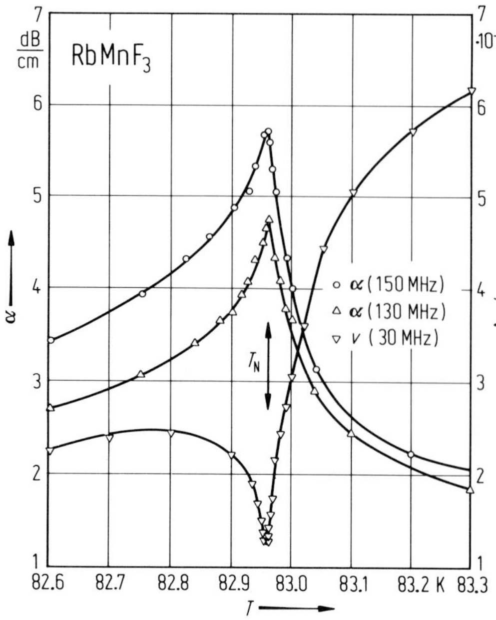

Fig. 146. $RbMnF_3$. Temperature dependence of critical ultrasonic attenuation and velocity of [0 0 1] propagating longitudinal waves [68G3].

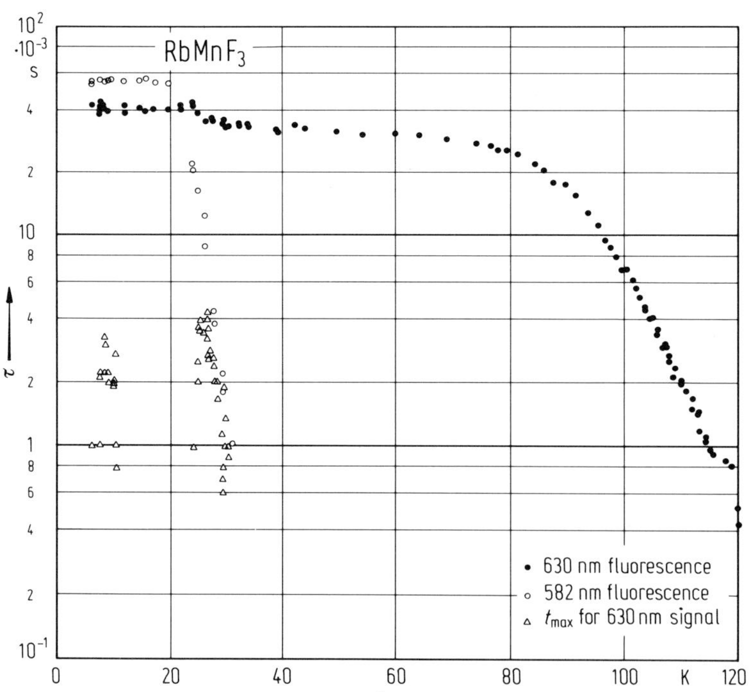

Fig. 147. $RbMnF_3$. The fluorescence lifetime at 0.5820 and 0.6300 μm [69G2]. t_{max}: time at which maximum in fluorescence signal occurs during decay process.

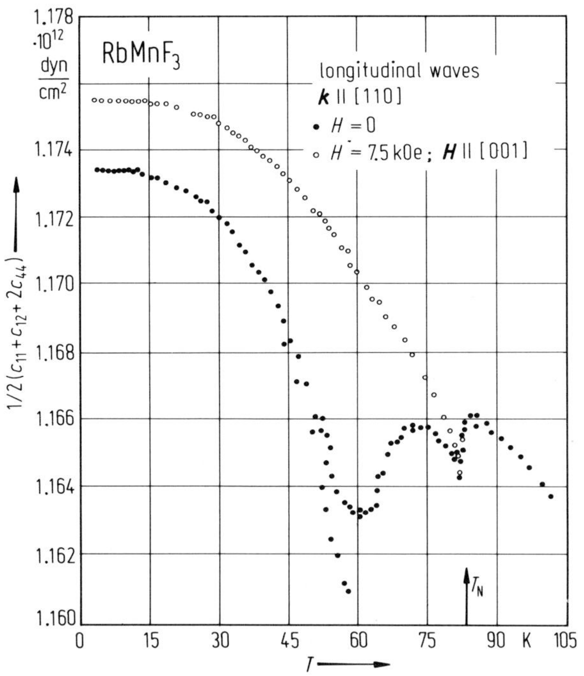

Fig. 148. $RbMnF_3$. Temperature dependence of elastic constant $c_L = \frac{1}{2}(c_{11} + c_{12} + 2c_{44})$ [69M1].

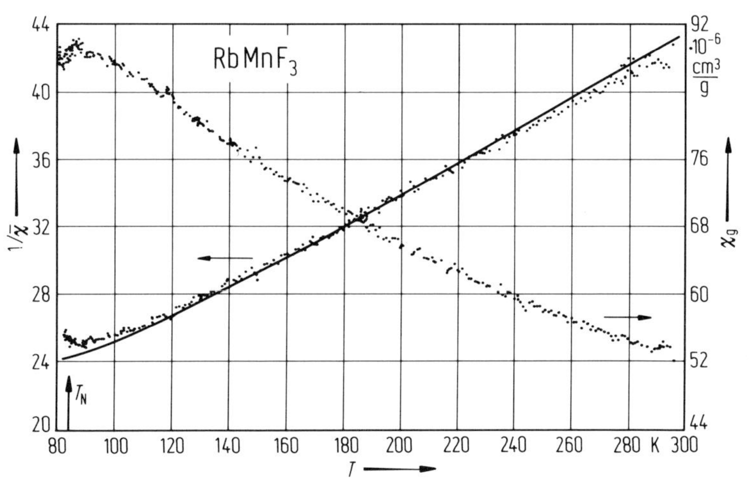

Fig. 149. $RbMnF_3$. Temperature dependence of χ_g and $(\bar{\chi})^{-1} = Ng^2\mu_B^2/\chi J$ for single crystal (χ is isotropic above T_N). The solid curve is theoretical. Experimental points are normalized using $J/k = 3.37$ K and $g = 1.975$ [72B2].

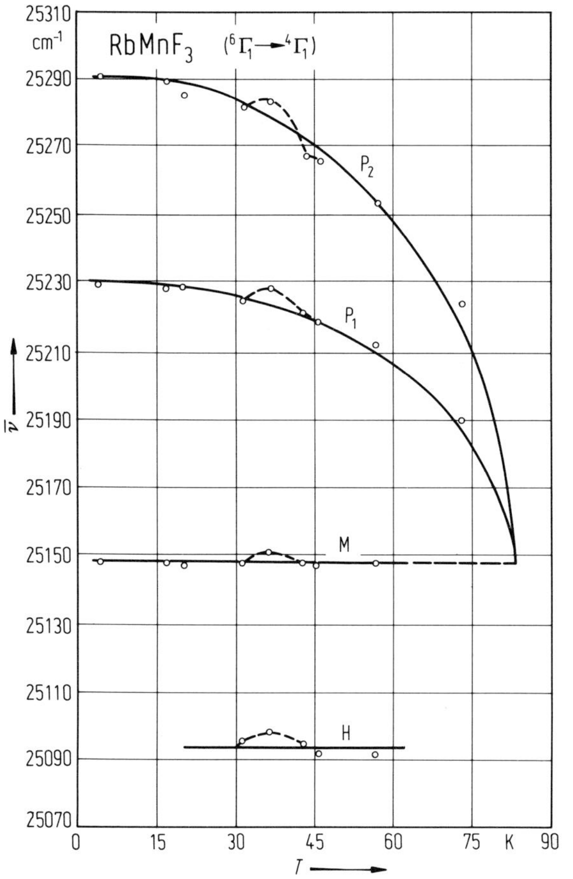

Fig. 150. $RbMnF_3$. Temperature dependence of the energies of the excitation line M, the magnon sidebands P_1 and P_2 and the magnon assisted hot band H, all associated with transition to the $^4\Gamma_1$ level. The solid curves for P_1 and P_2 have been calculated using formula $\bar{\nu}(T) = \bar{\nu}(M) + \Delta\bar{\nu}B_{5/2}(T/T_N)$, where $\Delta\bar{\nu}$ is separation from M line, $B_{5/2}$ = Brillouin function for $S = 5/2$ [72S3].

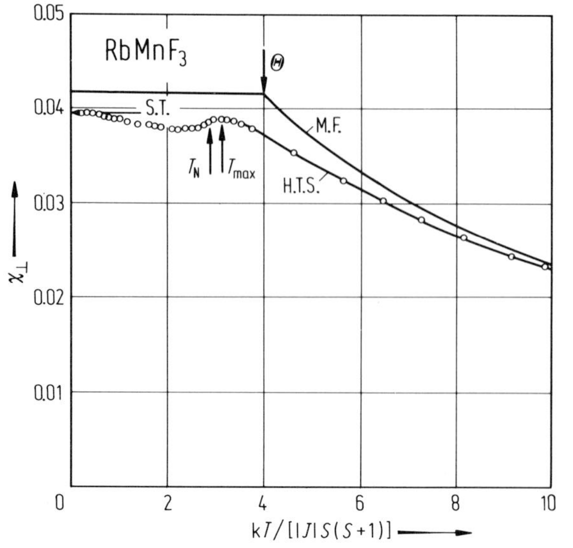

Fig. 151. $RbMnF_3$. Temperature dependence of $\chi_\perp$ for a single crystal. The measurements (open circles) are plotted on relative scales in order to facilitate a comparison with theoretical results for the simple cubic, $S = 5/2$, Heisenberg antiferromagnet. These include the high-temperature series prediction for the antiferromagnetic susceptibility (H.T.S.) and the spin wave prediction for $\chi_\perp$ at $T = 0$ (S.T.). The theoretical predictions for the transition temperature (T_N) and the position of the susceptibility maximum (T_{max}) have been indicated. The M.F. curve is the molecular field prediction for $\chi_\perp \cdot \bar{\chi} = \chi |J|/N_0 g^2 \mu_B^2$. $\chi_{max} = 87.2(1) \cdot 10^{-6}$ cm³/g, $T_{max} = 91.2$ K, $T_N = 84.2$ K, experimental data have been corrected for an estimated diamagnetic contribution of $\chi_{dia} = -0.32 \cdot 10^{-6}$ cm³/g [74J1]. $J/k = -3.40$ K, $H_A/H_E = 5 \cdot 10^{-6}$.

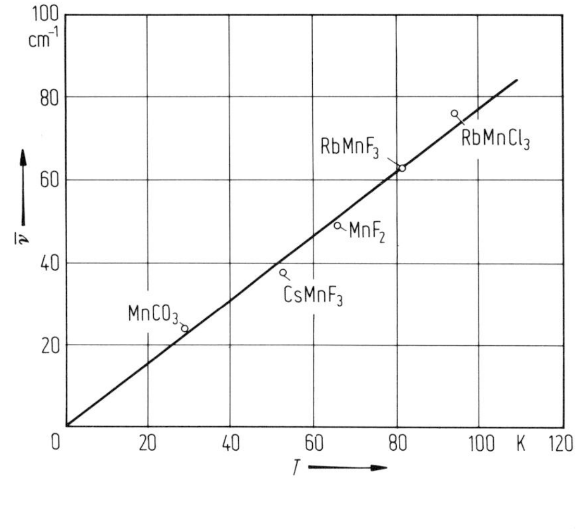

Fig. 152. $RbMnF_3$, $RbMnCl_3$, $CsMnF_3$. Dependence of the magnon frequencies on the boundary of the Brillouin zone versus magnetic ordering temperature [76B3].

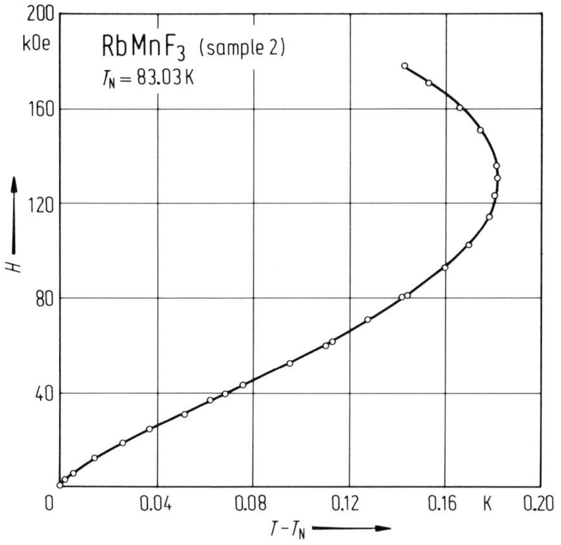

Fig. 153. $RbMnF_3$. Magnetic field dependence of the ordering temperature T_c. The solid line is a least-squares fit to the data according to equation given in Fig. 119 (see Fig. 154) [78S1].

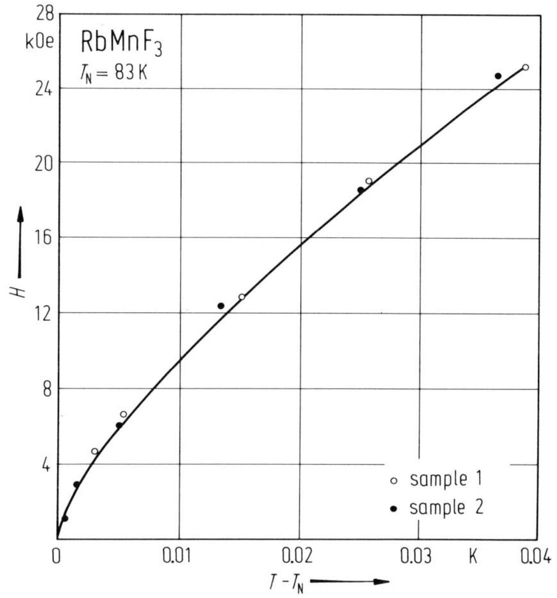

Fig. 154. $RbMnF_3$. Magnetic field dependence of the ordering temperature T_c for $H < 25$ kOe. Solid line is the low-H portion of the least-square fit shown in Fig. 153 [78S1].

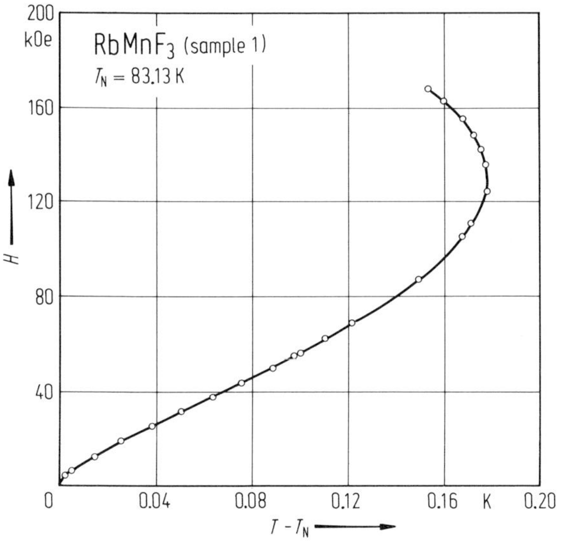

Fig. 155. $RbMnF_3$. Magnetic field dependence of the order-disorder transition. Data obtained from lambda peaks in the attenuation of a 45 MHz ultrasonic wave (see Fig. 154) [78S2]. See also [77S4]

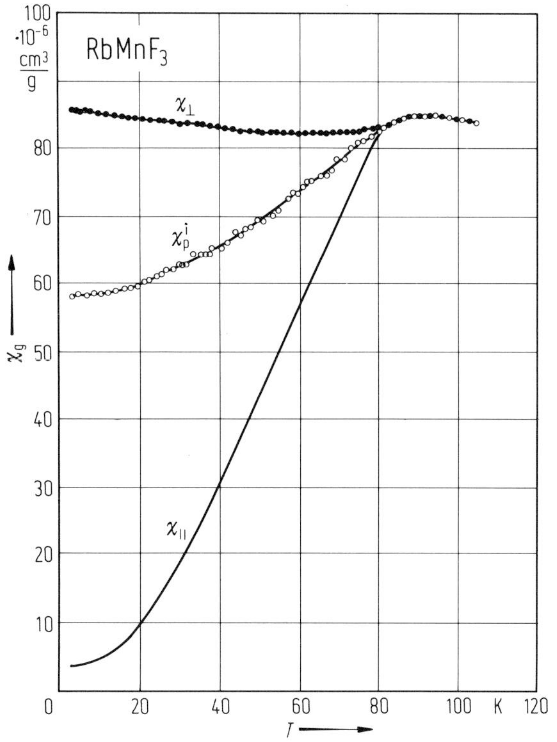

Fig. 156. $RbMnF_3$. Temperature dependence of $\chi_{g\perp}$ at 10 kOe, χ_p^i (initial, powder) at 64 Oe, and $\chi_{||}$ calculated according to Eq. $3\chi_p^i = \chi_{||} + 2\chi_\perp$ [82C1].

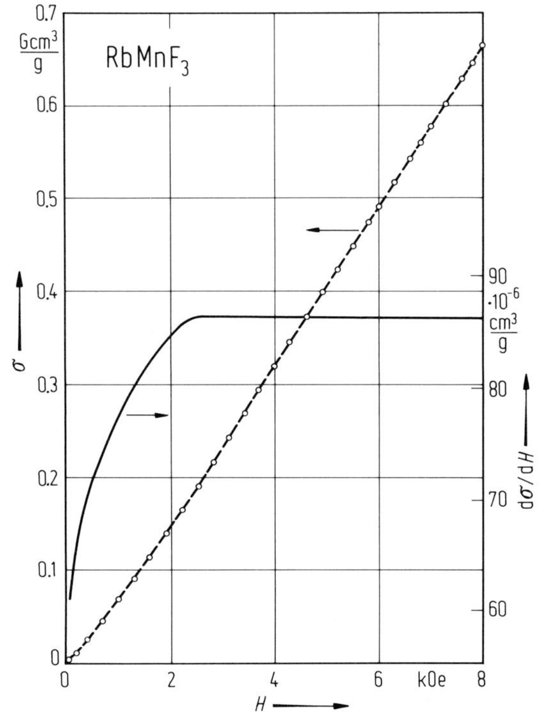

Fig. 157. $RbMnF_3$. Measured magnetization σ and computed $d\sigma/dH$ (solid line) vs. applied field for powder sample at 4.2 K. The dashed curve is drawn through the data points [82C1].

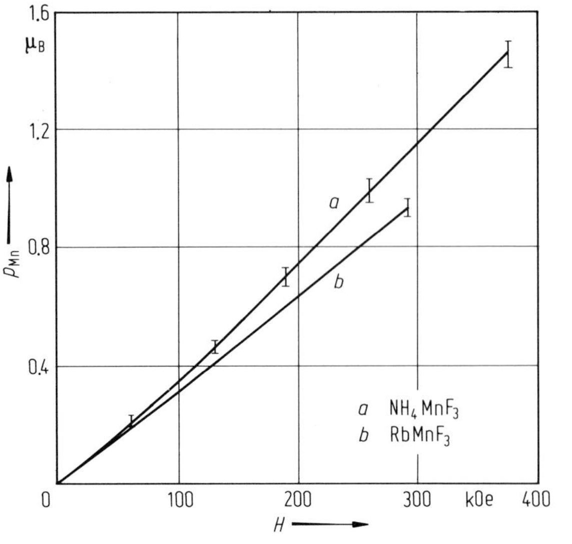

Fig. 158. NH_4MnF_3, $RbMnF_3$. Magnetic field dependence of the pulsed-field magnetic moment at 4.2 K for (*a*) randomly oriented single crystal, and (*b*) powder [83B2].

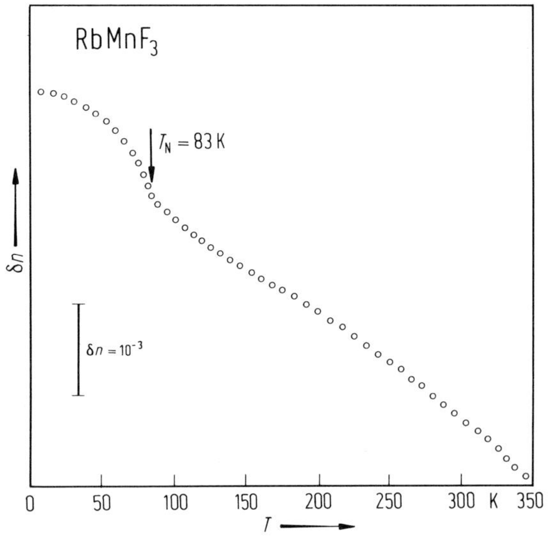

Fig. 159. $RbMnF_3$. Temperature dependence of the refraction index for $\lambda = 0.6328$ μm [84K4].

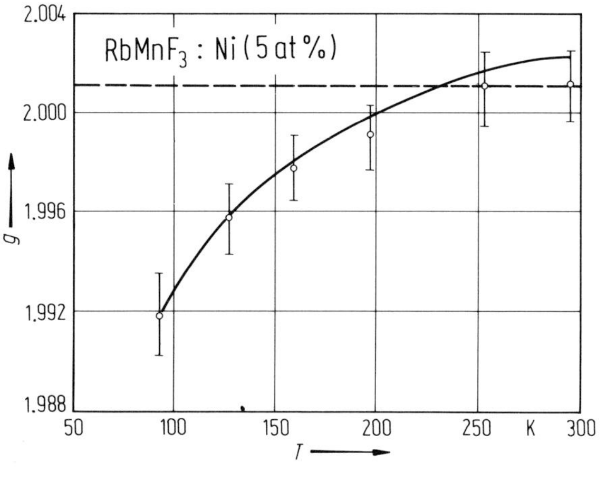

Fig. 160. $RbMnF_3$:Ni (5 at %). Temperature dependence of g at 21 GHz from EPR experiment. Dashed line: g for pure $RbMnF_3$ [72G1].

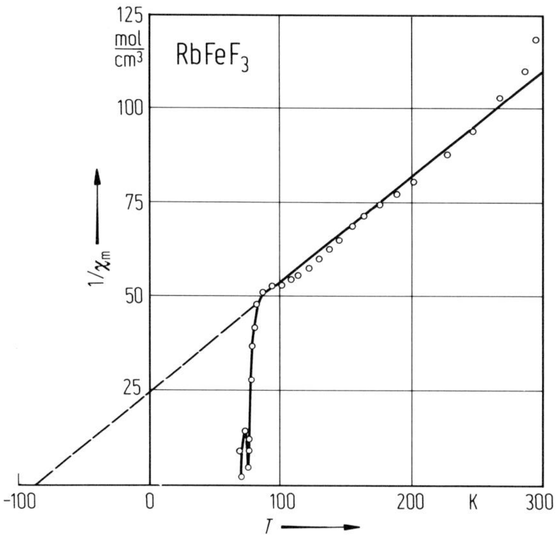

Fig. 161. $RbFeF_3$. Temperature dependence of χ_m^{-1} at 8.4 kOe [66W1].

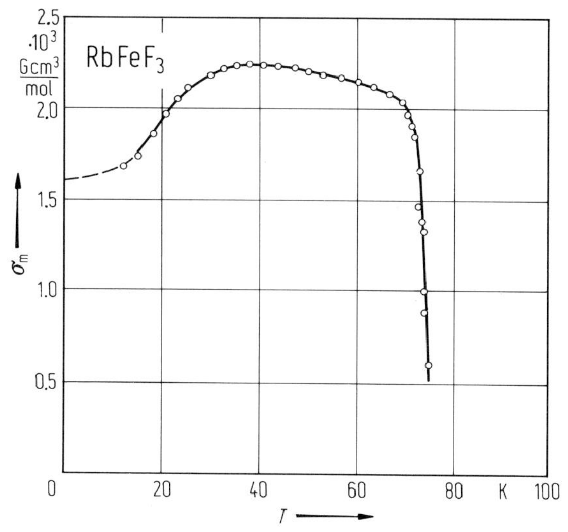

Fig. 162. $RbFeF_3$. Temperature dependence of σ_m for powdered sample at 8.4 kOe [66W1].

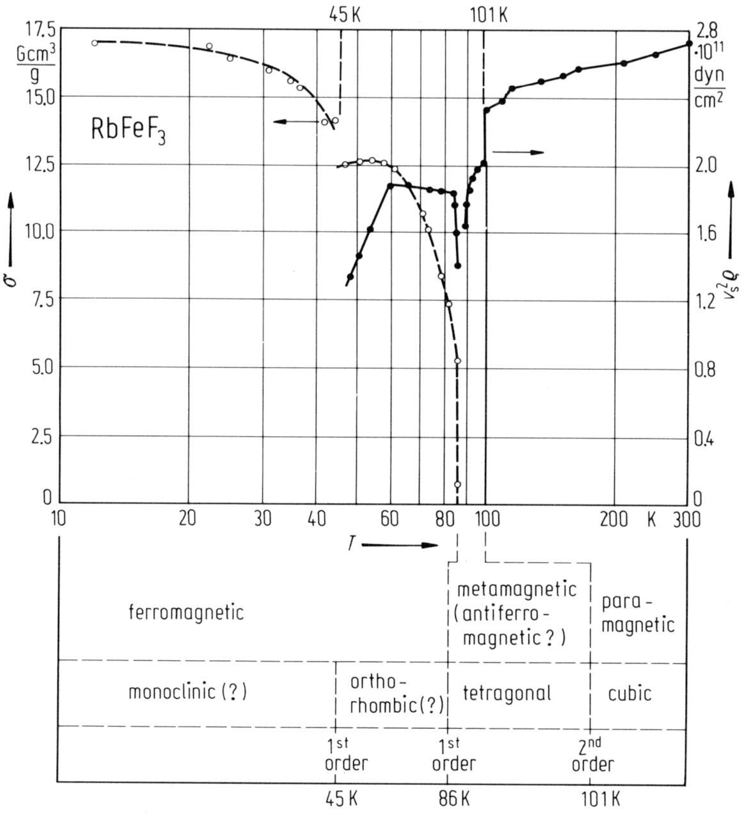

Fig. 163. $RbFeF_3$. Temperature dependence of σ (open circles) in zero applied field and shear modulus (full circles). ρ: density. Magnetic measurement was made on single crystalline long rod whose axis was within 10° of [1 1 0]. Material was found to be elastically isotropic to within 5% in the temperature range investigated [67T1].

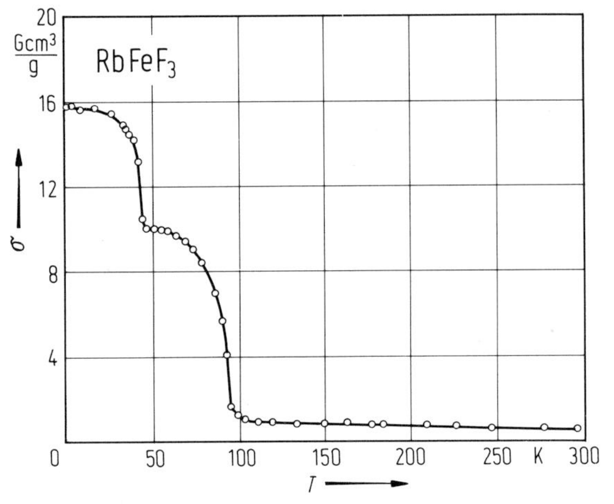

Fig. 164. $RbFeF_3$. Temperature dependence of σ at 14.24 kOe [67W1].

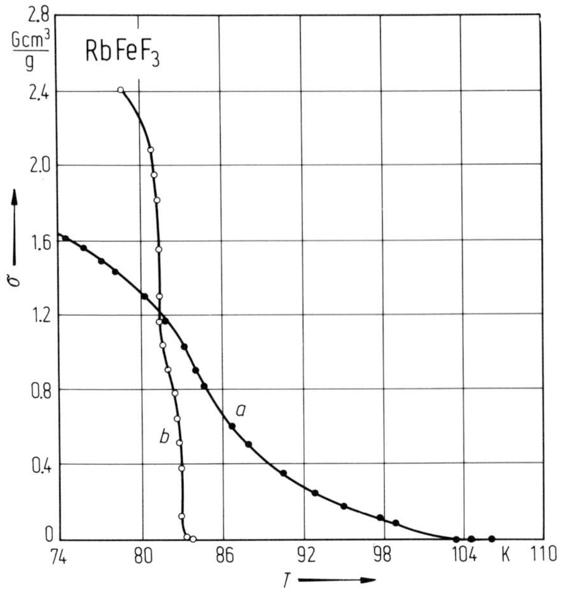

Fig. 165. $RbFeF_3$. Temperature dependence of σ at zero field near transition temperature for powder (a) and single crystal (b) [68G2].

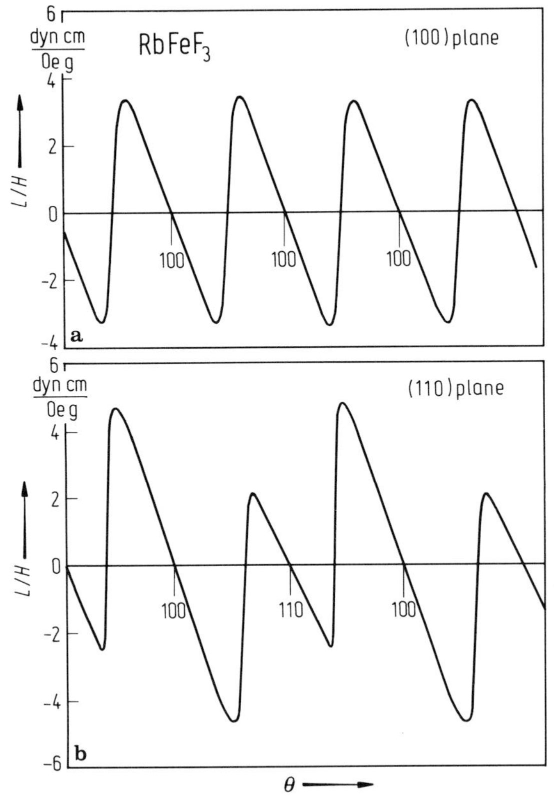

Fig. 166. $RbFeF_3$. Torque per magnetic field as a function of angle between the applied field and the [0 0 1] axis for (a) H in the (1 0 0) plane and (b) H in the (1 1 0) plane at 77 K and $H = 5$ kOe [69G1].

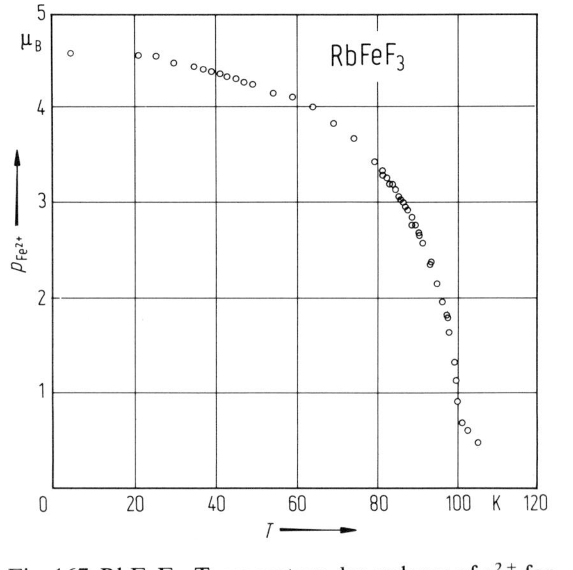

Fig. 167. $RbFeF_3$. Temperature dependence of p_{Fe}^{2+} for single crystal [71W1].

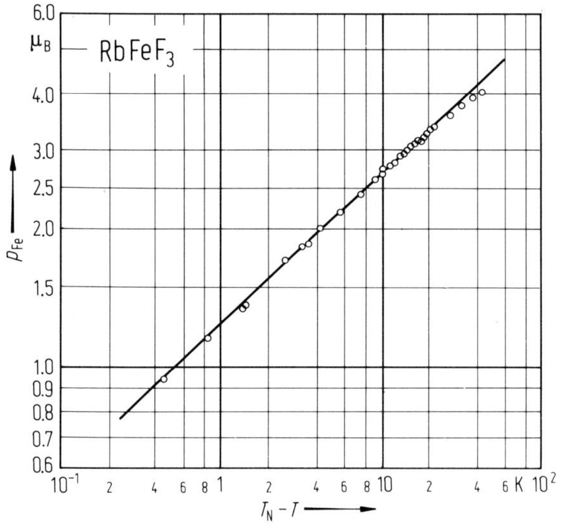

Fig. 168. $RbFeF_3$. Ordered moment p_{Fe} vs. $T_N - T$ for single crystal in the neighborhood of T_N. $T_N = 100.53\ (2)$ K, $p_{Fe} \propto (1 - T/T_N)^{\beta}$ with $\beta = 0.329$ [71W1].

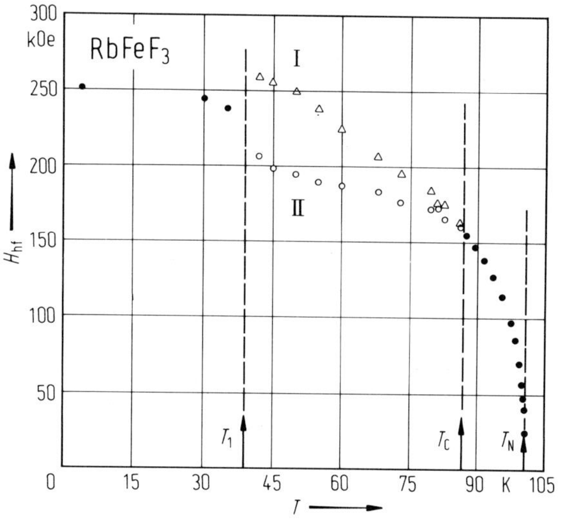

Fig. 169. $RbFeF_3$. Temperature dependence of hyperfine magnetic field H_{hf} from Mössbauer study. For $T_1 < T < T_C$ there are two values of H_{hf} corresponding to two sites I and II [81S1].

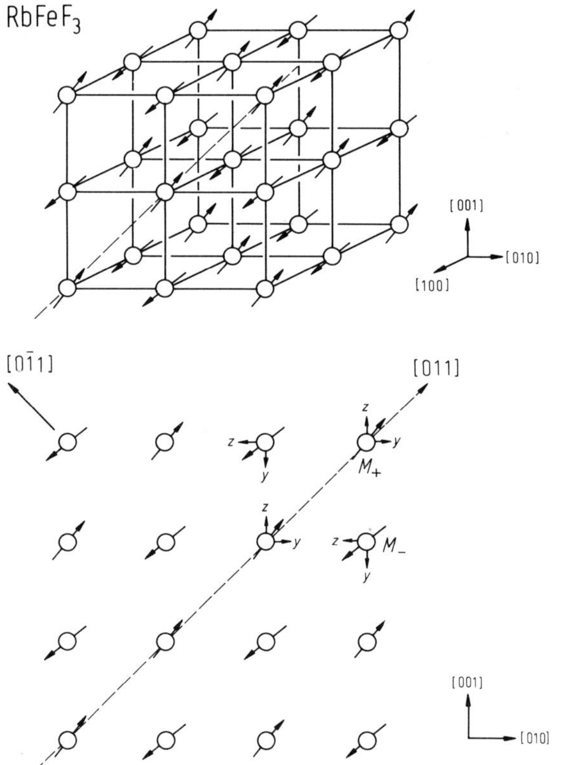

Fig. 170. $RbFeF_3$. Spin arrangement below T_1 [81S1].

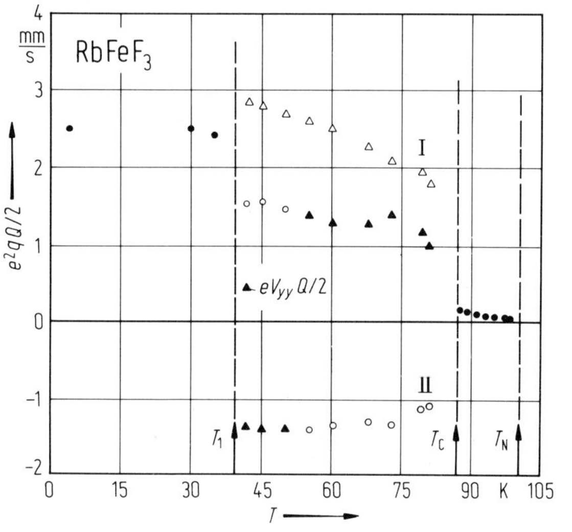

Fig. 171. $RbFeF_3$. Temperature dependence of $\frac{1}{2}e^2qQ$. In the region $T_1 < T < T_C$, $\frac{1}{2}eV_{yy}Q$ at the site II are also plotted [81S1].

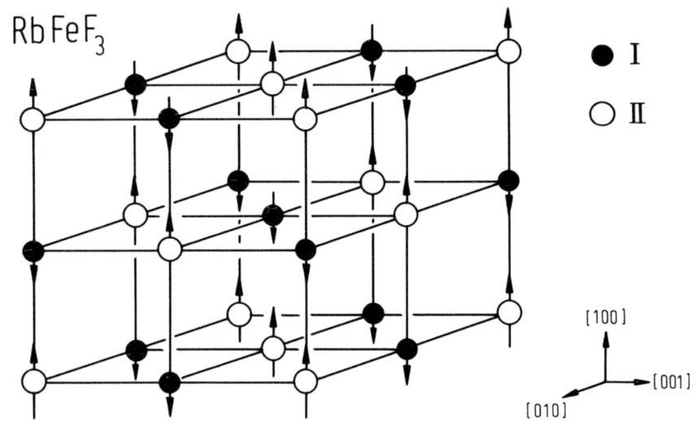

Fig. 172. $RbFeF_3$. Spin arrangement in the temperature range $T_1 < T < T_C$ at site I and site II, from Mössbauer experiment [81S2].

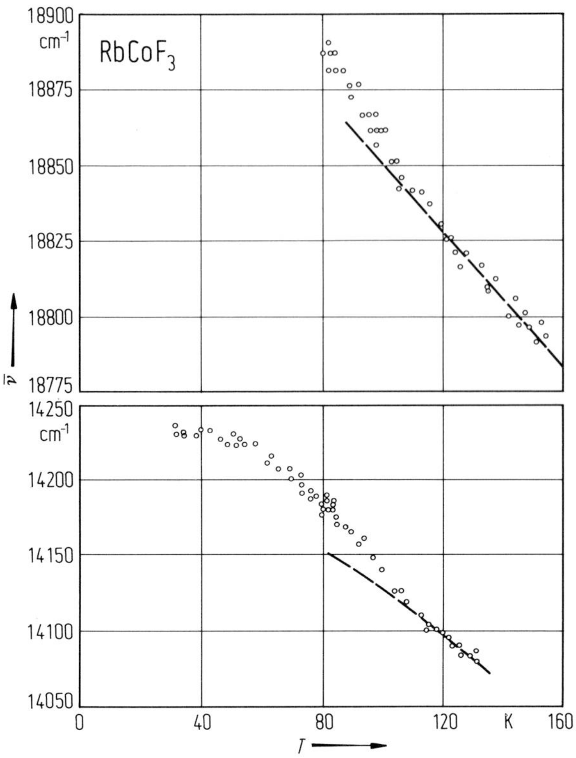

Fig. 173. $RbCoF_3$. Temperature dependence of ${}^4T_{1g} \rightarrow {}^4T_{1g}(P)$ ($\sim 19000\ cm^{-1}$) and the ${}^4T_{1g} \rightarrow {}^4A_{2g}$ ($\sim 14000\ cm^{-1}$) absorption band maxima of the Co^{2+} ion. The dashed lines indicate purely thermal behavior to emphasize the magnetically induced shifts [67H2].

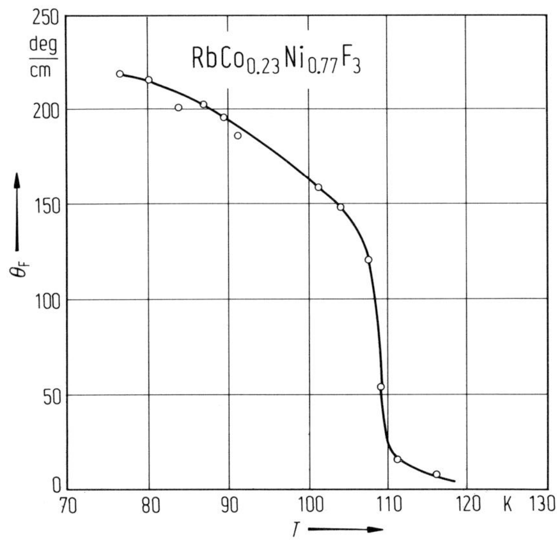

Fig. 174. $RbNiF_3$:23% Co^{2+}. Temperature dependence of the Faraday rotation θ_F for $\lambda = 0.5950\ \mu m$ at $H = 1$ kOe [68P1].

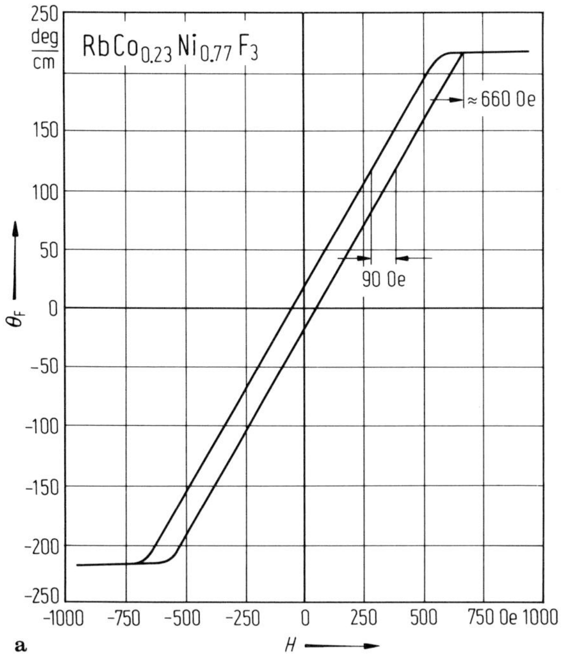

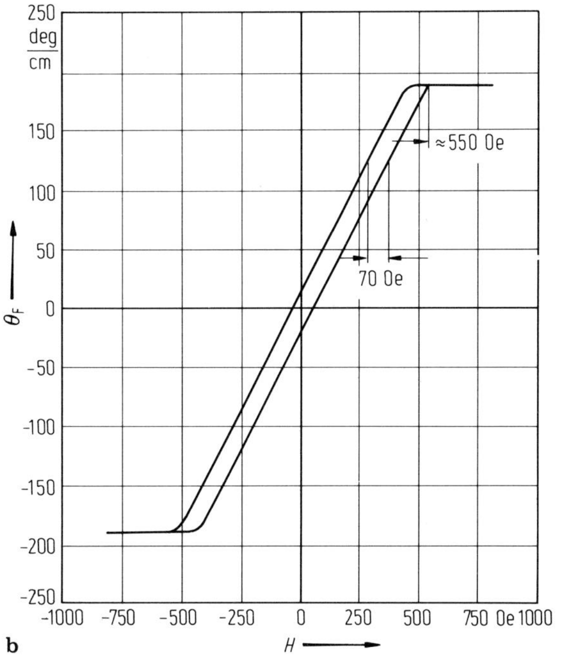

Fig. 175. $RbNiF_3$: 23% Co^{2+}. Magnetic field dependence of the Faraday rotation at 77 K (a) and 90 K (b) for $\lambda = 0.5950$ µm [68P1].

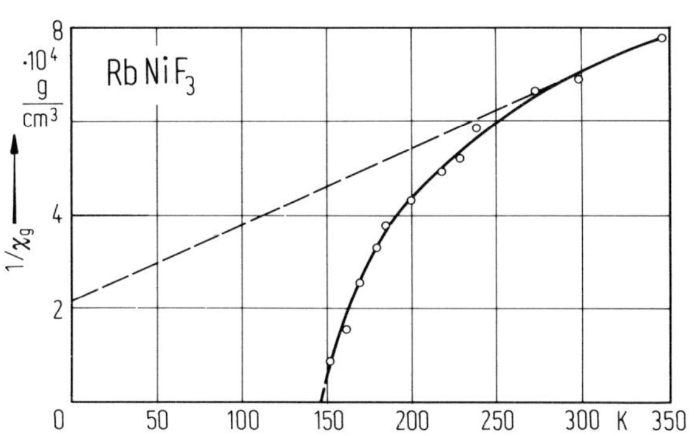

Fig. 176. $RbNiF_3$. Temperature dependence of χ_g^{-1} [66S1].

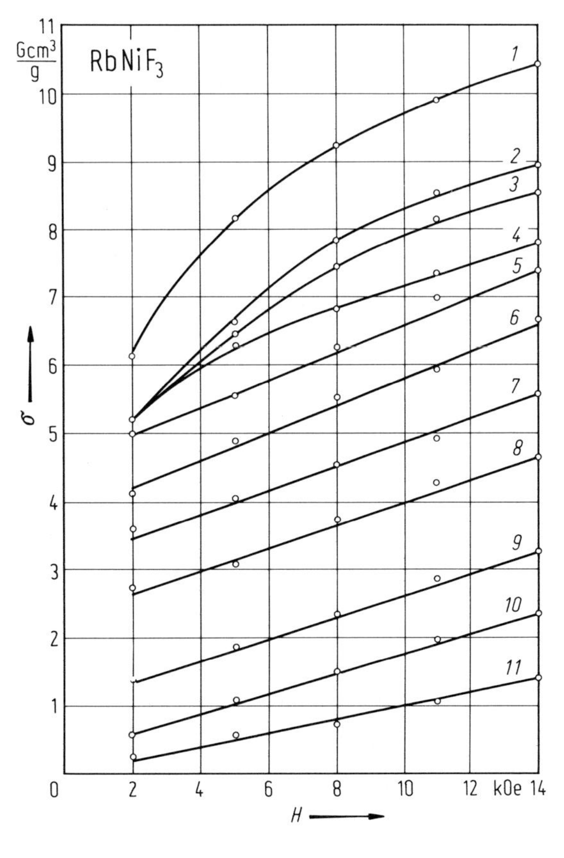

Fig. 177. $RbNiF_3$. Magnetic field dependence of σ in basal plane at (curve *1*) 77, (*2*) 90, (*3*) 100, (*4*) 110, (*5*) 120, (*6*) 125, (*7*) 130, (*8*) 135, (*9*) 140, (*10*) 145 and (*11*) 150 K [66S1].

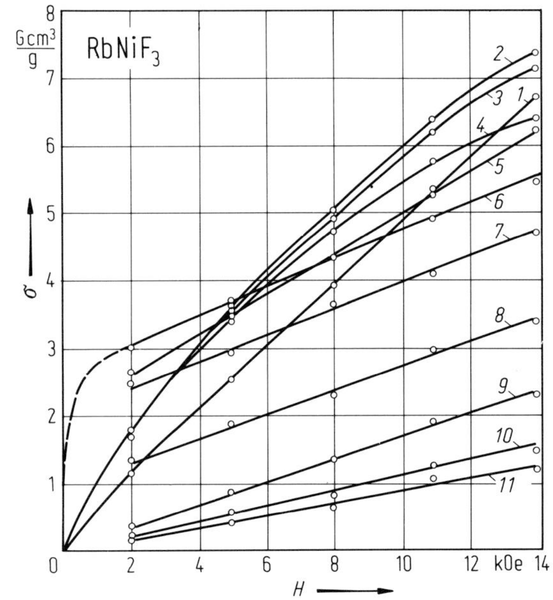

Fig. 178. $RbNiF_3$. Magnetic field dependence of σ along main axis at (curve *1*) 77, (*2*) 90, (*3*) 100, (*4*) 110, (*5*) 125, (*6*) 130, (*7*) 135, (*8*) 140, (*9*) 145, (*10*) 148 and (*11*) 155 K [66S1].

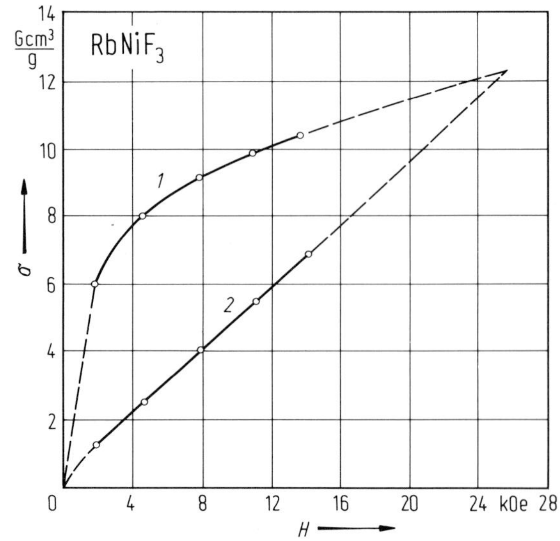

Fig. 179. $RbNiF_3$. Magnetic field dependence of σ at 77 K for $\sigma_\perp$(*1*) and $\sigma_{||}$(*2*) (see Fig. 180) [66S1].

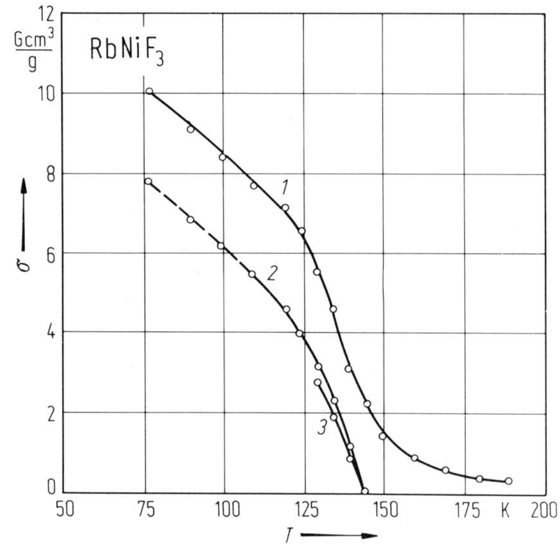

Fig. 180. $RbNiF_3$. Temperature dependence of σ. (curve *1*) magnetization in basal plane $\sigma_\perp$ at 14 kOe; (*2*) spontaneous magnetization in basal plane $\sigma_{0\perp}$ obtained by extrapolation to $H = 0$ according to $\sigma_\perp = \sigma_{0\perp} + \chi H$; (*3*) spontaneous magnetization along main axis $\sigma_{0||}$ obtained according to $\sigma_{||} = \sigma_{0||} + \chi H$ extrapolated to $H = 0$ [66S1].

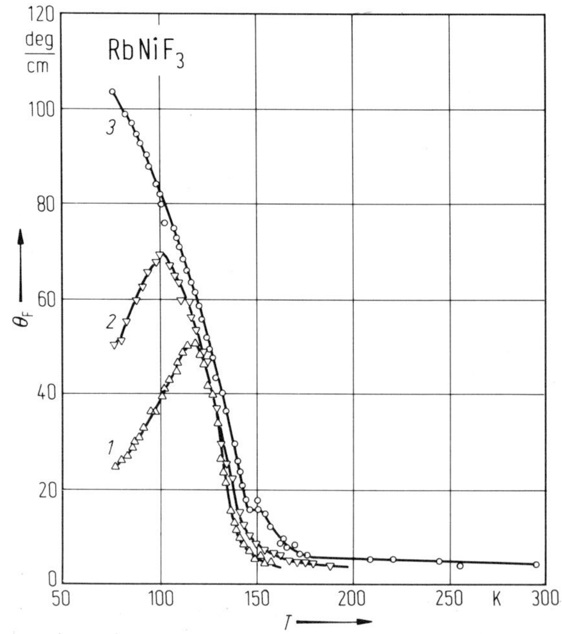

Fig. 181. $RbNiF_3$. Temperature dependence of Faraday rotation at 6.5 (curve *1*), 11.3 (*2*) and 25 kOe (*3*) for $\lambda = 0.55$ μm [67P1].

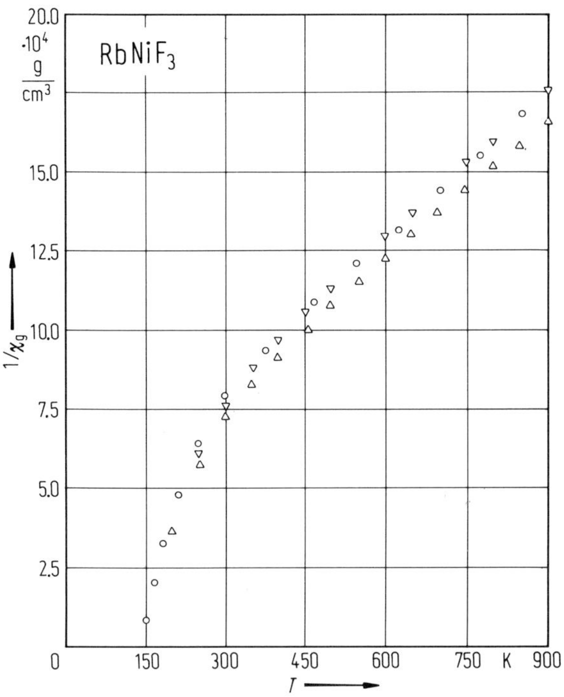

Fig. 182. $RbNiF_3$. Temperature dependence of χ_g^{-1} (○) [67S2] and calculated for $J_{AB}/k = -95.1$, $J_{BB}/k = 21.5$ K with $g = 2.35$ (△) and $g = 2.29$ (▽) [80K1]. J_{AB}: exchange interaction constants between nearest neighbors A(Rb) and B(Ni).

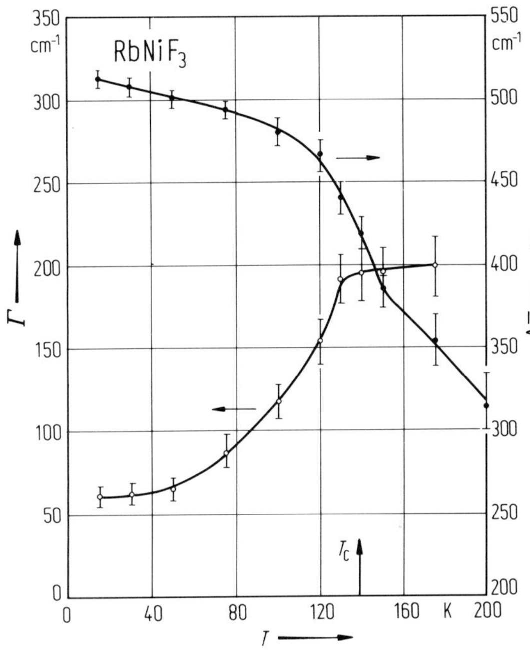

Fig. 183. $RbNiF_3$. Temperature dependence of second-order magnon Raman shift $\Delta\nu$ and linewidth Γ [69C1].

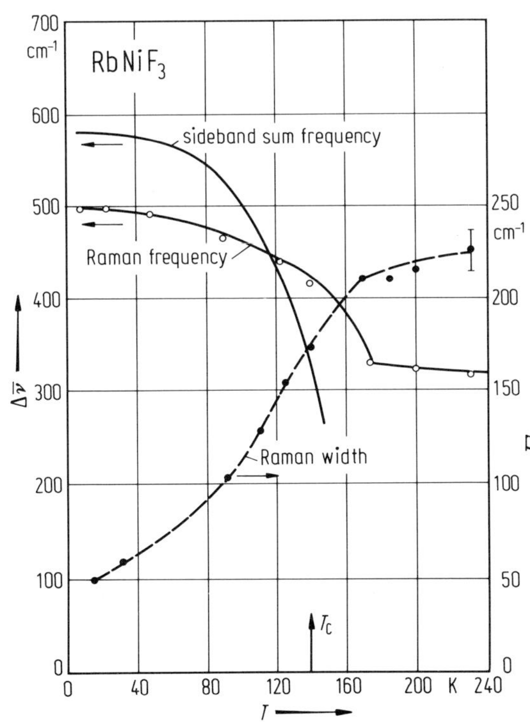

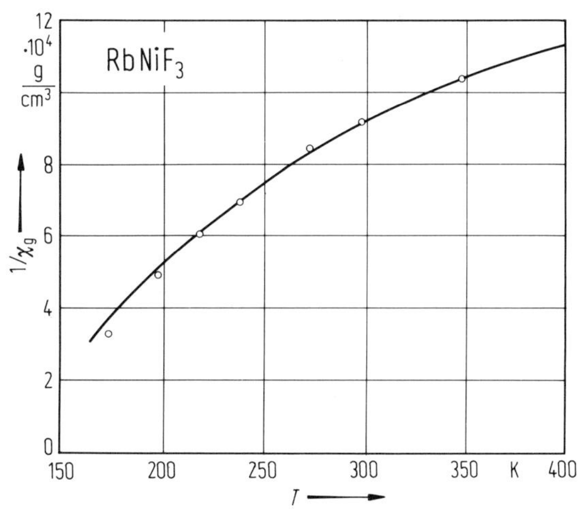

Fig. 185. $RbNiF_3$. Temperature dependence of χ_g^{-1}. The drawn line from NMR experiment and circles from [67S2]; [69S5].

←

Fig. 184. $RbNiF_3$. Temperature dependence of the frequency and of the full width at half-maximum of the two-magnon peak [69F1]. Sum of the magnon sideband frequencies is also shown (from other literature).

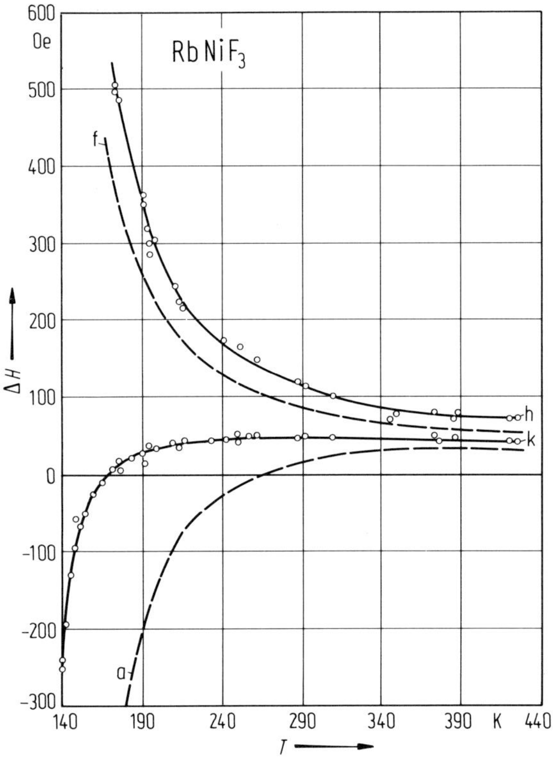

Fig. 186. $RbNiF_3$. Temperature dependence of the NMR line shift of ^{19}F in two different lattice positions (h and k, Fig. 187b) for $\boldsymbol{H} \| c$ at $\nu_0 = 25.7$ MHz. Solid lines are calculated magnetizations of sublattices in relative units. Broken lines are due to NMR experiment for Ni in a, f sites (Fig. 187b). Experimental points are taken from [67S2], [67P1, 69S5].

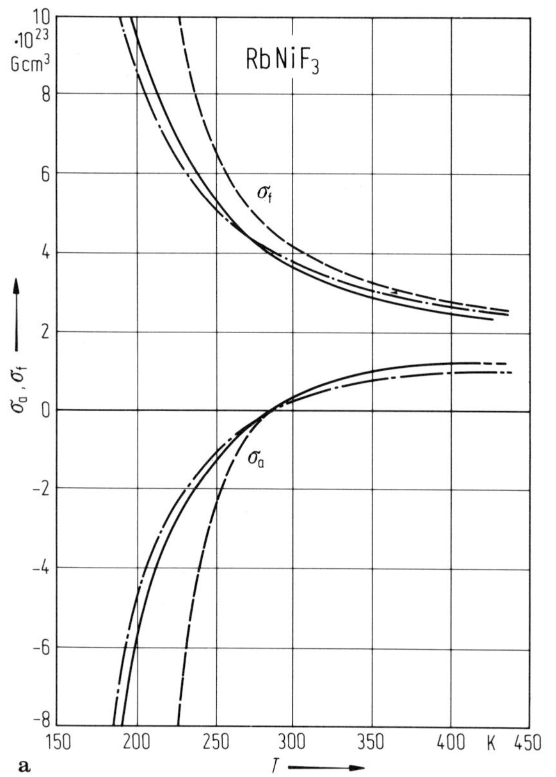

a

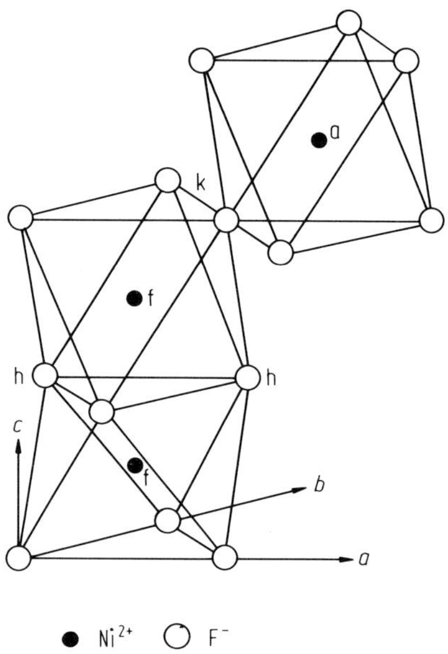

b

Fig. 187. $RbNiF_3$. (a) Temperature dependence of sublattice magnetization σ_a and σ_f per one ion. Solid line: NMR data, broken line: conventional molecular theory ($J_{af}/k = 36$ K), dashed-dotted line: the constant coupling method ($J_{af}/k = 39$ K, $J_{ff} = 0$). (b) Part of the unit cell [70P1].

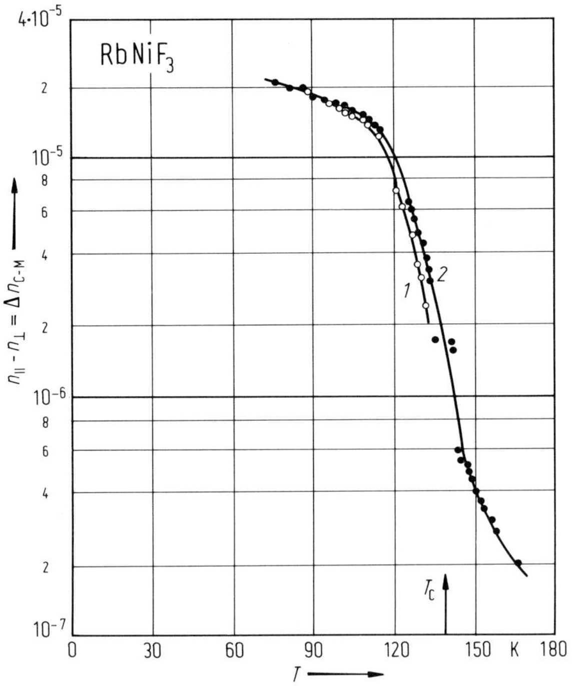

Fig. 188. $RbNiF_3$. Temperature dependence of the Cotton-Mouton effect for $\lambda = 0.63$ µm at 1.5 kOe (○) and 10 kOe (●) [70S3].

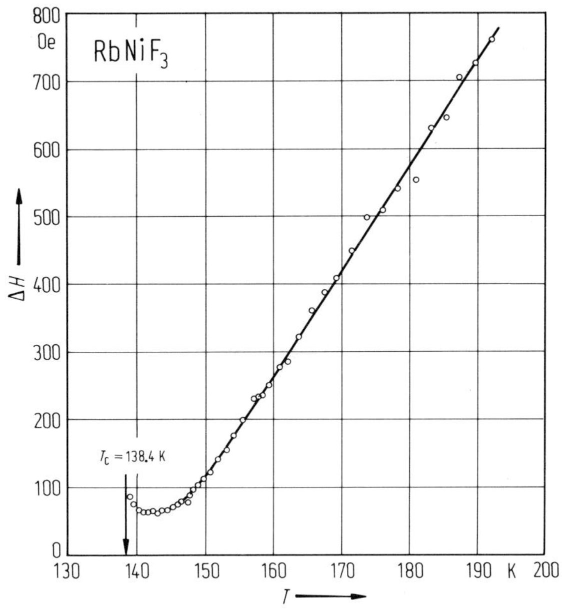

Fig. 189. $RbNiF_3$. Temperature dependence of ESR linewidth ΔH [71T1].

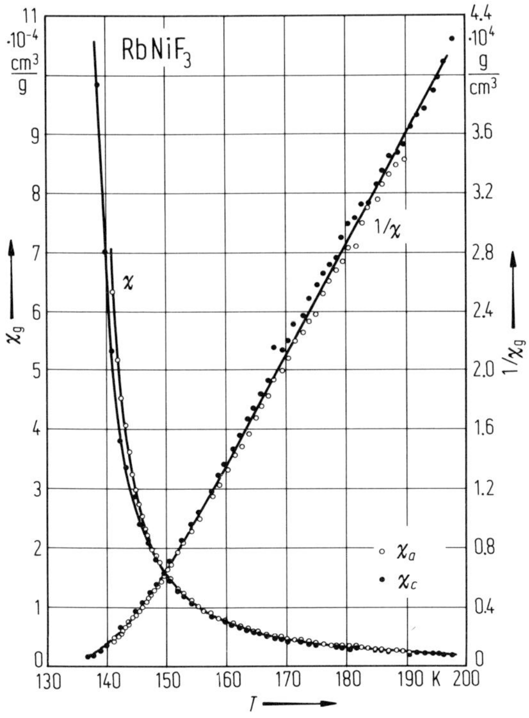

Fig. 190. $RbNiF_3$. Temperature dependence of χ_g and χ_g^{-1} along a and c axes [71T1].

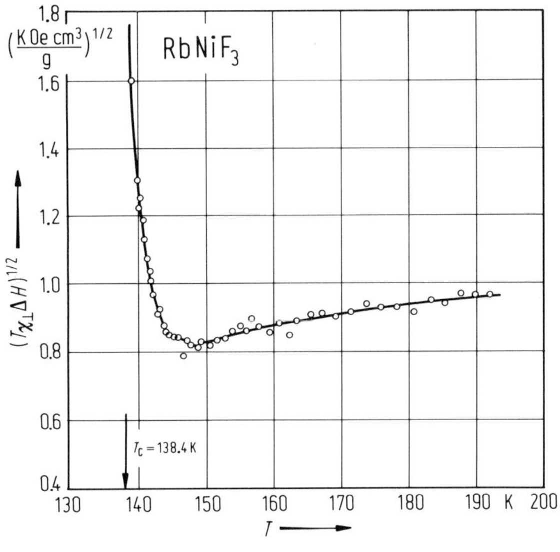

Fig. 191. $RbNiF_3$. Temperature dependence of $(T\chi_{\perp}\Delta H)^{1/2}$ where ΔH is ESR halfwidth of the maximum derivatives, $\chi_{\perp}$ is the susceptibility normal to the static field. [71T1].

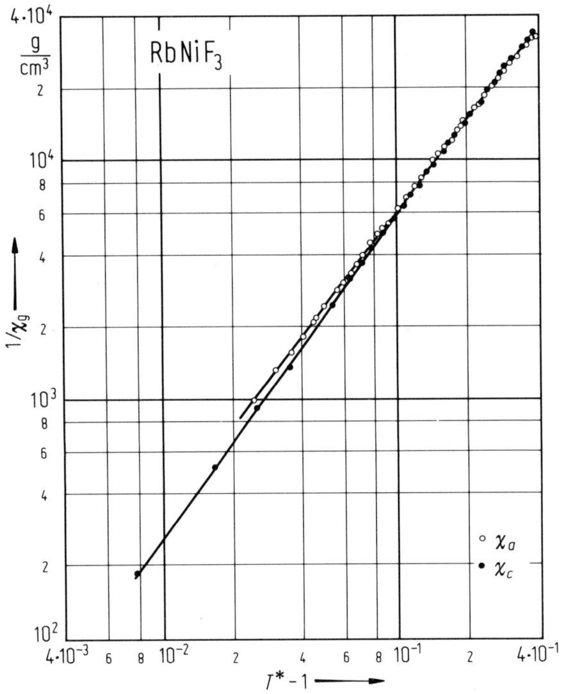

Fig. 192. $RbNiF_3$. Logarithmic plot of χ_g^{-1} as a function of temperature. The straight lines are for $\chi_a^{-1} \propto (T^* - 1)^{-1.29}$ and $\chi_c^{-1} \propto (T^* - 1)^{-1.35}$ where $T^* = T/T_c$ [71T1].

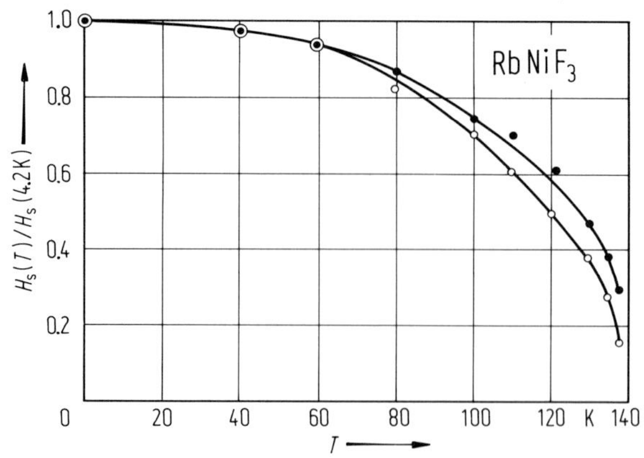

Fig. 193. $RbNiF_3$. Temperature dependence of $H_s(T)/H_s$ (4.2 K). H_s: field at which domain structure disappears. ●: $\boldsymbol{H} \| \boldsymbol{C}_2' \| \boldsymbol{k}$, ○: $\boldsymbol{H} \| \boldsymbol{C}_2 \perp \boldsymbol{k}$ [80B2].

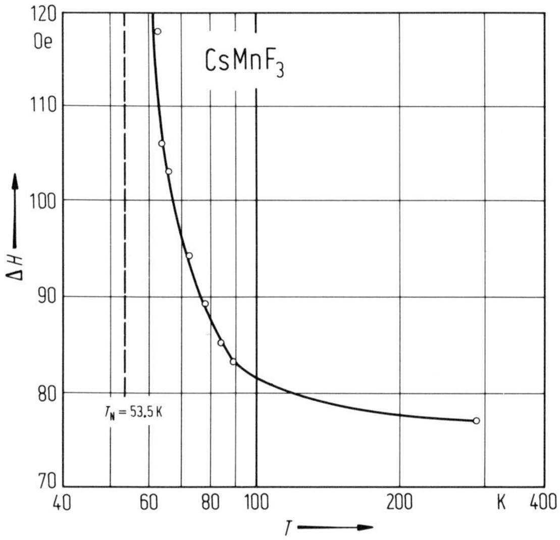

Fig. 194. $CsMnF_3$. Temperature dependence of the halfwidth of EPR line, $\nu = 9.66$ GHz [63L1].

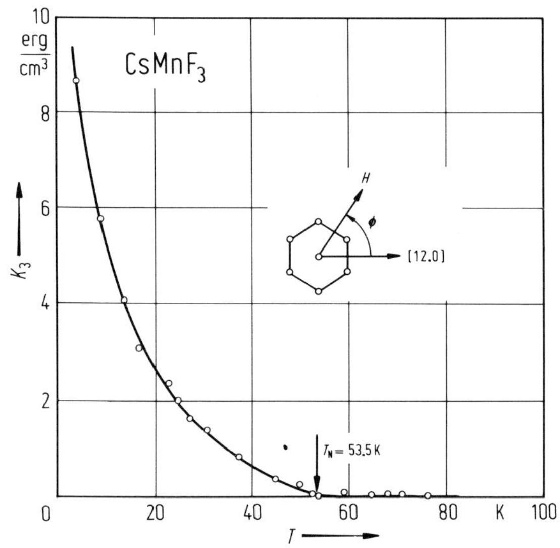

Fig. 195. $CsMnF_3$. Temperature dependence of the sixfold anisotropy energy. $H = 6$ kOe applied at 45° to the [12.0] direction [63L1].

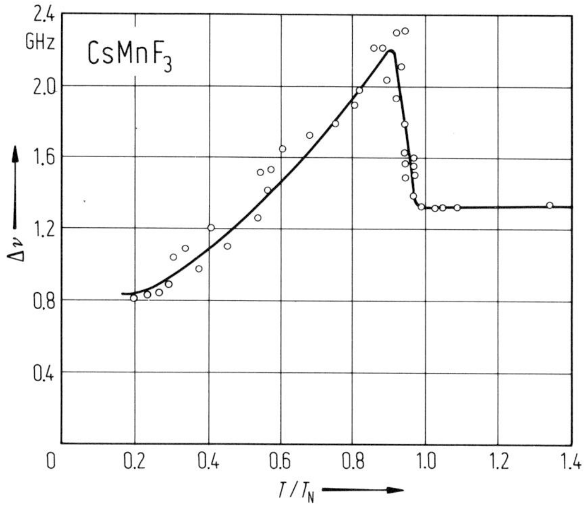

Fig. 196. $CsMnF_3$. Temperature dependence of high frequency AFMR branch [70B2]. $T_N = 53.6$ K.

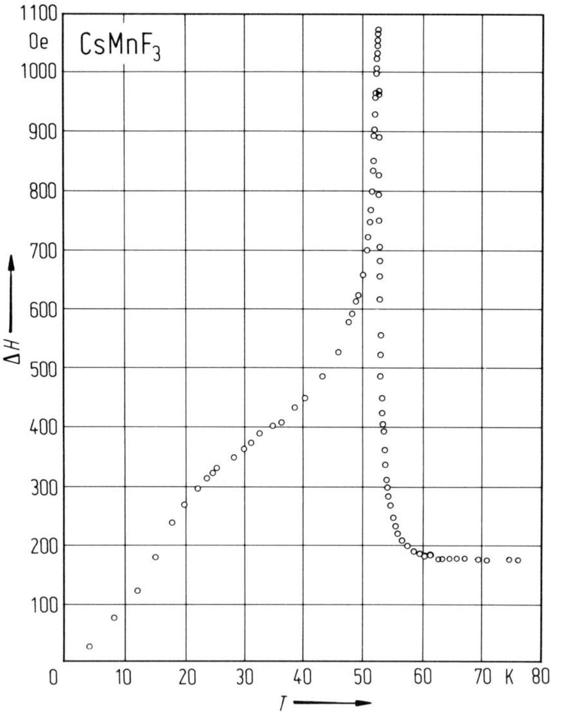

Fig. 197. $CsMnF_3$. Temperature dependence of EPR linewidth for $\nu = 10.032$ GHz and $\boldsymbol{H} \perp c$ [72W1].

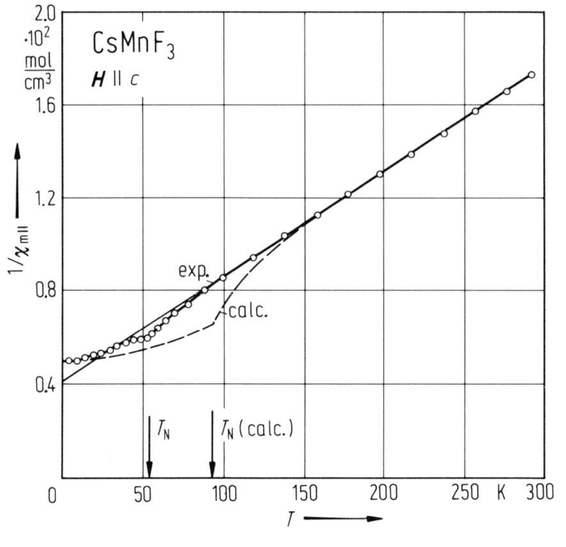

Fig. 198. $CsMnF_3$. Temperature dependence of initial χ_m^{-1} parallel to the c axis [75Y2].

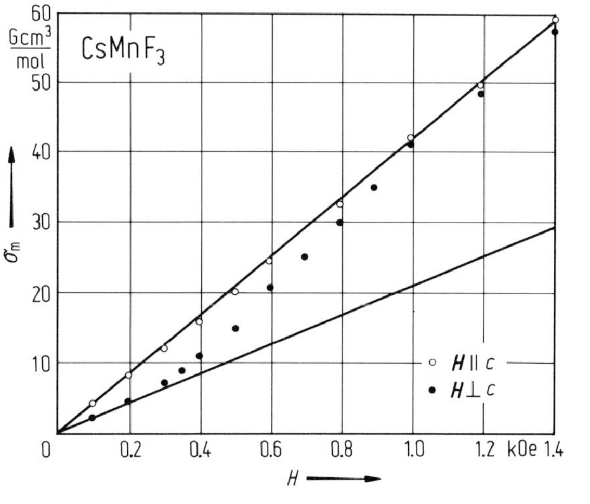

Fig. 199. $CsMnF_3$. Magnetic field dependence of σ_m at 4.2 K [75Y2].

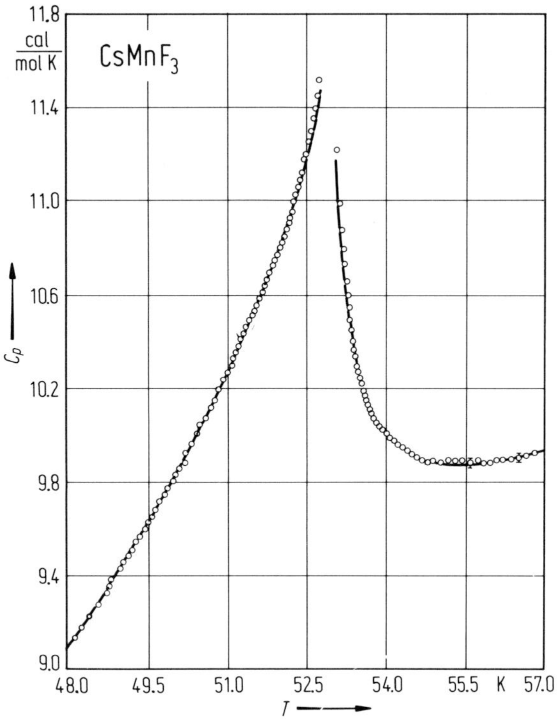

Fig. 200. $CsMnF_3$. Temperature dependence of heat capacity C_p for a second run [77I1].

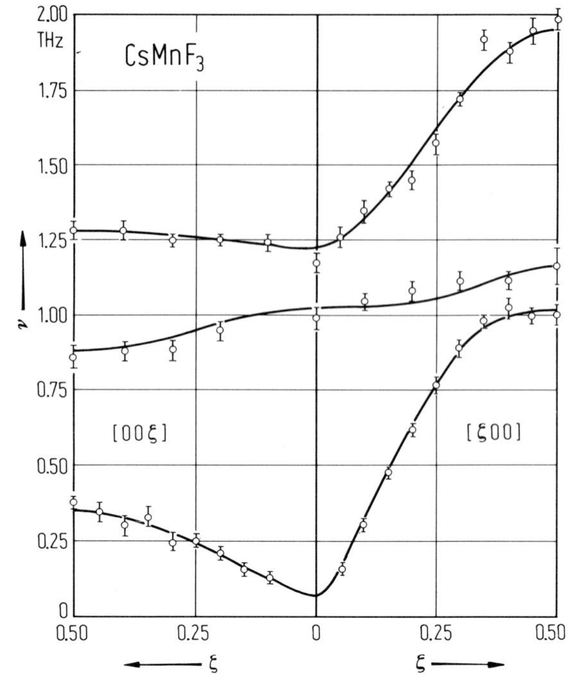

Fig. 201. $CsMnF_3$. Magnon dispersion curves at 4.2 K along the [0 0 ζ] and [ζ 0 0] directions. The solid curves are obtained from the theory with dipole-dipole interactions included [77K4].

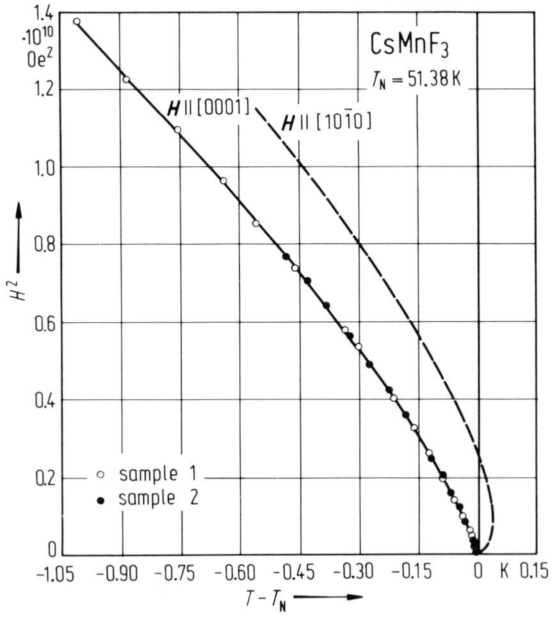

Fig. 202. $CsMnF_3$. Temperature dependence of the phase boundaries for $H \parallel [0001]$ and $H \parallel [10\bar{1}0]$. The data points for $H \parallel [10\bar{1}0]$ are neglected. The solid line is a fit to Eq. $T_c - T_N = A[(H^2 + h_0^2)^{1/\varphi} - h_0^{2/\varphi}] - BH^2$; $\varphi = 1.25$, A and B are positive constants and h_0 magnetic field obtained by fitting [79S1].

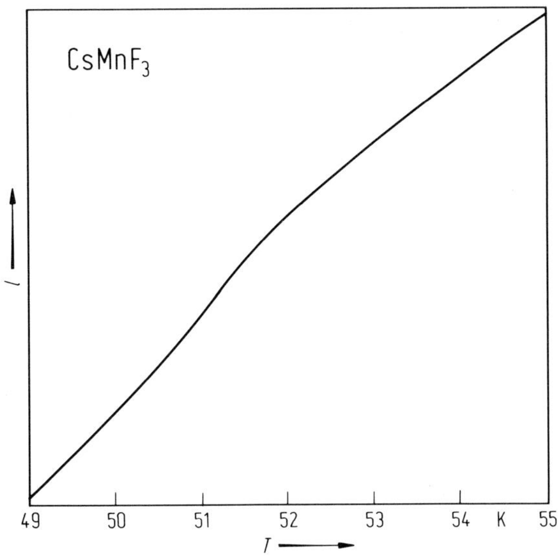

Fig. 203. $CsMnF_3$. Temperature dependence of the thermal expansion at $H = 0$, along the [1 1 2 0] direction relative to that of the copper capacitance cell [80S2].

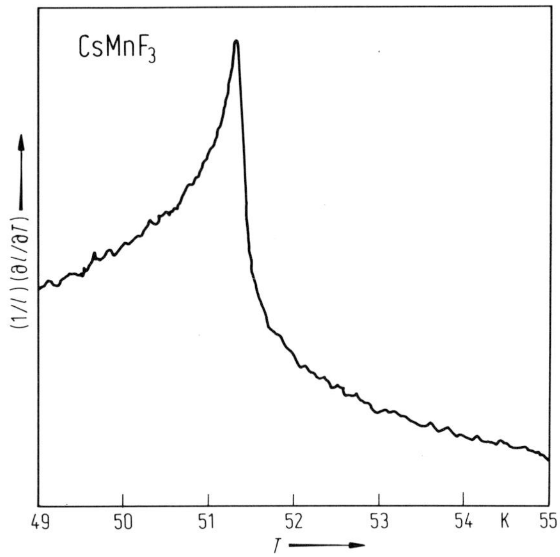

Fig. 204. $CsMnF_3$. Temperature dependence of the thermal expansion coefficient $\left(\frac{1}{l}\frac{\partial l}{\partial T}\right)$ [80S2].

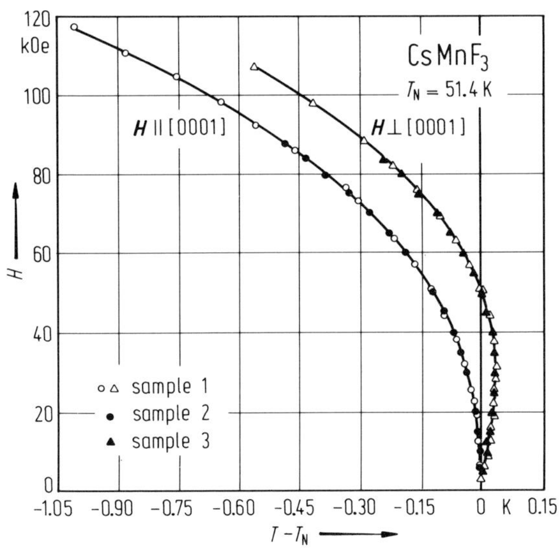

Fig. 205. $CsMnF_3$. Phase boundaries, $T_c(T_c(0) = T_N)$ vs. H for $\boldsymbol{H}$ parallel and perpendicular to the hexagonal axis [0 0 0 1] for different samples [80S2]. See also [79O1].

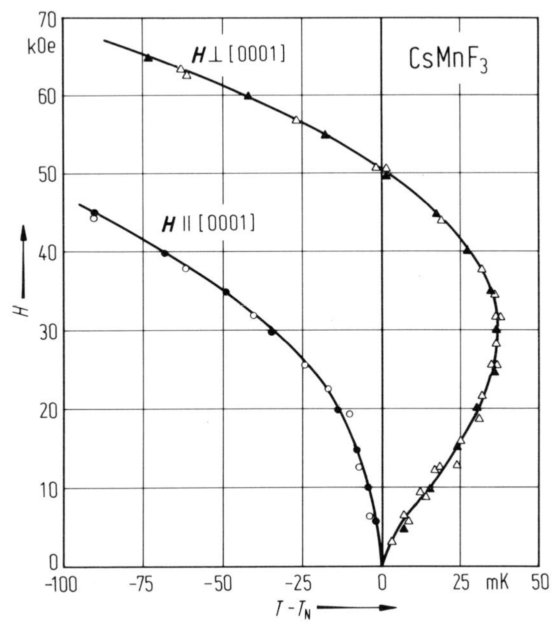

Fig. 206. $CsMnF_3$. Expanded view of the low-H portion of Fig. 205 [80S2]. See also [79O1].

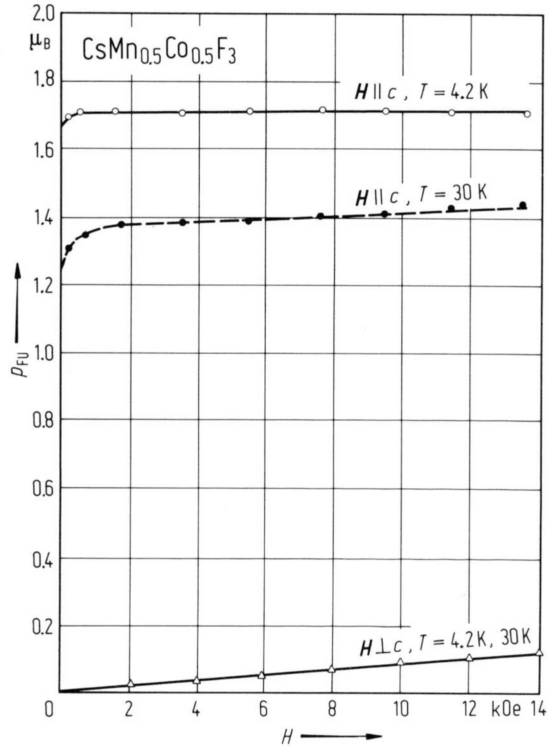

Fig. 207. $CsMn_{0.5}Co_{0.5}F_3$. Magnetic field dependence of magnetic moment per formula unit at 4.2 and 30 K. Demagnetizing field has been corrected for in the H axis [77Y1].

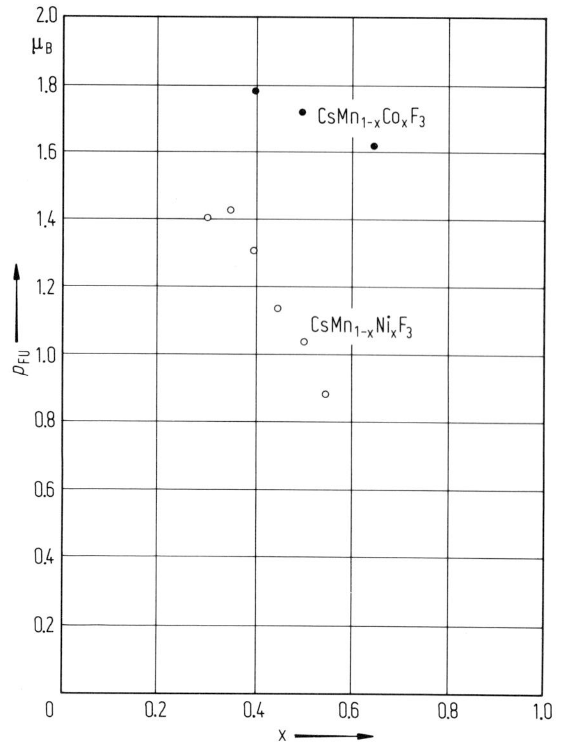

Fig. 208. $CsMn_{1-x}Ni_xF_3$, $CsMn_{1-x}Co_xF_3$. Concentration dependence of p_{FU}. The magnetic moment has been extrapolated from magnetization between 4 and 14 kOe at 4.2 K to that in $H = 0$ [77Y1].

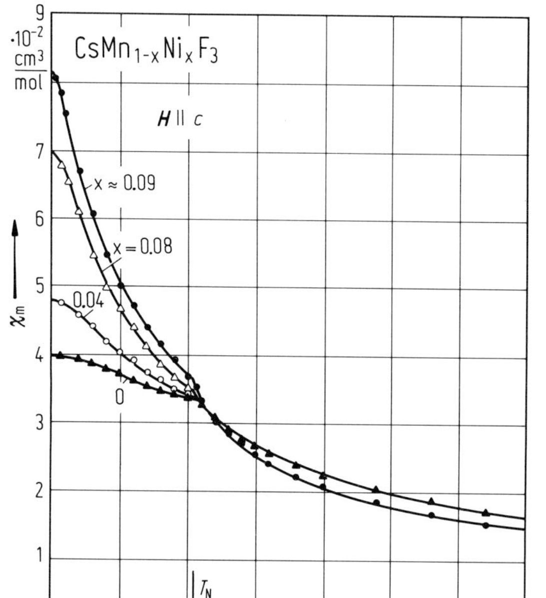

Fig. 209. $CsMn_{1-x}Ni_xF_3$. Temperature dependence of initial χ_m for different x. $H \parallel c$ [75Y2]. For x = 0.09: $T_N = 51.5(5)$ K, for x = 0: $T_N = 53.5(5)$ K.

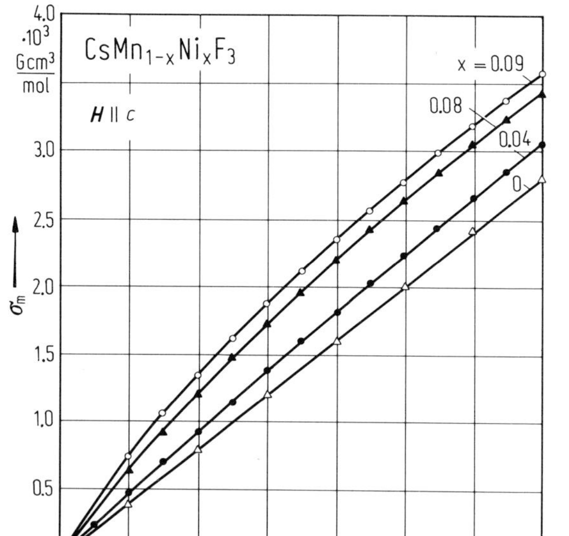

Fig. 210. $CsMn_{1-x}Ni_xF_3$. Magnetic field dependence of σ_m for different x at 4.2 K [75Y2].

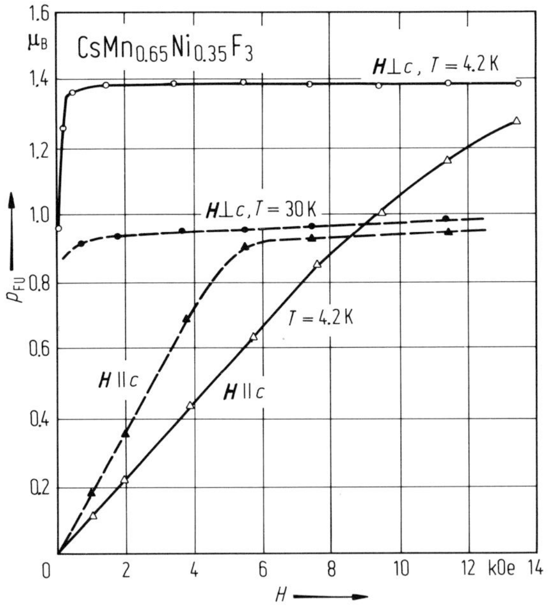

Fig. 211. $CsMn_{0.65}Ni_{0.35}F_3$. Magnetic field dependence of p_{FU} at 4.2 and 30 K. Demagnetization field has been corrected for in the H axis [77Y1].

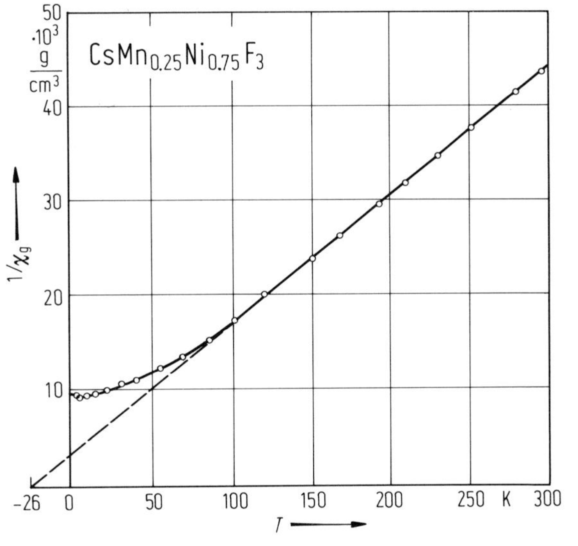

Fig. 212. $CsMn_{0.25}Ni_{0.75}F_3$. Temperature dependence of χ_g^{-1} for 9H phase [79D2].

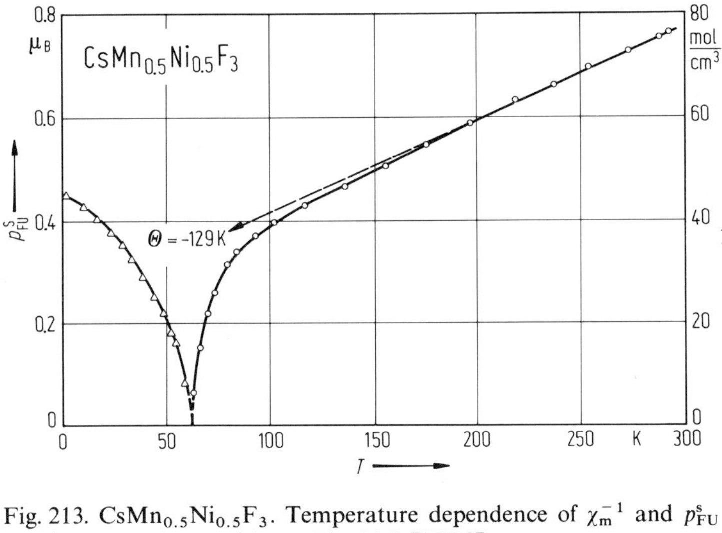

Fig. 213. $CsMn_{0.5}Ni_{0.5}F_3$. Temperature dependence of χ_m^{-1} and p_{FU}^s for high-pressure phase (6H) (see Fig. 216) [79D2].

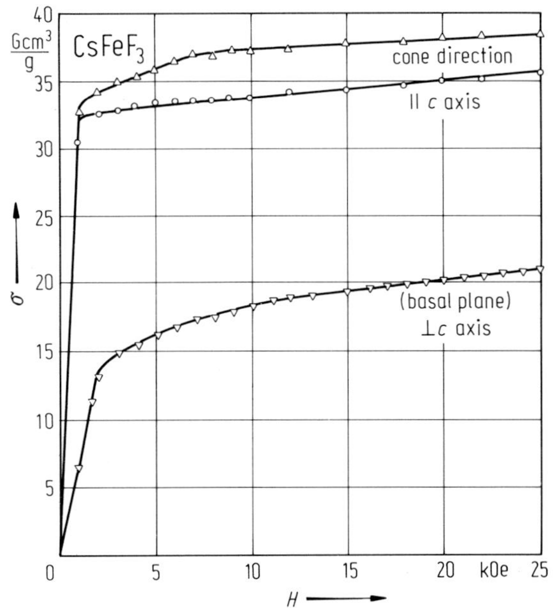

Fig. 214. $CsFeF_3$. Magnetic field dependence of σ at 4.2 K for a single crystal in two principal crystallographic directions, and in easy magnetic direction which is 25° from the c axis [70M3].

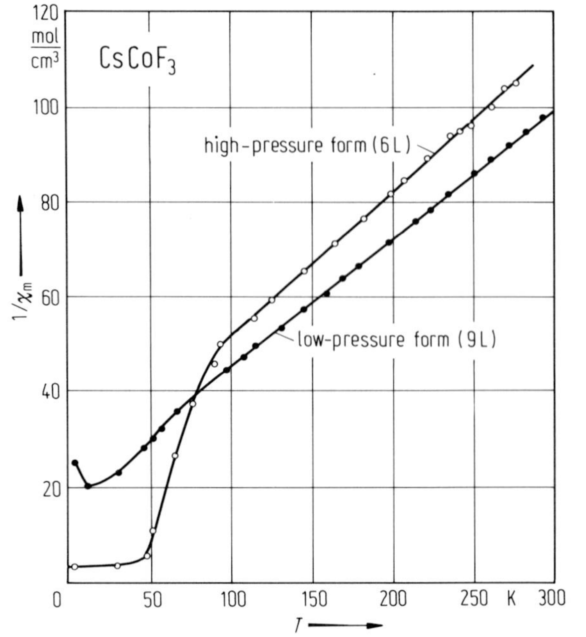

Fig. 215. $CsCoF_3$. Temperature dependence of χ_m^{-1} for nine-layer (9L) and six-layer (6L) structures (see Fig. 216) [70L1].

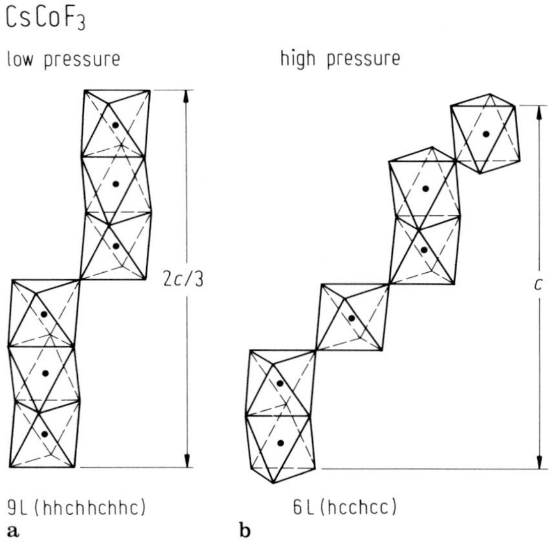

Fig. 216. $CsCoF_3$. Octahedral linkages for the two structural forms [70L1].

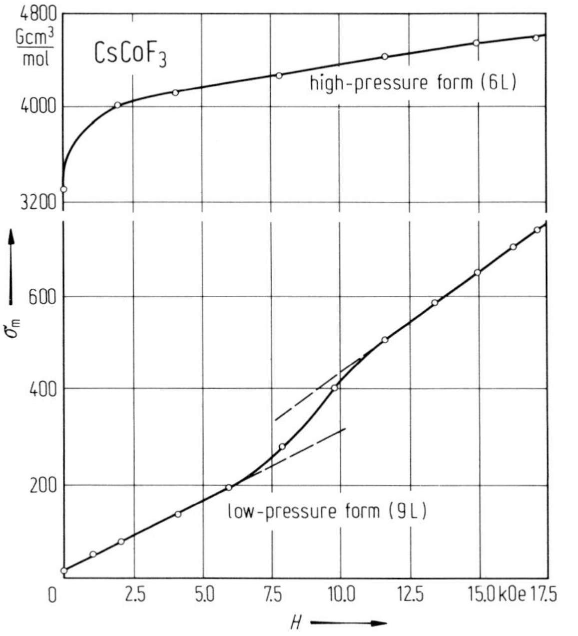

Fig. 217. $CsCoF_3$. Magnetic field dependence of σ_m for six-layer (6L) and nine layer (9L) structures at 4.2 K (see Fig. 216) [70L1].

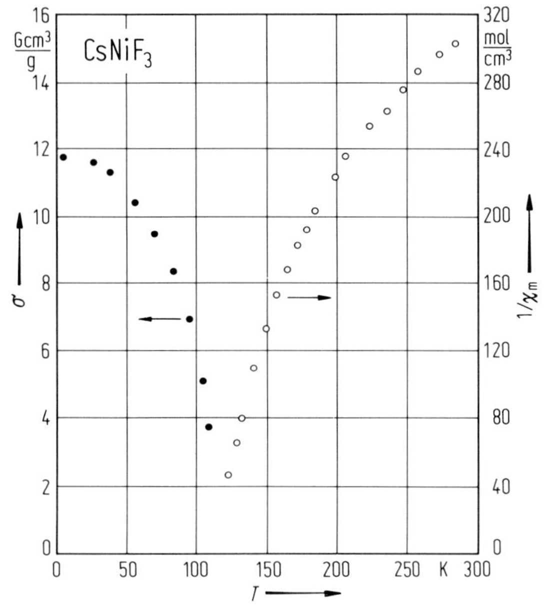

Fig. 218. $CsNiF_3$. Temperature dependence of χ_m^{-1} and σ at 10 kOe [69L1]. Sample is 6L modification (high-pressure form).

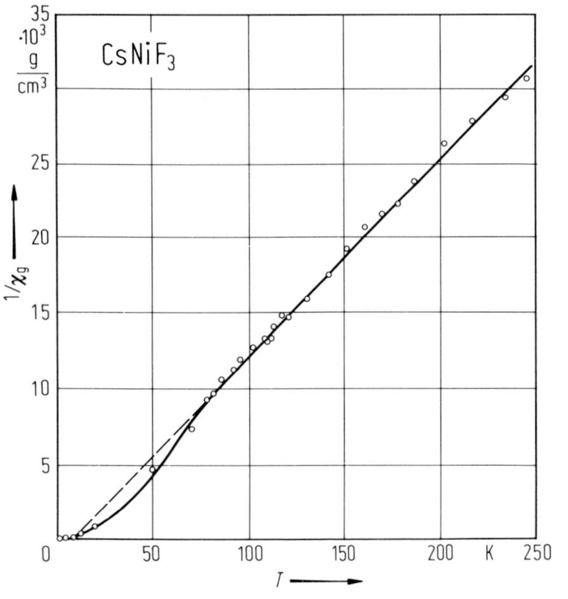

Fig. 219. $CsNiF_3$. Temperature dependence of χ_g^{-1} for powder sample [71S2].

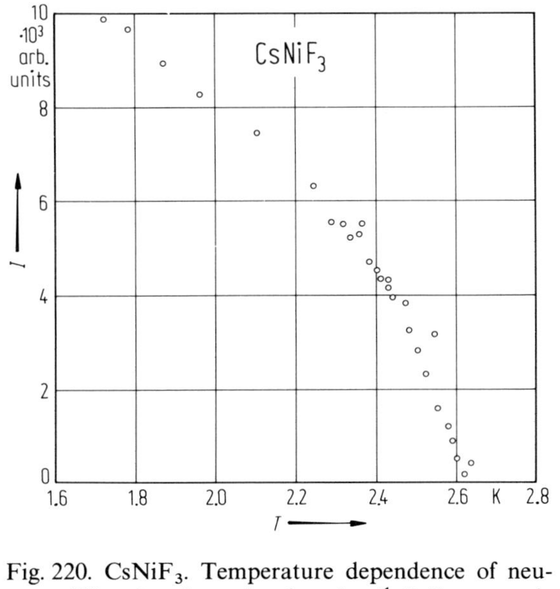

Fig. 220. $CsNiF_3$. Temperature dependence of neutron diffraction intensity for the $(\frac{1}{2}, 0, 0)$ magnetic Bragg peak [72S1].

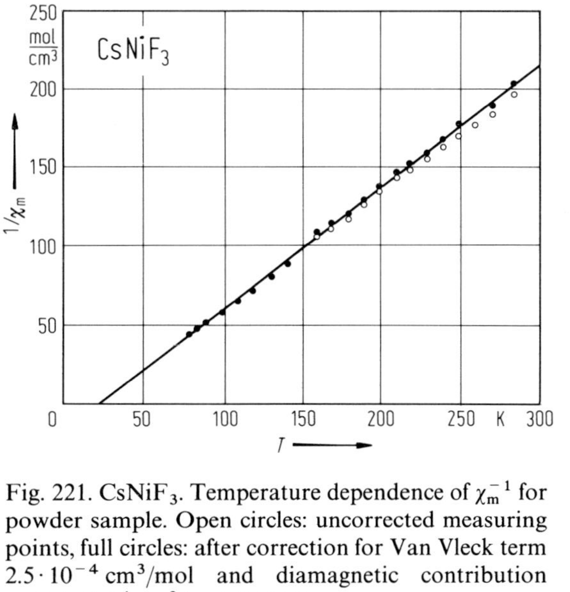

Fig. 221. $CsNiF_3$. Temperature dependence of χ_m^{-1} for powder sample. Open circles: uncorrected measuring points, full circles: after correction for Van Vleck term $2.5 \cdot 10^{-4}$ cm^3/mol and diamagnetic contribution $-0.8 \cdot 10^{-4}$ cm^3/mol [73L1].

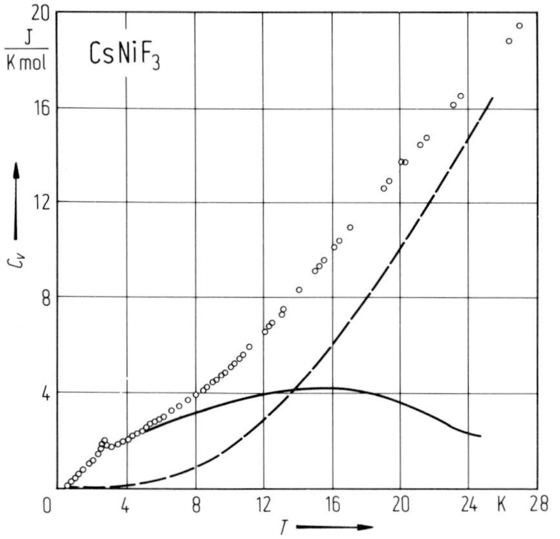

Fig. 222. $CsNiF_3$. Temperature dependence of heat capacity C_v. Dashed curve: lattice contribution, full curve: difference between total and lattice heat capacity [73L1].

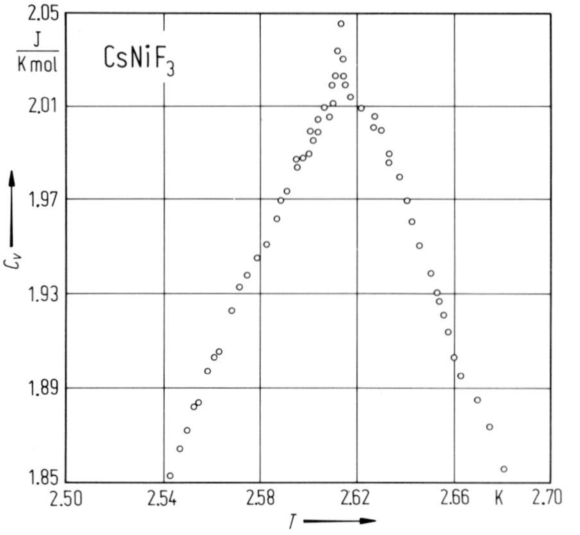

Fig. 223. $CsNiF_3$. Temperature dependence of heat capacity C_v. Experimental points in the region around transition temperature (2.613 K) (see Fig. 222) [73L1].

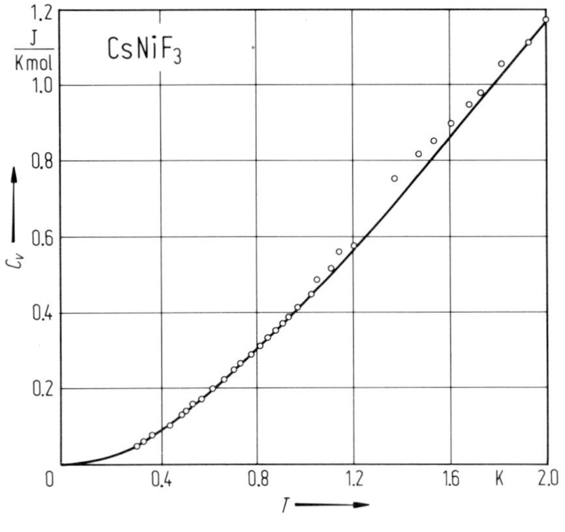

Fig. 224. $CsNiF_3$. Temperature dependence of magnetic heat capacity C_v. Experimental points below 2 K (see Fig. 222) [73L1].

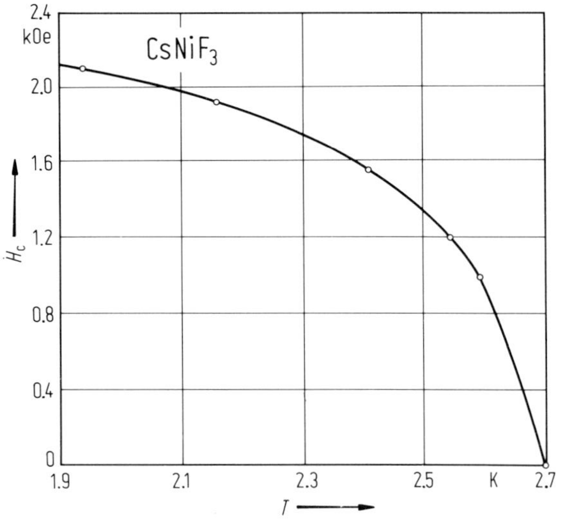

Fig. 225. $CsNiF_3$. Phase diagram as measured with $(\frac{1}{2}, 0, 0)$ neutron magnetic reflection [74S2].

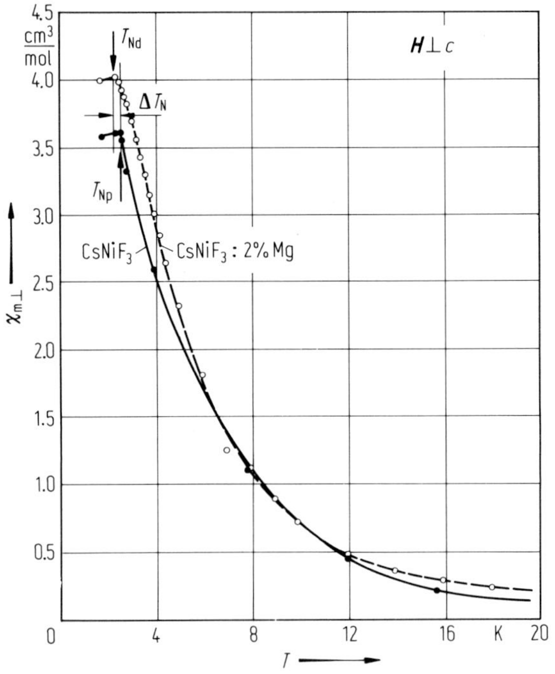

Fig. 226. $CsNiF_3$, $CsNiF_3$:2%Mg. Temperature dependence of $\chi_{m\perp}$ ($\boldsymbol{H}\perp c$) at $H = 1$ kOe [76S1]. $T_{Np} = 2.61$ K, $T_{Nd} = 2.32$ K, $\Delta T_N = 0.29$ K.

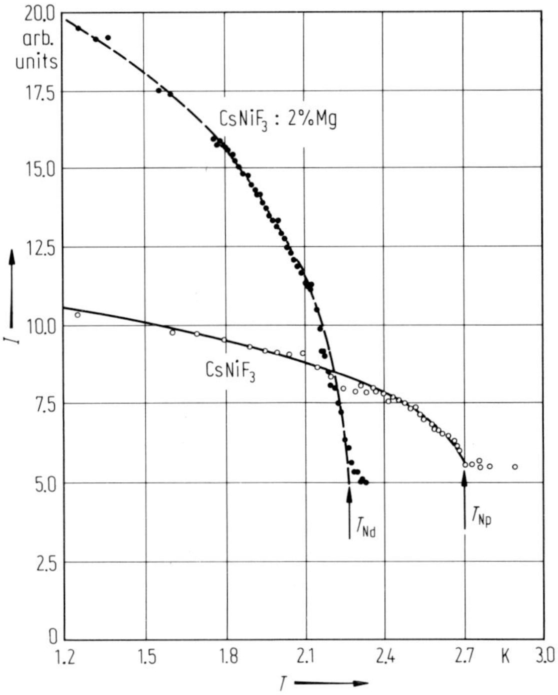

Fig. 227. $CsNiF_3$, $CsNiF_3$:2% Mg. Temperature dependence of ($\frac{1}{2}$, 0, 0) neutron magnetic Bragg peak intensity [76S1]. $T_{Np} = 2.70$ K, $T_{Nd} = 2.37$ K, $\Delta T_N = 0.33$ K.

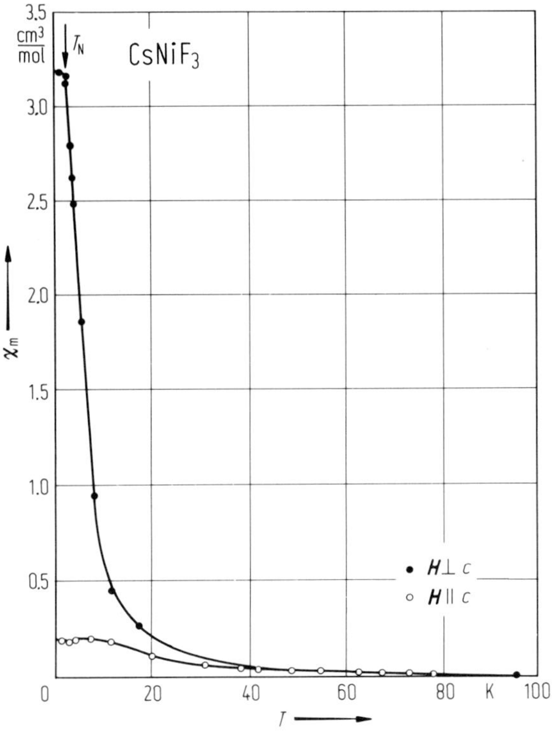

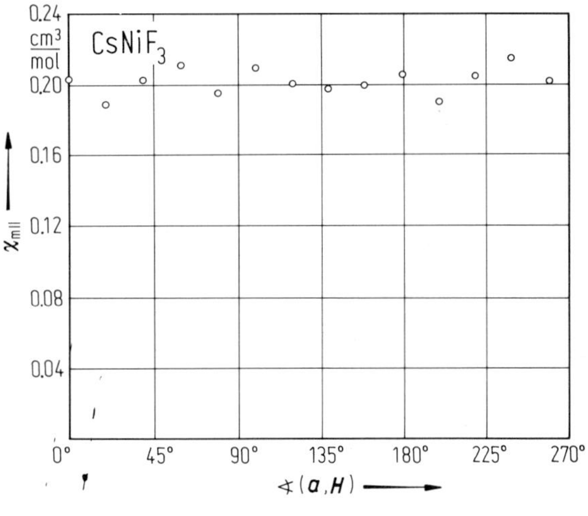

Fig. 229. $CsNiF_3$. Angle dependence of the single crystal $\chi_{m\parallel}$ at 1.4 K and 500 Oe; see Fig. 228 [76S2].

Fig. 228. $CsNiF_3$. χ_m vs. temperature for a single crystal with external field in easy plane, $\boldsymbol{H}\perp c$, and perpendicular to this plane, $\boldsymbol{H}\parallel c$ [76S2].

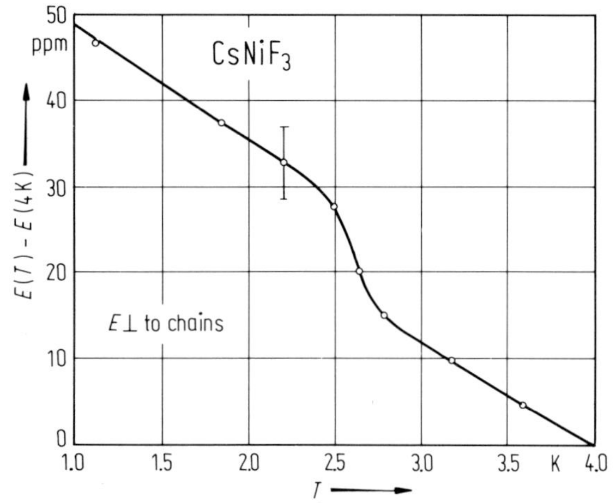

Fig. 230. $CsNiF_3$. Temperature dependence of the Young modulus $E_{\perp}$ perpendicular to the linear chain near T_N (see Fig. 231). ν (4 K) = 1.25 kHz, E (4 K) = $7.4 \cdot 10^{11}$ dyn cm^{-2}. $CsNiF_3$ with hexagonal structure (space group $P6_3/mmc$) consists of (NiF_6) octahedra sharing common faces along the c direction thus forming ferromagnetic chains along the c direction. These chains are widely separated by intermediate Cs ions [77B2].

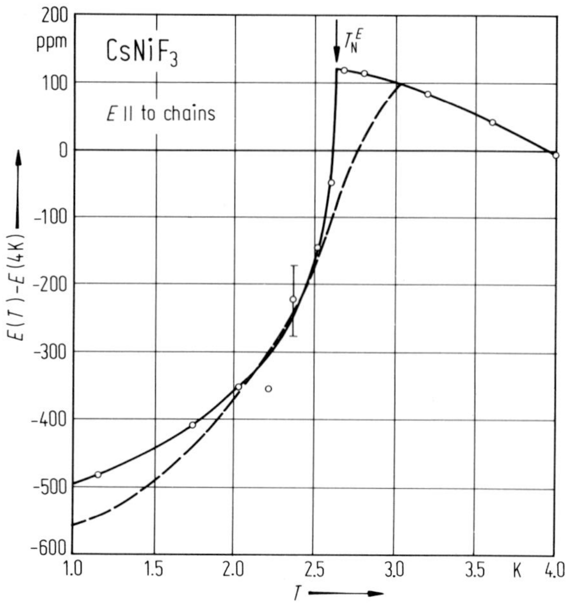

Fig. 231. $CsNiF_3$. Temperature dependence of the Young modulus $E_{\parallel}$ parallel to the linear chain (see caption Fig. 230) near T_N. The dashed line is theoretical due to the thermodynamical analysis. The elastic measurements were carried out in the frequency range of 1 ··· 4kHz [77B2]. E(4K) = $3.1 \cdot 10^{11}$ dyn cm^{-2}.

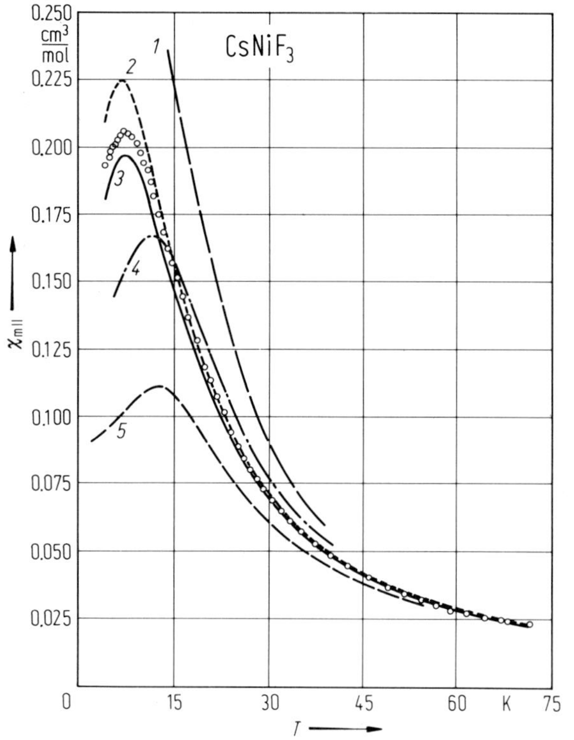

Fig. 232. $CsNiF_3$. Temperature dependence of $\chi_{m\parallel}$ parallel to the chain (circles). Lines *2* and *3* are theoretical after [75N1] for $J/k = 10$ K, $D/k = 8$ K and $J/k = 10$ K, $D/K = 9$ K, respectively. Lines *1*, *4* and *5* are calculated after [75L2] for *1*: $J/k = 11.8$ K and $D/k = 4.5$ K, *4*: $J/k = 10$ K and $D/k = 7.5$ K and *5*: $J/k = 8$ K and $D/k = 11$ K, respectively [77D1]. For chain axis see caption Fig. 230. J: intrachain exchange parameter, D: single-ion anisotropy constant.

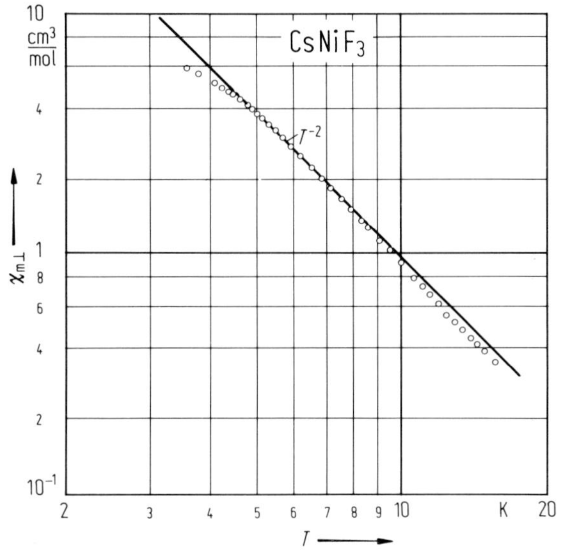

Fig. 233. $CsNiF_3$. Log-log plot of the temperature dependence of χ_m perpendicular to the chain axis (see Fig. 230). Solid straight line represents T^{-2} law [77D1].

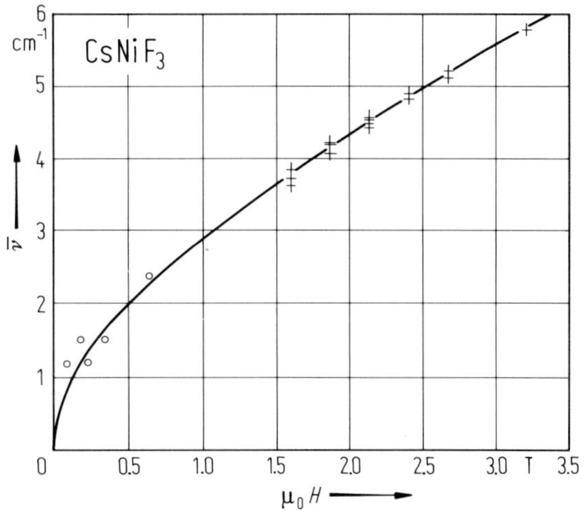

Fig. 234. $CsNiF_3$. Magnetic field dependence of the FMR at 4.2 K. The solid line is calculated for $D = 3.3\ cm^{-1}$ and $g_\perp = 2.28$ (circles: results of [77R1]) [77G1].

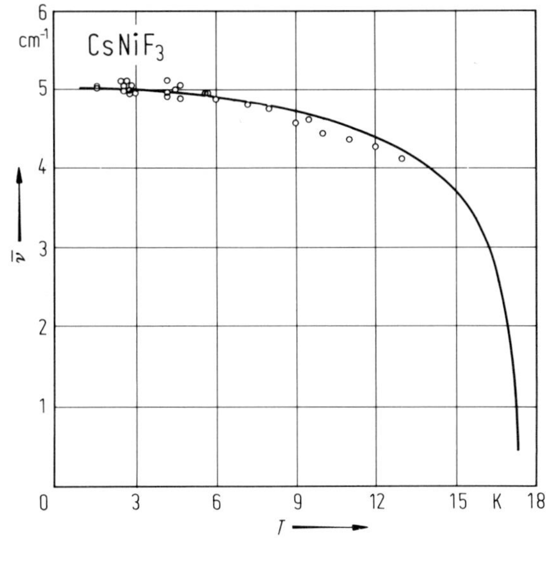

Fig. 235. $CsNiF_3$. Temperature dependence of FMR at $\mu_0 H_\perp = 2.675$ T. The solid line shows the result of a spin wave renormalization calculation [77G1].

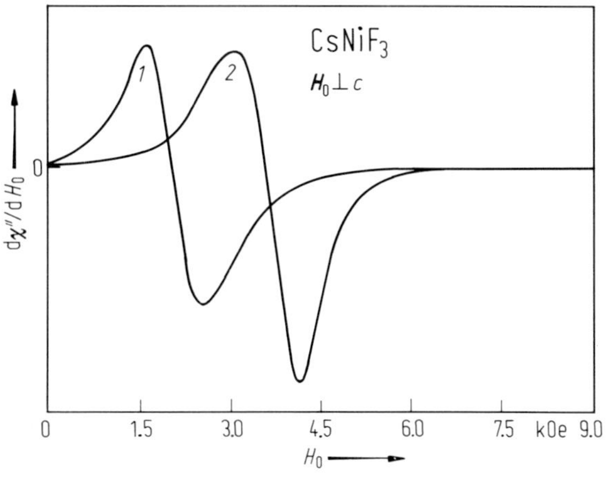

Fig. 236. $CsNiF_3$. Magnetic field dependence of $d\chi''/dH_0$, measured on thin disc at 4.2 K and $\nu = 35$ GHz with $\boldsymbol{H}_0 \perp c$, $\boldsymbol{H}_1 \perp \boldsymbol{H}_0$, and $\boldsymbol{H}_1 \perp c$, for $\boldsymbol{H}_0 \perp$ to the disk plane (*2*) and within the disk plane (*1*) [77R1]. H_0: external field, H_1: rf field.

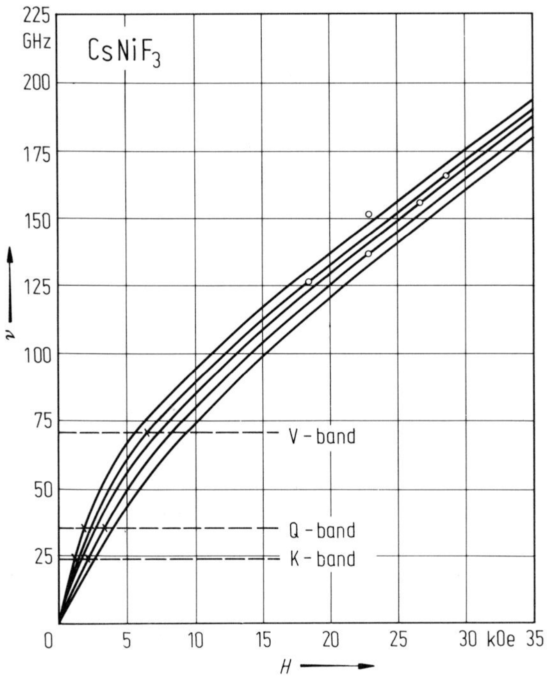

Fig. 237. $CsNiF_3$. Magnetic field dependence of FMR frequency at 4.2 K and $\boldsymbol{H}_0 \perp c$ axis for different sample shapes. The middle curve represents a sphere, the two outer ones the bordering case of infinitely thin disks [77R1].

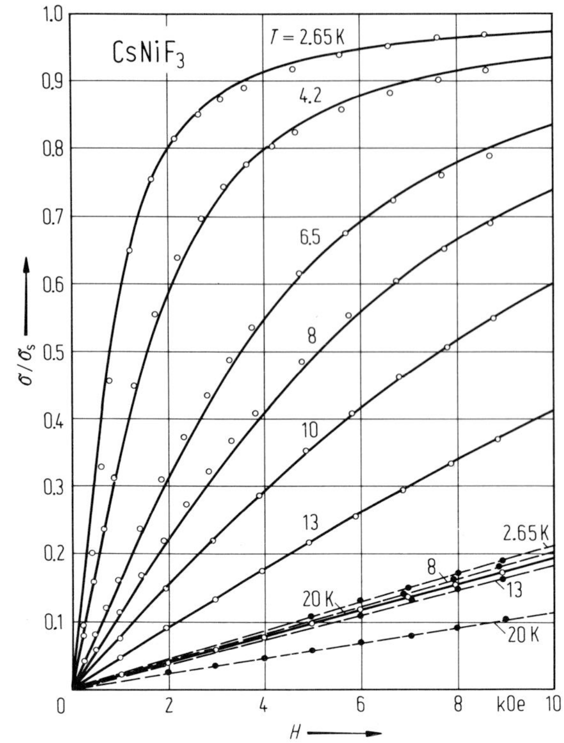

Fig. 238. $CsNiF_3$. Internal magnetic field dependence ($H_{int} = H_0 - NM$) of σ, open circles: experimental points, $\boldsymbol{H}_{int} \perp c$, full circles: experimental points, $\boldsymbol{H}_{int} \| c$, full and broken lines: theoretical curves [77R1].

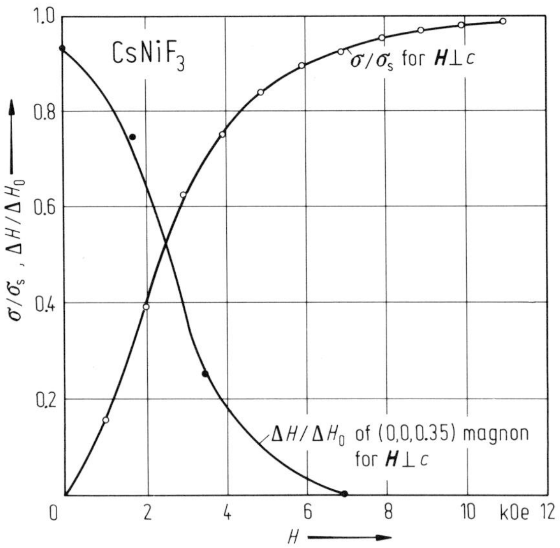

Fig. 239. $CsNiF_3$. Magnetic field dependence of the linewidth of the (0, 0, 0.35) magnon and homogeneous magnetization at 4.2 K. Linewidth corrected for resolution of the experiment [77S2]. σ_s: saturation magnetization.

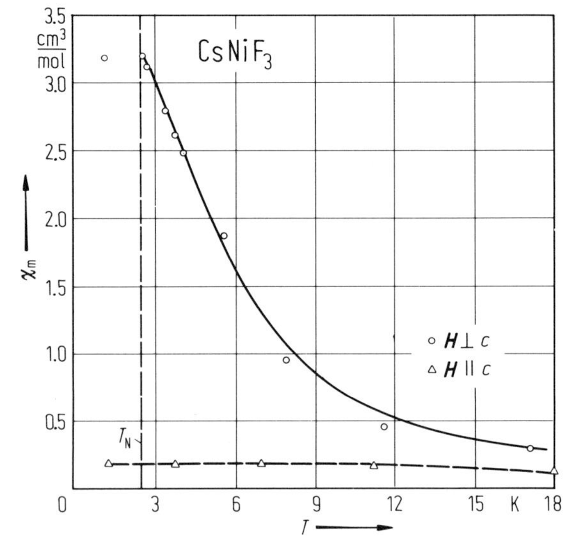

Fig. 240. $CsNiF_3$. Temperature dependence of χ_m. Triangles: $\boldsymbol{H} \| c$, circles: $\boldsymbol{H} \perp c$ [77S3].

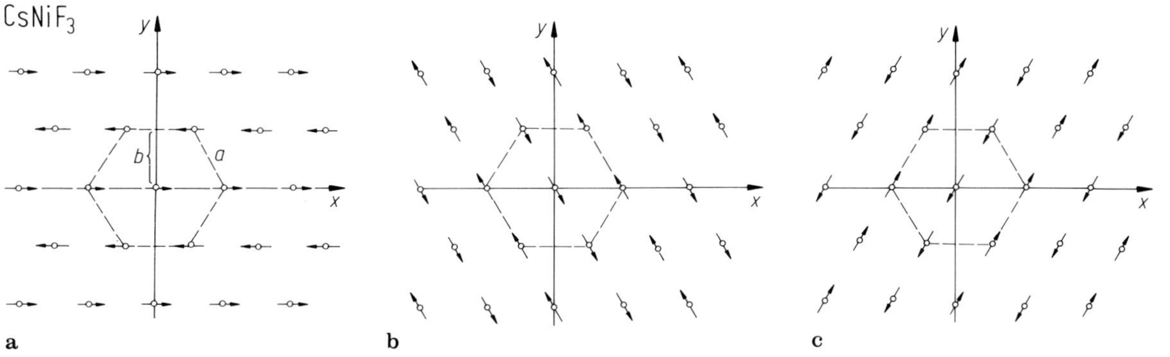

Fig. 241. $CsNiF_3$. Three equivalent states of order for local magnetic moments. Each arrow represents a chain of spins, perpendicular to the plane of the paper [77S3].

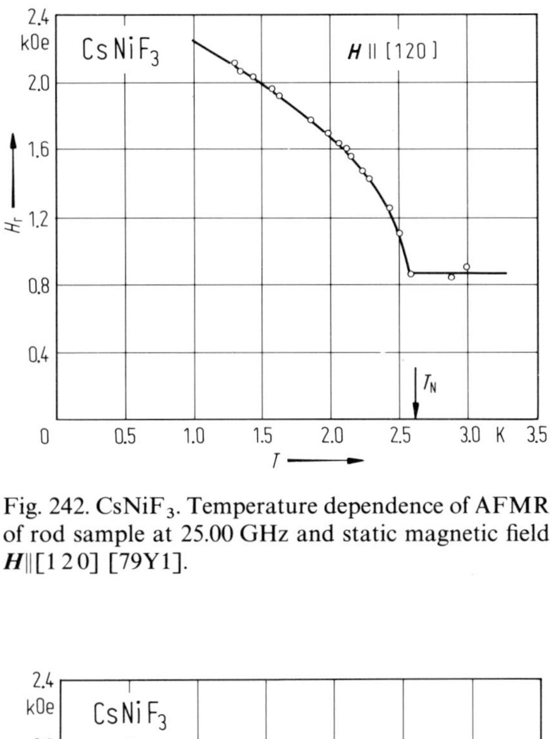

Fig. 242. $CsNiF_3$. Temperature dependence of AFMR of rod sample at 25.00 GHz and static magnetic field $\boldsymbol{H} \| [1\,2\,0]$ [79Y1].

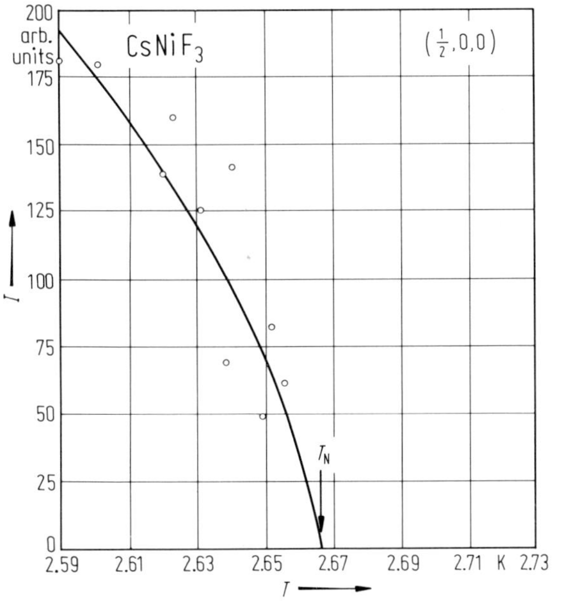

Fig. 244. $CsNiF_3$. Temperature dependence of $(\frac{1}{2}, 0, 0)$ neutron magnetic Bragg peak intensity corresponding to $\langle \sigma_s^2 \rangle$. The solid line is the result of a least squares fit to the experimental points with the parameters $T_N = 2.665$ K, $\beta = 0.34$ [80S3].

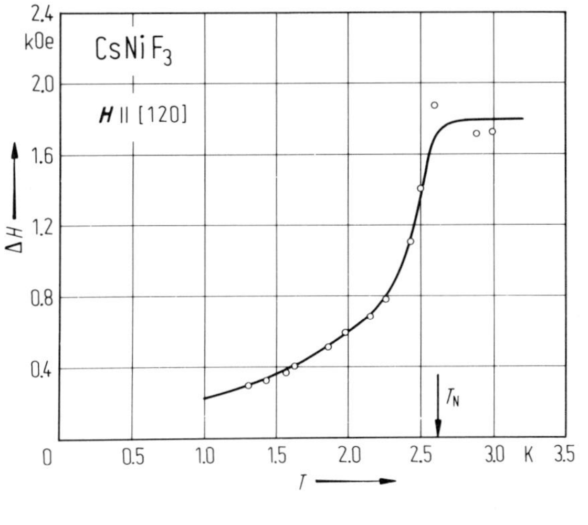

Fig. 243. $CsNiF_3$. Temperature dependence of linewidth ΔH of rod sample at 25.00 GHz and external field $\boldsymbol{H} \| [1\,2\,0]$ [79Y1].

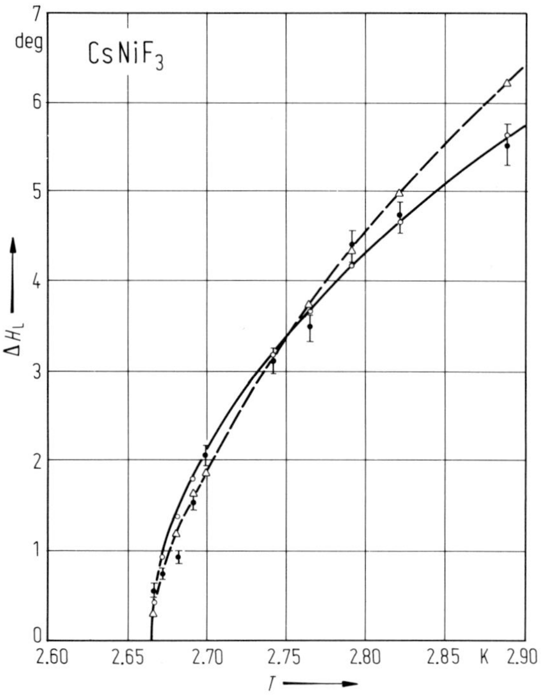

Fig. 245. $CsNiF_3$. Temperature dependence of the intrinsic halfwidth of the Lorentzian line ΔH_L. Full circles: experimental points; open circles: calculated with $\nu = 0.54$; dashed line: calculated with $\nu = 0.68$; ν: critical exponent, according to $2\pi\zeta = \frac{1}{\kappa} \propto \left(\frac{T - T_N}{T_N}\right)^{-\nu}$, ζ = correlation length [80S3].

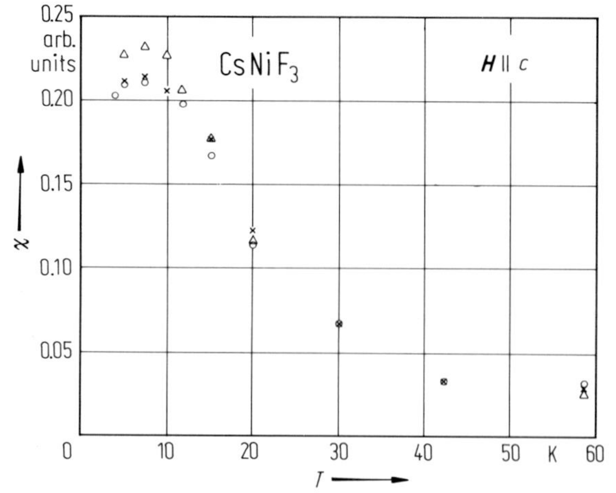

Fig. 246. $CsNiF_3$. Temperature dependence of χ from Faraday rotation from [84I1] (crosses), from [71S1] (open circles) and from [77D1] (open triangles).

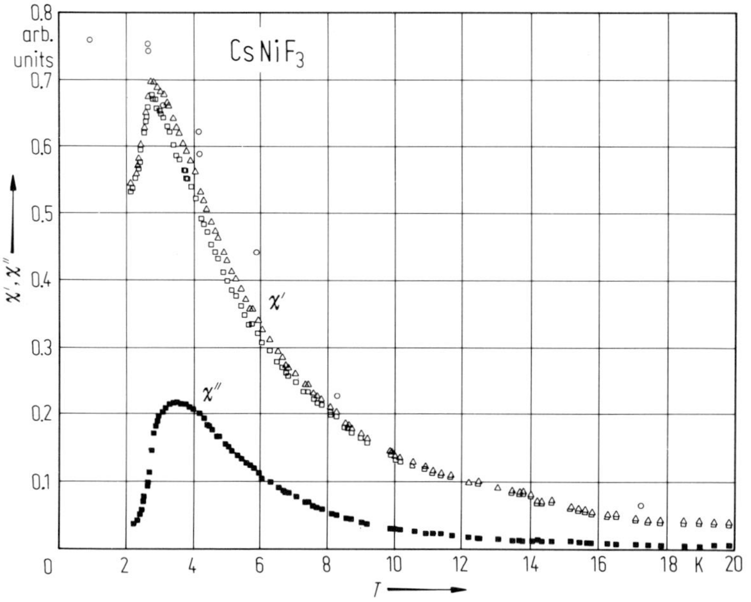

Fig. 247. $CsNiF_3$. Temperature dependence of χ at 9.5 GHz in the easy plane perpendicular to the c axis: (open squares) real $\chi'(\omega, T)$ and (full squares) imaginary $\chi''(\omega, T)$ parts. Static susceptibility $\chi(0, T)$: triangles: calculated, (circles) results from [71S1]; [86C1].

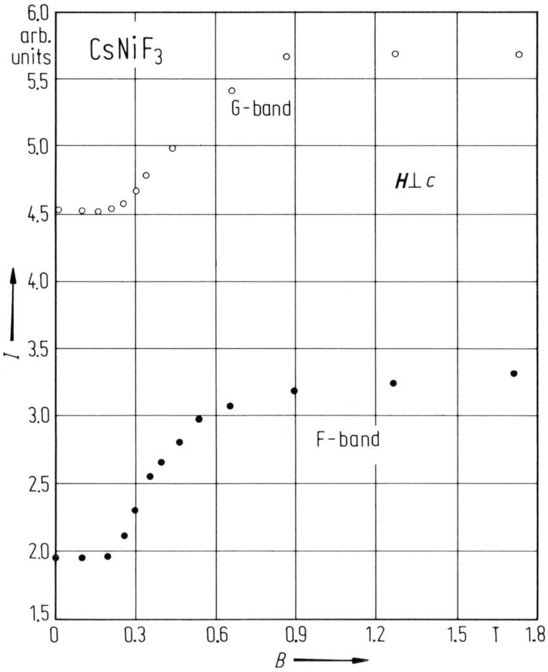

Fig. 248. $CsNiF_3$. Magnetic field dependence of the optical absorption intensity I of the exciton-phonon bands along the easy plane at 2 K. Appearance of a disordered spin structure at $0.2\ T < B < 0.55\ T$ presumably of an incommensurate magnetic phase suggested [86L1].

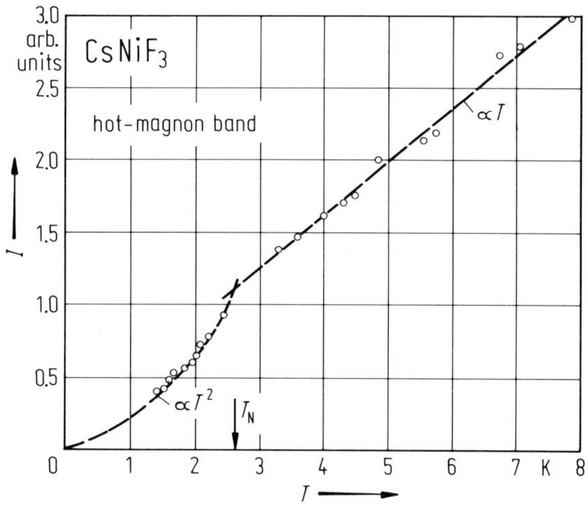

Fig. 249. $CsNiF_3$. Temperature dependence of the intensity of the hot magnon A band optical absorption [86L1].

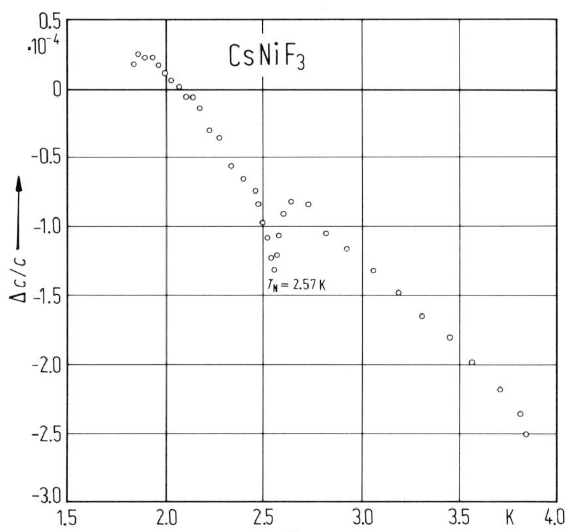

Fig. 250. $CsNiF_3$. Temperature dependence of the slower "bar mode" near T_N determined by ultrasonic measurements. This elasticity is believed to mostly represent c_{44} elastic constant. Measurements were made on small needles cleaved from large parent crystal [87S1].

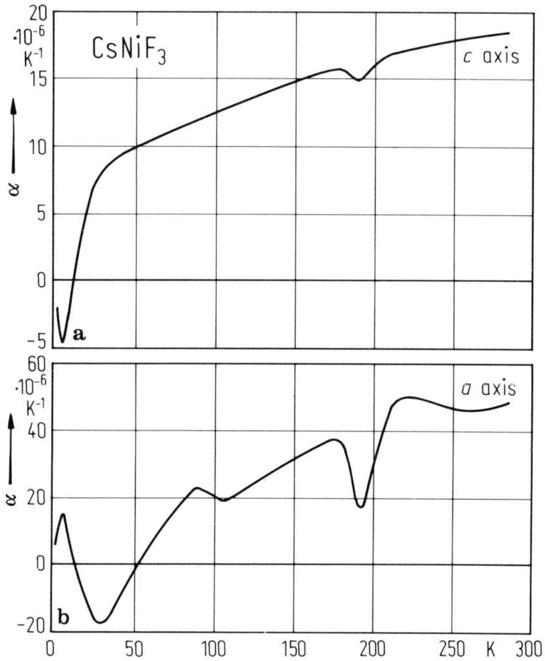

Fig. 251. $CsNiF_3$. Temperature dependence of the c axis (a) and a axis (b) expansion coefficient [87S1].

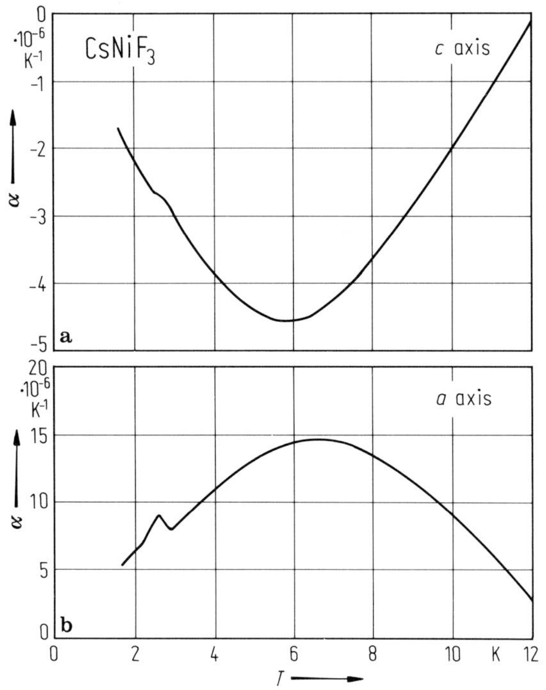

Fig. 252. $CsNiF_3$. Low temperature dependence of the c axis (a) and a axis (b) expansion coefficient [87S1].

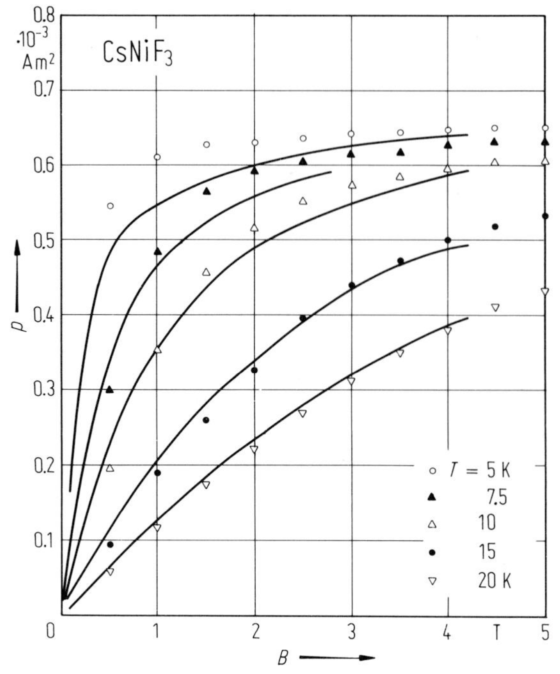

Fig. 253. $CsNiF_3$. Magnetic field dependence of in-plane magnetic moment at different temperatures. Solid lines are experimental. Symbols are calculated [91C1].

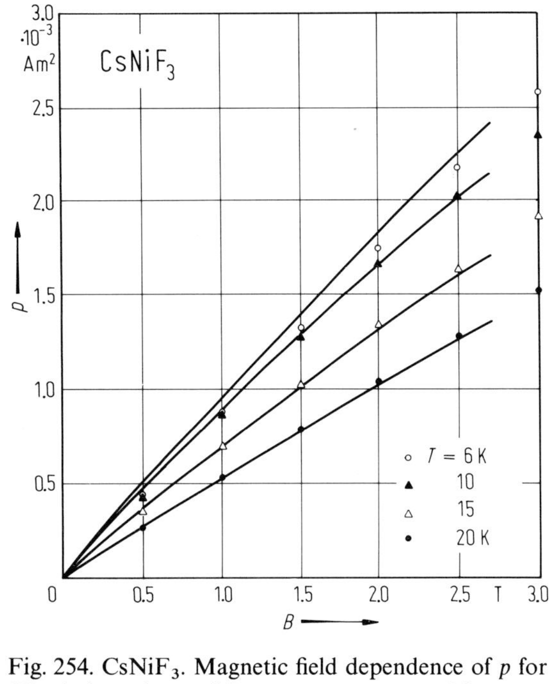

Fig. 254. $CsNiF_3$. Magnetic field dependence of p for $\boldsymbol{H}\|$ chain axis at different temperatures. Lines are experimental and symbols calculated [91C1].

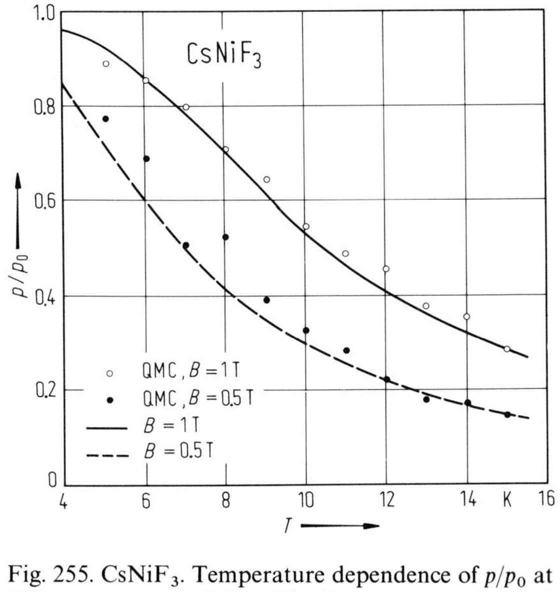

Fig. 255. $CsNiF_3$. Temperature dependence of p/p_0 at different magnetic fields. The lines are experimental and symbols are calculated [91C1].

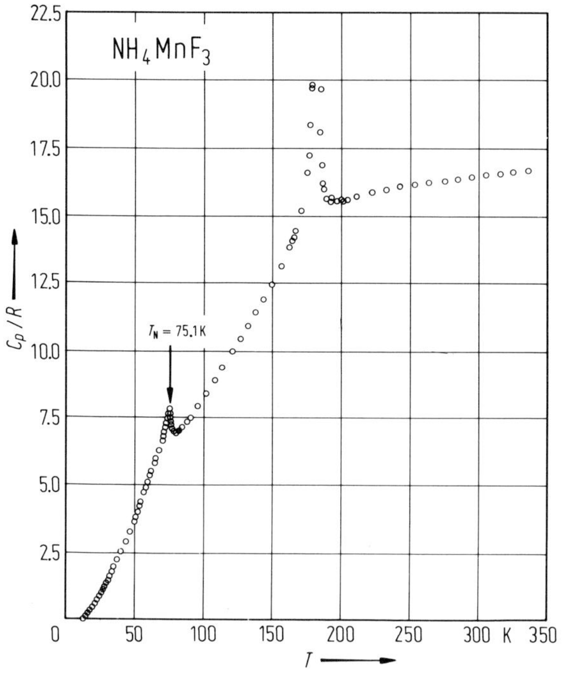

Fig. 256. NH_4MnF_3. Temperature dependence of heat capacity C_p [83B2].

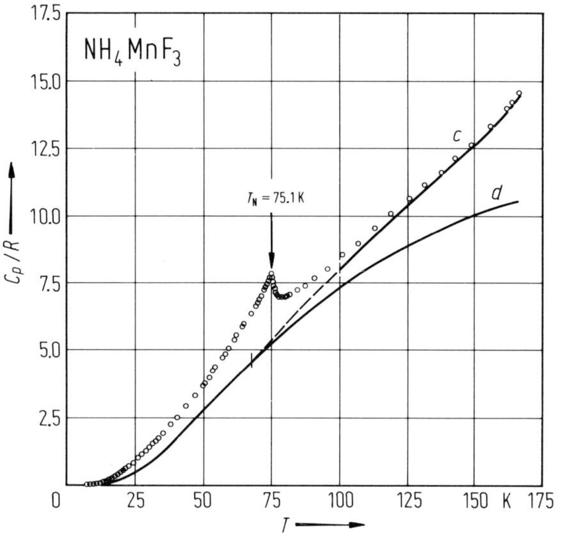

Fig. 257. NH_4MnF_3. Temperature dependence of heat capacity C_p near T_N. Curve *c*: non-magnetic base line; curve *d*: phonon contribution as obtained from corresponding states law from C_p ($KZnF_3$) [83B2].

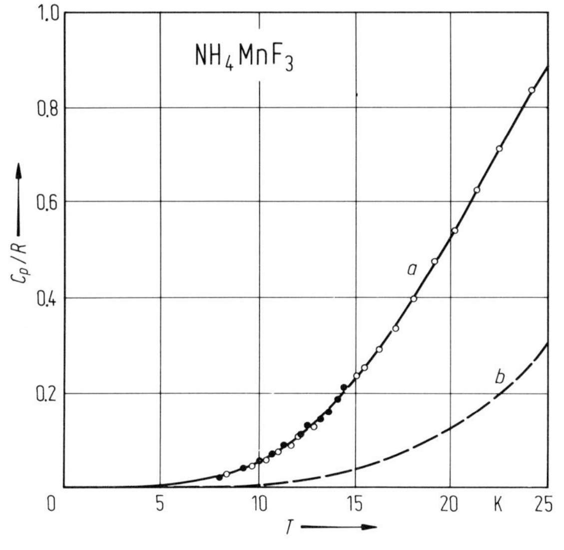

Fig. 258. NH_4MnF_3. Low temperature dependence of heat capacity C_p; curve *a*: A T^3 fit of the experimental data; curve *b*: calculated spin-wave contribution (see Fig. 257) [83B2].

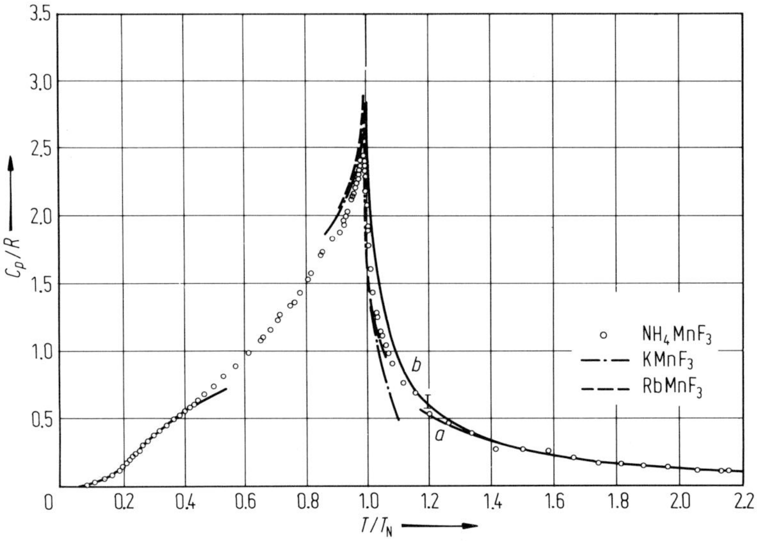

Fig. 259. NH_4MnF_3. Temperature dependence of heat capacity in reduced temperature scale, compared to available data of $KMnF_3$ and $RbMnF_3$. The H.T.S. theoretical prediction (full line) is included for (curve *a*) $S = 5/2$ and (*b*) $S = \infty$ extrapolated to T_N. The spin-wave prediction for $S = 5/2$ at $T < 0.5\ T_N$ is also shown [83B2].

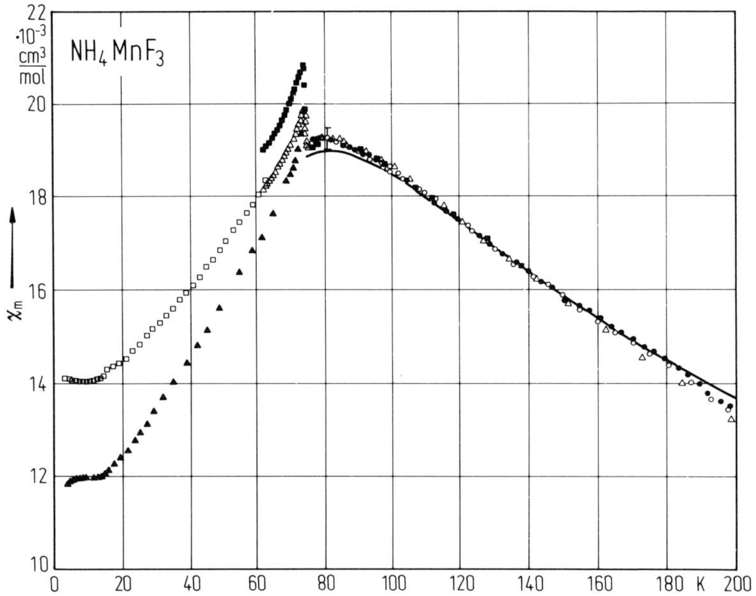

Fig. 260. NH_4MnF_3. Temperature dependence of zero field χ_m. △, ○, ●: different series on milled crystals. Full and open squares: randomly oriented single crystals. The two series differ in sample mass, $m = 0.6398$ and 0.2804 g, respectively. Full triangles: stacked single crystals with two faces perpendicular to the *c* axis, $m = 0.1453$ g. Full line: theoretical prediction for the Heisenberg antiferromagnetic simple cubic $S = 5/2$ model calculated by the Pade-approximation analysis of the H.T.S. for $J/k = -3.11$ K [83B2].

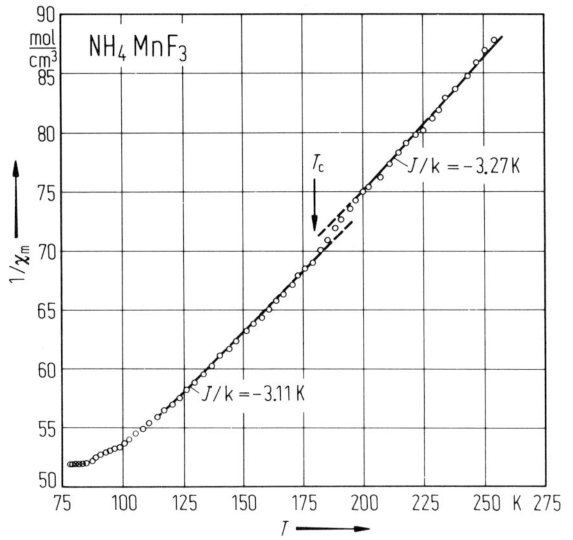

Fig. 261. NH_4MnF_3. Temperature dependence of zero-field χ_m^{-1}. The full lines represent the theoretical fits above and below T_c [83B2].

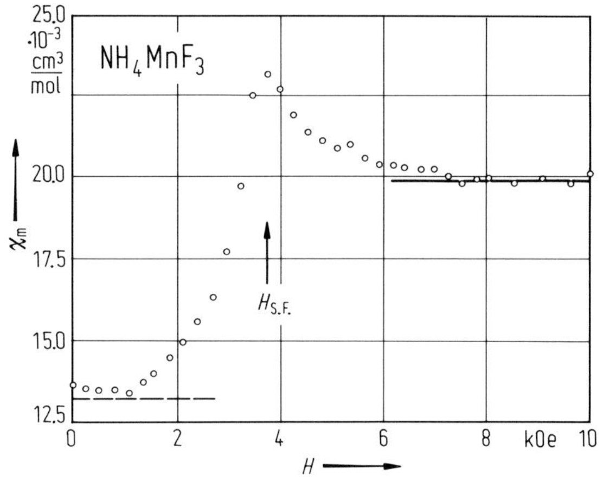

Fig. 263. NH_4MnF_3. Magnetic field dependence χ_m at $T = 4.22$ K for H up to 10 kOe on randomly stucked single crystals. The spin-flop transition is detected at $H_{SF} = 3.7(2)$ kOe. The full line corresponds to the experimental tendency to $\chi_\perp$, while the dashed line to $\frac{2}{3}\chi_\perp$ [83B2].

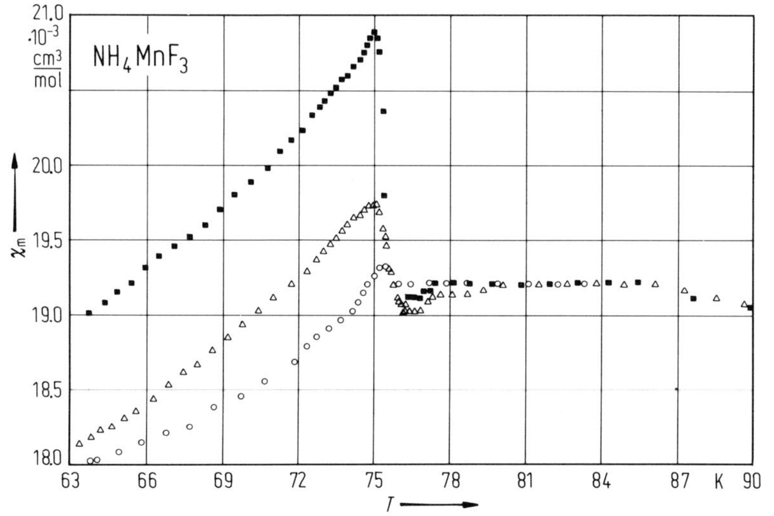

Fig. 262. NH_4MnF_3. Weak-ferromagnetic anomaly in the zero-field χ_m. Dependence on particle size: full squares: 1 mm wide cubic single crystals, open triangles: 100 μm wide milled crystals, open circles: 0.05 μm wide powder [83B2].

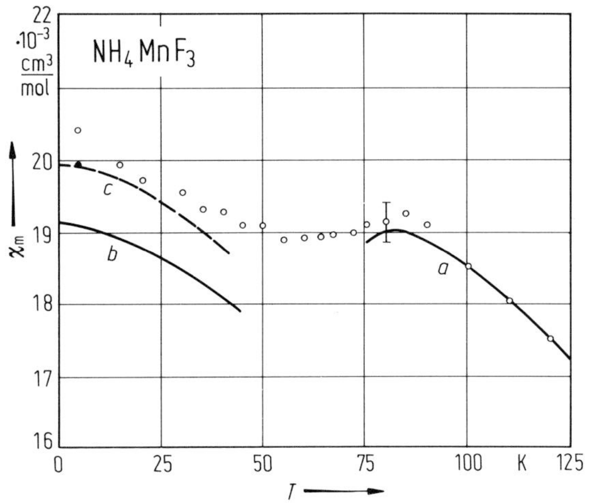

Fig. 264. NH_4MnF_3. Temperature dependence of χ_m at $H = 10$ kOe. Full circle: value obtained from measurements shown in Fig. 263. Curve *a*: H.T.S. prediction of χ for $J/k = -3.11$ K, curve *b*: spin wave prediction for $J/k = -3.11$ K, curve *c*: idem for $J/k = -2.98$ K [83B2].

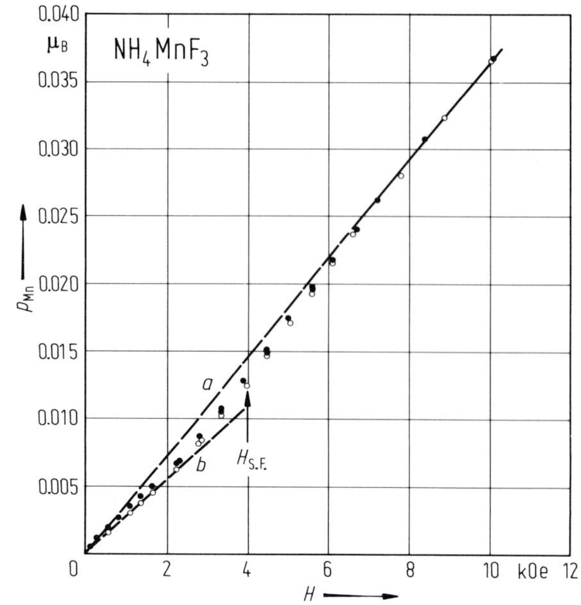

Fig. 265. NH_4MnF_3. Magnetic field dependence of p_{Mn} for randomly oriented single crystals at 4.2 K. Full line: the pulsed-field measurements, dashed line *a*: extrapolation to $H = 0$. Dashed line *b*: the low-field tangent. H_{SF} is assigned to the inflexion point [83B2].

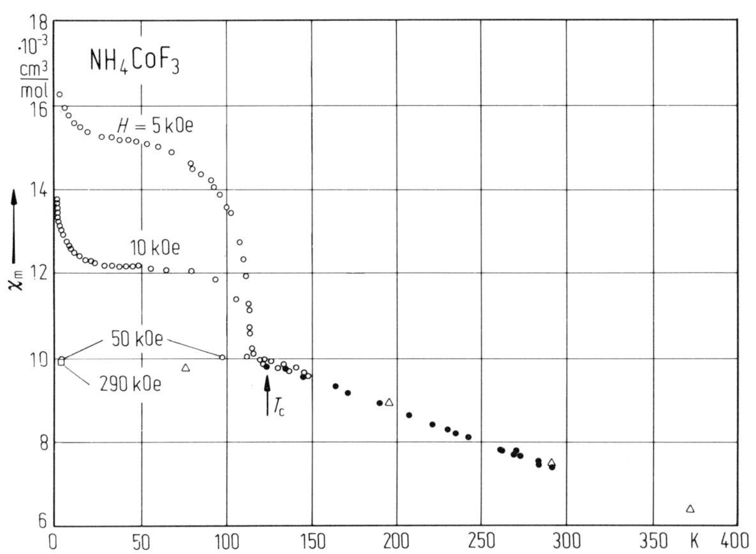

Fig. 266. NH_4CoF_3. Temperature dependence of χ_m for different H. Open circles: Foner magnetometer, full circles: Faraday balance, open triangles: data of [63R1], square: derived from magnetization at $H = 290$ kOe. T_c: transition temperature found in the heat capacity (124.5 K) [77B3].

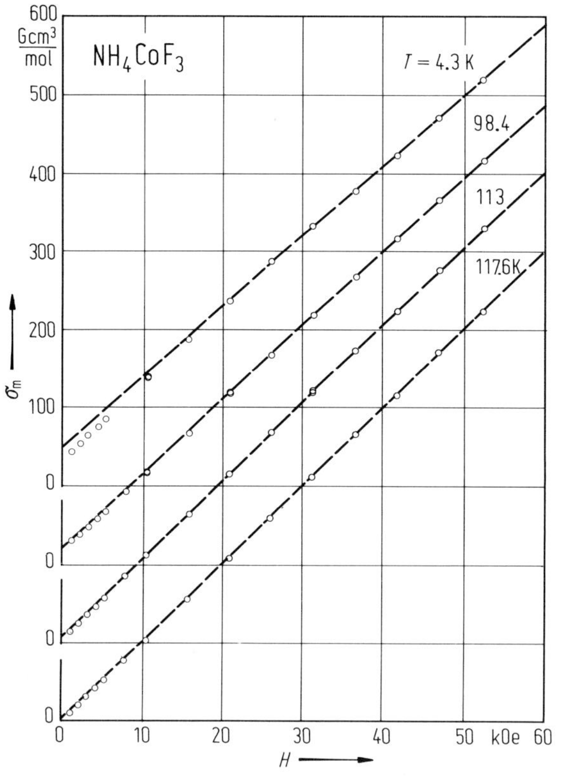

Fig. 267. NH_4CoF_3. Magnetic field dependence of σ_m measured with the Foner magnetometer at different temperatures [77B3].

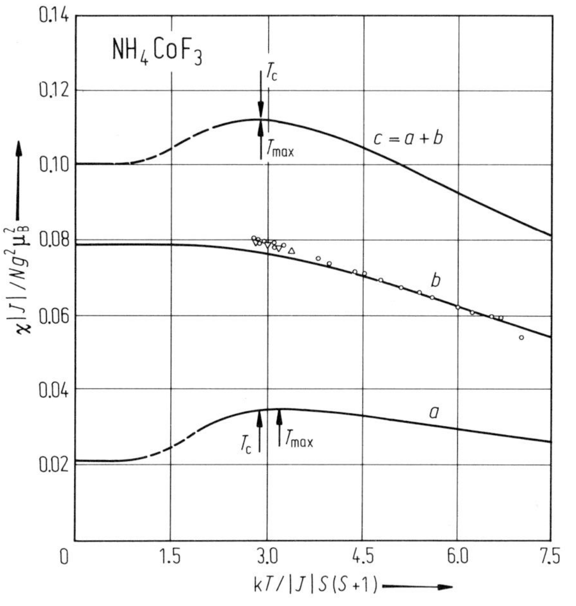

Fig. 269. NH_4CoF_3. Temperature dependence of the susceptibility compared with theoretical calculations. *a*: estimated behavior of the powder χ of the simple cubic Heisenberg antiferromagnet with $S = 1/2$; *b*: contribution to the χ arising from the higher-energy levels neglecting exchange effects [77B3].

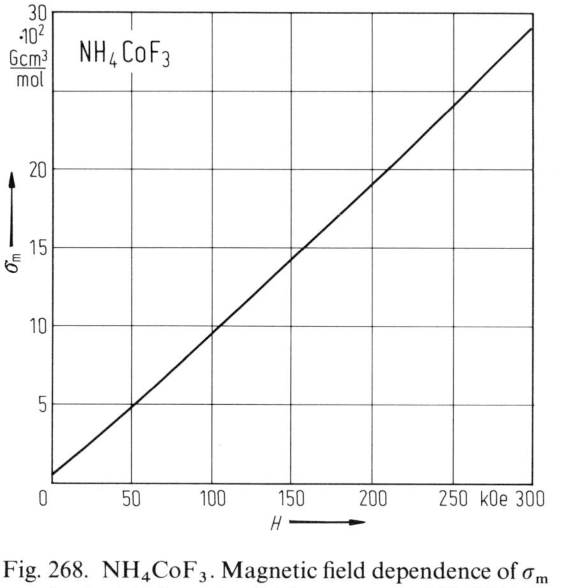

Fig. 268. NH_4CoF_3. Magnetic field dependence of σ_m obtained with the pulse field set-up at 4.2 K [77B3].

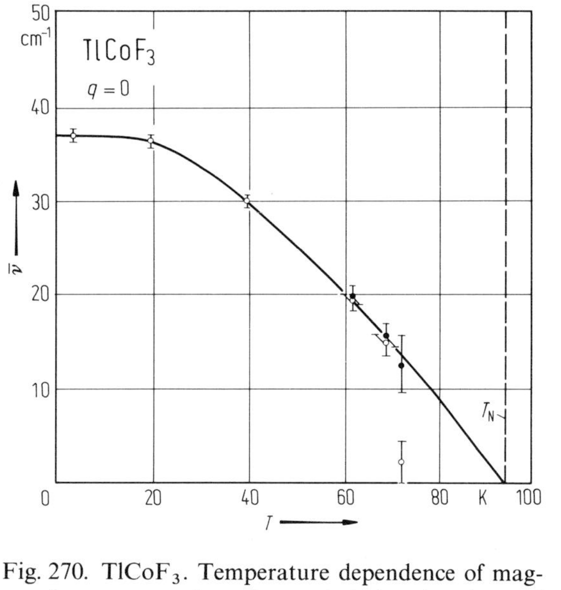

Fig. 270. $TlCoF_3$. Temperature dependence of magnon frequency peak ω_p (open circles) and undamped quasi-harmonic frequency ω_0 (full circles) given as wavenumbers $\bar{\nu}$. $\omega_o^2 = \omega_p^2 + \Gamma^2/2$. Γ: intrinsic half-width at half-height [73R1].

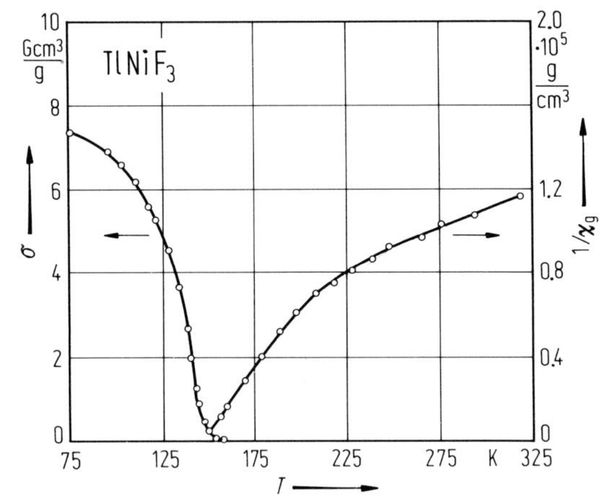

Fig. 271. $TlNiF_3$. Temperature dependence of χ_g^{-1} and σ at 7.6 kOe [67K1].

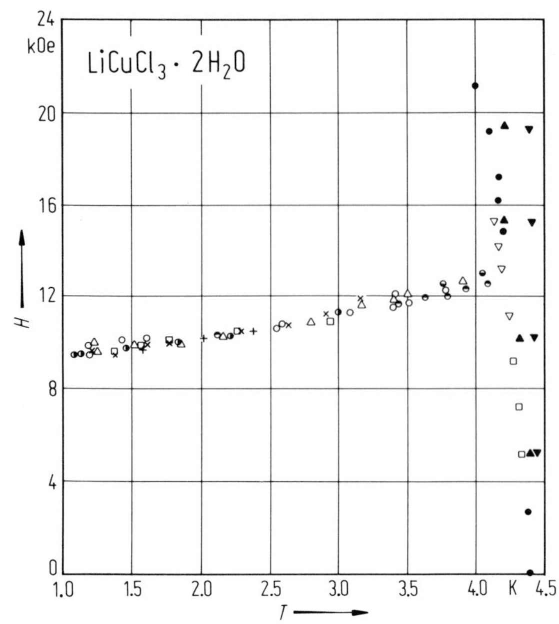

Fig. 272. $LiCuCl_3 \cdot 2H_2O$. Magnetic phase diagram. All the data were taken with $\boldsymbol{H} \parallel c'$, except ▲and ▼which correspond to $\boldsymbol{H} \perp c'$ axis [71F1]. c' axis is perpendicular to ab plane.

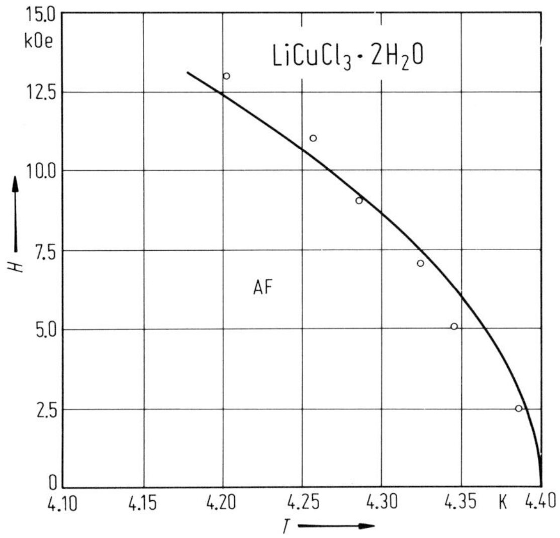

Fig. 273. $LiCuCl_3 \cdot 2H_2O$. AF-P boundary of the phase diagram. Solid curve indicates least-squares fit to the data [71F1].

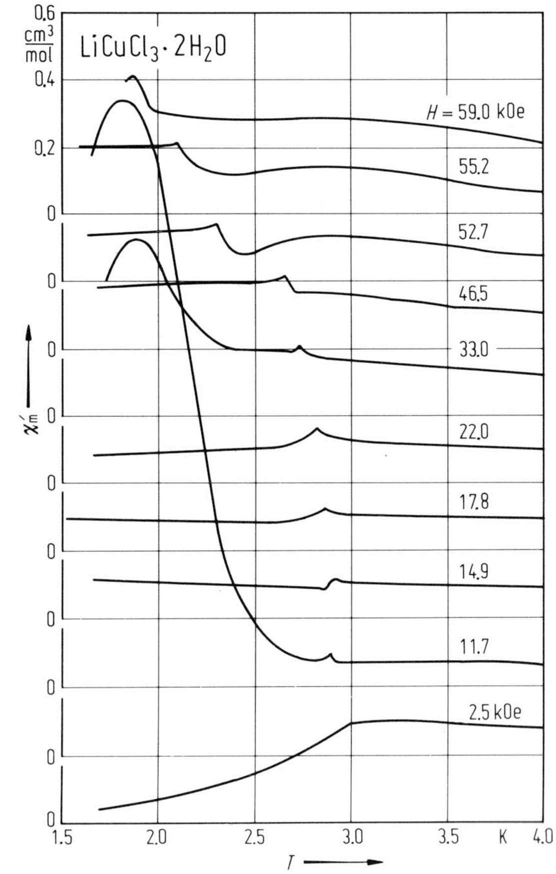

Fig. 274. $LiCuCl_3 \cdot 2H_2O$. Temperature dependence of χ'_m (of 225 Hz) at various fields along [101] axis [73M3].

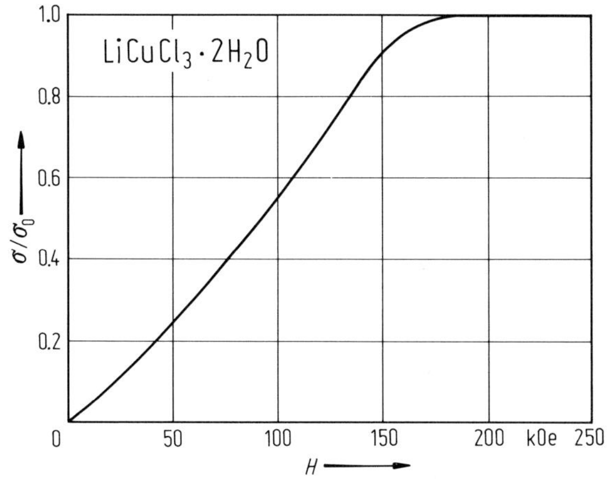

Fig. 275. $LiCuCl_3 \cdot 2H_2O$. Magnetic field dependence of σ at 1.55 K. The measurements have been performed in high pulsed magnetic fields [73M3].

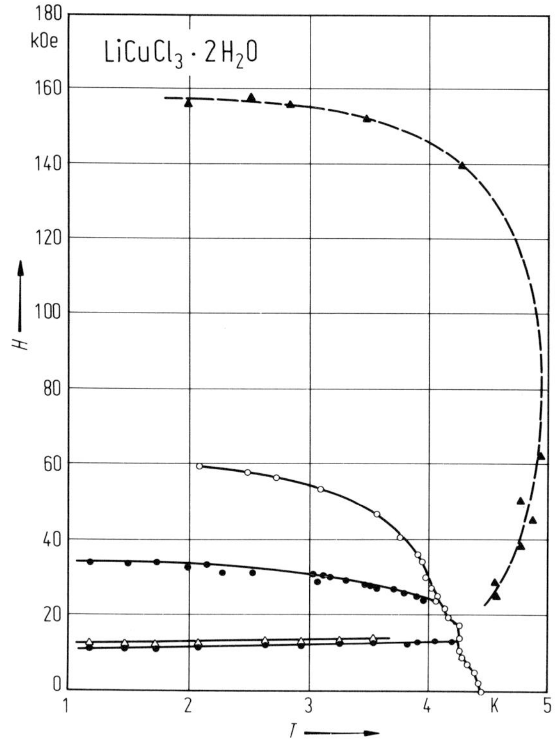

Fig. 276. $LiCuCl_3 \cdot 2H_2O$. Magnetic phase diagram for the easy spin direction [101] axis. Open circles: second order phase transitions; full triangles: anomalies which are, for the lower temperatures, related to saturation; full circles: first-order phase transitions; open triangles: weak anomaly in the susceptibility [73M3].

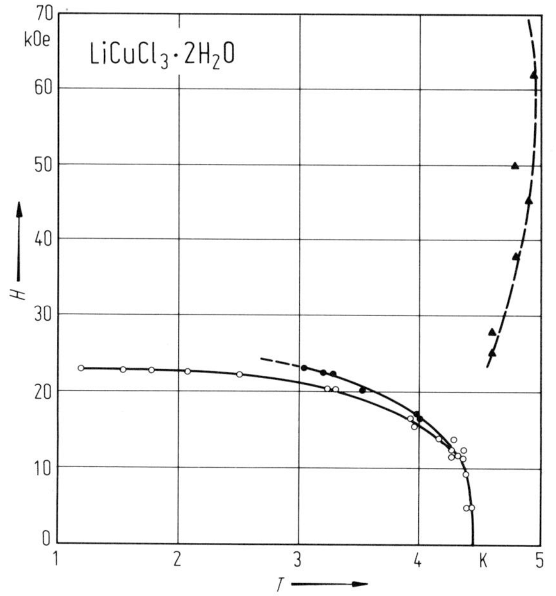

Fig. 277. $LiCuCl_3 \cdot 2H_2O$. Magnetic phase diagram for the *b* direction. Full triangles: broad anomalies in the susceptibility; open circles: first-order phase transitions; full circles: weak anomalies in the susceptibility [73M3].

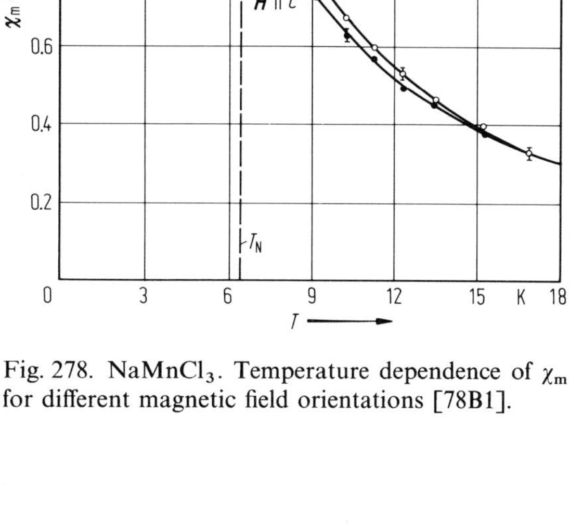

Fig. 278. $NaMnCl_3$. Temperature dependence of χ_m for different magnetic field orientations [78B1].

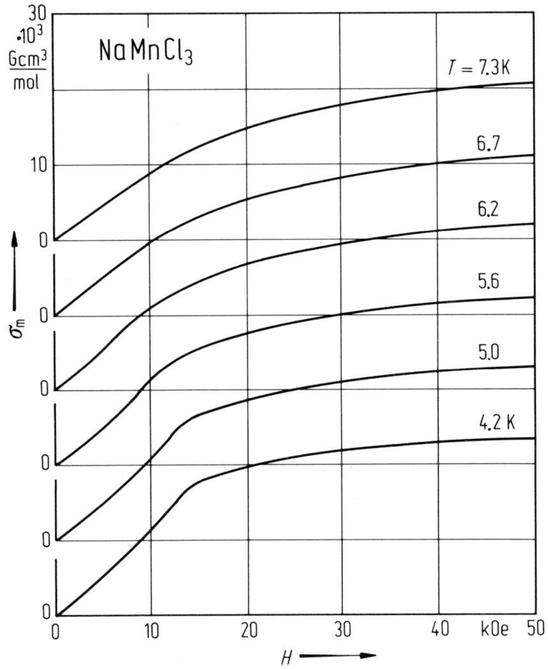

Fig. 279. $NaMnCl_3$. Magnetic field dependence of σ_m at different temperatures for $H \perp c$ [78B1].

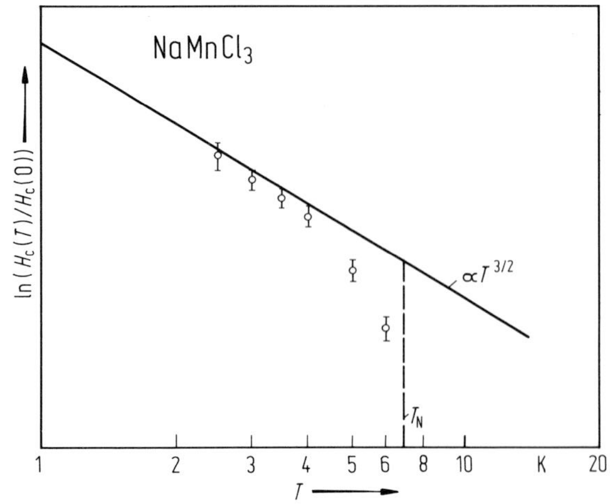

Fig. 280. $NaMnCl_3$. Temperature dependence of H_c in logarithmic scales [78B1].

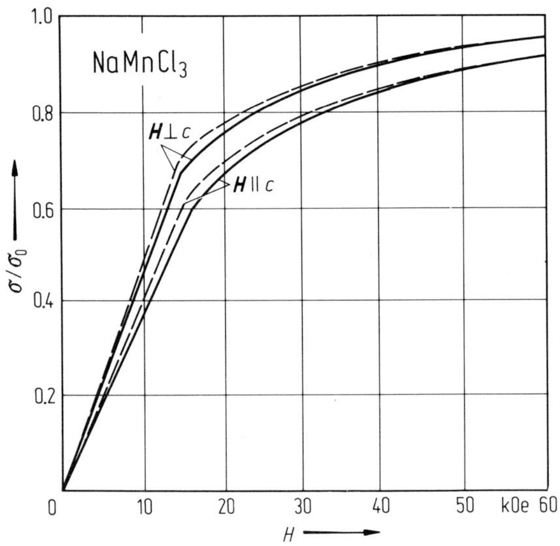

Fig. 281. $NaMnCl_3$. Magnetic field dependence of magnetization $\sigma_m(H)/\sigma_m(0)$ at 4.2 K for $H \perp c$, and $H \parallel c$ [80F1]. Dashed lines: see Fig. 282. See also [78B1].

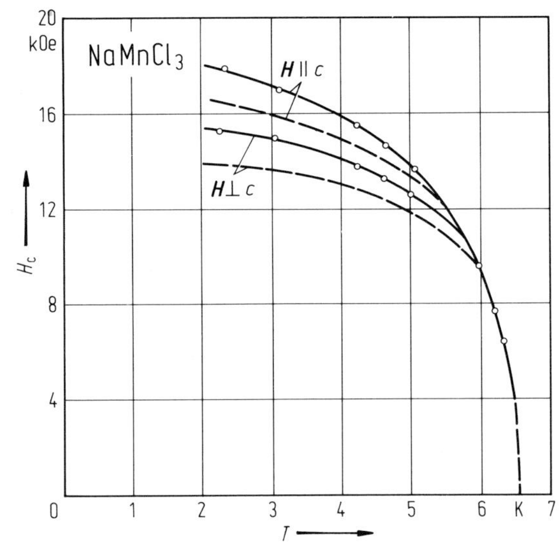

Fig. 282. $NaMnCl_3$. Temperature dependence of the spin-flip transition field. The dashed lines correspond to the curves corrected for the internal demagnetizing field [80F1]. See also [78B1].

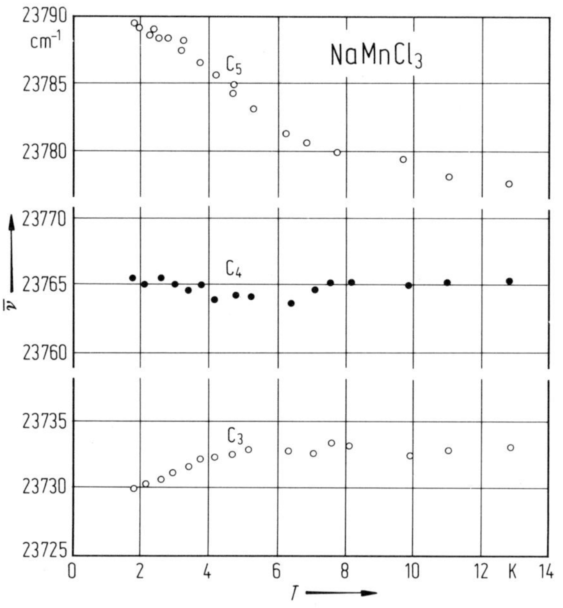

Fig. 283. $NaMnCl_3$. Temperature dependence of the energies of the optical absorption bands C_3 to C_5 [85P1]. C_5 is exciton band and C_3, C_4 its magnon side bands.

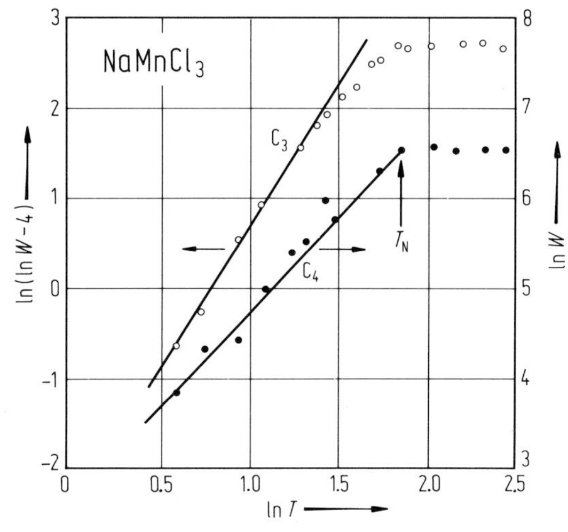

Fig. 284. $NaMnCl_3$. Temperature dependence of the integral intensities W of the bands C_3 and C_4 (see Fig. 283) [85P1]. T in K.

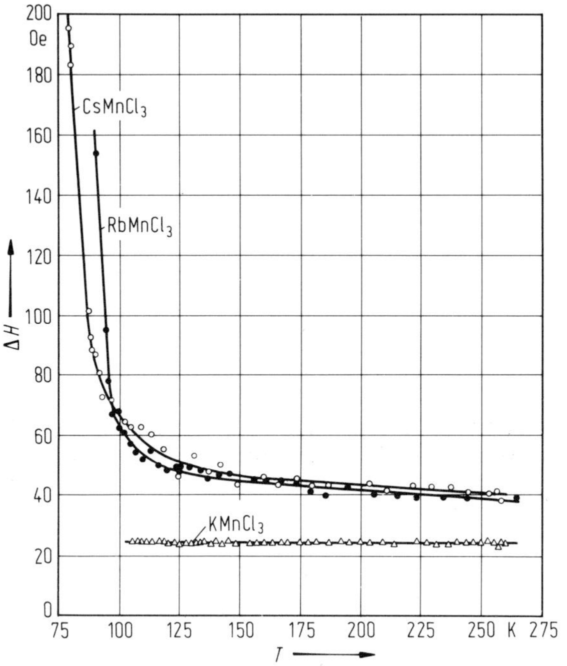

Fig. 285. $KMnCl_3$, $RbMnCl_3$, $CsMnCl_3$. Temperature dependence of the EPR linewidth at K band (25 GHz) [65K1].

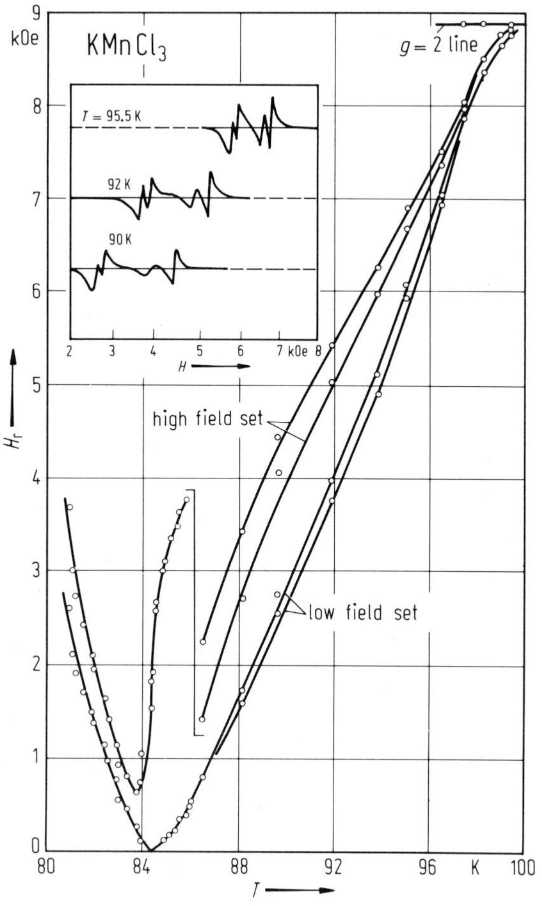

Fig. 286. $KMnCl_3$. Temperature dependence of the magnetic field for AFMR observed at 24.87 GHz [65K1].

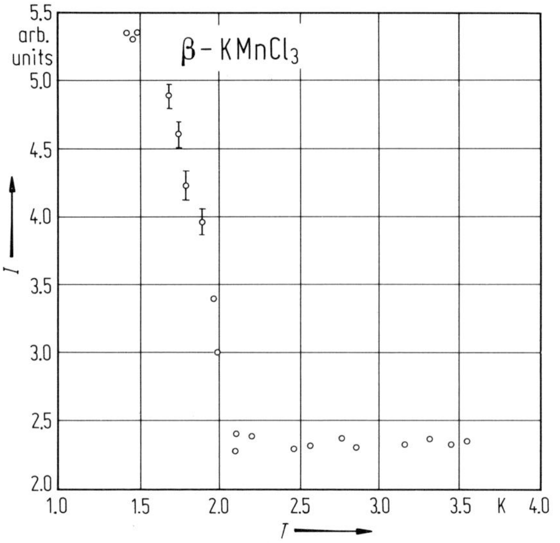

Fig. 287. β-$KMnCl_3$. Temperature dependence of neutron $(000)^{\pm}$ magnetic satellite reflection of powder sample (β-phase, isostructural with $KCdCl_3$) [79G1].

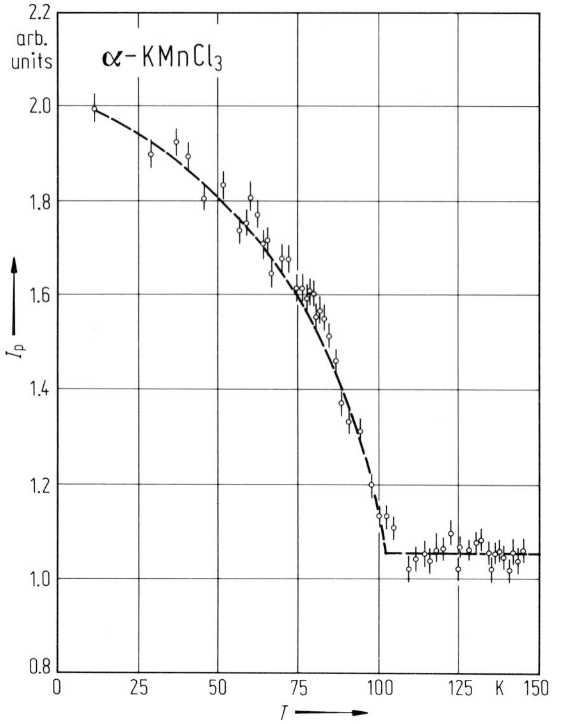

Fig. 288. α-$KMnCl_3$. Temperature dependence of the neutron peak intensity of (1, 0, 1) magnetic reflection from powder sample [82G1].

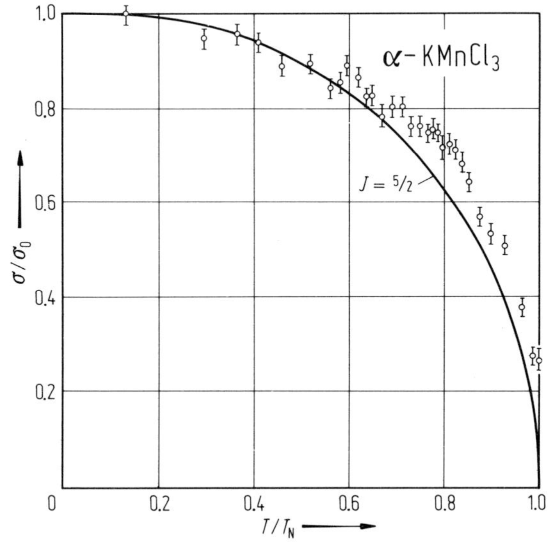

Fig. 289. α-$KMnCl_3$. Reduced temperature dependence of reduced magnetization calculated from the (101) reflection. The line corresponds to Brillouin curve for $J = 5/2$ [82G1].

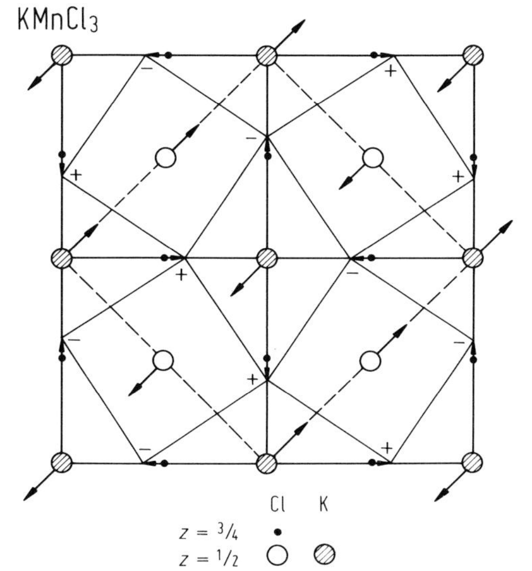

Fig. 290. $KMnCl_3$. Schematic drawing of the two rotations of the ions induced on the Cl^- octahedra (thin lines) and the K^+ ions. The respective displacements of these ions from their ideal positions (full circles and hatched circles) are indicated by arrows in the *ab* plane and ± in the *z* direction. The heavy lines are the ideal perovskite unit cell and the dashed lines are the orthorhombic unit cell [82G1].

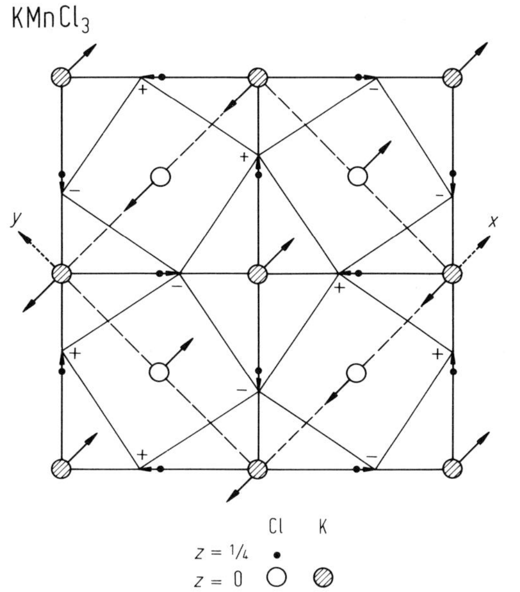

Fig. 291. $KMnCl_3$. For explanation see Fig. 290 [82G1].

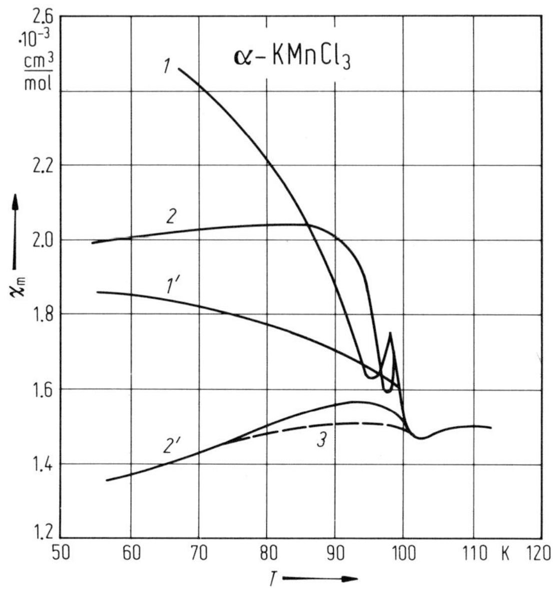

Fig. 292. α-$KMnCl_3$. Temperature dependence of χ_m in the T_N vicinity. Curves: *1*, *1′* in [1 1 0] direction; *2*, *2′* in [0 0 1] direction; *3*: powder, $H = 0$; *1*, *2*: $H =$ 0.1 kOe; *1′*, *2′*: $H = 3.3$ kOe [89F1].

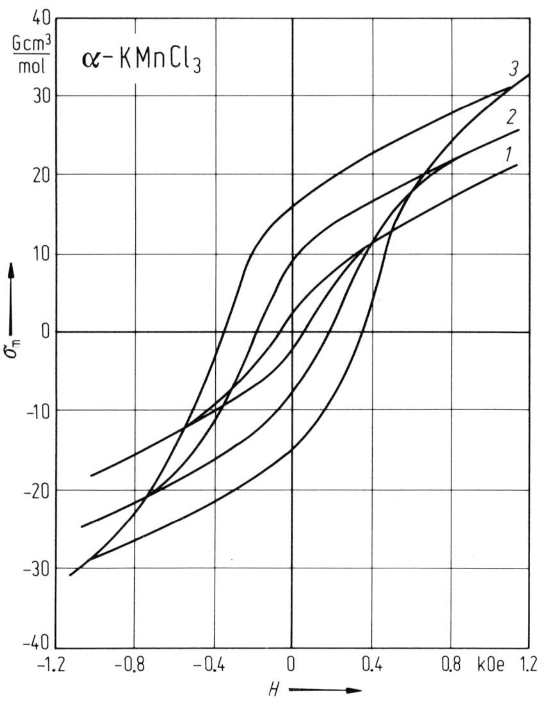

Fig. 293. α-$KMnCl_3$. Hysteresis loops for $\boldsymbol{H} \parallel [110]$ at (curve *1*) $T = 96$ K, (*2*) 86 K and (*3*) 33.4 K [89F1].

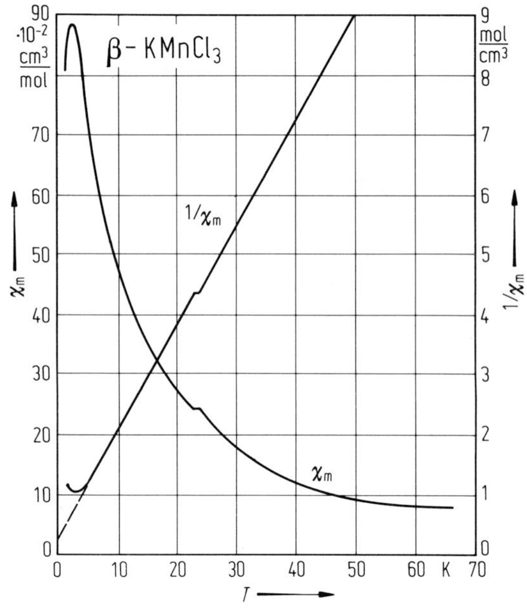

Fig. 294. β-$KMnCl_3$. Temperature dependence of χ_m and χ_m^{-1} at 178 Oe [83V3].

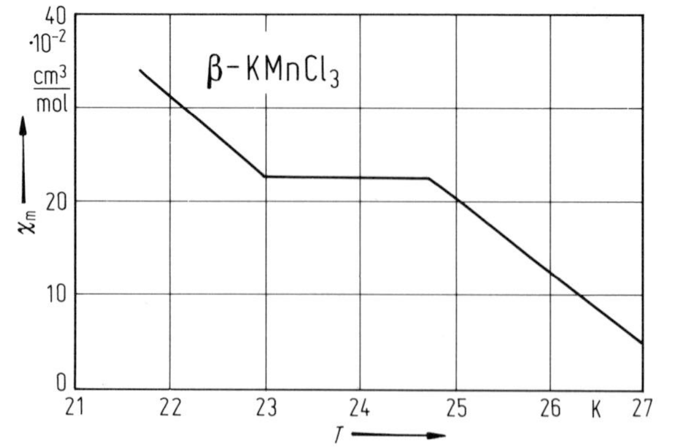

Fig. 295. β-$KMnCl_3$. Temperature dependence of χ_m at 178 Oe in the region of structural transition [83V3].

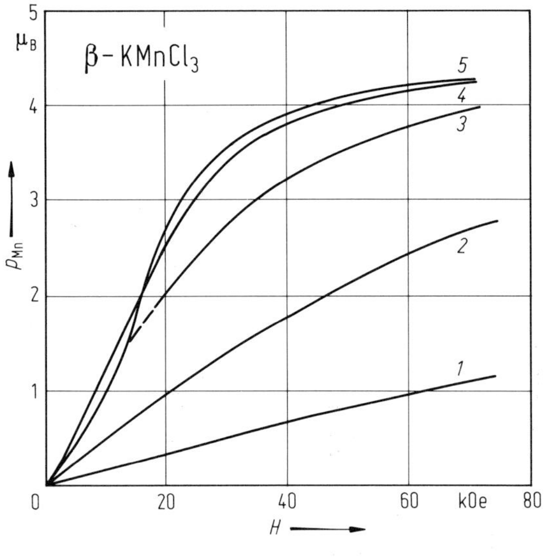

Fig. 296. β-$KMnCl_3$. Magnetic field dependence of p_{Mn} for (curve *1*) $T = 41.1$; (*2*) 13.5; (*3*) 4.2; (*4*) 1.85; (*5*) 1.6 K [83V3, 84F1].

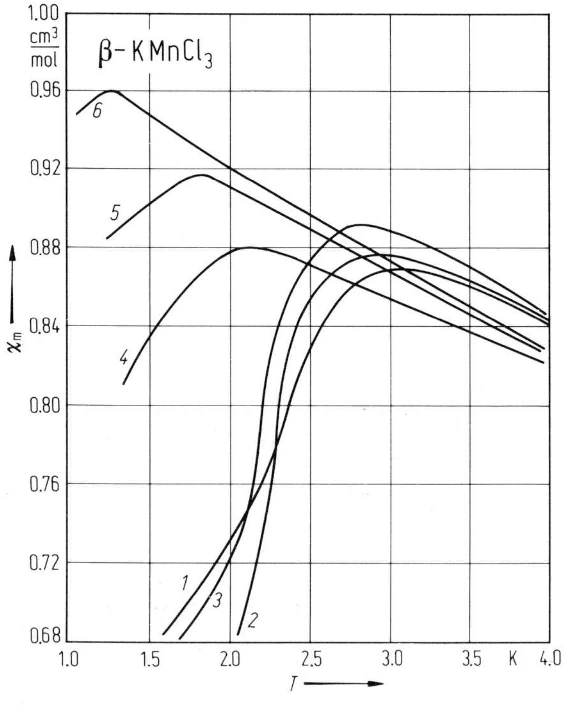

Fig. 297. β-$KMnCl_3$. Temperature dependence of χ_m in the magnetic phase transition range, at (curve *1*) $H = 1 \cdot 10^{-2}$; (*2*) 0.2; (*3*) 3.3; (*4*) 10; (*5*) 16.6; (*6*) 20 kOe [83V3, 84F1].

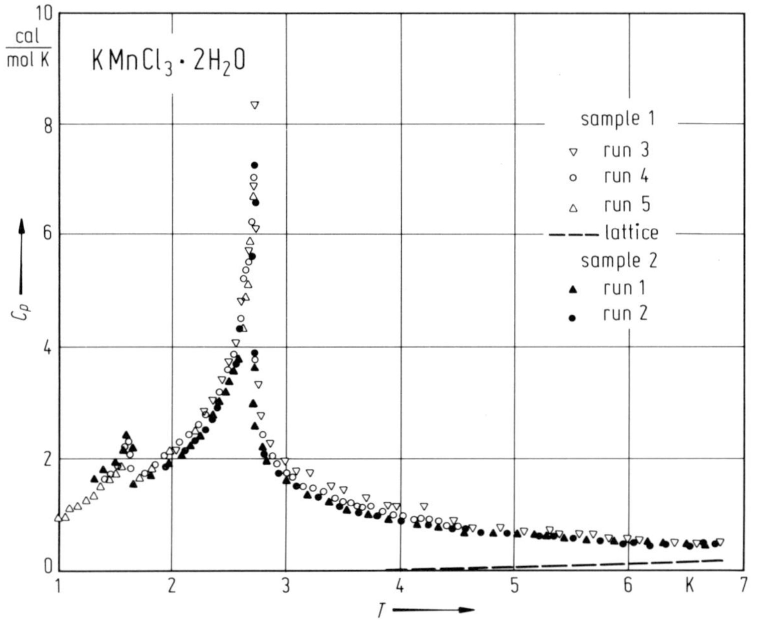

Fig. 298. $KMnCl_3 \cdot 2H_2O$. Temperature dependence of heat capacity C_p, for different runs and samples. The dashed line is the lattice contribution to C_p. The second peak at 1.6 K is due to the $MnCl_2 \cdot 4H_2O$ impurity [68F3].

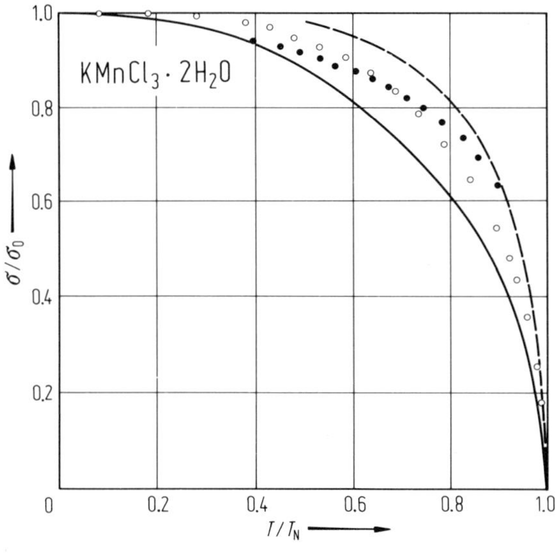

Fig. 299. $KMnCl_3 \cdot 2H_2O$. Relative temperature dependence of sublattice magnetization obtained from C_p (open circles), NMR (full circles) and calculated from molecular field approximation (full line) and three dimensional bcc Ising lattice approximation (dashed line) [68F3].

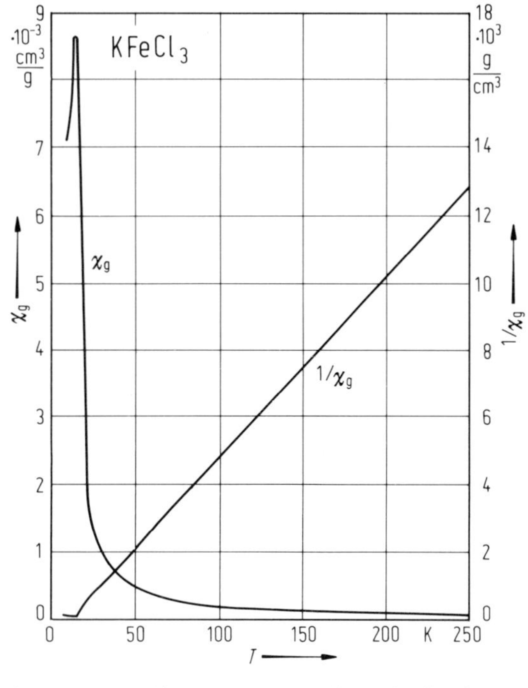

Fig. 300. $KFeCl_3$. Temperature dependence of χ_g and χ_g^{-1} at 1 kOe [74G1].

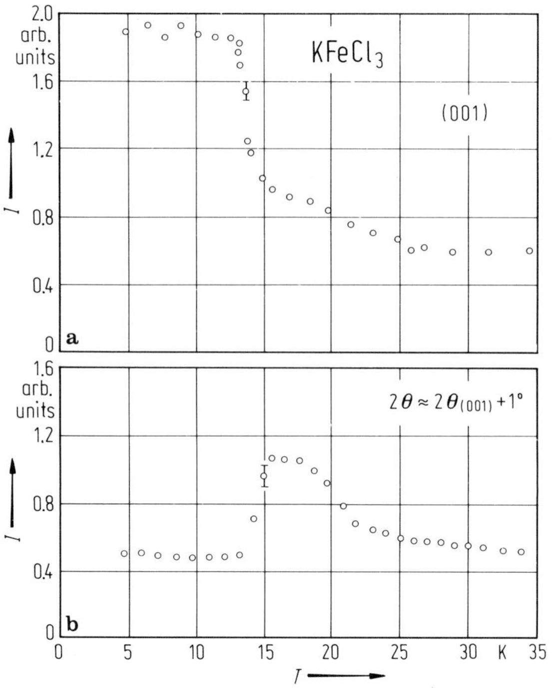

Fig. 301. $KFeCl_3$. Temperature dependence of neutron magnetic reflection (a) at (0, 0, 1); (b) at a ridge taken at $2\theta \approx 2\theta_{(001)} + 1°$ [74G1].

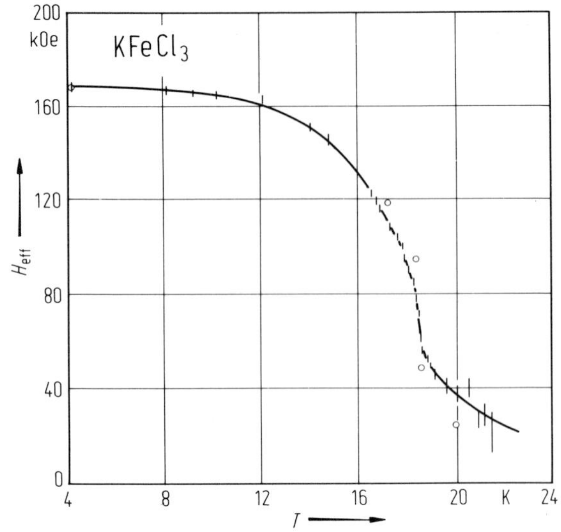

Fig. 302. $KFeCl_3$. Temperature dependence of H_{eff} from Mössbauer experiment. Open circles represent exchange field values normalized with respect to the effective field value at 4.2 K [74P2].

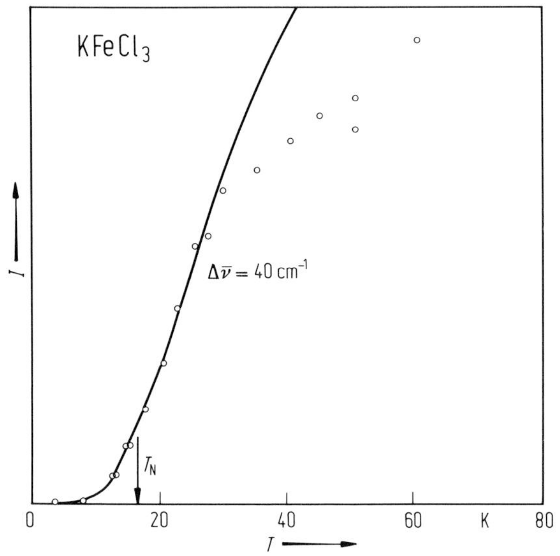

Fig. 303. $KFeCl_3$. Temperature dependence of the 23490 cm^{-1} (hot) band intensity from optical absorption. Circles are experimental and full curve is calculated for two level system with energy separation of 40 cm^{-1} [77K2].

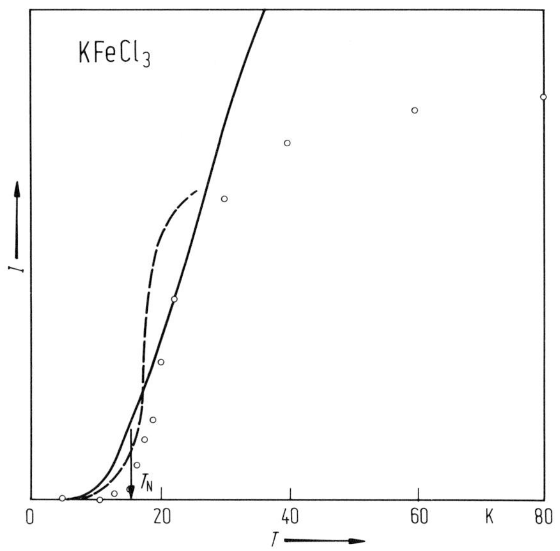

Fig. 304. $KFeCl_3$. Temperature dependence of the 19400 cm^{-1} hot band intensity. Full line calculated for two-level model system with energy separation of 40 cm^{-1}. Dashed line is reduced sublattice magnetization from [75P1] [77K2].

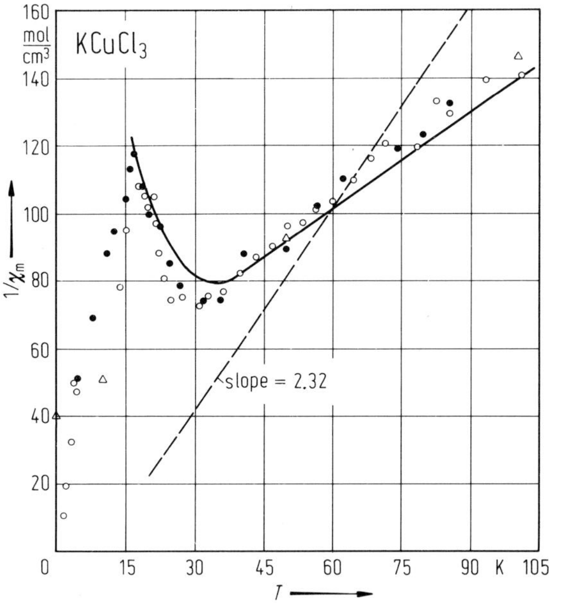

Fig. 305. $KCuCl_3$. Temperature dependence of χ_m^{-1}, open circles: cooling data, full circles: heating data, triangles: least-squares points. Solid line is theoretical curve based on $\Delta E/k = 55$ K. ΔE: energy difference between the singlet and triplet states, χ_m is taken per mol of $K_2Cu_2Cl_6$ [67M1].

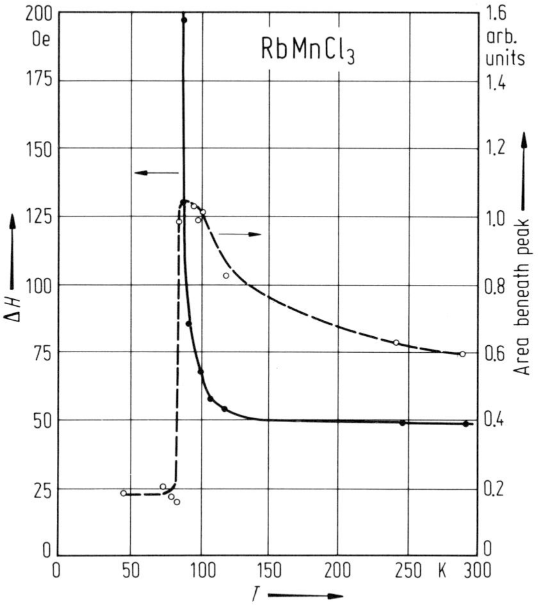

Fig. 306. $RbMnCl_3$. Temperature dependence of EPR peak half-width ΔH [79P1].

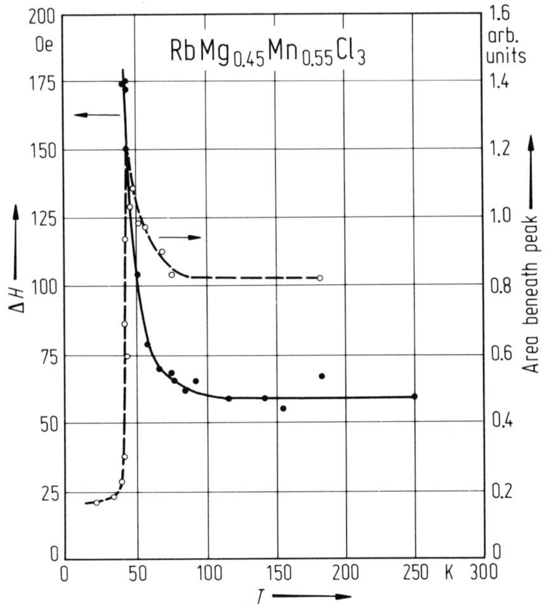

Fig. 307. $RbMg_{0.45}Mn_{0.55}Cl_3$. Temperature dependence of EPR peak half-width ΔH [79P1].

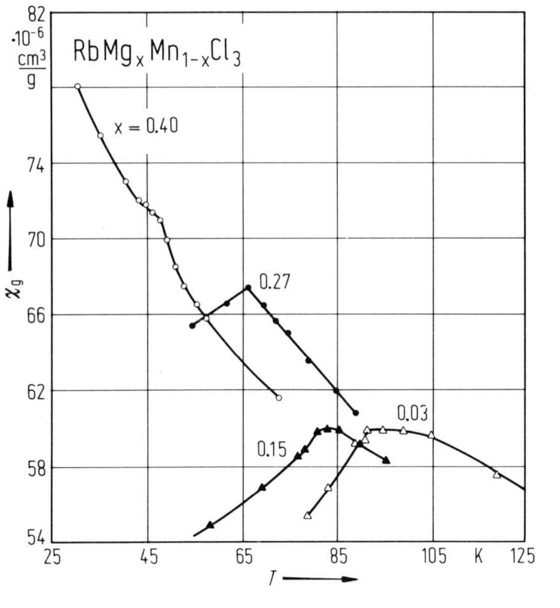

Fig. 308. $RbMg_xMn_{1-x}Cl_3$. Temperature dependence of χ_g for various x [79P1].

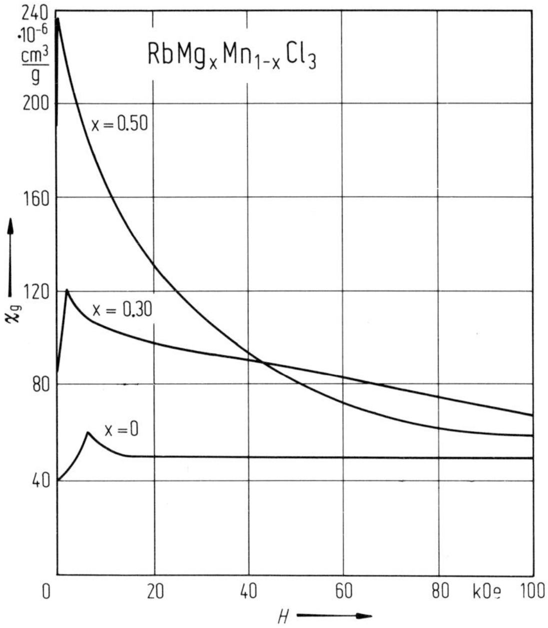

Fig. 309. $RbMg_xMn_{1-x}Cl_3$. Magnetic field dependence of χ_g at 4.2 K for different x [79P1].

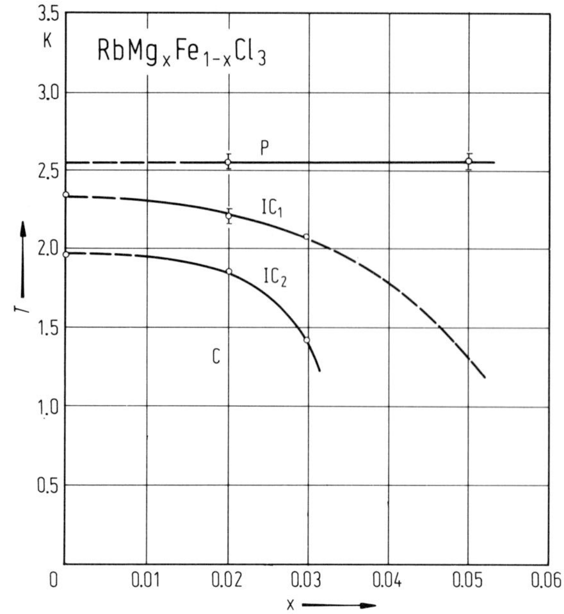

Fig. 311. $RbMg_xFe_{1-x}Cl_3$. Temperature dependence of the periodicity of magnetic correlations for x = 0, 0.02, 0.03 and 0.05. Symbols IC_1, IC_2 and C refer to incommensurate and commensurate periodicities. Two lower phase boundaries are not seen down to 1.3 K for x = 0.05 sample [90H1].

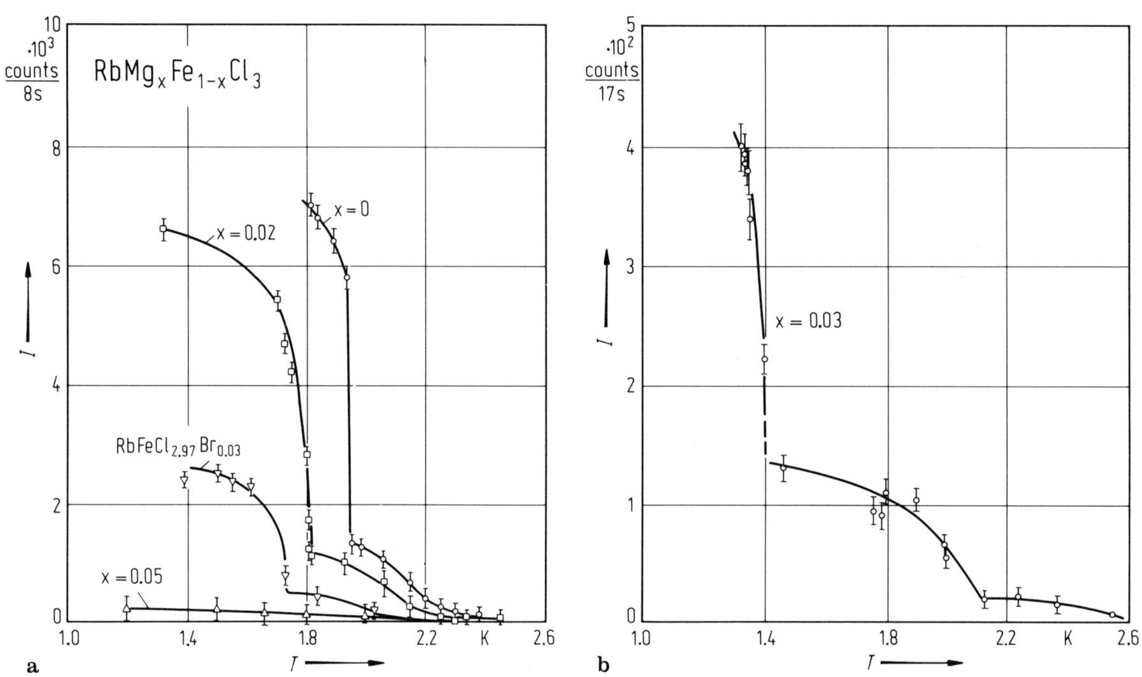

Fig. 310. $RbMg_xFe_{1-x}Cl_3$. Temperature dependence of neutron intensity measured at the K point $[\boldsymbol{Q} = (\frac{1}{3}, \frac{1}{3}, 0)_N]$ for (a) x = 0 (circles), x = 0.02 (squares), 0.05 (triangles) and (b) x = 0.03. The data for x = 0.02 and 0.05 are scaled against each other using the nuclear (100) and (220) reflections intensities. For comparison similar results on $RbFeCl_{3-x}Br_x$ (x = 0.03) is included in (a) [90H1].

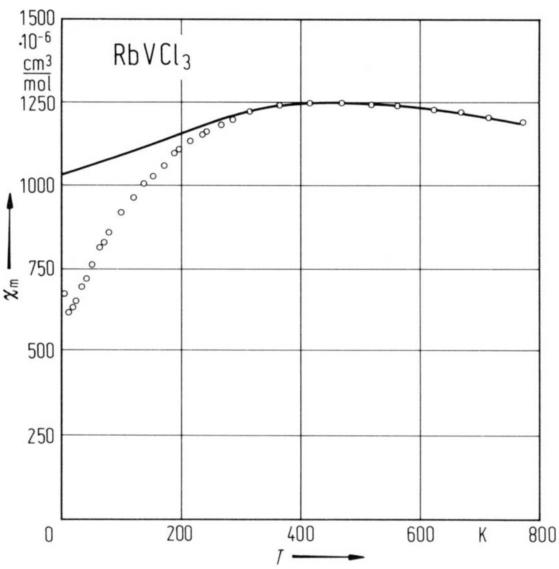

Fig. 312. $RbVCl_3$. Temperature dependence of χ_m reduced to spin contribution. Solid line is a theoretical fit according to Smith and Friedberg [68S1] for $J/k = -123$ K [77N2].

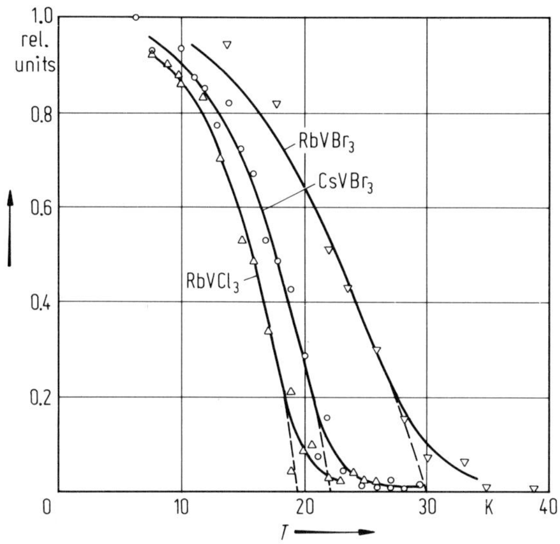

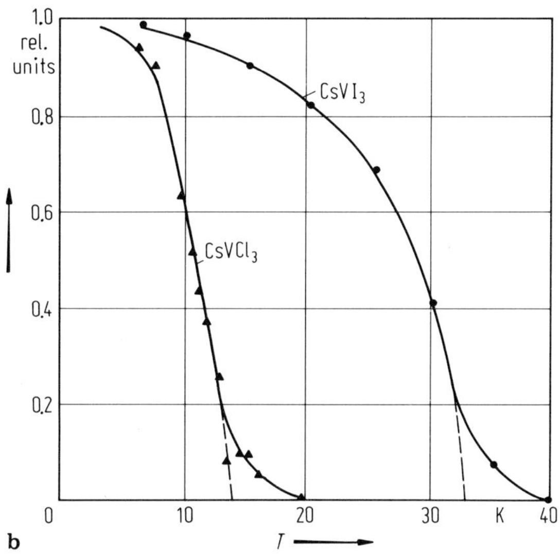

b

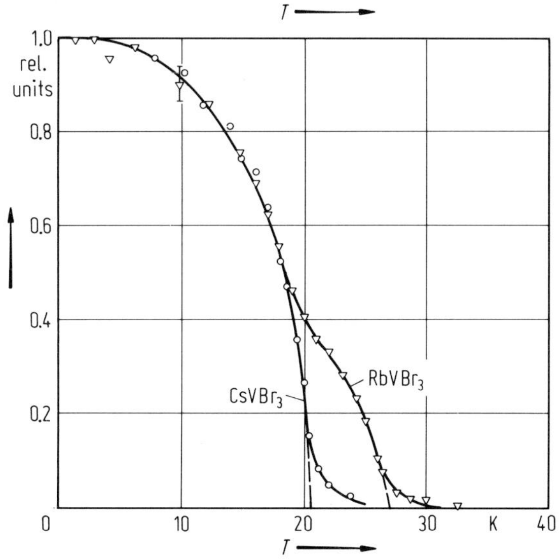

Fig. 313. $RbVCl_3$, $RbVBr_3$, $CsVCl_3$, $CsVBr_3$, $CsVI_3$. Temperature dependence of neutron magnetic reflection $(\frac{1}{3},\frac{1}{3},1)$. (a) Powder data; (b), (c) single crystal data [83H1, 85H1].

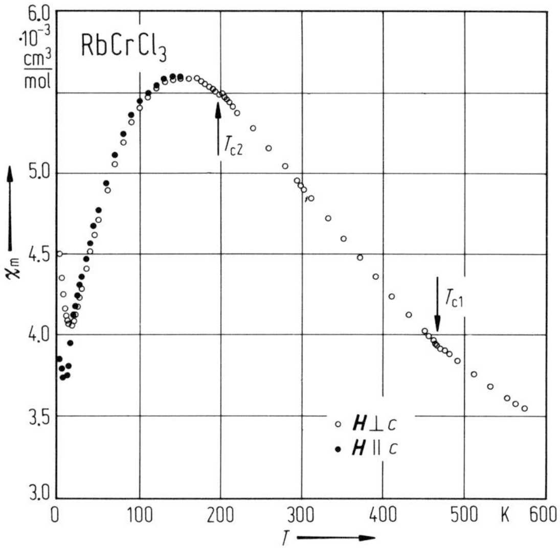

Fig. 314. $RbCrCl_3$. Temperature dependence of χ_m at 11.9 kOe after the correction of core diamagnetism ($-1.12 \cdot 10^{-4}$ cm^3/mol) [84T1].

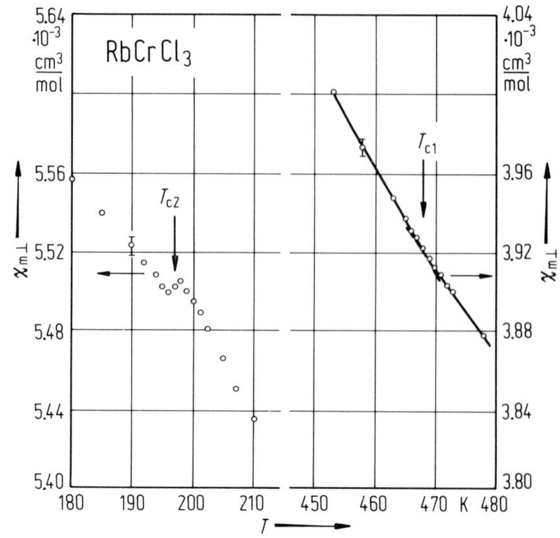

Fig. 315. $RbCrCl_3$. Temperature dependence of χ_m for $\boldsymbol{H}$ perpendicular to the c axis around the crystallographic phase transition temperatures T_{c1} and T_{c2} [84T1].

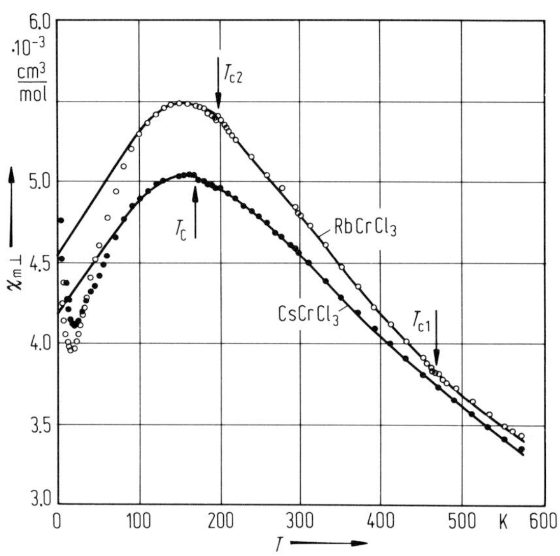

Fig. 316. $RbCrCl_3$, $CsCrCl_3$. Temperature dependence of $\chi_{m\perp}$, with $\boldsymbol{H} \perp c$, after the correction of the Van Vleck paramagnetism. Solid lines calculated [84T1].

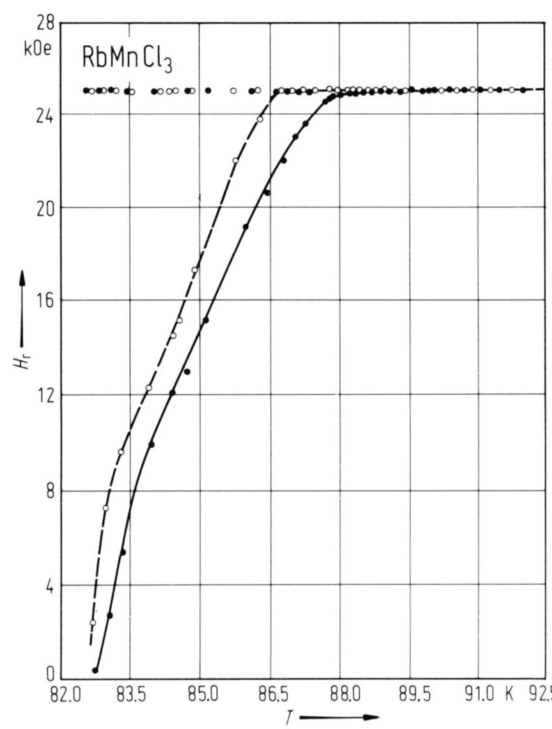

Fig. 317. $RbMnCl_3$. Temperature dependence of the field for resonance at 69.8 GHz. --o--: increasing field sweep, —•—: decreasing field sweep [66S2].

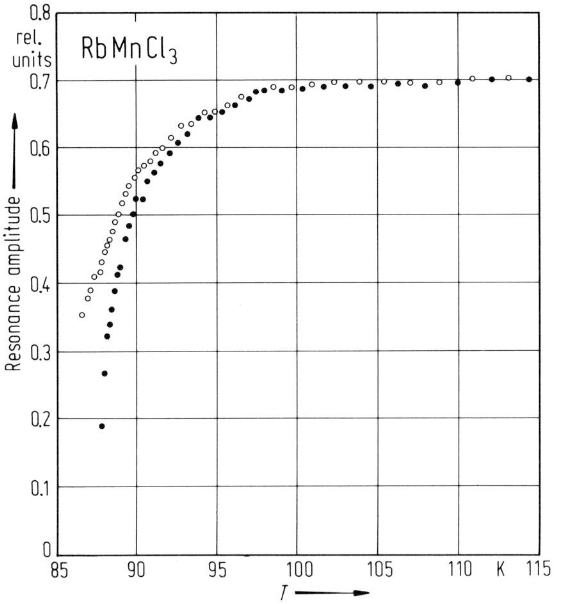

Fig. 318. $RbMnCl_3$. Temperature dependence of the resonance amplitude above the ordering temperature at 69.8 GHz. Open circles: for increasing, and full circles: decreasing field sweep, respectively [66S2].

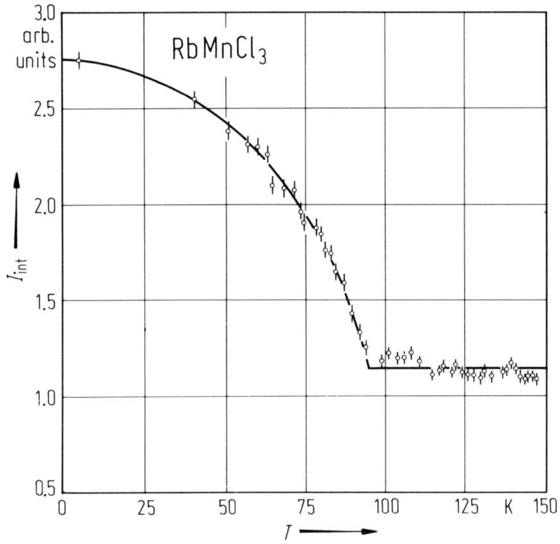

Fig. 319. $RbMnCl_3$. Temperature dependence of integrated intensity of the (1,0,0) + (0,0,3) + (1,0,1) neutron magnetic line [71M2].

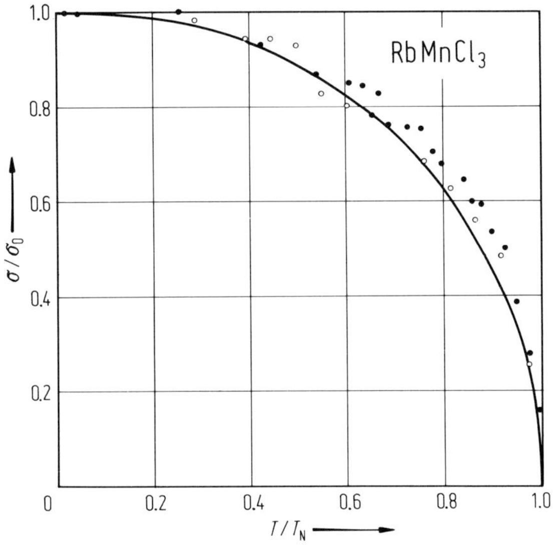

Fig. 320. $RbMnCl_3$. Reduced temperature dependence of reduced sublattice magnetization derived from the magnetic lines (1,0,0) + (0,0,3) + (1,0,1) (full circles) and (1,0,2) (open circles) compared with the result of molecular field model for Mn^{2+} ($S = 5/2$) (solid line) [71M2].

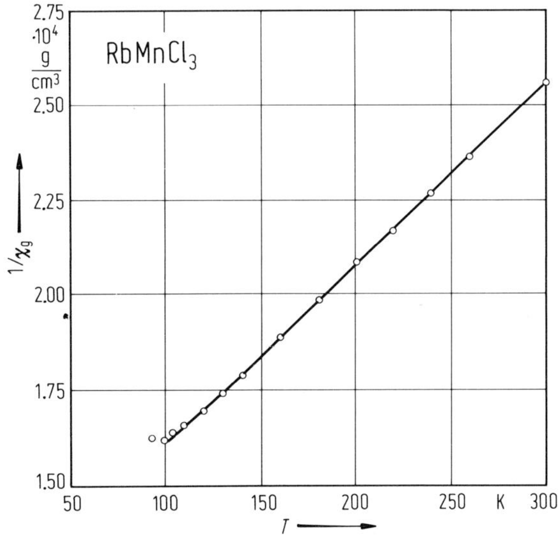

Fig. 321. $RbMnCl_3$. Temperature dependence of χ_g^{-1} [75F3].

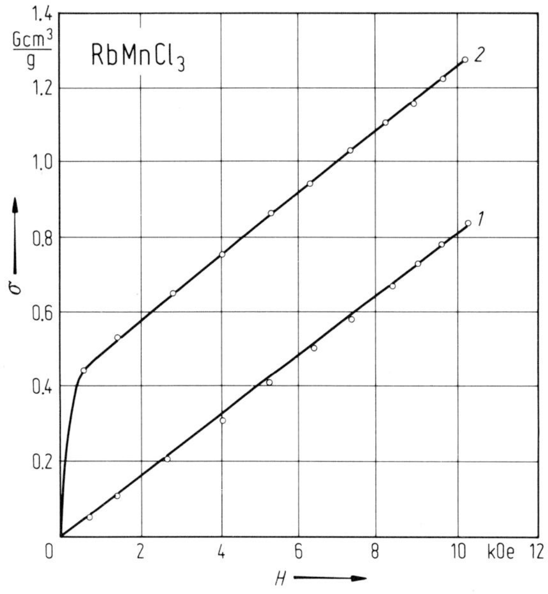

Fig. 322. $RbMnCl_3$. Magnetic field dependence of σ at 78 K for (curve *1*) $\boldsymbol{H} \perp [001]$, (*2*) $\boldsymbol{H} \parallel [001]$ [75F3].

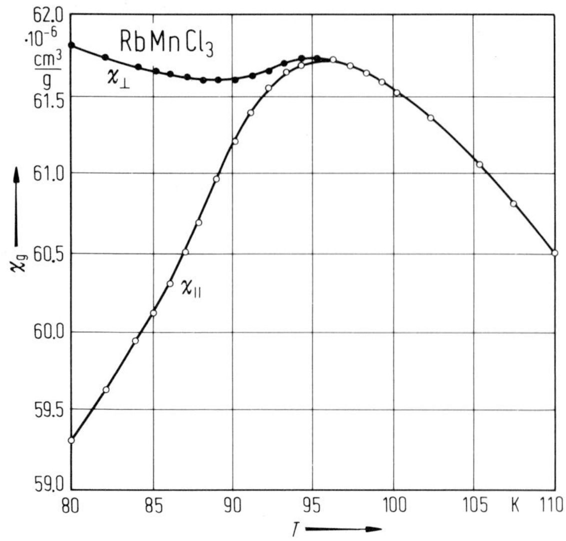

Fig. 323. $RbMnCl_3$. Temperature dependence of χ_g for $\boldsymbol{H} \parallel [001]$ and $\boldsymbol{H} \perp [001]$ [75F3].

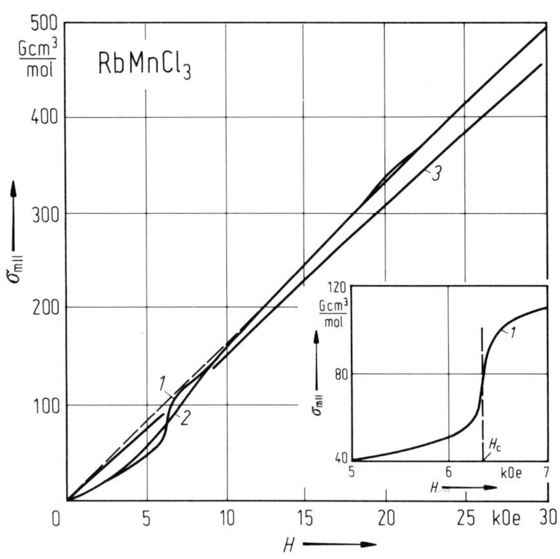

Fig. 324. $RbMnCl_3$. Magnetic field dependence of $\sigma_{m\parallel}$ for a single crystal. Curve *1*: $\boldsymbol{H}$ along one of three C_2; *2*: $\boldsymbol{H}$ oriented at 30° to C_2; *3*: $\boldsymbol{H} \parallel c$ [78B5].

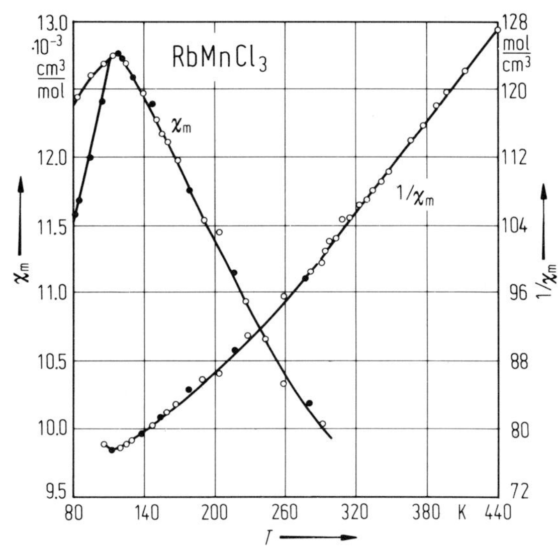

Fig. 325. $RbMnCl_3$. Temperature dependence of χ_m and χ_m^{-1} of the cubic phase. The full and open circles indicate the experimental points for two directions of the crystal [79F1].

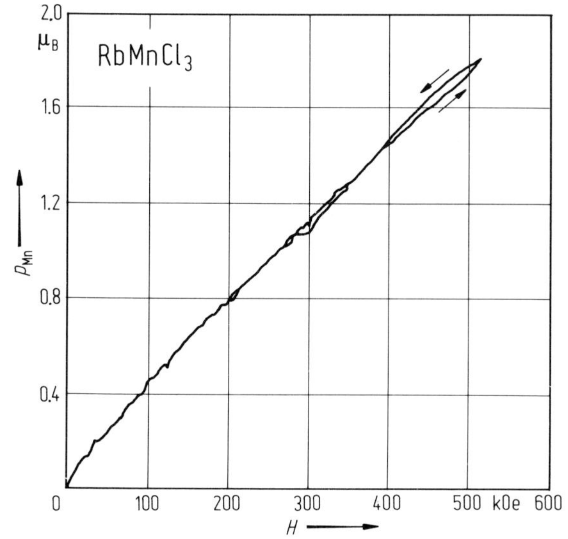

Fig. 326. $RbMnCl_3$. Magnetic field dependence of magnetic moment p_{Mn} at 4.2 K, for $\boldsymbol{H} \parallel c$ up to 520 kOe [81M1].

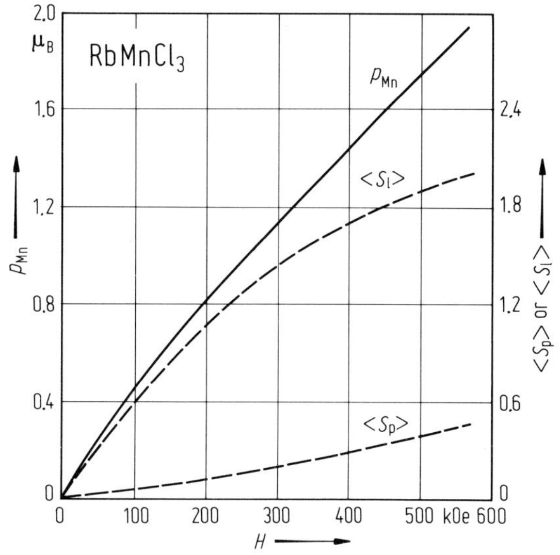

Fig. 327. $RbMnCl_3$. Field induced spin moment p_{Mn} calculated as a function of external field for $J_2/J_0 = 0.1$, $J_1/k = -21$ K and $g = 2$ (solid curve). J_1 and J_2 are strong and weak exchange coupling constants of the system. Dashed lines are calculated thermal averages of pair-spin and lone-spin, respectively [81M1].

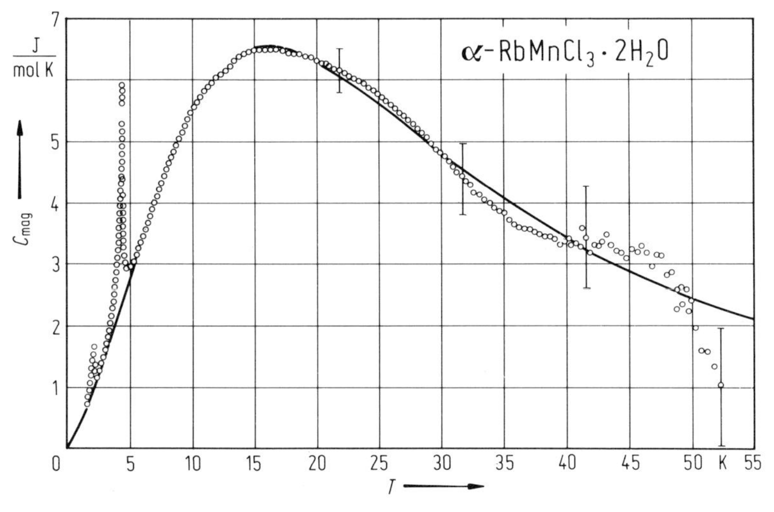

Fig. 328. α-$RbMnCl_3 \cdot 2H_2O$. Temperature dependence of magnetic heat capacity. The anomaly at 2.19 K is due to a small fraction of β-$RbMnCl_3 \cdot 2H_2O$. For explanation see Fig. 434 [77K3].

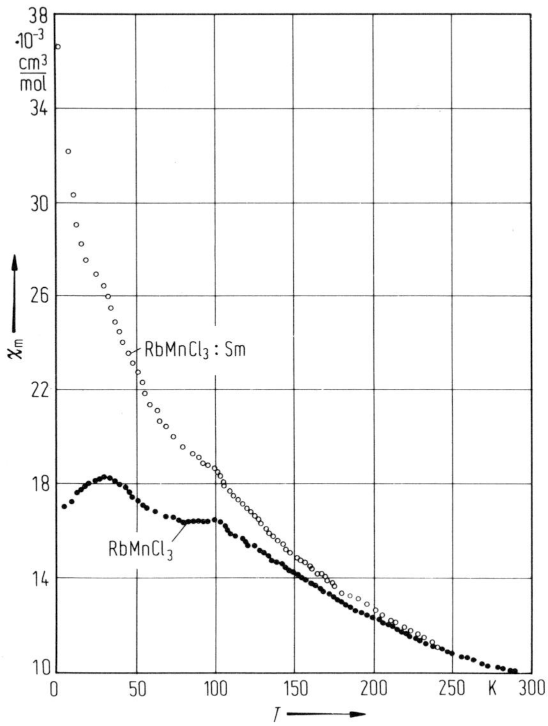

Fig. 329. $RbMnCl_3$, $RbMnCl_3$: Sm. Temperature dependence of χ_m [88P1].

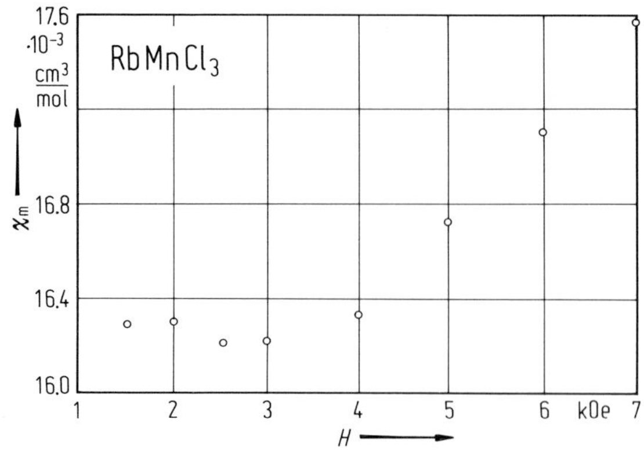

Fig. 330. $RbMnCl_3$. Magnetic field dependence of χ_m at 4.2 K [88P1].

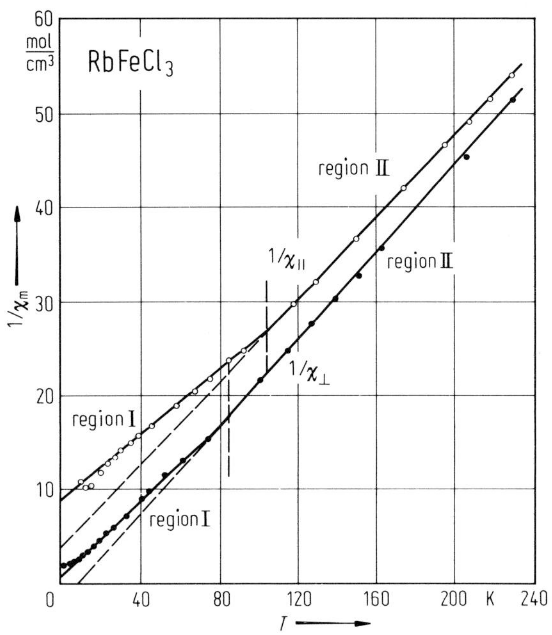

Fig. 331. $RbFeCl_3$. Temperature dependence of χ_m^{-1} for a single crystal, having different Curie constants in the two temperature regions. $\chi_\parallel$ is taken along [0001] and $\chi_\perp$ along [11$\bar{2}$0] [69A1].

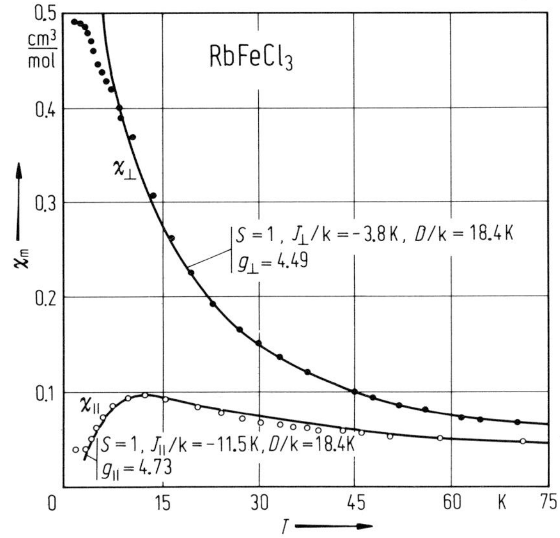

Fig. 332. $RbFeCl_3$., Temperature dependence of single crystal χ_m parallel and perpendicular to the c axis. The solid curves indicate the tabulated values from the equations

$$\chi_\parallel = \frac{2Ng_\parallel^2 \mu_B^2}{kT[\exp(D/kT) + 2] - 8J};$$

$$\chi_\perp = \frac{2Ng_\perp^2 \mu_B^2}{kT[\exp(-D/2kT) + 2] - 8J}$$ [69A1]. See also [73M1].

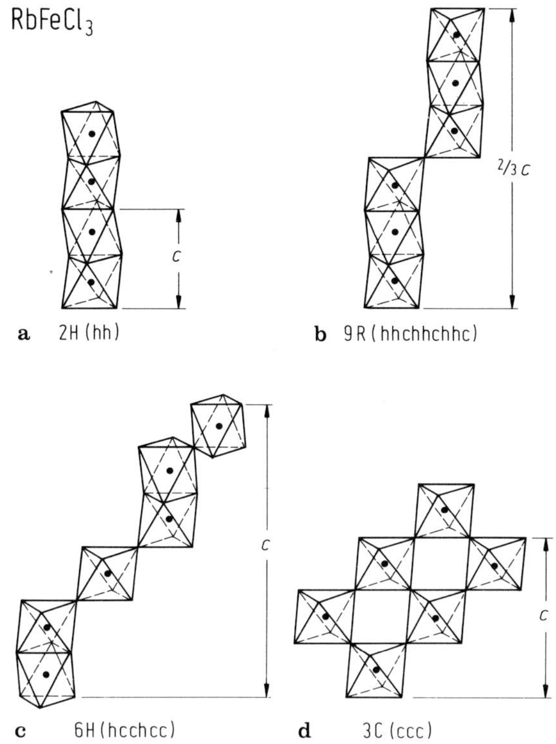

Fig. 333. $RbFeCl_3$. Octahedra linkage of four polytypes suggested [71L1].

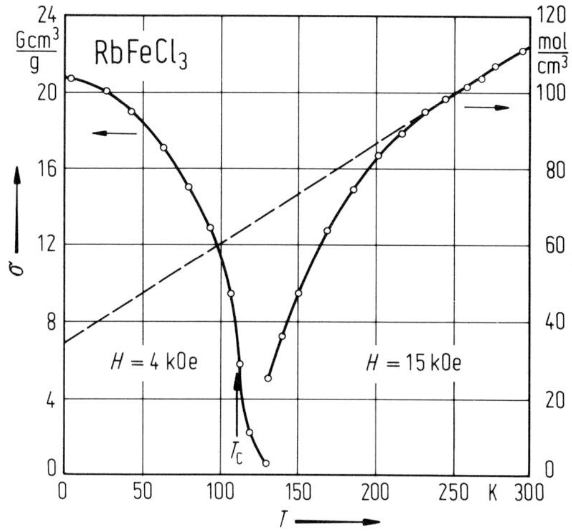

Fig. 334. $RbFeCl_3$. Temperature dependence of σ and χ_m^{-1} for polycrystalline sample, 6H form suggested (see Fig. 333) [71L1]. $\Theta = -134$ K, $C_m = 3.89$ cm^3 K/mol.

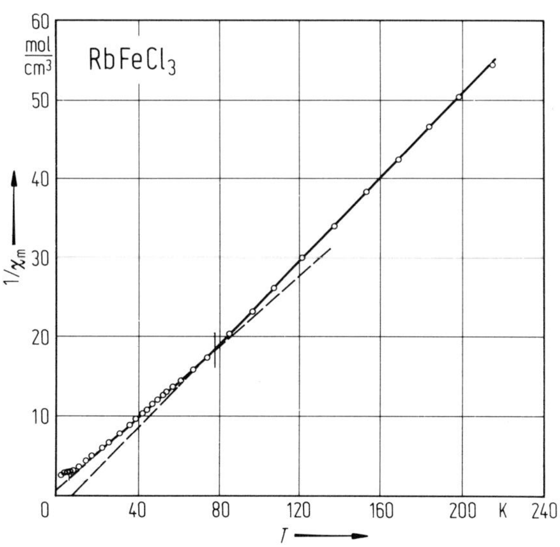

Fig. 335. $RbFeCl_3$. Temperature dependence of χ_m^{-1} for polycrystal [73W1].

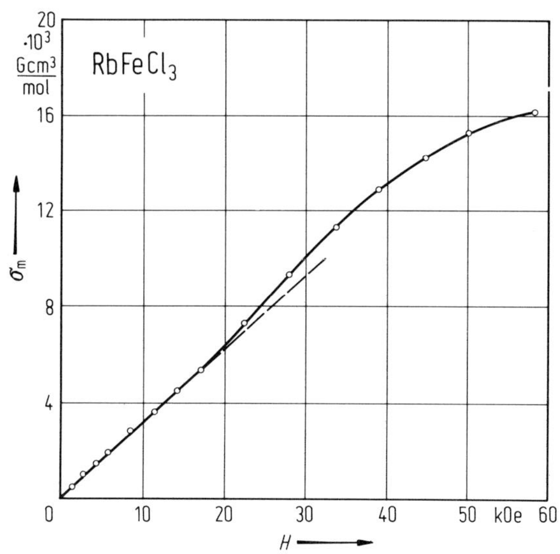

Fig. 336. $RbFeCl_3$. Magnetic field dependence of σ_m at 2.3 K [73W1].

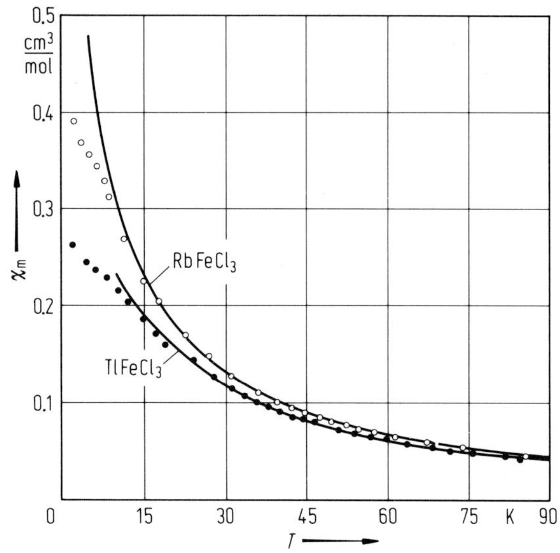

Fig. 337. $RbFeCl_3$, $TlFeCl_3$. Temperature dependence of χ_m. Solid curves: theoretical fit [73W1].

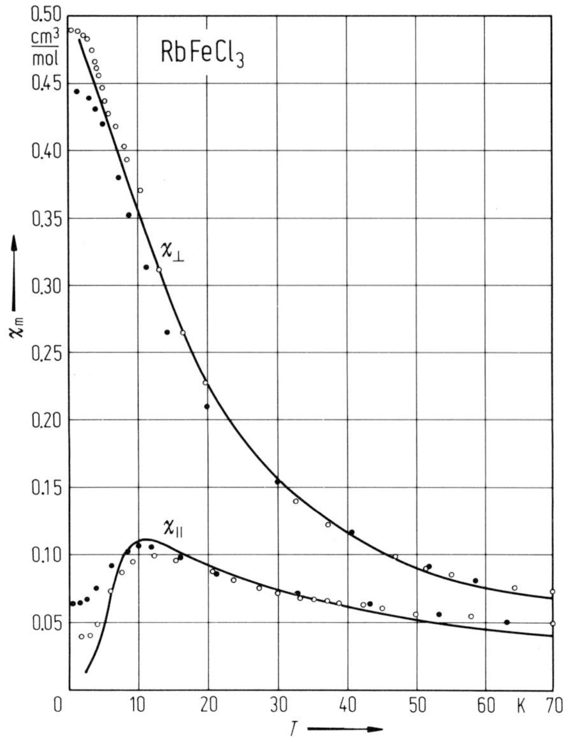

Fig. 339. $RbFeCl_3$. Temperature dependence of χ_m. Full line the best fit with CEF values for the isolated-chain susceptibility for $\Delta'/|\lambda| = 1.3$, $J/|\lambda| = 0.015$, $2z'J'/|\lambda| = -0.038$; where Δ': trigonal distortion, λ: spin-orbit coupling parameter, J: intrachain isotropic Heisenberg exchange between real spins, J': interchain Heisenberg exchange, z': number of interchain nearest-neighbor spins [75E1]. Full circles [75E1], open circles [69A1].

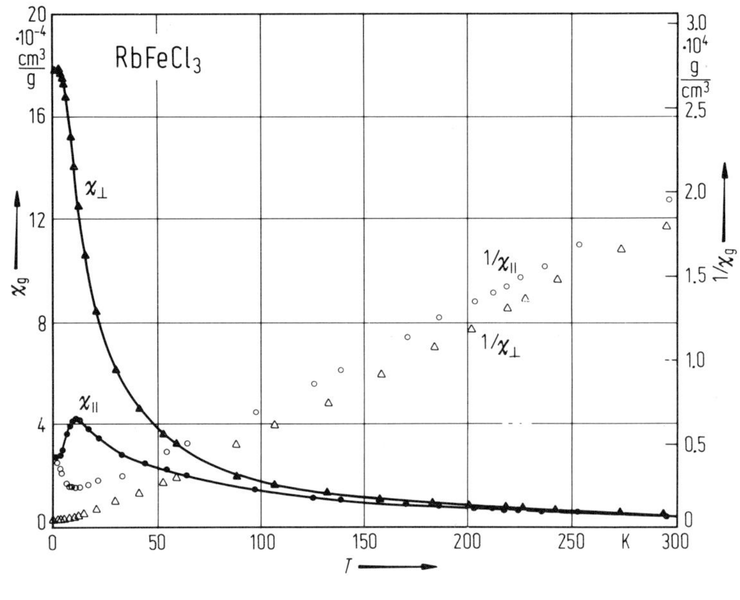

Fig. 338. $RbFeCl_3$. Temperature dependence of χ_g and χ_g^{-1} at $H = 15.3$ kOe for single crystals, parallel and perpendicular to the c axis [75E1].

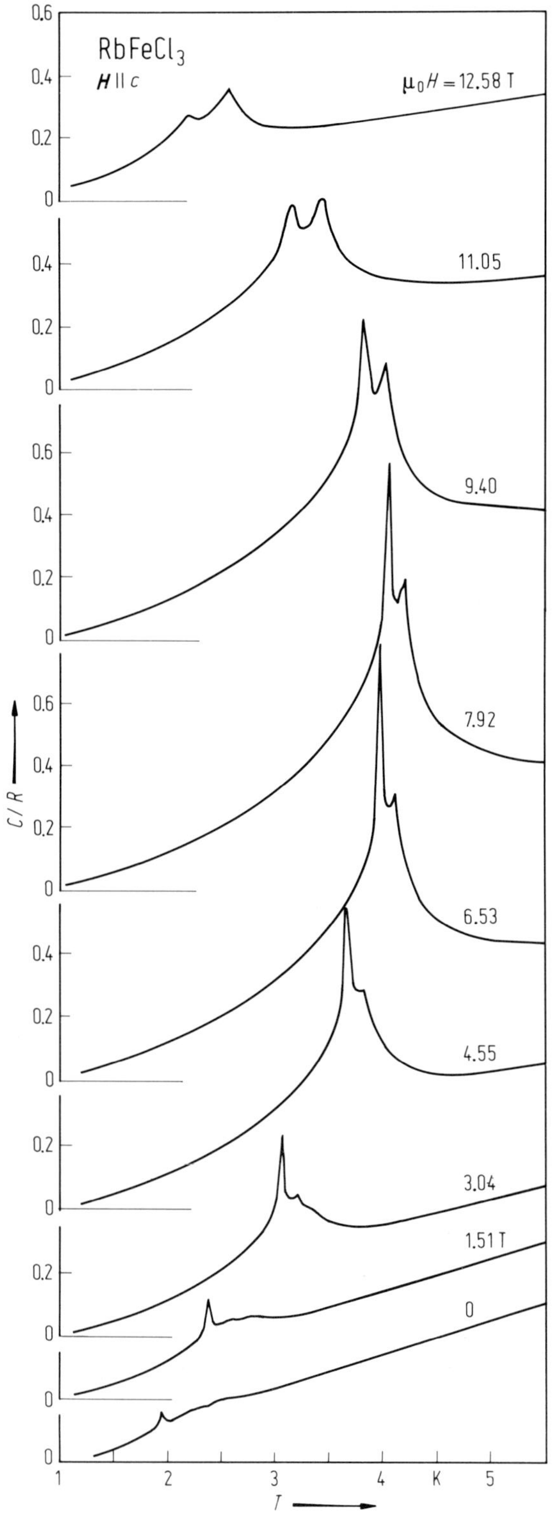

Fig. 340. $RbFeCl_3$. Heat capacity under several magnetic fields for ***H*** parallel to the *c* axis [81H1].

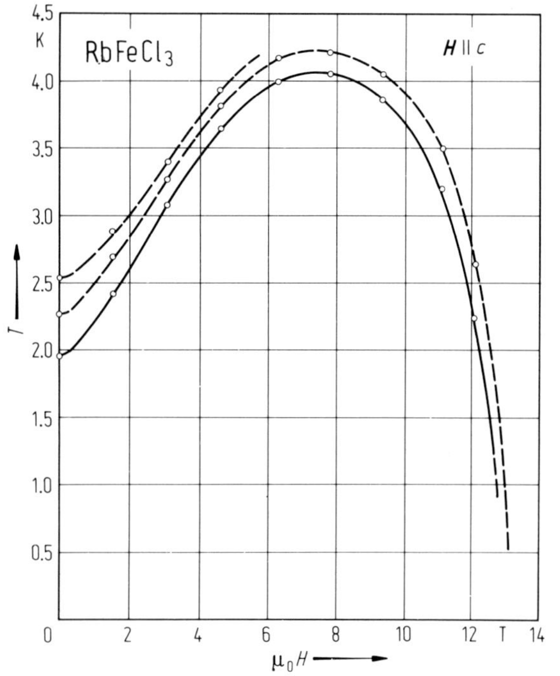

Fig. 341. $RbFeCl_3$. *T*–*H* phase boundary [81H1]. Curves correspond to maxima of heat capacity vs. *H* plots of Fig. 340.

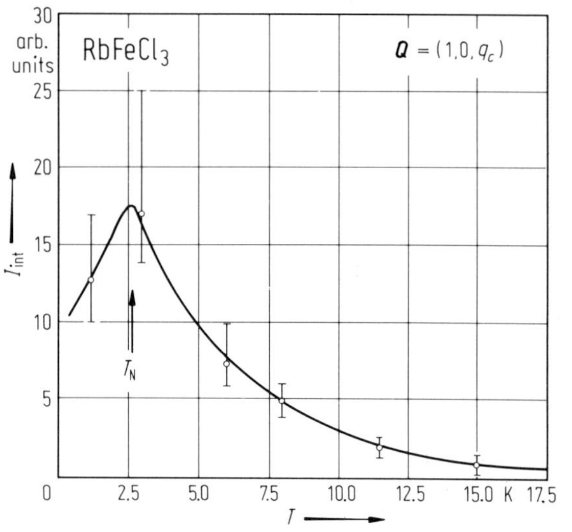

Fig. 342. $RbFeCl_3$. Temperature dependence of neutron integrated intensity for $q_c = 0.1$ [81Y1].

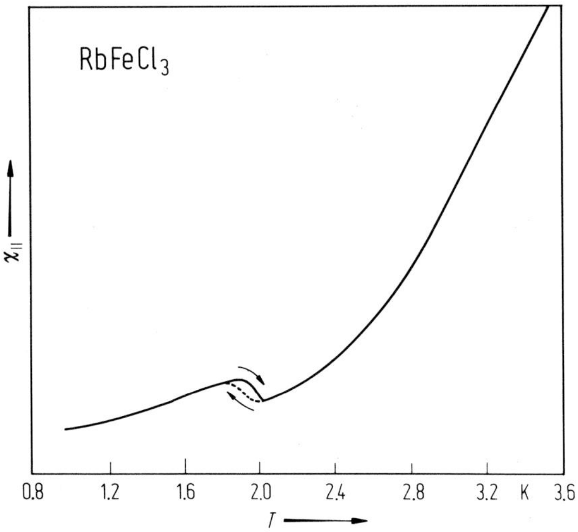

Fig. 343. $RbFeCl_3$. Temperature dependence of χ along the c axis ($H = 0$). Around 1.95 K a hysteresis is observed [82W1].

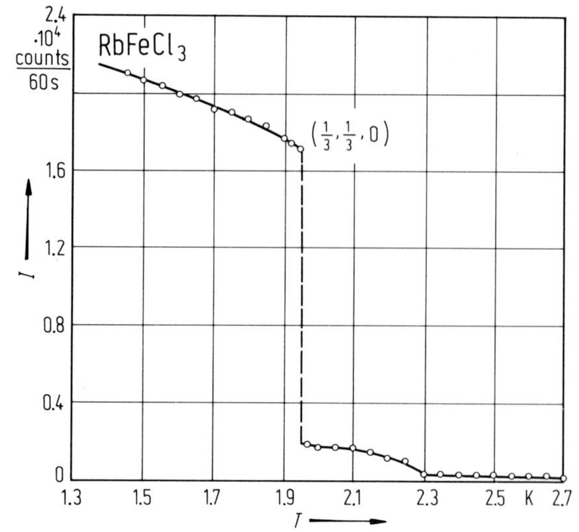

Fig. 344. $RbFeCl_3$. Temperature dependence of neutron scattering intensity at the $(\frac{1}{3},\frac{1}{3},0)$ reciprocal lattice point for a single crystal [82W1].

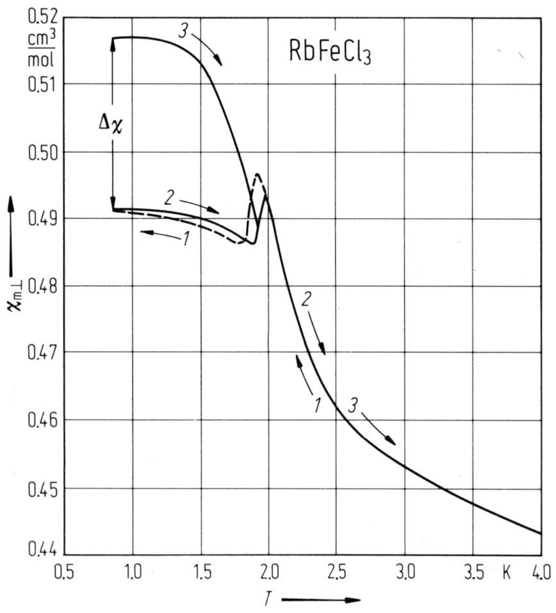

Fig. 345. $RbFeCl_3$. Temperature dependence of $\chi_{m\perp}$ perpendicular to the c axis at $H = 0$ changing the temperature under three conditions. Curve *1*: T decreased in $H = 0$; *2*: T increased after decreasing along the line *1*; *3*: T increased just after the external field is applied in the c plane at $T = 0.85$ K up to ≈ 1 kOe and demagnetized to zero field again. $\Delta\chi$ difference between the susceptibilities before and after the field applications [83W1].

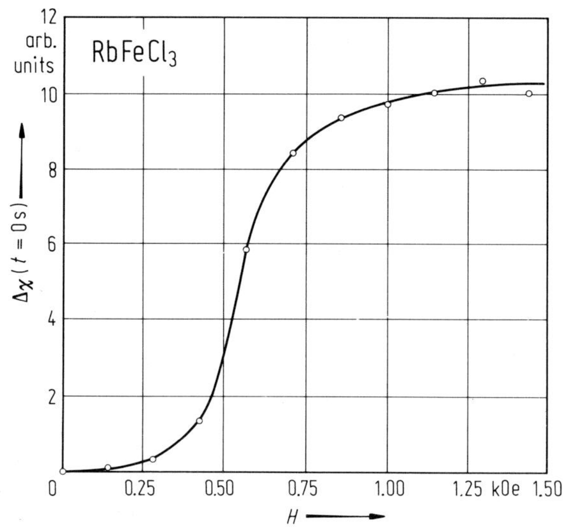

Fig. 346. $RbFeCl_3$. The dependence of $\Delta\chi$ just after the demagnetization ($t = 0$ s) at $T = 1.35$ K on the starting field H of the demagnetization [83W1].

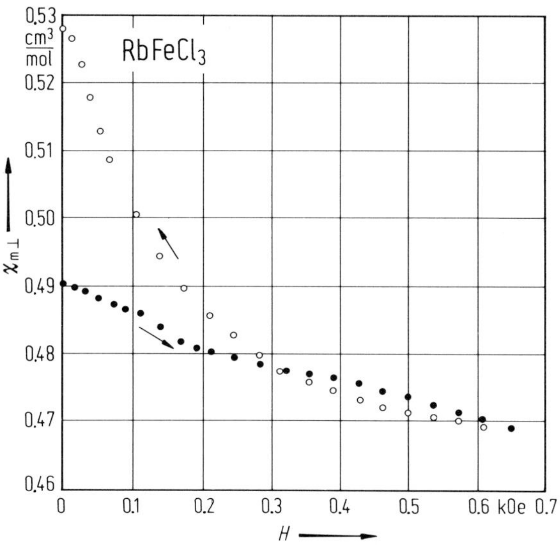

Fig. 347. $RbFeCl_3$. The change of the susceptibility ($\boldsymbol{H} \perp c$) at $T = 1.2$ K when the external field applied up to 0.65 kOe and demagnetized again. The arrows show the sweep directions [83W1].

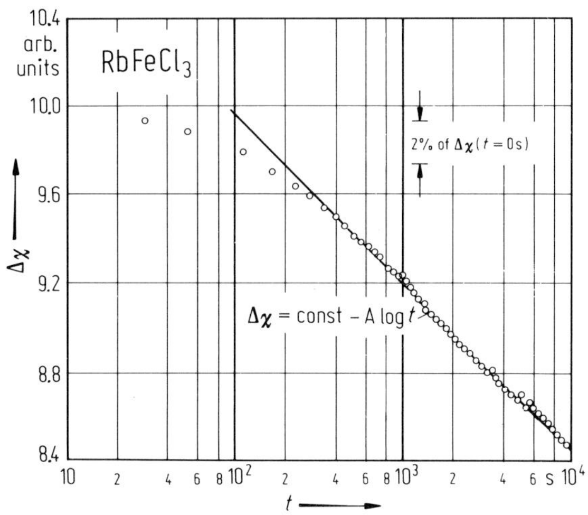

Fig. 348. $RbFeCl_3$. The dependence of $\Delta\chi$ on the time t after demagnetization ($T = 1.25$ K and $H_0 = 0.93$ kOe). After $t \approx 400$ s $\Delta\chi$ decays as $\Delta\chi(t) = \text{const} - A \log t$ [83W1]. H_0: starting field of demagnetization.

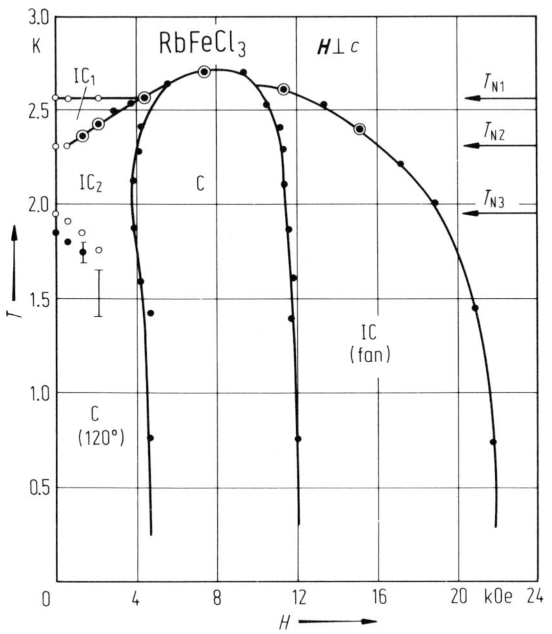

Fig. 349. $RbFeCl_3$. T–H phase diagram for $\boldsymbol{H} \perp c$ ($\boldsymbol{H} \parallel [110]$) determined from the anomalies in heat capacity and susceptibility at the points marked with open and closed circles respectively [83W1].

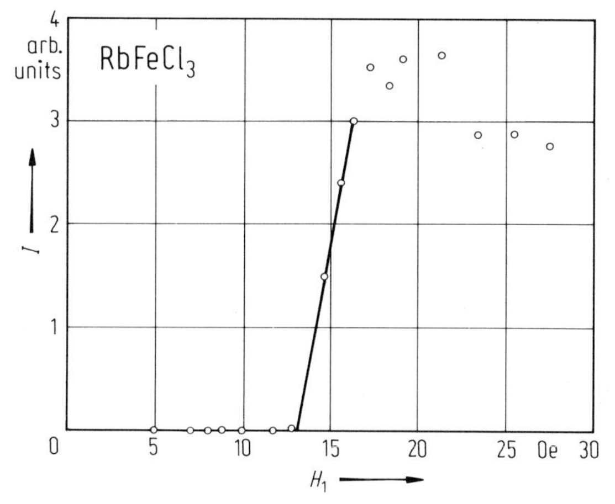

Fig. 350. $RbFeCl_3$. The rf-field H_1 dependence of the unusual NMR echo intensity of ^{35}Cl at 1.3 K (below T_{N3}). $\tau = 40$ µs (τ: double pulse duration). $w_1 = w_2 = 5$ µs ($w_{1,2}$: widths of the double pulses). rf-field $H_1 \geq 13$ Oe. Repetition rate 50 Hz, $H_0 = 6$ kOe (H_0: starting field) [83W2].

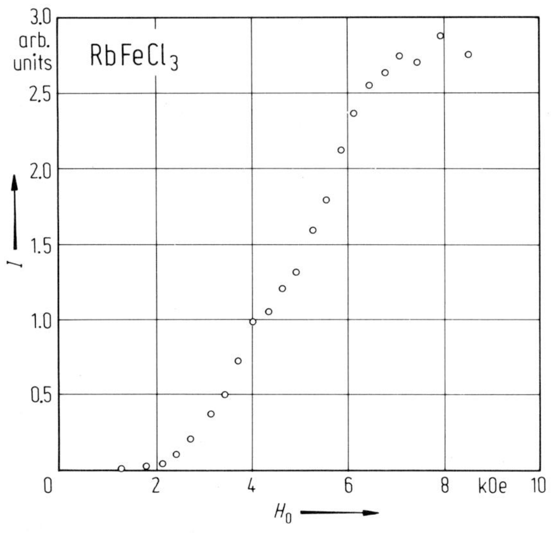

Fig. 351. $RbFeCl_3$. The echo intensity vs. the starting field H_0 of demagnetization at $T = 1.3$ K (below T_{N3}) [83W2]. $\tau = 40$ µs, $w_1 = w_2 = 5$ µs, repetition rate 50 Hz, $H_1 = 15$ Oe (see Fig. 350).

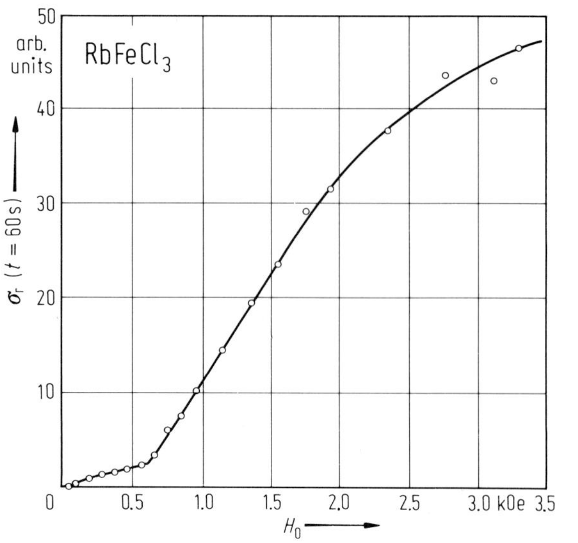

Fig. 353. $RbFeCl_3$. Magnetic field dependence of remanent magnetization σ_r at 1.2 K. p_A at 3.3 kOe and 1.2 K is $3 \cdot 10^{-4} \mu_B$ per magnetic ion [85K2]. H_0: starting field for demagnetization.

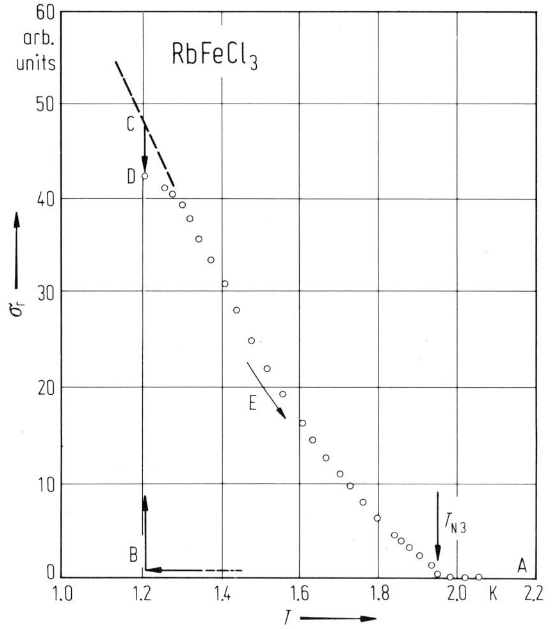

Fig. 352. $RbFeCl_3$. Temperature dependence of the remanent magnetization σ_r in the following procedures. (A → B): Zero field cooling from 4.2 to 1.2 K. (C): After H application up to $H_0 = 3.1$ kOe in the c plane at 1.2 K. (D): About half an hour after the demagnetization. (D → E → A): Zero field heating from 1.2 to 4.2 K [85K2].

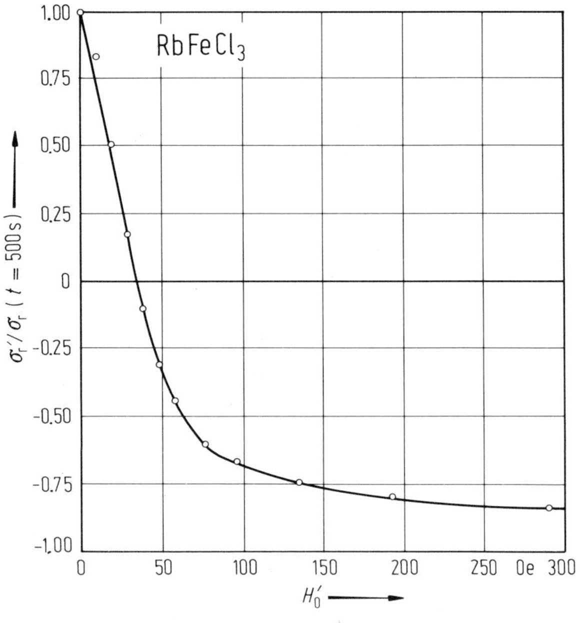

Fig. 354. $RbFeCl_3$. Magnetic field dependence of the remanent magnetization σ_r' at $T = 1.2$ K (below T_{N3}) after following procedure: H_0 was applied up to 3.4 kOe, and reduced to zero. After ~ 500 s H_0' was applied opposite to H_0 direction. The magnetization σ_r' was measured at zero field [85K2]. H_0: see Fig. 353.

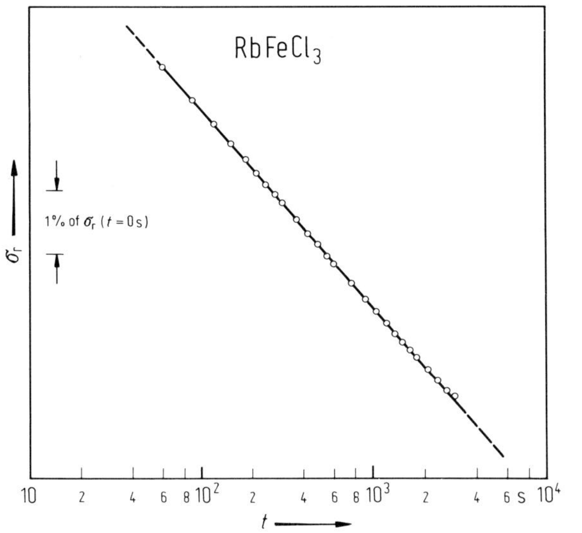

Fig. 355. $RbFeCl_3$. Remanent magnetization σ_r vs. time t, just after the demagnetization ($T = 1.2$ K, $H_0 = 3.3$ kOe), σ_r decreases logarithmically in time [85K2]. H_0: see Fig. 353.

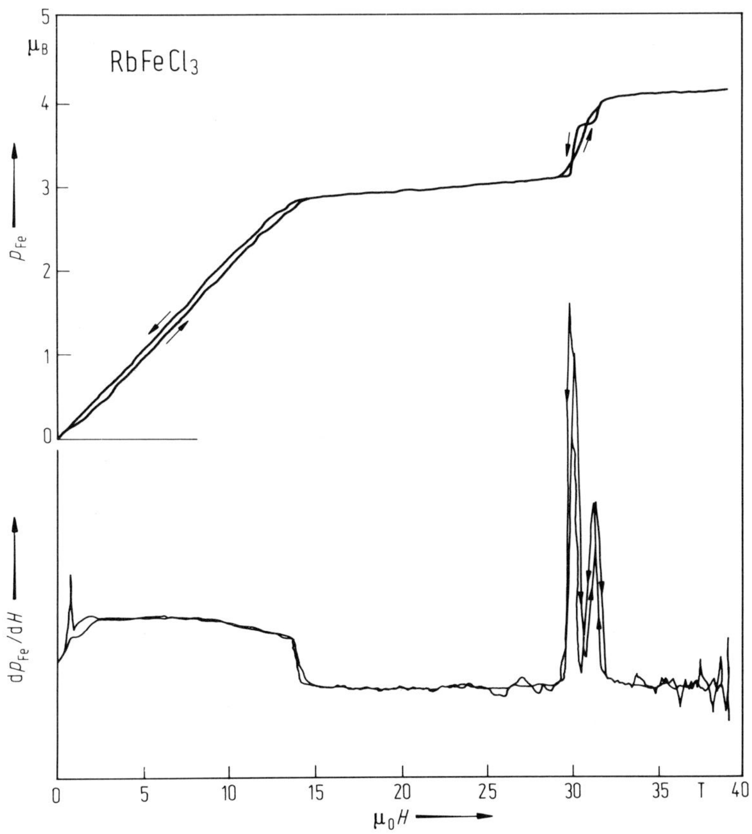

Fig. 356. $RbFeCl_3$. Magnetic field dependence of p_{Fe} and dp_{Fe}/dH for $\boldsymbol{H} \parallel c$ at 1.7 K. The hysteresis corresponding to the increasing and decreasing mode of the pulse field is shown by the arrows [88A2].

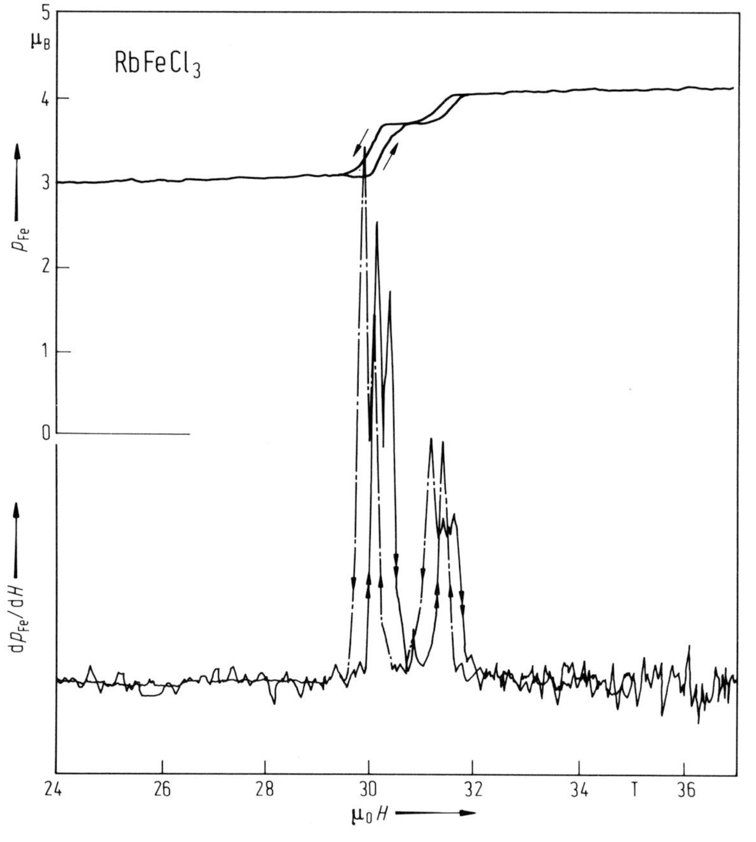

Fig. 357. $RbFeCl_3$. Magnetic moment p_{Fe} vs. H and dp_{Fe}/dH vs. H; $\boldsymbol{H} \parallel c$, $T = 1.7$ K. The horizontal H axis in Fig. 356 is expanded to increase resolution [88A2].

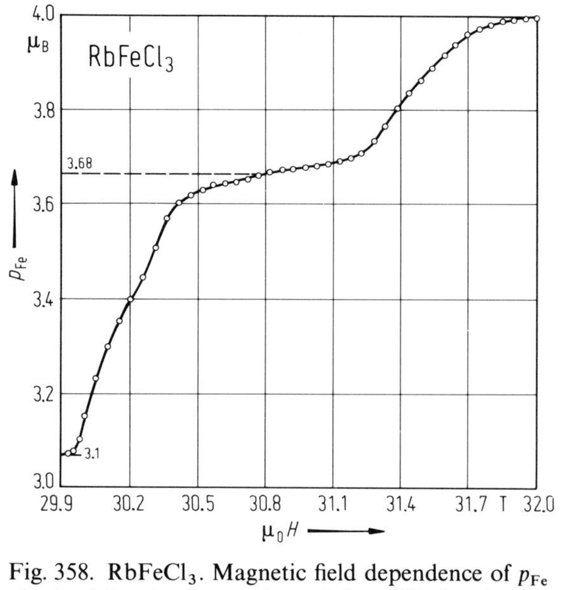

Fig. 358. $RbFeCl_3$. Magnetic field dependence of p_{Fe} obtained by the integration of dp_{Fe}/dH signal with respect to H [88A2].

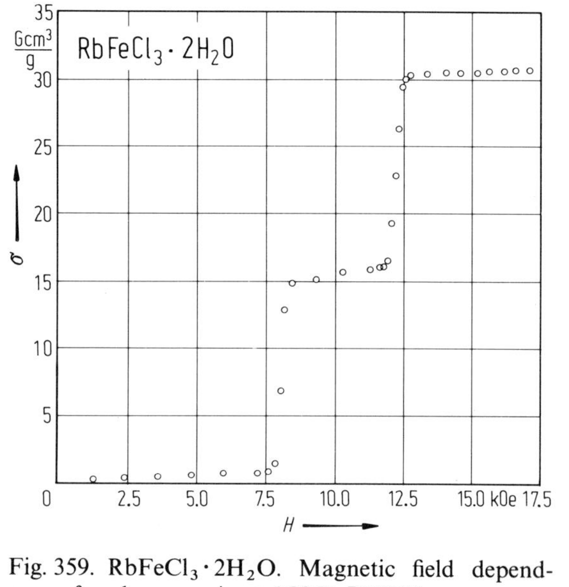

Fig. 359. $RbFeCl_3 \cdot 2H_2O$. Magnetic field dependence of σ along c axis at 4.22 K [77K1].

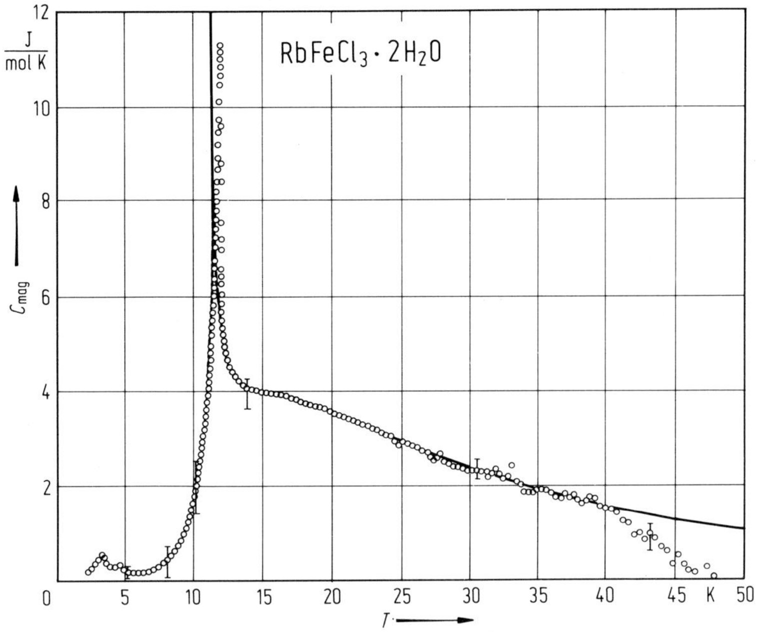

Fig. 360. $RbFeCl_3 \cdot 2H_2O$. Temperature dependence of magnetic heat capacity. Open circles are the experimental points corrected for lattice contribution. The solid curve is calculated for Ising $S = 1/2$ model with $|J_1/k| = 39$ K, $|J_2/k| = 0.7$ K. Anomalies at 3.3 and 4.6 K are due to impurities [77K1].

For Fig. 361, see next page.

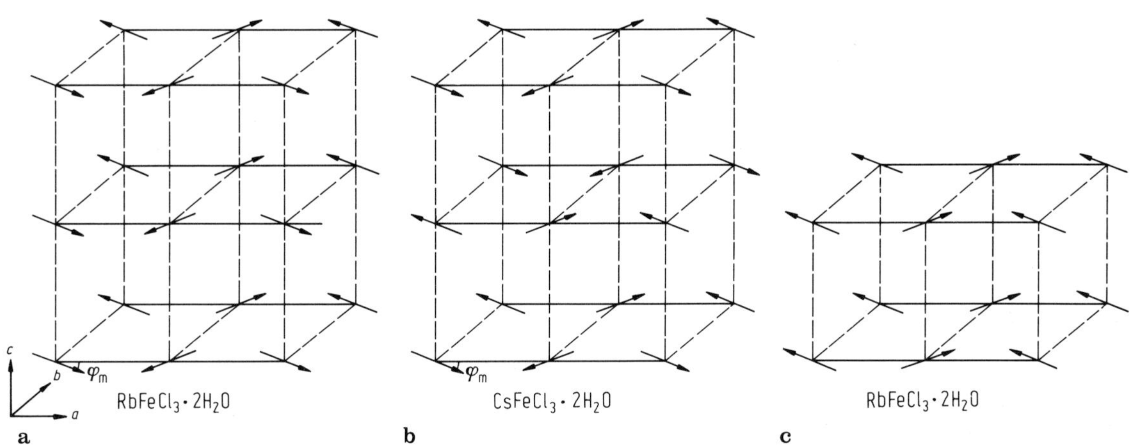

Fig. 362. (a) $RbFeCl_3 \cdot 2H_2O$ (RFC), (b) $CsFeCl_3 \cdot 2H_2O$ (CFC). Magnetic moments array in the ordered state at $H = 0$ and $T = 0$ K. Moments located in the *ca* plane at an angle φ_m from the *a* axis [78B4]. (c) For (RFC) proposed magnetic structure at $B_c \geq 1.24$ T and $T = 0$ K [85S1].

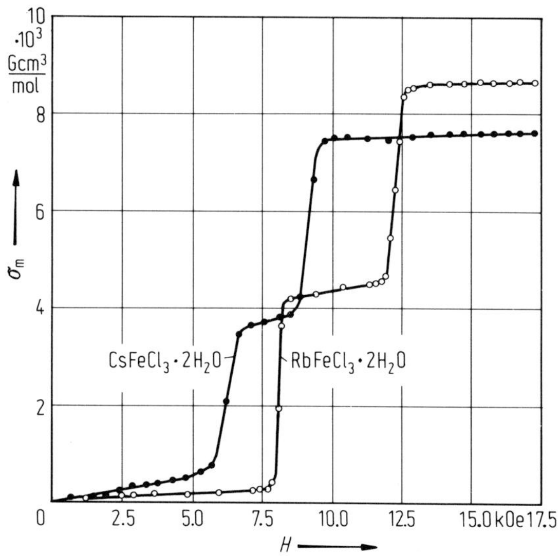

Fig. 361. $RbFeCl_3 \cdot 2H_2O$, $CsFeCl_3 \cdot 2H_2O$. Magnetic field dependence of σ_m along c axis at 4.2 K [78B4].

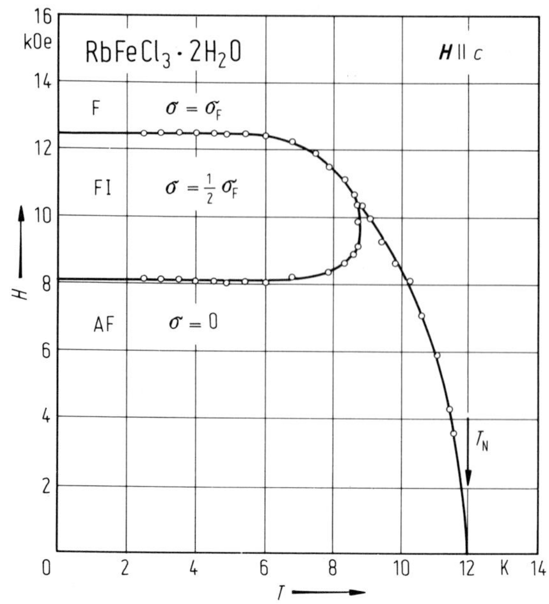

Fig. 364. $RbFeCl_3 \cdot 2H_2O$. Magnetic phase diagram for $\boldsymbol{H} \parallel c$. AF: antiferromagnetic state, FI and F: ferrimagnetic and ferromagnetic ordering between ferromagnetic moments of the chains, respectively. σ_F: magnetization in the ferromagnetic phase [79V1, 80V2].

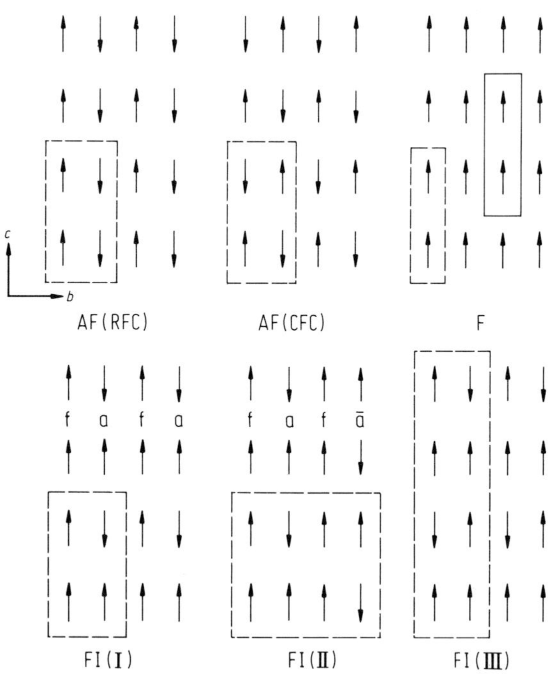

Fig. 363. $RbFeCl_3 \cdot 2H(D)_2O$ (RFC), $CsFeCl_3 \cdot 2H(D)_2O$ (CFC). Stacking in the bc plane of the net magnetic moments of the chains for various structures. For each structure, the magnetic unit cell is indicated by dashed lines, whereas the crystallographic unit cell is indicated by solid lines. The FI(I) and FI(II) structures are characterised by "antiferromagnetic" (a, ā) and "ferromagnetic" (f) stacking in adjacent ac planes [78B4, 83S3, 85S2].

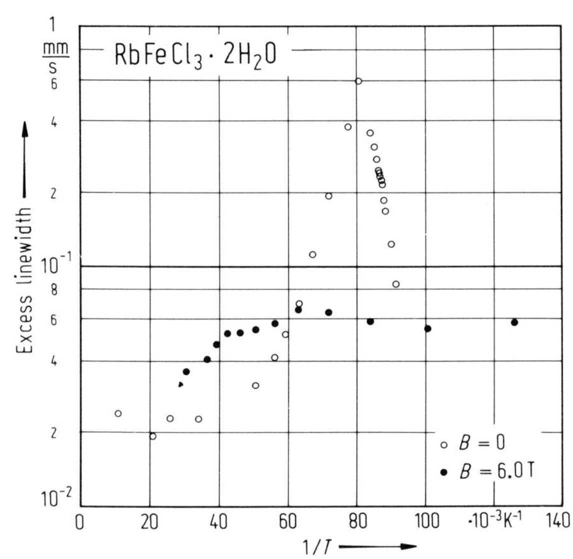

Fig. 365. $RbFeCl_3 \cdot 2H_2O$. Inverse temperature dependence of excess linewidth obtained from Mössbauer relaxation measurements in $B = 0$ and 6.0 T [85S1].

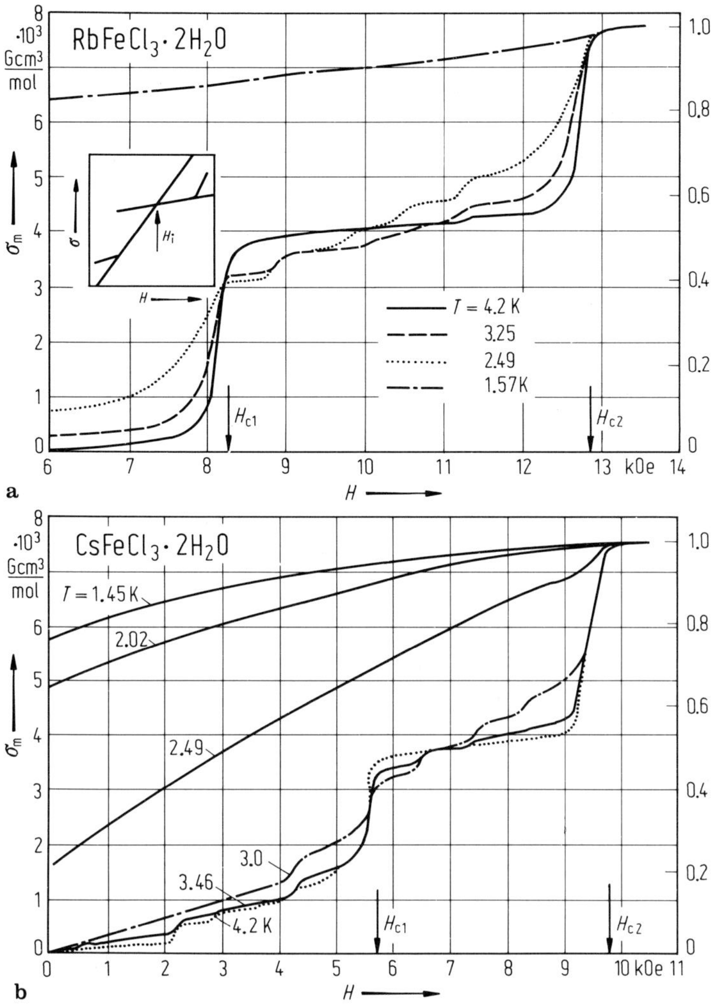

Fig. 366. $RbFeCl_3 \cdot 2H_2O$, $CsFeCl_3 \cdot 2H_2O$. Variation of $\sigma_m(H)$ in decreasing field scans (700 Oe/min) at various temperatures. The transition fields H_{c1}, H_{c2}, and the fields H_i are determined by linearizing the curves, as is shown schematically in the insert [85S2].

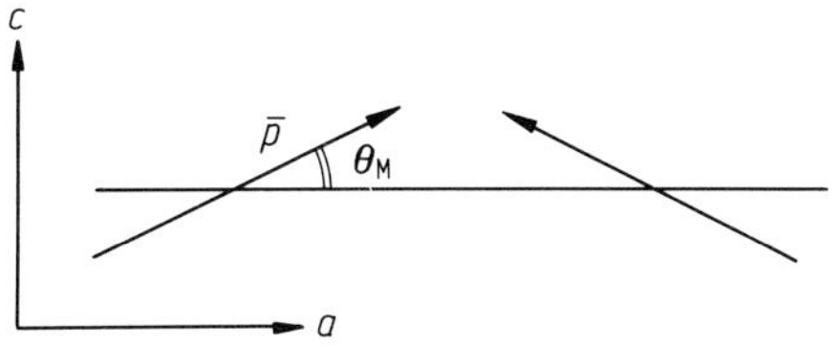

Fig. 367. (a) $CsFeCl_3 \cdot 2H(D)_2O$, (b) $RbFeCl_3 \cdot 2H(D)_2O$. The orientation in the *ac* plane of the canted magnetic moment in a chain. $\theta_M \cong 19°$ for (b) and $\theta \cong 15°$ for (a) [85S2].

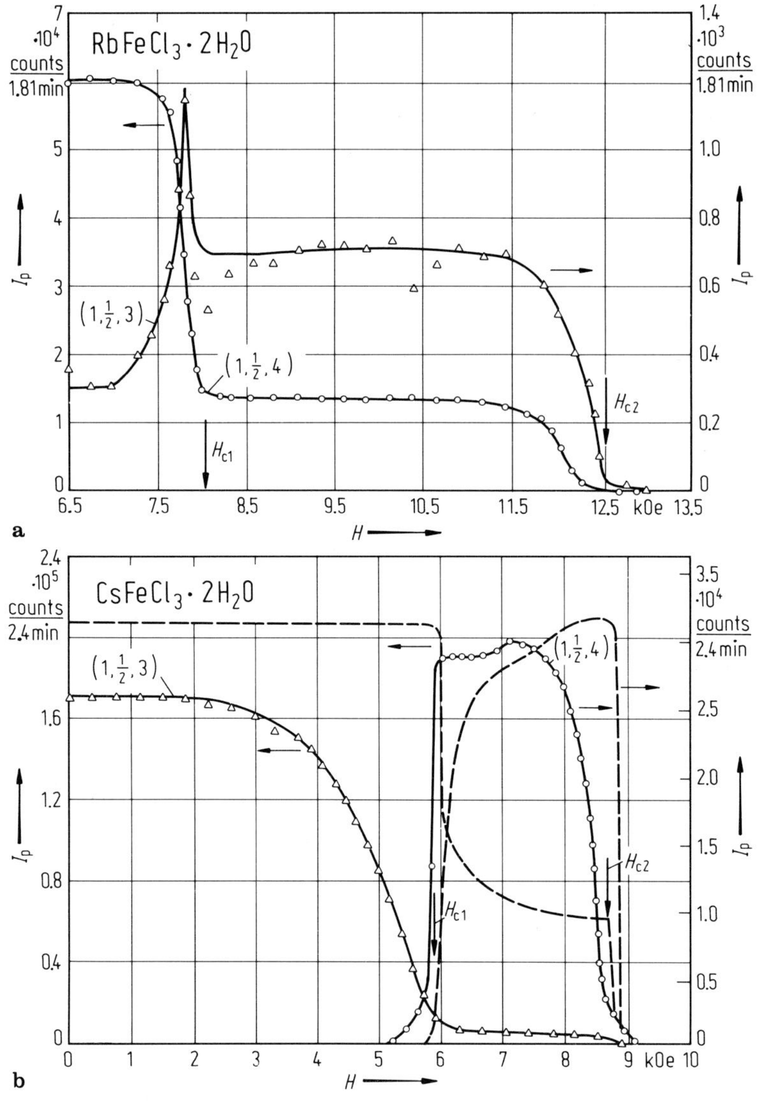

Fig. 368. $RbFeCl_3 \cdot 2H_2O$ (RFC), $CsFeCl_3 \cdot 2H_2O$ (CFC). Variation of the $(1,\frac{1}{2},3)$ and $(1,\frac{1}{2},4)$ magnetic peak intensities at 4.2 K. The data points represent the observations in a decreasing field scan (100 Oe/min). For (CFC) the variations observed in an increasing field scan (100 Oe/min) are also shown (dashed curves). Note the different field scales for RFC and CFC and the different intensity scales for the various reflections [85S2].

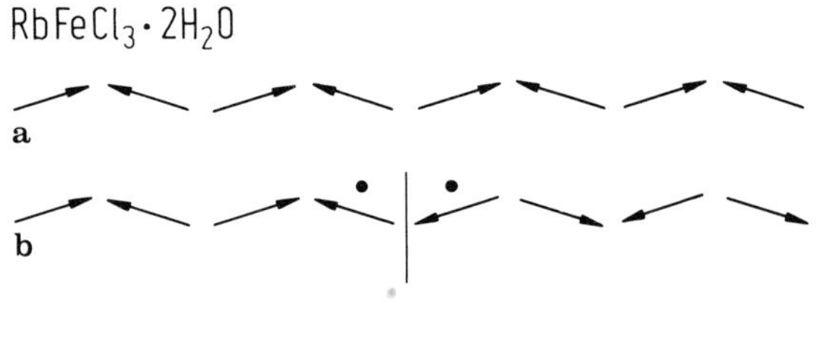

Fig. 369. $RbFeCl_3 \cdot 2H_2O$. (a) Antiferromagnetic chain of canted Ising spins in AF phase (schematic representation) (b) The soliton. The arrows represent Ising spins along the chain and the vertical bar the soliton [90T2].

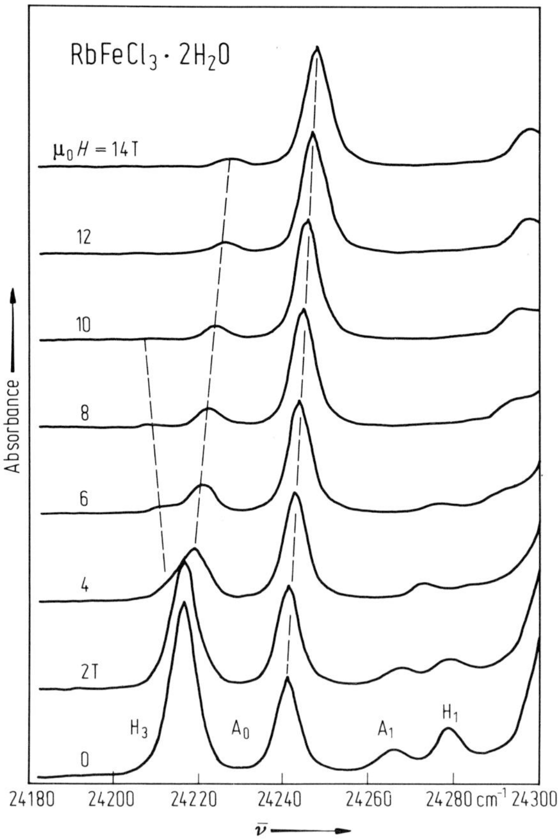

Fig. 370. $RbFeCl_3 \cdot 2H_2O$. Fine structure of the magnetoabsorption spectra at $14\ K > T_N (= 11.96\ K)$. The incident light electric field vector $\tilde{\boldsymbol{E}} \parallel a$; $\boldsymbol{H} \parallel c$. Even at $T > T_N$ there exist antiferromagnetic clusters in the linear chain. A_0 is due to the usual exciton in AF cluster, H_3-line is assigned as the exciton at the sites on the soliton, A_1, H_1 are magnon sidebands of the A_0 and H_3 line [90T2].

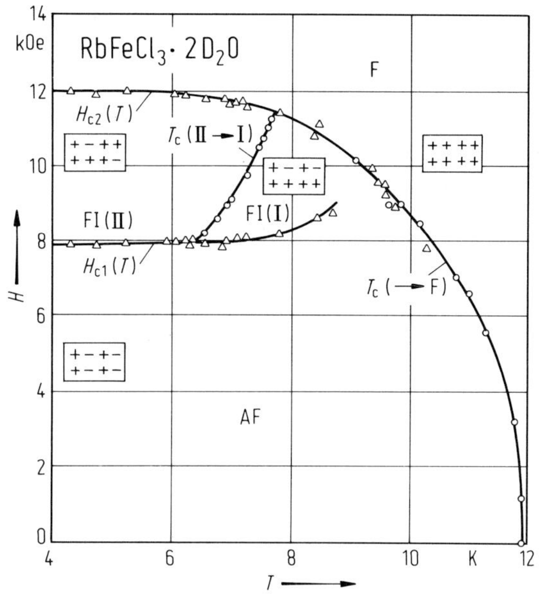

Fig. 371. $RbFeCl_3 \cdot 2D_2O$. Magnetic (H, T) phase diagram with $\boldsymbol{H} \parallel c$ as determined from neutron-diffraction experiments. The data points denote transitions in increasing field (triangles) and increasing temperature (circles) [85S2].

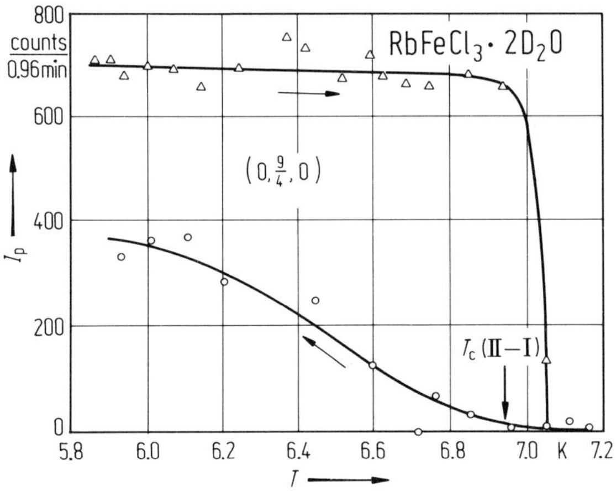

Fig. 372. $RbFeCl_3 \cdot 2D_2O$. Typical variations of magnetic peak intensities across the phase boundary $T_c(\mathrm{II} \rightarrow \mathrm{I})$. Temperature scans at constant field ($H = 9.38$ kOe) [85S2].

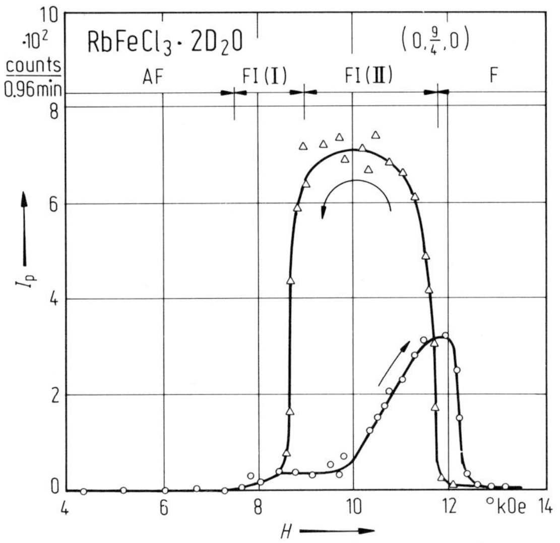

Fig. 373. $RbFeCl_3 \cdot 2D_2O$. Typical variations of magnetic peak intensities across the phase boundary $T_c(II \to I)$. Field scans at constant temperature ($T = 6.92$ K). The indicated F, FI(II), FI(I) and AF phases refer to the field scan [85S2].

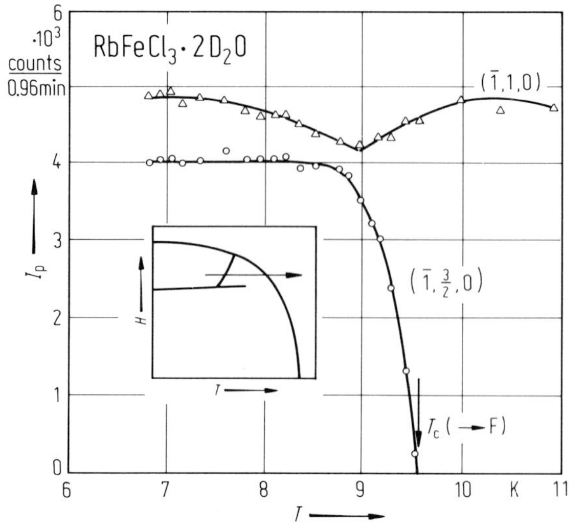

Fig. 374. $RbFeCl_3 \cdot 2D_2O$. Temperature dependence of the $(\bar{1},1,0)$ and $(\bar{1},\frac{3}{2},0)$ intensities at constant field $H = 9.38$ kOe. The insert shows the path followed in the H-T phase diagram [85S2].

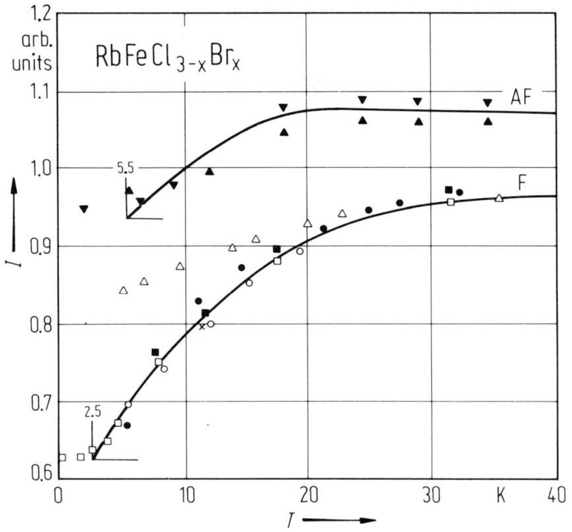

Fig. 375. $RbFeCl_{3-x}Br_x$. Temperature dependence of absorption intensities. ▲ and ▼: bands a and b for x = 3, □ and ■: bands a and b for x = 0, ×: sharp line for x = 0, ○: band a for x = 1, ●: band a for x = 0.2, △: band a for x = 2. The full lines are theoretical for ferro- (F) and antiferromagnetic (AF) coupling [82B1].

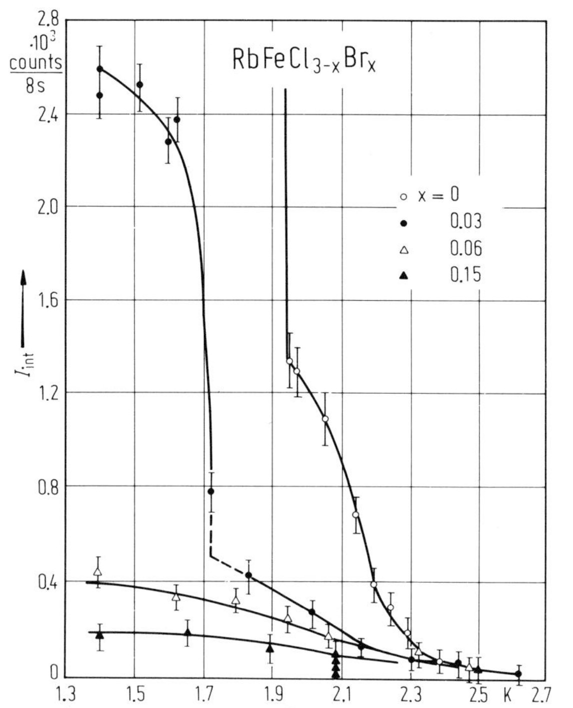

Fig. 376. $RbFeCl_{3-x}Br_x$. Temperature dependence of the integrated neutron scattering intensity for small values of x measured at the K point [83S2] ($\boldsymbol{Q} = (\frac{1}{3}, \frac{1}{3}, 0)_N$). The data have been scaled against the relative intensities of the nuclear (1,1,0) and (2,2,0) reflections [89H1].

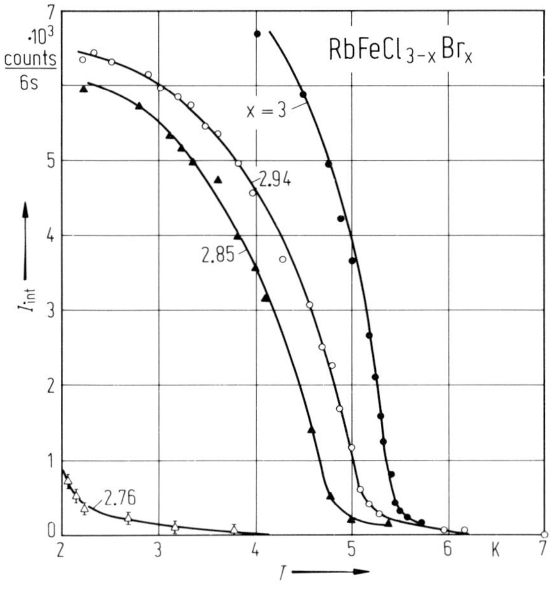

Fig. 377. $RbFeCl_{3-x}Br_x$. Temperature dependence of the integrated neutron scattering intensity for large values of x (x ≥ 2.76), measured at the H point ($\mathbf{Q} = (\frac{1}{3}, \frac{1}{3}, 1)_N$). The data have been scaled against the relative intensities of the nuclear (1,1,0) and (2,2,0) reflections [89H1].

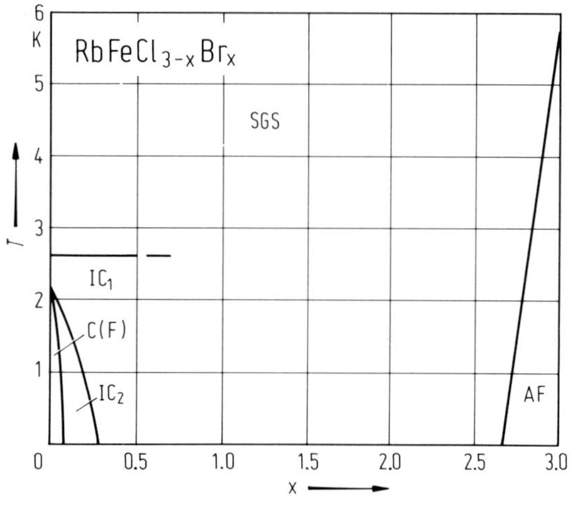

Fig. 379. $RbFeCl_{3-x}Br_x$. Schematic magnetic phase diagram. Only single ground-state (SGS) behavior at intermediate composition is shown [89H1].

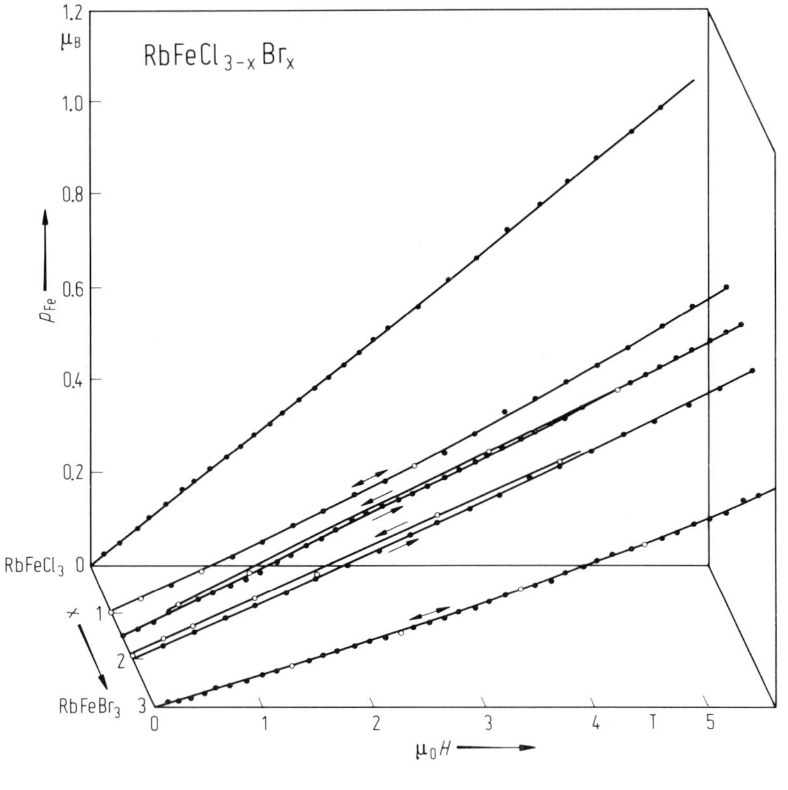

Fig. 378. $RbFeCl_{3-x}Br_x$. Magnetic field dependence of magnetic moment p_{Fe} for $\mathbf{H} \| c$ at 2 K for various x in increasing and decreasing magnetic field [89H1].

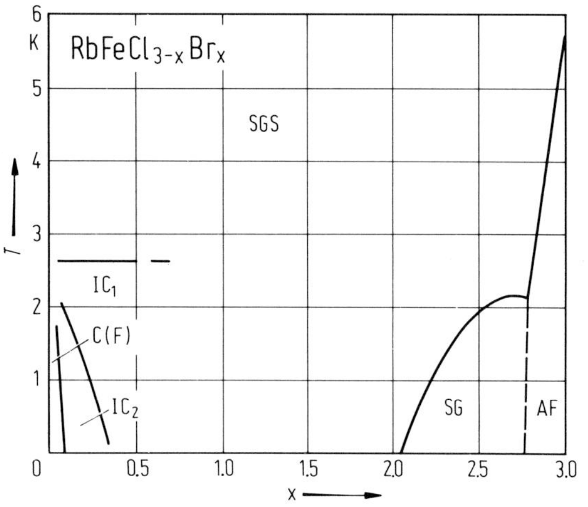

Fig. 380. $RbFeCl_{3-x}Br_x$. Schematic magnetic phase diagram for the case where there is a spin glass (SG) phase on the Br side of the phase diagram [89H1].

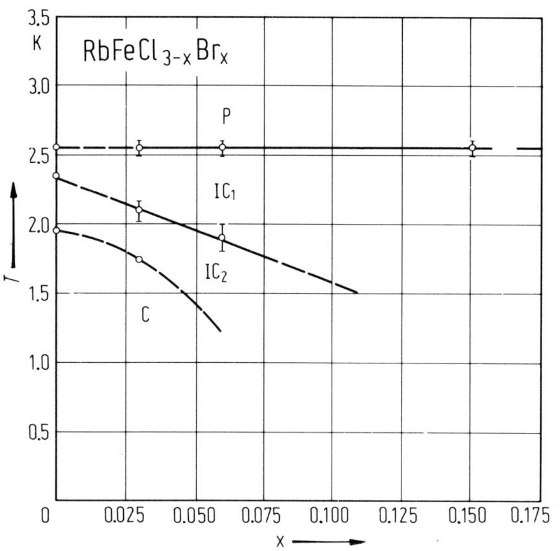

Fig. 381. $RbFeCl_{3-x}Br_x$. Temperature dependence of the periodicity of magnetic correlations. Symbols IC_1, IC_2 and C refer to the incommensurate and commensurate periodicities. The two lower phase boundaries were not seen down to 1.3 K for the x = 0.15 sample [89H1].

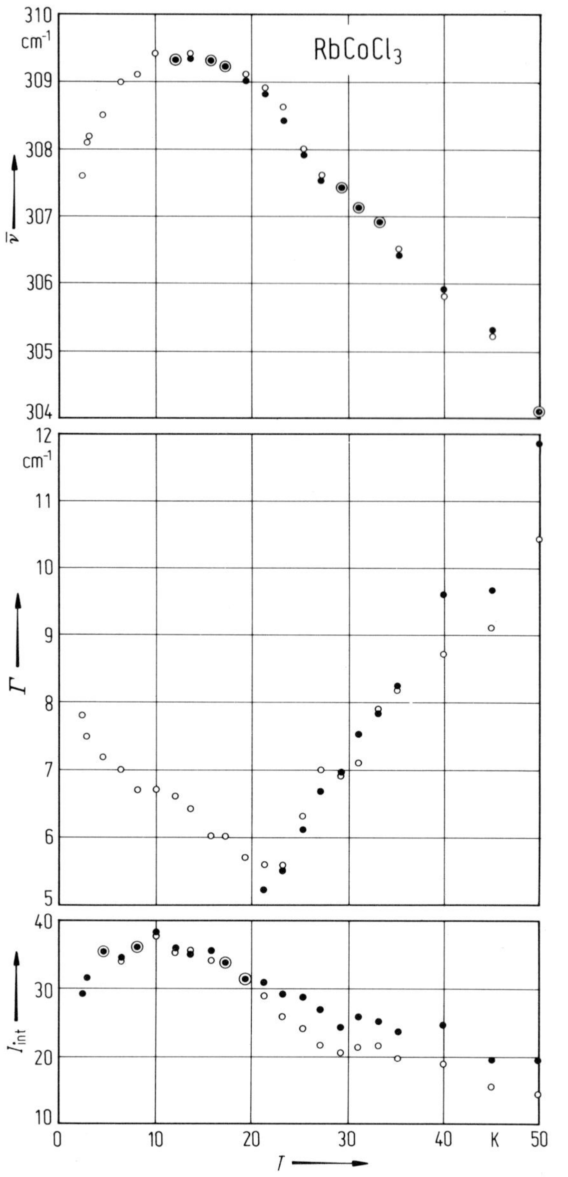

Fig. 382. $RbCoCl_3$. Temperature dependence of the band parameters for the exciton frequency near 300 cm^{-1} obtained from direct Raman measurement (open circles) and computer curve-fitting (full circles) [83L1].

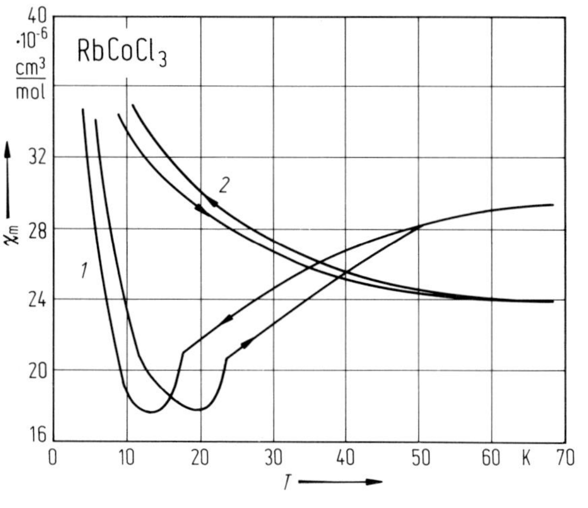

Fig. 383. RbCoCl$_3$. Temperature dependence of χ_m at 1.2 kOe for $H \| c$ (*1*) and $H \perp c$ (*2*) [89P1].

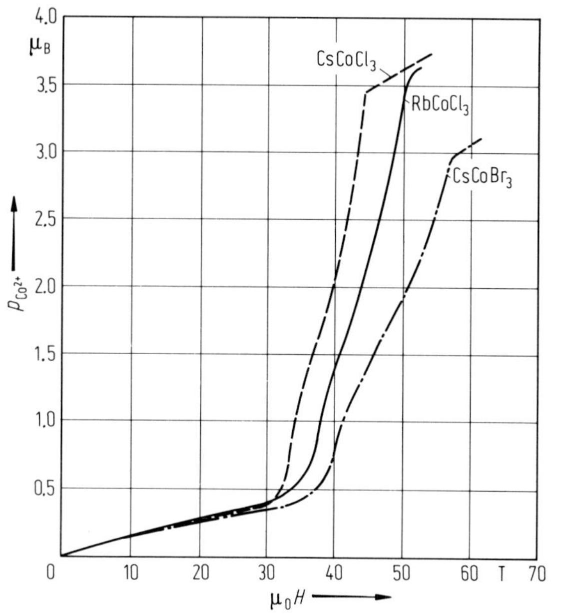

Fig. 384. RbCoCl$_3$, CsCoCl$_3$, CsCoBr$_3$. Magnetic moment $p_{Co^{2+}}$ vs. pulsed magnetic field at 4.2 K and for $H \| c$ [90H2].

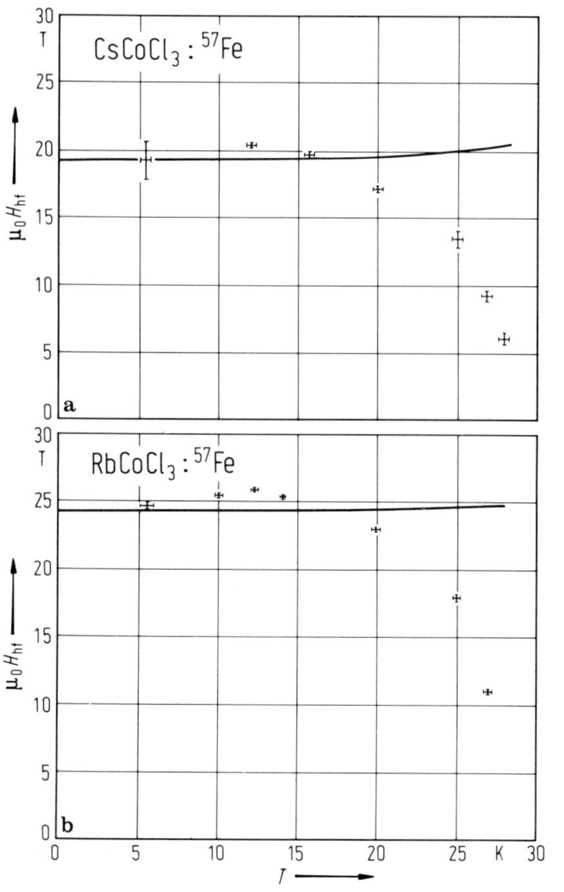

Fig. 385. (a) CsCoCl$_3$:^{57}Fe, (b) RbCoCl$_3$:^{57}Fe. Magnetic hyperfine field at ^{57}Fe. Full curves are from the two-level relaxation model and the crystal field fit [88B1].

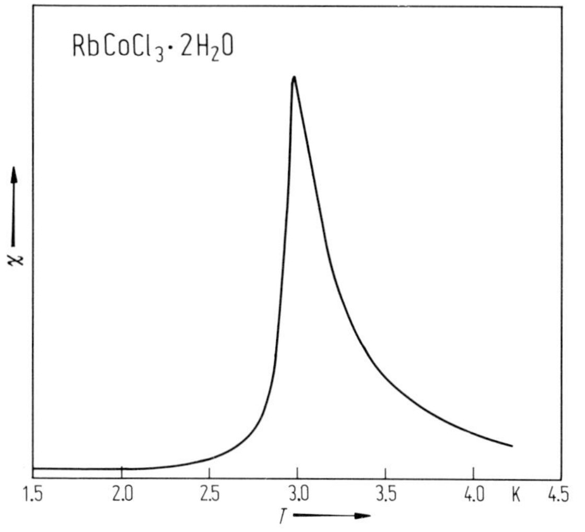

Fig. 386. RbCoCl$_3$·2H$_2$O. Temperature dependence of χ for a single crystal in c direction at $H = 0$ and frequency of 275 Hz [75F2].

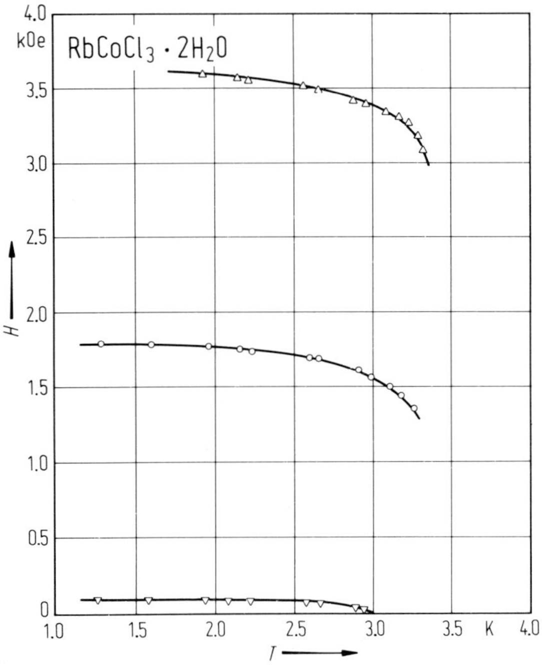

Fig. 387. $RbCoCl_3 \cdot 2H_2O$. H–T phase diagram. Transitions in the c direction [75F2]. See Fig. 632

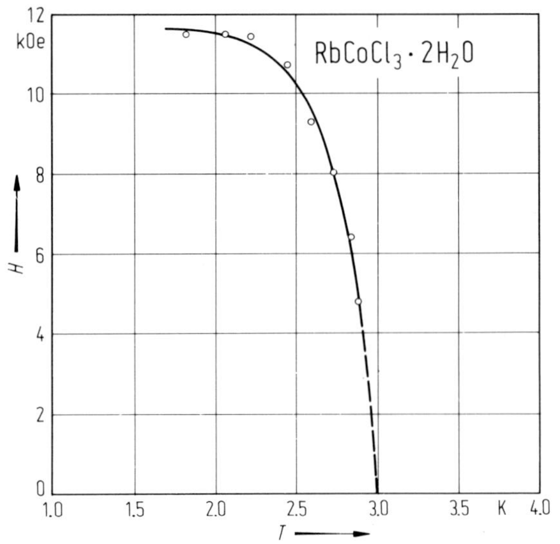

Fig. 388. $RbCoCl_3 \cdot 2H_2O$. H–T phase diagram. Transition in the b direction [75F2].

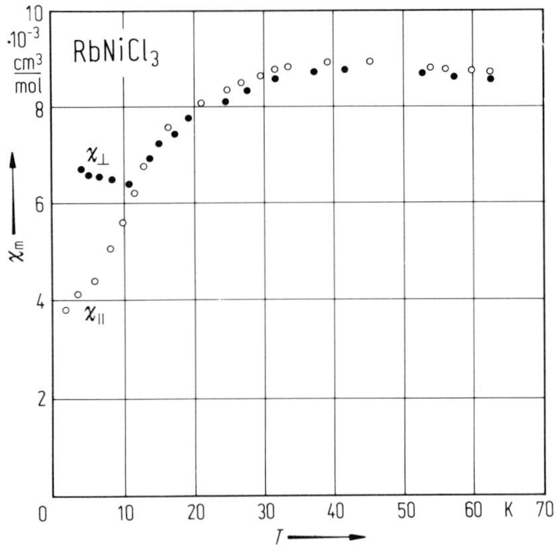

Fig. 389. $RbNiCl_3$. Temperature dependence of χ_m parallel and perpendicular to the c axis [69A1].

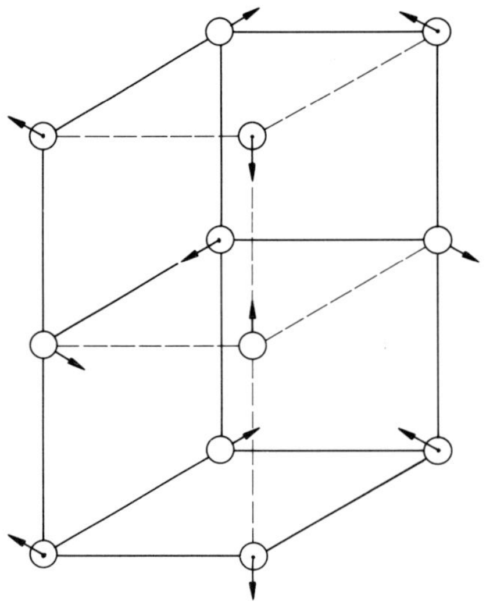

Fig. 390. $RbNiCl_3$, $CsNiCl_3$. Magnetic structure consists of an antiferromagnetic sequence of moments along c axis, and a screw spiral propagating in the basal plane with wavelength $\frac{3}{2}a$ as one proceeds along ac plane [70M2].

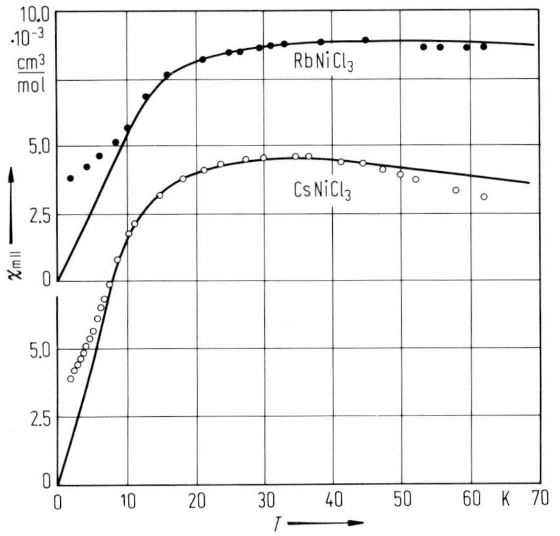

Fig. 391. $CsNiCl_3$, $RbNiCl_3$. Temperature dependence of χ_m parallel to the c axis (experimental points from [69A1]). The solid line is a theoretical fit with $D = 0.25\ cm^{-1}$, $J = 11.1\ cm^{-1}$, and $D = 1.1\ cm^{-1}$, $J = 13.2\ cm^{-1}$ for Cs and Rb compounds, respectively [72M3].

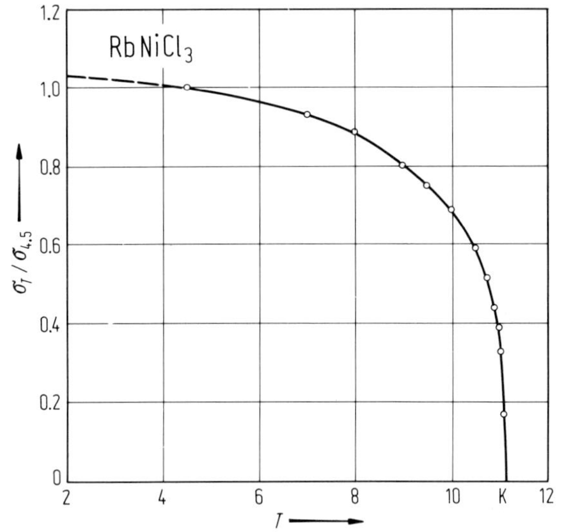

Fig. 392. $RbNiCl_3$. Temperature dependence of σ determined from the intensity of the $(\frac{1}{3}, \frac{1}{3}, 1)$ reflection. Extrapolation to 0K yields a Ni moment of $1.3\mu_B$ [72Y1].

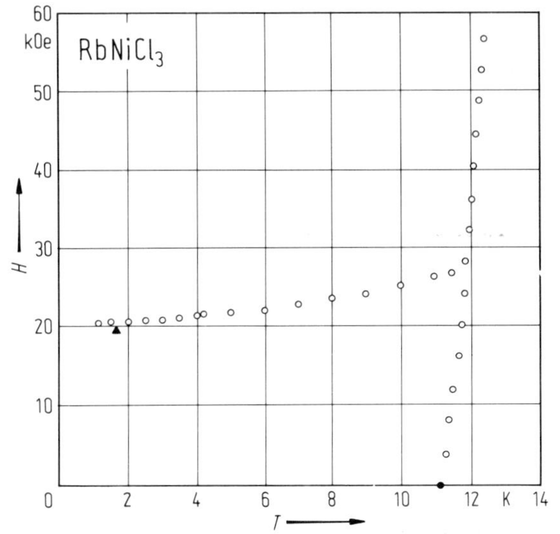

Fig. 393. $RbNiCl_3$. H–T phase diagram with $\boldsymbol{H} \parallel c$; full triangle: from [77S1], full circle from [72Y1]; [79J1].

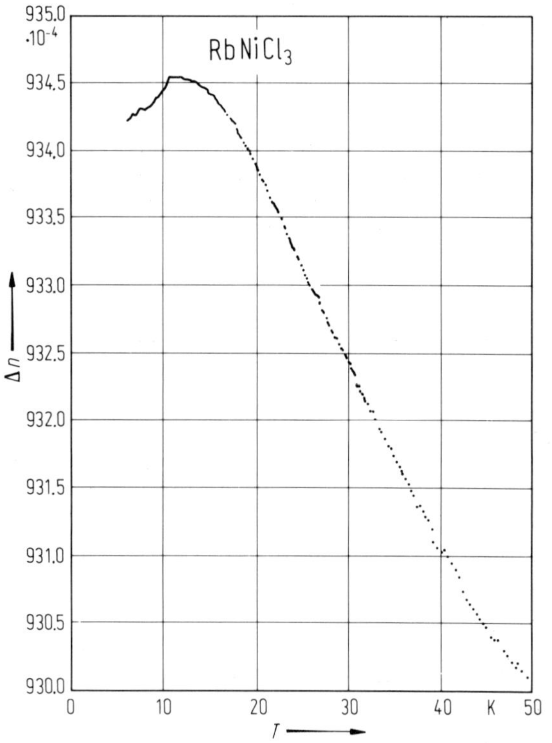

Fig. 394. $RbNiCl_3$. Temperature dependence of the birefringence $\Delta n = n_c - n_a$ at $0.6328\mu m$ wavelength. A sudden decrease of Δn below $T_N = 11$ K is observed [83I1].

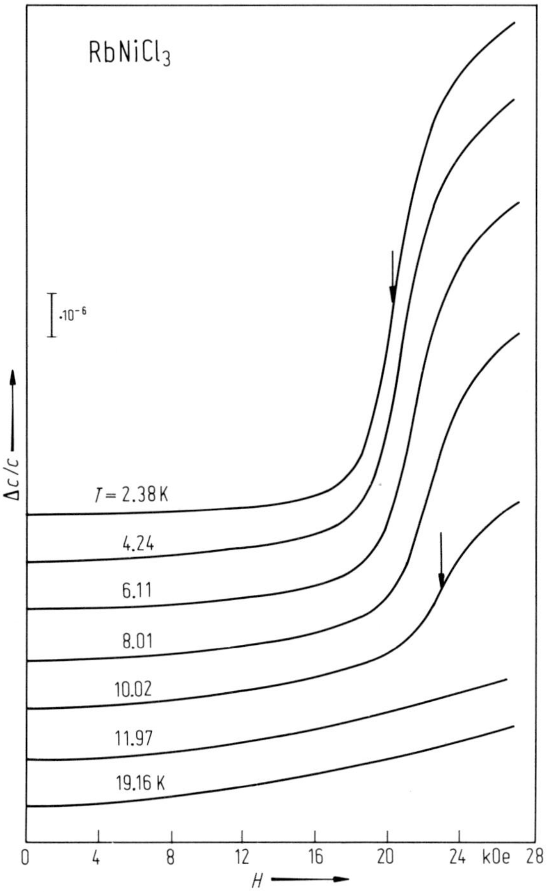

Fig. 395. $RbNiCl_3$. Magnetic field dependence of magnetostriction, for applied field parallel to the c axis. Arrows correspond to spin-flop at 10.02 and 2.38 K, respectively [84R1].

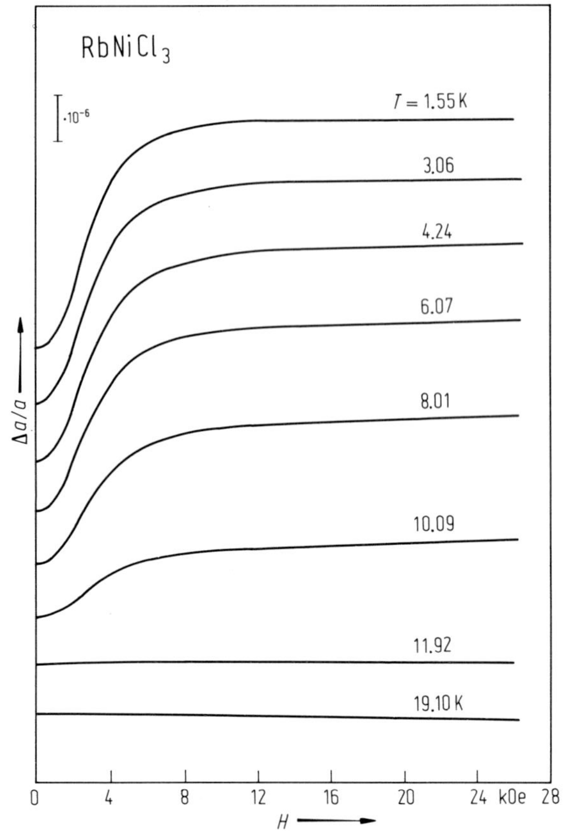

Fig. 396. $RbNiCl_3$. Magnetic field dependence of magnetostriction, for applied field normal to c axis. Above $T_N = 11.1$ K negative effect is observed [84R1].

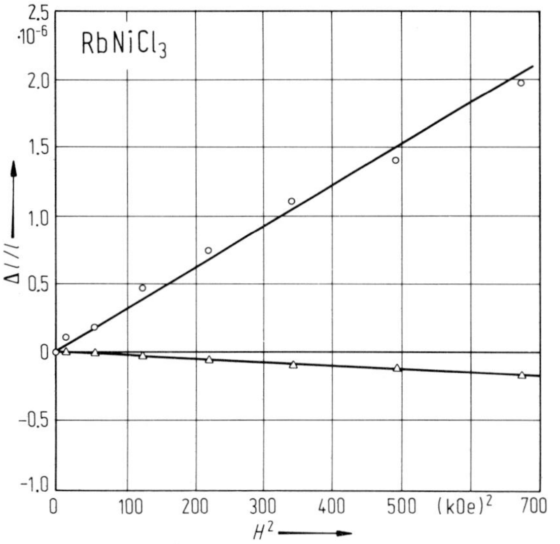

Fig. 397. $RbNiCl_3$. Magnetostriction $\Delta l/l$ vs. H^2 at 1.9 K for $\boldsymbol{H} \parallel c$ axis (circles), $\boldsymbol{H} \perp c$ axis (triangles) [84R1].

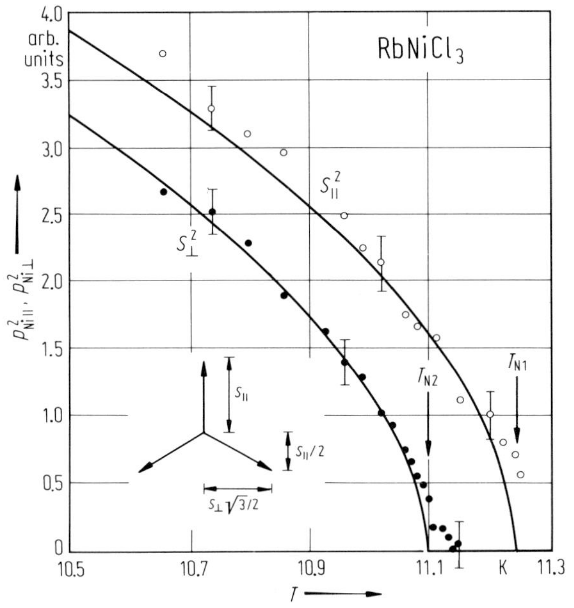

Fig. 398. $RbNiCl_3$. Temperature dependence of squares of sublattice moments perpendicular ($p^2_{Ni\perp}$) and parallel ($p^2_{Ni\parallel}$) to the c axis. Solid lines are fits to $p^2_{\parallel} \propto (T_{N1} - T)^{2\beta\parallel}$, $p^2_{\perp} \propto (T_{N2} - T)^{2\beta\perp}$. In the inset the three sublattice spin structure in the c plane and definitions of $p_{\parallel}$ and $p_{\perp}$ are shown [91O1].

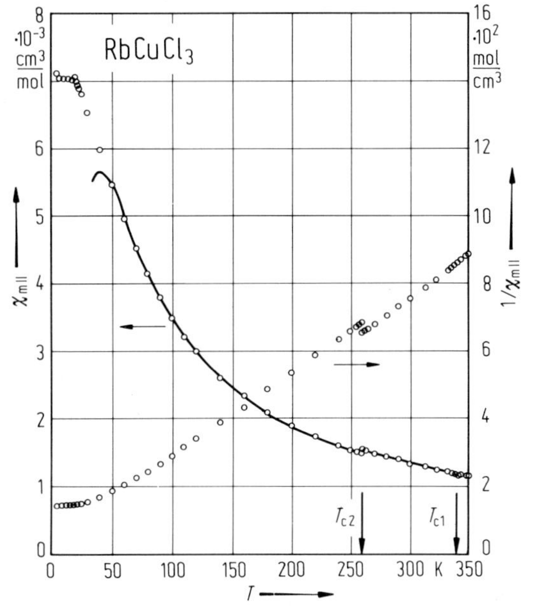

Fig. 400. $RbCuCl_3$. Temperature dependence of $\chi_{m\parallel}$ and $\chi^{-1}_{m\parallel}$ measured at 16.38 kOe parallel to the c axis with increasing temperature. Correction for inner-core diamagnetism and Van Vleck paramagnetism have already been made. Solid line shows the calculated values [86T1].

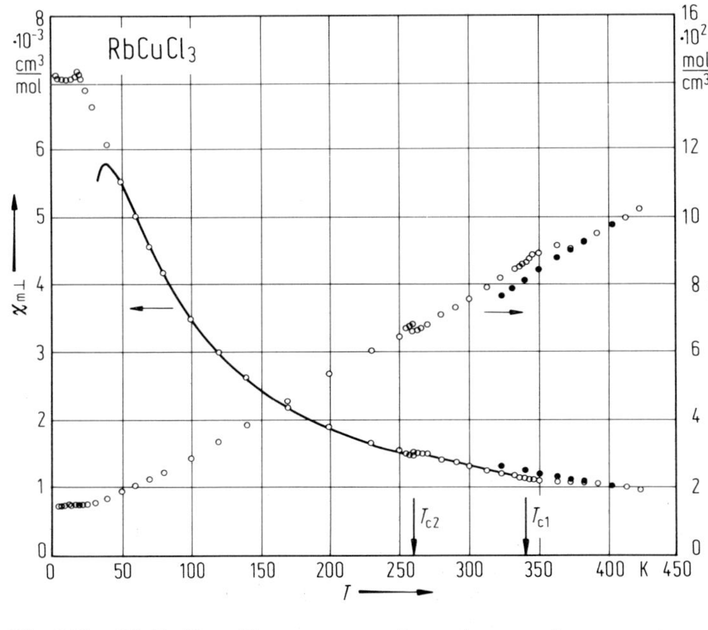

Fig. 399. $RbCuCl_3$. Temperature dependence of $\chi_{m\perp}$ and $\chi^{-1}_{m\perp}$ perpendicular to the c axis, at 16.38 kOe. Open circles: increasing temperature, full circles: decreasing temperature. Corrections for inner-core diamagnetism and Van Vleck paramagnetism have been made. Solid line calculated [86T1].

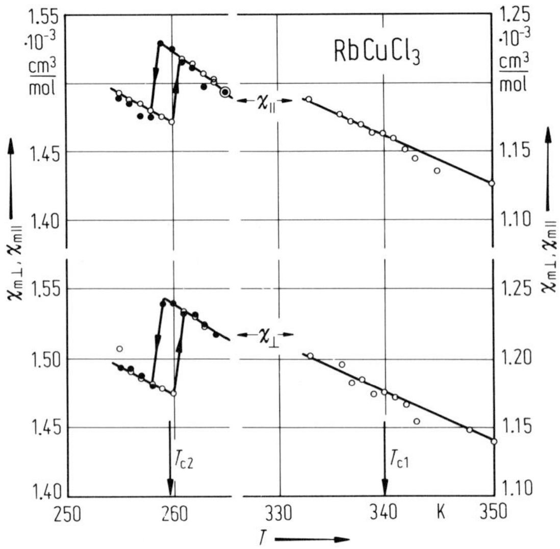

Fig. 401. $RbCuCl_3$. Temperature dependence of χ_m parallel and perpendicular to the c axis around the two phase transition temperatures, T_{c1} and T_{c2}. Open circles indicate measurements with increasing and full circles with decreasing temperature [86T1].

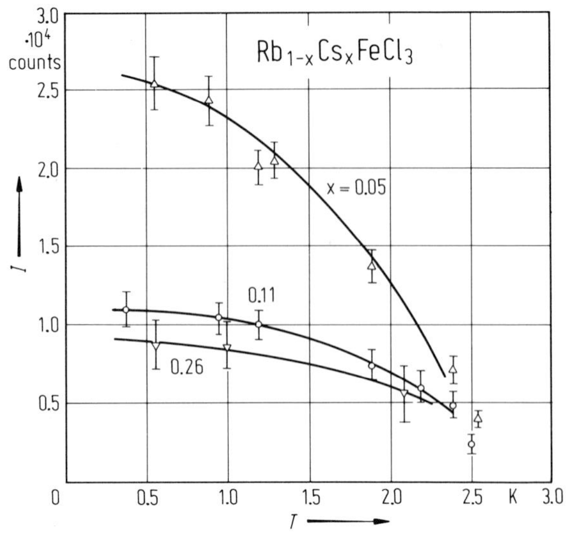

Fig. 402. $Rb_{1-x}Cs_xFeCl_3$. Temperature dependence of the intensity of the diffuse neutron scattering. Scans were made along the [110] direction (y) through the intensity maximum over 20 s. Consequently, the position along the [110] (x) direction of the centre of the scan varied with the changing position of the intensity maximum. The integrated intensity of such a scan was corrected for a background scattering at 6.0 K and scaled against the nuclear (1,1,0) reflection of each sample. [86H1].

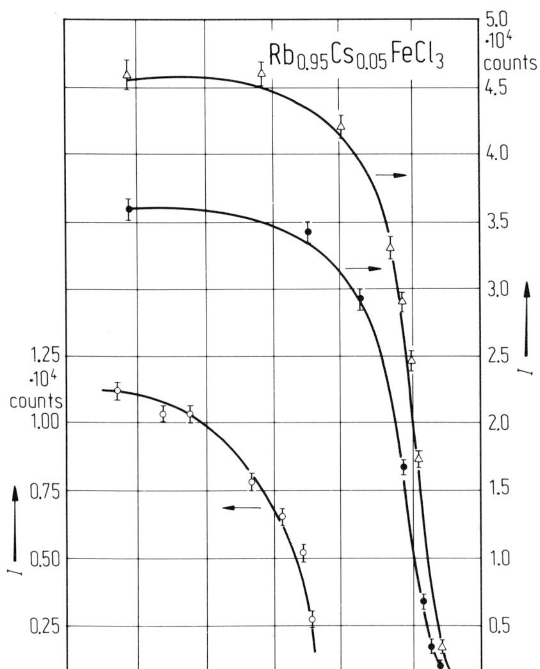

Fig. 403. $Rb_{0.95}Cs_{0.05}FeCl_3$. Temperature dependence of magnetization curves obtained by scanning along x through the K point ($\boldsymbol{Q} = (\frac{1}{3}, \frac{1}{3}, 0)$) as the temperature was lowered from 6 K in zero magnetic field (open circles, left-hand axis) and in a magnetic field of 4.9 T or 4.5 T (right-hand axis, triangles and full circles, respectively) [86H1].

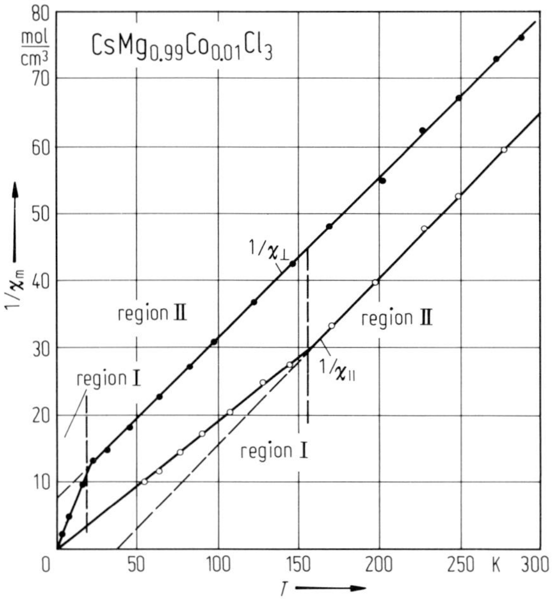

Fig. 404. $CsMg_{0.99}Co_{0.01}Cl_3$. Temperature dependence of χ_m parallel and perpendicular to the c axis, having different Curie constants in two temperature regions [69A1].

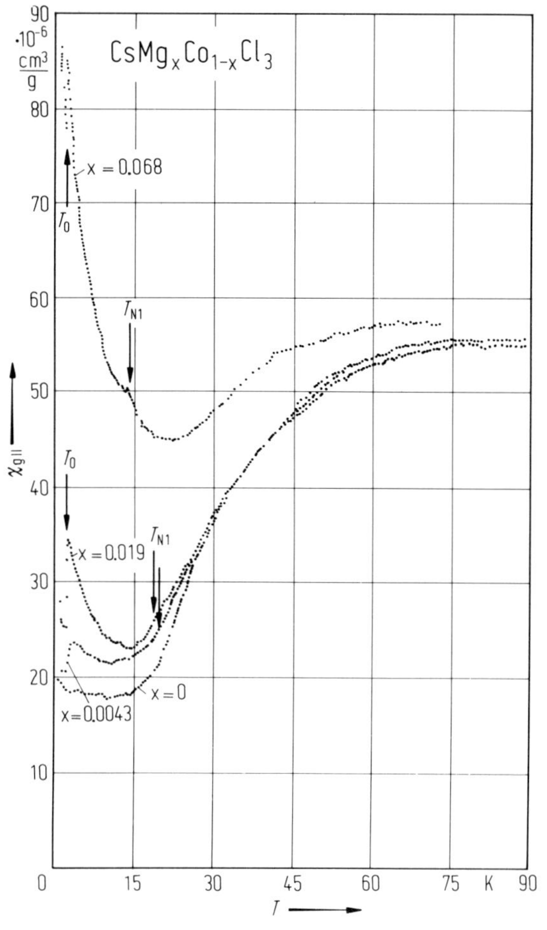

Fig. 406. $CsMg_xCo_{1-x}Cl_3$. Temperature dependence of $\chi_{g\parallel}$ parallel to c axis at 5 kOe for different x [87M3].

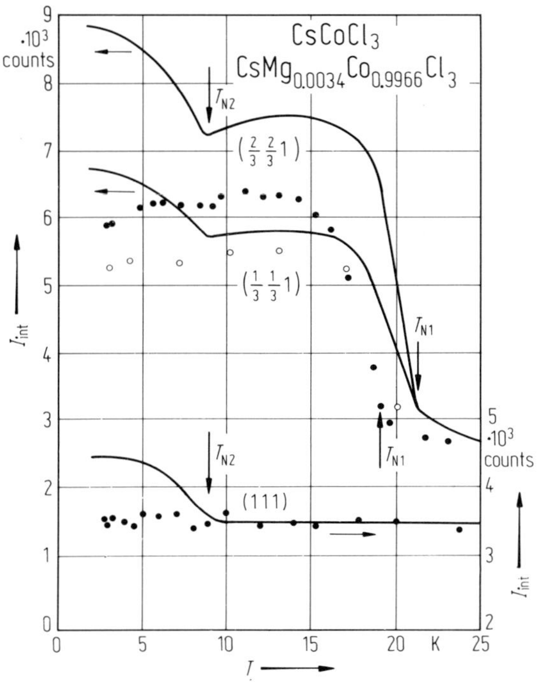

Fig. 405. $CsCoCl_3$, $CsMg_{0.0034}Co_{0.9966}Cl_3$. Temperature dependence of the integrated intensity of elastic neutron scattering. Solid curves for $CsCoCl_3$ [87M3].

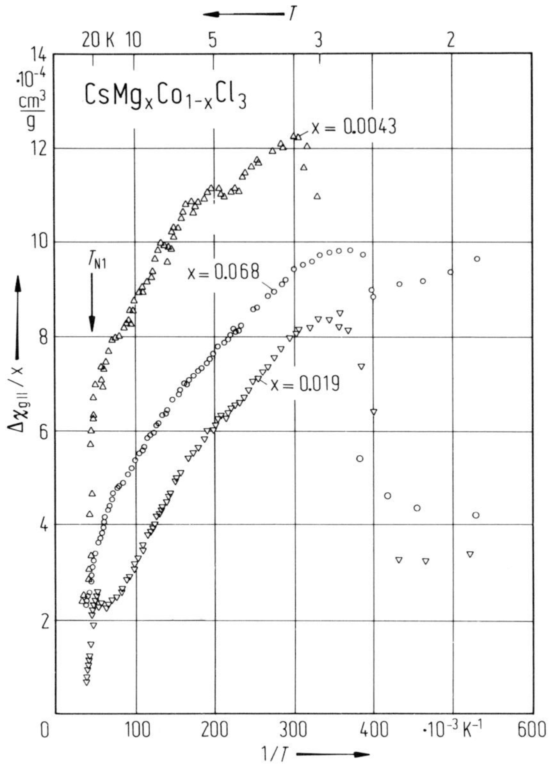

Fig. 407. $CsMg_xCo_{1-x}Cl_3$. Inverse of temperature dependence of incremental magnetic susceptibility from that of pure $CsCoCl_3$, $\Delta\chi_{g\parallel}/x$ for $\chi \parallel c$ [87M3].

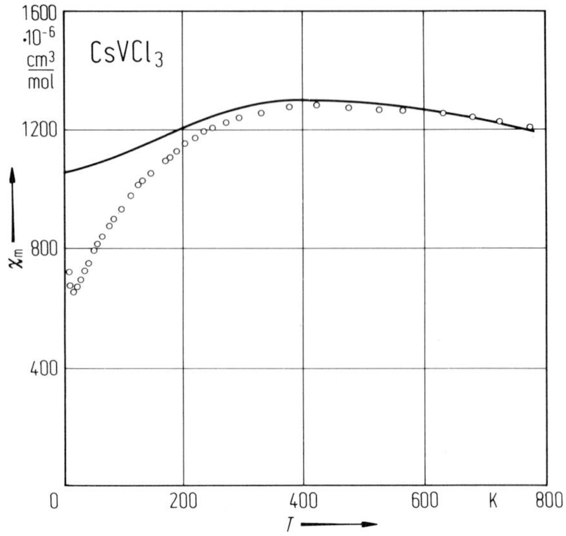

Fig. 408. $CsVCl_3$. Temperature dependence of χ_m. Solid line is theoretical fit [77N1].

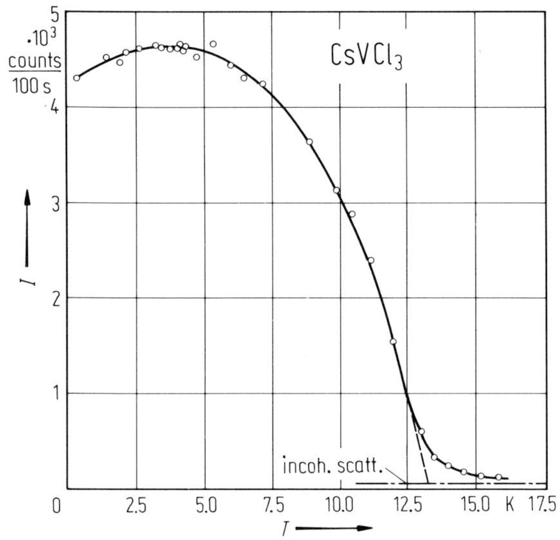

Fig. 409. $CsVCl_3$. Temperature dependence of the magnetic neutron Bragg scattering intensity measured at $(\frac{1}{3}, \frac{1}{3}, 1)$ with incident neutron energy of 13.7 meV [82H1].

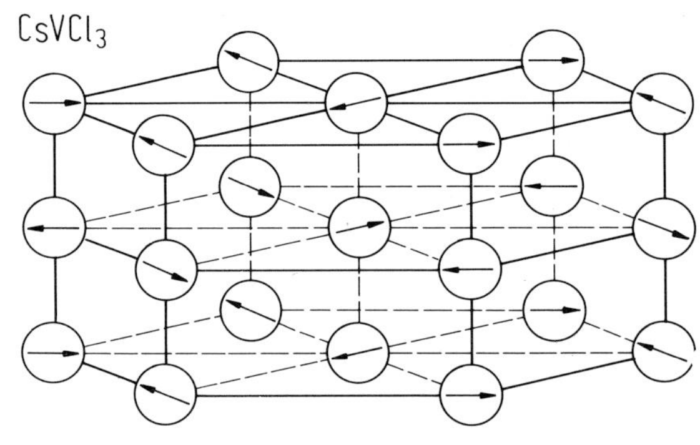

Fig. 410. $CsVCl_3$. Supposed spin structure from neutron Bragg scattering at low temperature [82H1].

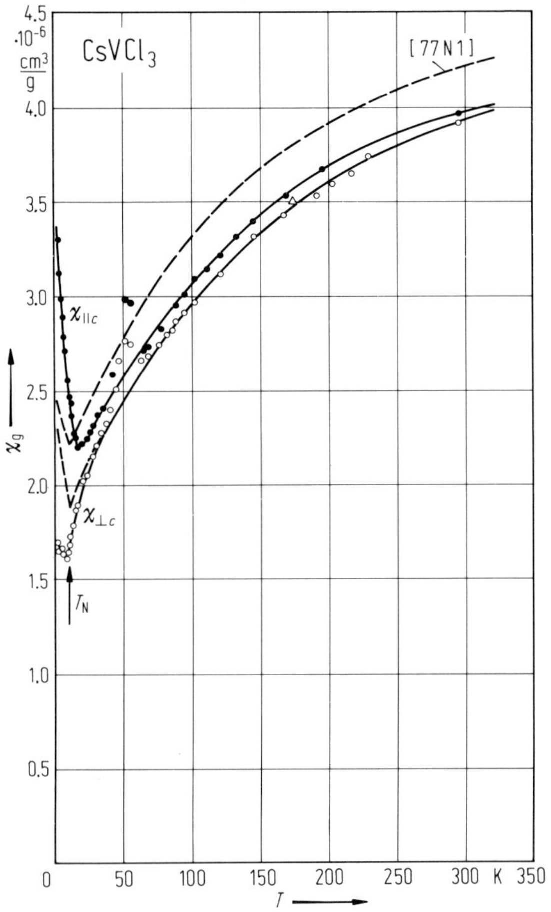

Fig. 411. $CsVCl_3$. Temperature dependence of χ_g parallel (full circles) and perpendicular (open circles) to c axis at 4 kOe, triangle: χ_c by SQUID at 50 Oe [83H2]. Results from [77N1] for powder sample for comparison. See also [82H1].

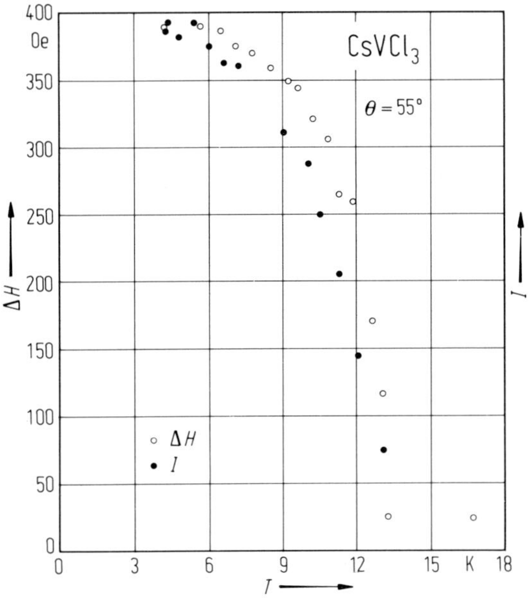

Fig. 412. $CsVCl_3$. Temperature dependence of the linewidth of the ^{133}Cs NMR at the magic angle (open circles) measured at 8.0 MHz and of the intensity of the $(\frac{1}{3},\frac{1}{3},1)$ peak of the magnetic neutron Bragg scattering (full circles) normalized at 4.2 K [85T1].

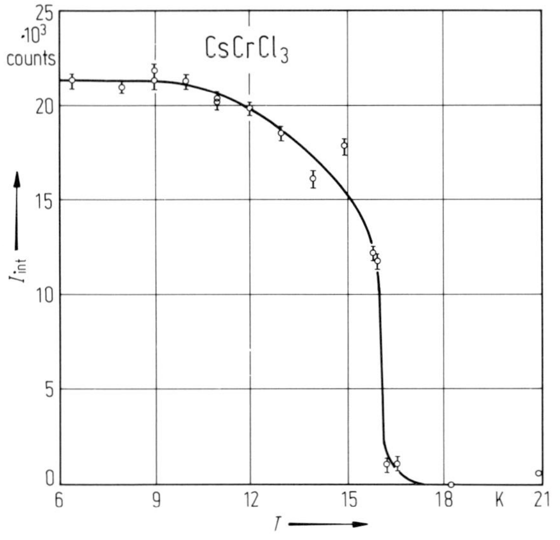

Fig. 413. $CsCrCl_3$. Temperature dependence of integrated intensity of $(\bar{1},\frac{1}{2},\frac{1}{2},1)$ magnetic peak in single-crystal neutron diffraction [79D1].

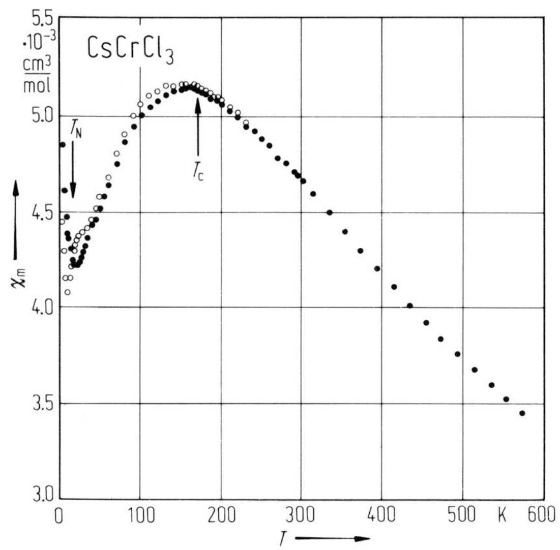

Fig. 414. $CsCrCl_3$. Temperature dependence of χ_m at 11.9 kOe after the correction of the core diamagnetism ($-1.23 \cdot 10^{-4}$ cm^3/mol). Full circles: $\boldsymbol{H} \perp c$, open circles: $\boldsymbol{H} \| c$. T_c: crystallographic phase transition [84T1].

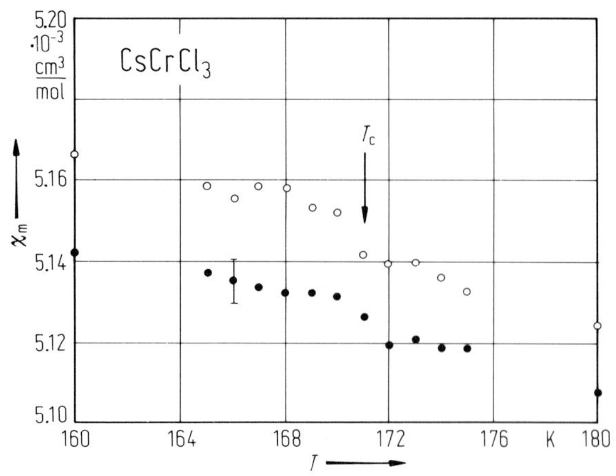

Fig. 415. $CsCrCl_3$. Temperature dependence of χ_m around crystallographic phase transition T_c [84T1]. For symbols see Fig. 414.

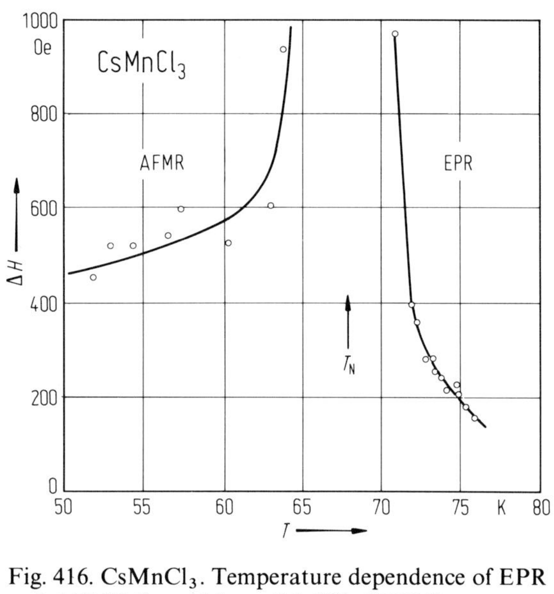

Fig. 416. $CsMnCl_3$. Temperature dependence of EPR and AFMR linewidths at 9.3 GHz [65K1].

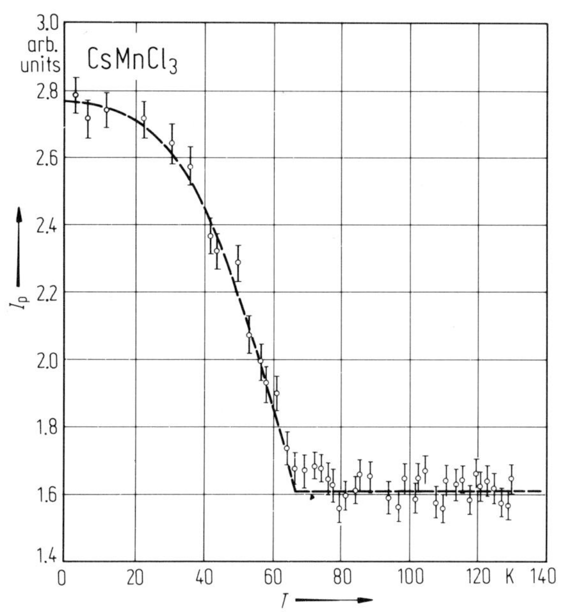

Fig. 417. $CsMnCl_3$. Temperature dependence of the peak intensity of the (0, 1, 1) + (0, 0, 9) magnetic line [71M3].

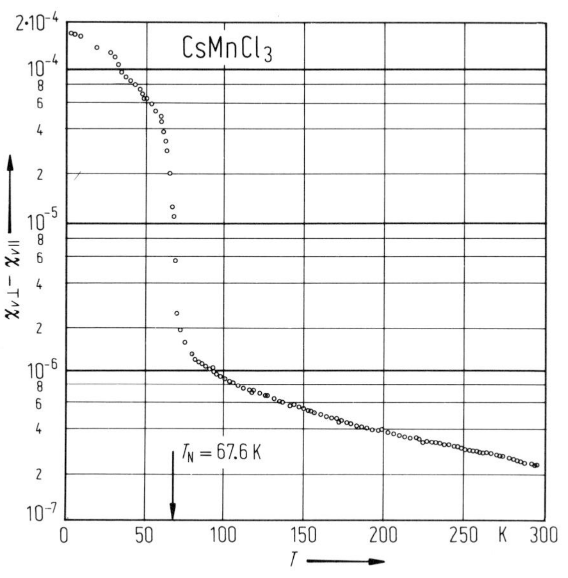

Fig. 418. $CsMnCl_3$. Temperature dependence of $\chi_{v\perp} - \chi_{v\parallel}$ [78Y2]. χ_v: volume susceptibility.

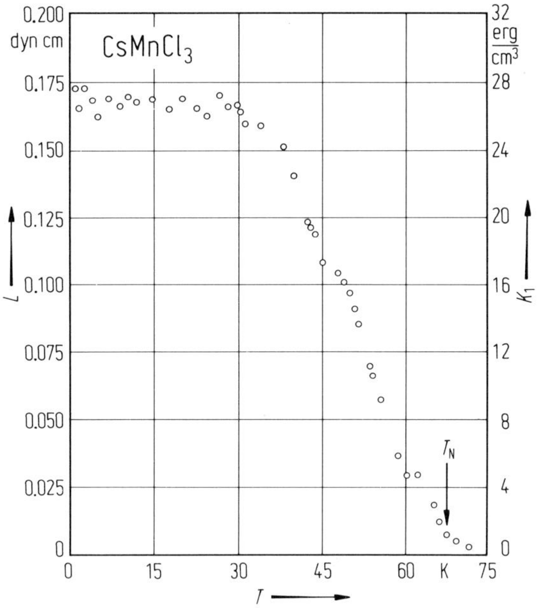

Fig. 419. $CsMnCl_3$. Temperature dependence of anisotropy constant and of torque amplitude in the *ab* plane at 8.0 kOe [78Y2]. Unit of *L* as given in the original paper.

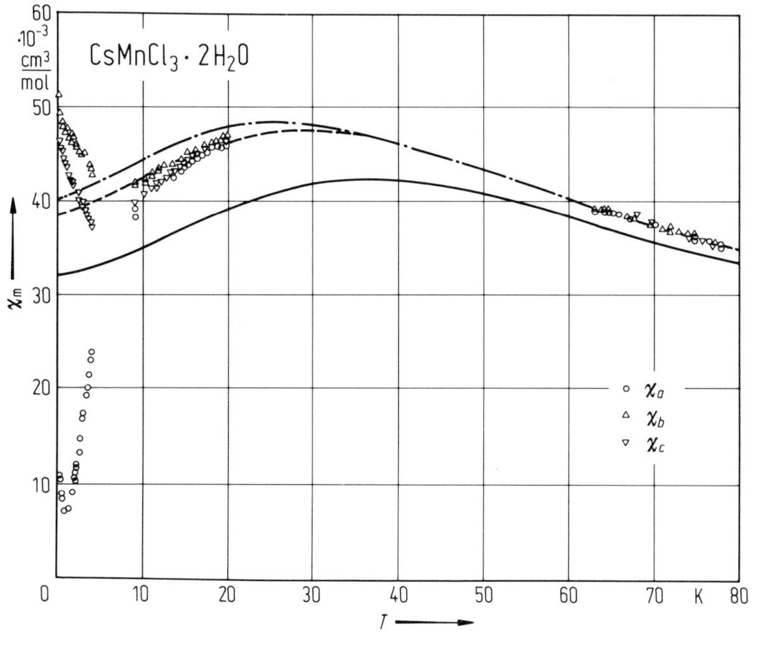

Fig. 420. $CsMnCl_3 \cdot 2H_2O$. Temperature dependence of χ_m along the *a*, *b* and *c* axes. Dot-dashed line: Fisher's curve for $S = 5/2$, $J/k = -3.115$ K; dashed line: Weng's interpolation estimate for $S = \frac{5}{2}$ curve with $J/k = -3.00$ K; solid line: Weng's $S = 1$ curve renormalized to $S = 5/2$, $J/k = -3.115$ K [68S1].

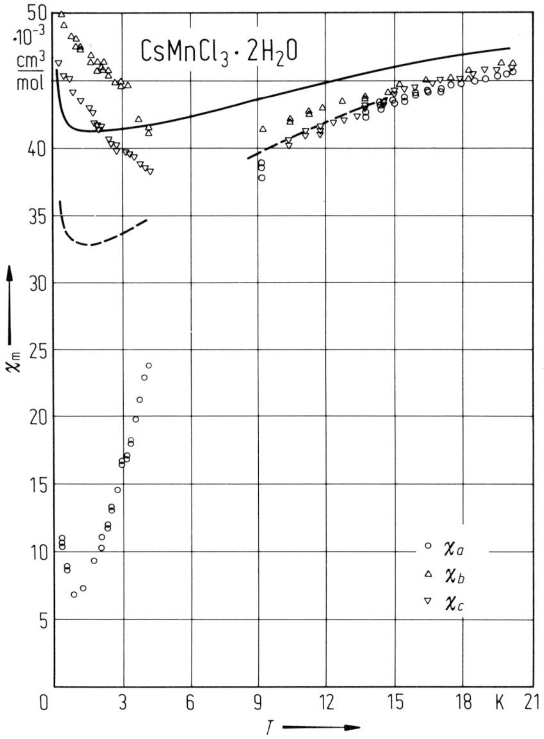

Fig. 421. $CsMnCl_3 \cdot 2H_2O$. Temperature dependence of the low temperature χ_m. Dashed line: average over 3 directions, solid line: Fisher's model [68S1].

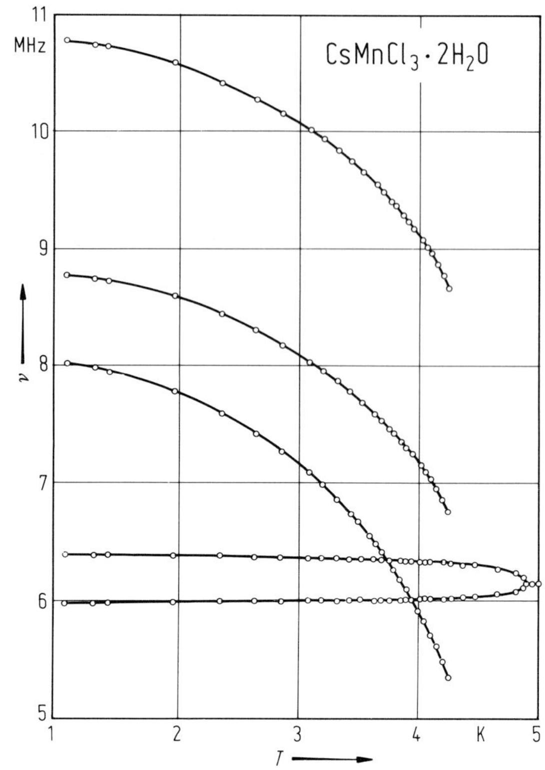

Fig. 423. $CsMnCl_3 \cdot 2H_2O$. Temperature dependence of NMR frequency of ^{35}Cl lines [69S1].

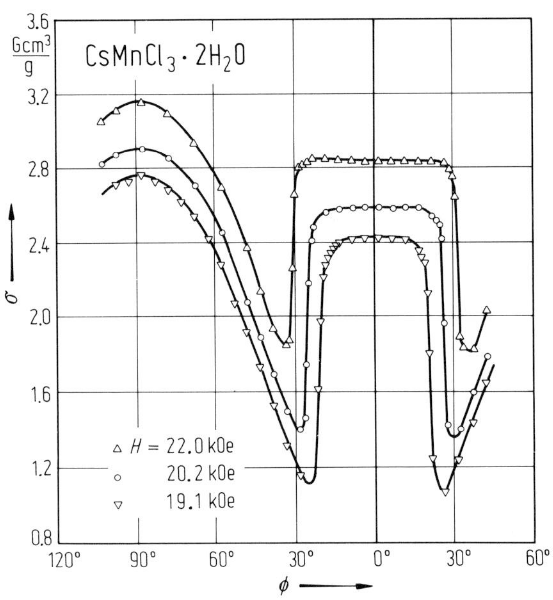

Fig. 422. $CsMnCl_3 \cdot 2H_2O$. Magnetization σ in the *ab* plane as a function of the angle ϕ between $\boldsymbol{H}$ and *b* axis at 1.07 K in different magnetic fields [69B1].

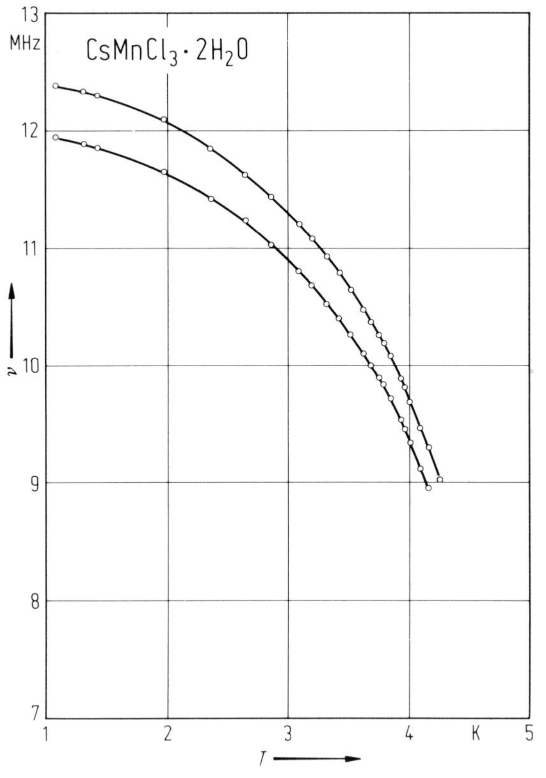

Fig. 424. $CsMnCl_3 \cdot 2H_2O$. Temperature dependence of proton frequency in zero applied field [69S1].

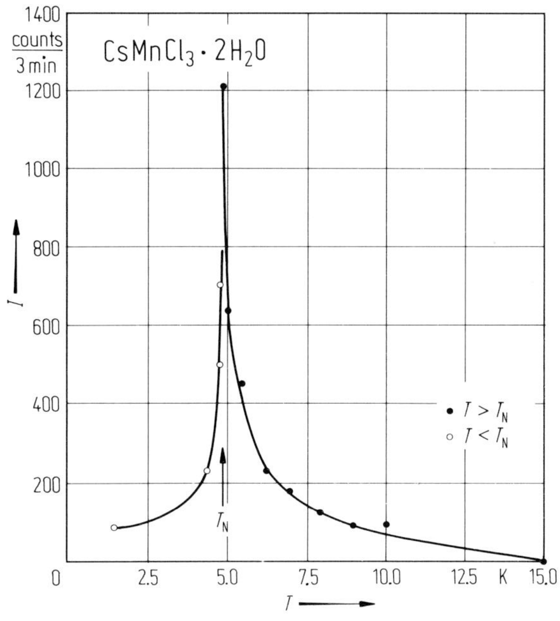

Fig. 425. $CsMnCl_3 \cdot 2H_2O$. Temperature dependence of the $(3,\frac{\bar{1}}{2},1)$ peak intensities. The background has been subtracted [70S2].

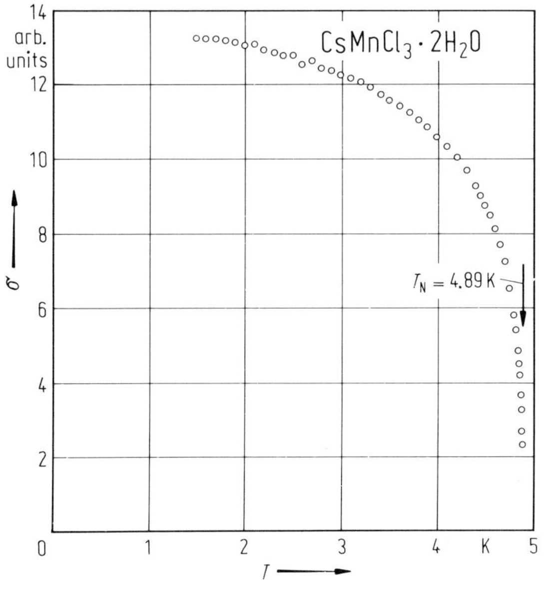

Fig. 426. $CsMnCl_3 \cdot 2H_2O$. Temperature dependence of relative magnetization carried out at $(3,\frac{\bar{1}}{2},1)$ peak [70S2].

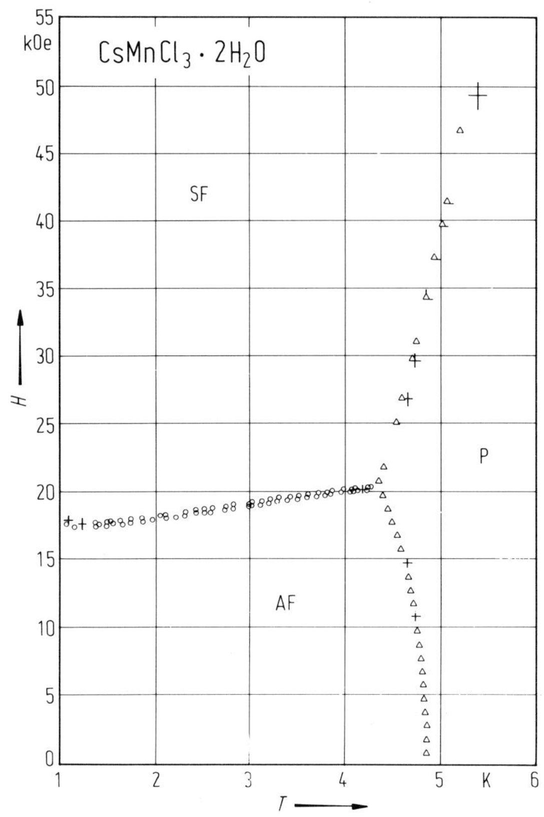

Fig. 427. $CsMnCl_3 \cdot 2H_2O$. Magnetic phase diagram, circles: magnetothermal data, triangles: heat capacity maxima, crosses: magnetostriction data [73B2].

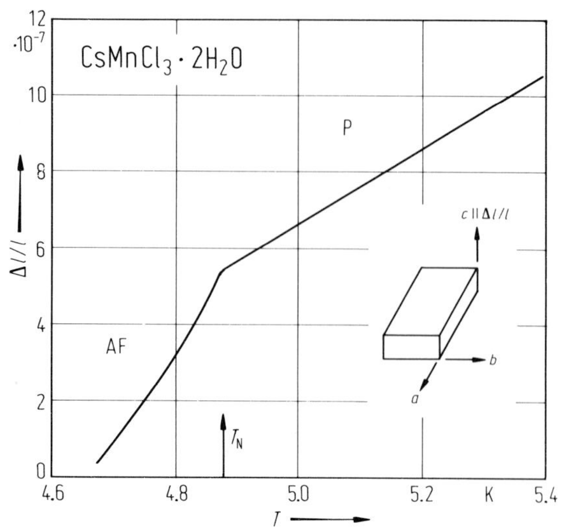

Fig. 428. $CsMnCl_3 \cdot 2H_2O$. Temperature dependence of the thermal expansion $\Delta l/l$ parallel to the c axis. [73B2].

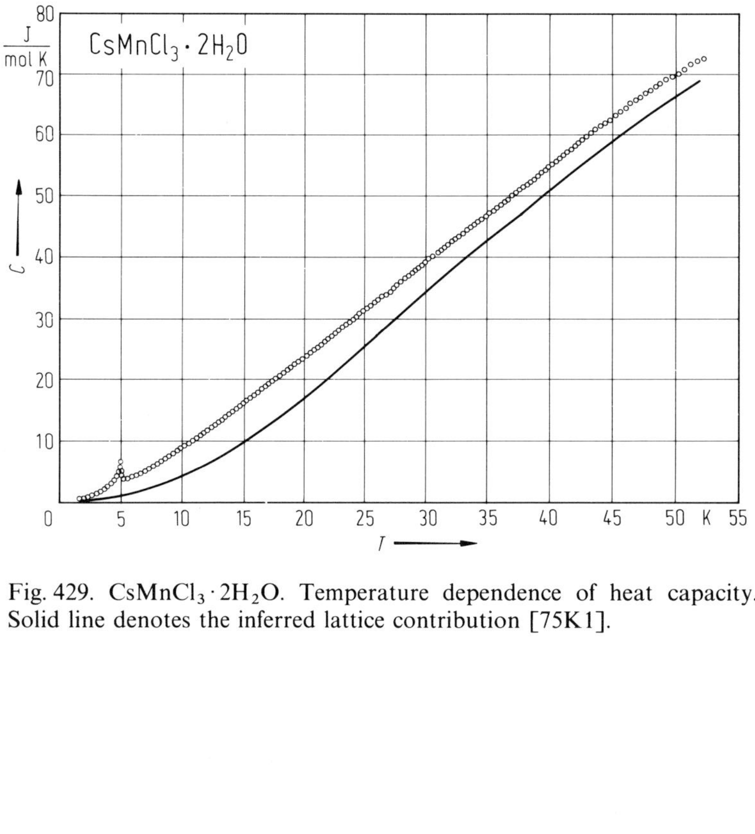

Fig. 429. $CsMnCl_3 \cdot 2H_2O$. Temperature dependence of heat capacity. Solid line denotes the inferred lattice contribution [75K1].

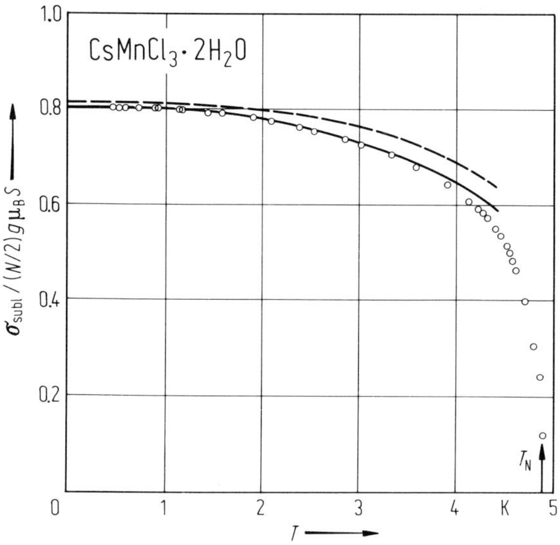

Fig. 430. $CsMnCl_3 \cdot 2H_2O$. Sublattice magnetization. Open circles: experimental deduced from NMR. Dashed line: spin wave prediction calculated with $J_1/k = -3.0$ K, $J_2/J_1 = 8 \cdot 10^{-3}$, $J_3 = 0$ or $J_1/k = -3.2$ K, $J_2/J_1 = 7 \cdot 10^{-3}$, $J_3 = 1/5\, J_2$. Solid line: calculated with $J_1/k = -3.0$ K, $J_2/J_1 = 4 \cdot 10^{-3}$, $J_3 = 0$. J_1: intrachain exchange constant along a; J_2, J_3: interchain exchange constants [76J1].

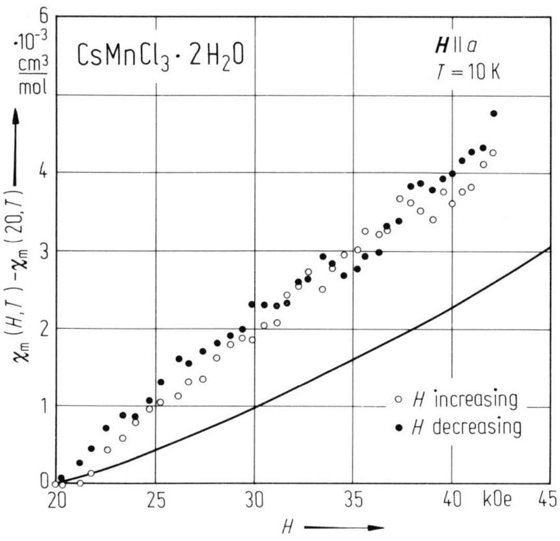

Fig. 431. $CsMnCl_3 \cdot 2H_2O$. Magnetic field dependence of $\chi_m(H, T) - \chi_m(20, T)$ at 10.0 K for $\boldsymbol{H} \| a$. Solid curve is calculated [76K4].

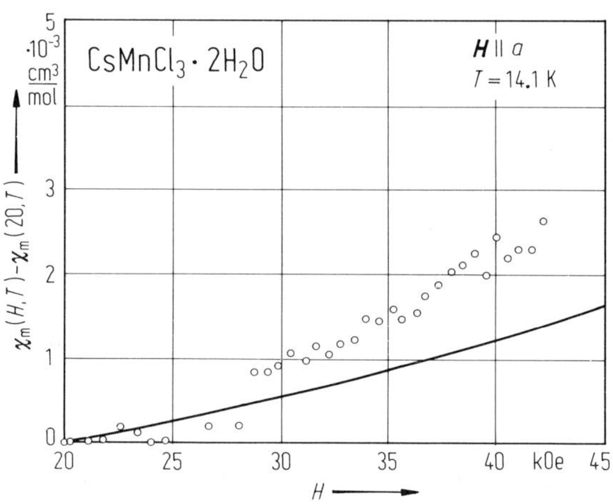

Fig. 432. $CsMnCl_3 \cdot 2H_2O$. Magnetic field dependence of $\chi_m(H, T) - \chi_m(20, T)$ for $\boldsymbol{H} \| a$ at 14.1 K. Solid curve is calculated [76K4].

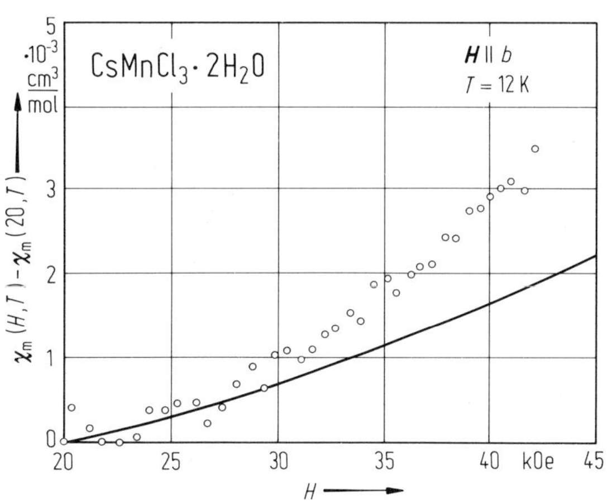

Fig. 433. $CsMnCl_3 \cdot 2H_2O$. Magnetic field dependence of $\chi_m(H, T) - \chi_m(20, T)$ for $\boldsymbol{H} \| b$ at 12 K. Solid curve is calculated [76K4].

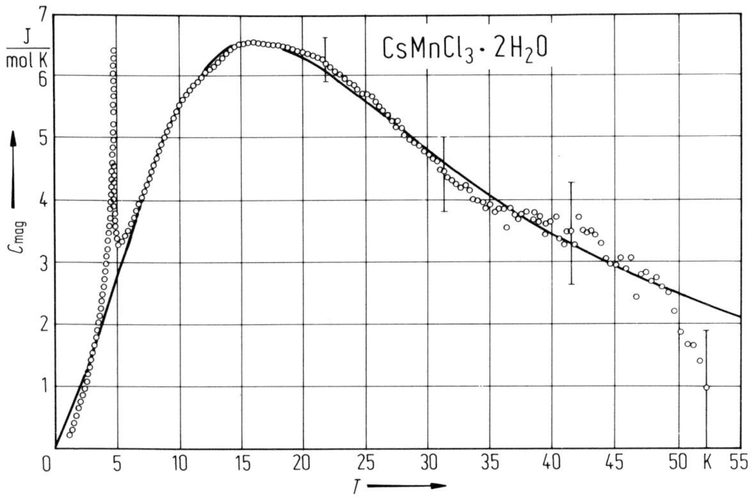

Fig. 434. $CsMnCl_3 \cdot 2H_2O$. Temperature dependence of magnetic heat capacity. Open circles correspond to $C_{mag} = C_{exp} - C_{L,calc}$, the error bars reflect the uncertainty in C_{exp}. The solid line denotes the theoretical estimate for a $S = 5/2$ Heisenberg linear chain system with $J/k = -3.0$ K [75K1, 77K3]. $C_{L,calc}$: calculated lattice contribution of heat capacity.

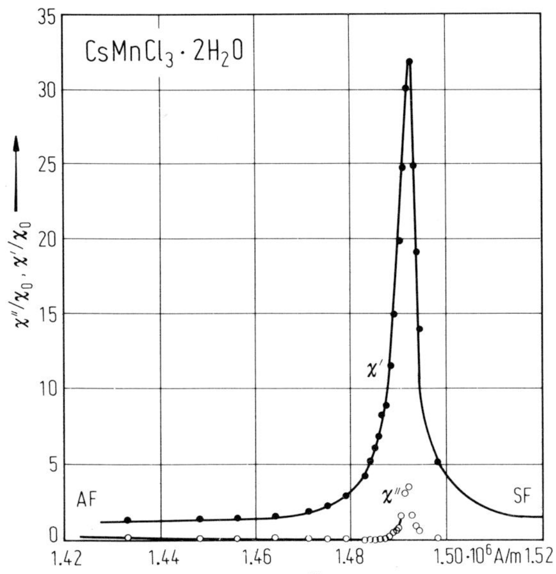

Fig. 435. $CsMnCl_3 \cdot 2H_2O$. Magnetic field dependence of the dispersion χ' and absorption χ'' relative to χ' at $H = 0$ (χ_0), showing the AF-SF phase transition at 2.776 K for frequency $\omega/2\pi = 127$ Hz and the amplitude of oscillating field $H_p = 33$ A/m [85C1].

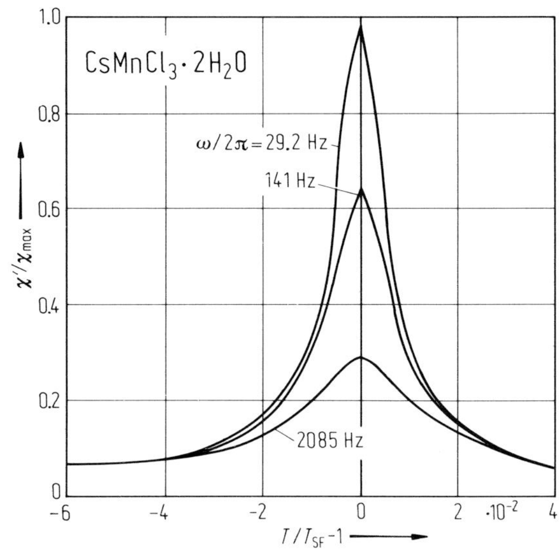

Fig. 436. $CsMnCl_3 \cdot 2H_2O$. Temperature dependence of the dispersion χ' near the spin-flop transition $H_{SF} = 1.44 \cdot 10^6$ A/m, $T_{SF} = 2.00$ K. χ_{max} is the maximum isothermal susceptibility at the phase transition [85C1].

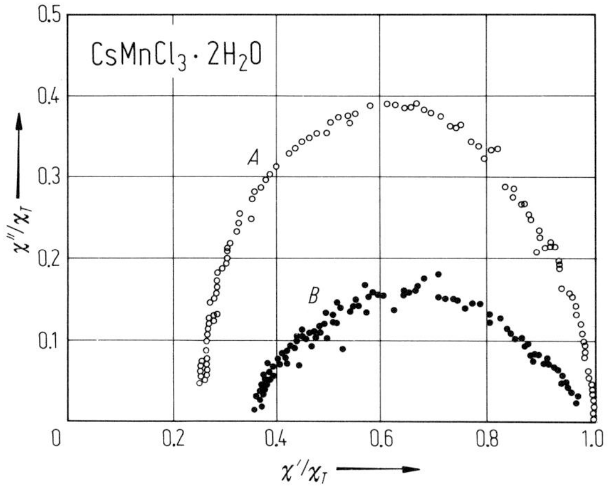

Fig. 437. $CsMnCl_3 \cdot 2H_2O$. Typical Argand (χ'' vs. χ') diagrams for the relaxation at the spin-flop transition obtained from measurements on a single crystal in contact with liquid helium (A) at $T = 1.894$ K, $H = 1.4288 \cdot 10^6$ A/m and (B) at $T = 2.358$ K, $H = 1.4624 \cdot 10^6$ A/m, below and above T_λ respectively [85C1]. T_λ: lambda transition of liquid He ($= 2.17$ K); χ_T: isothermal susceptibility in low frequency region.

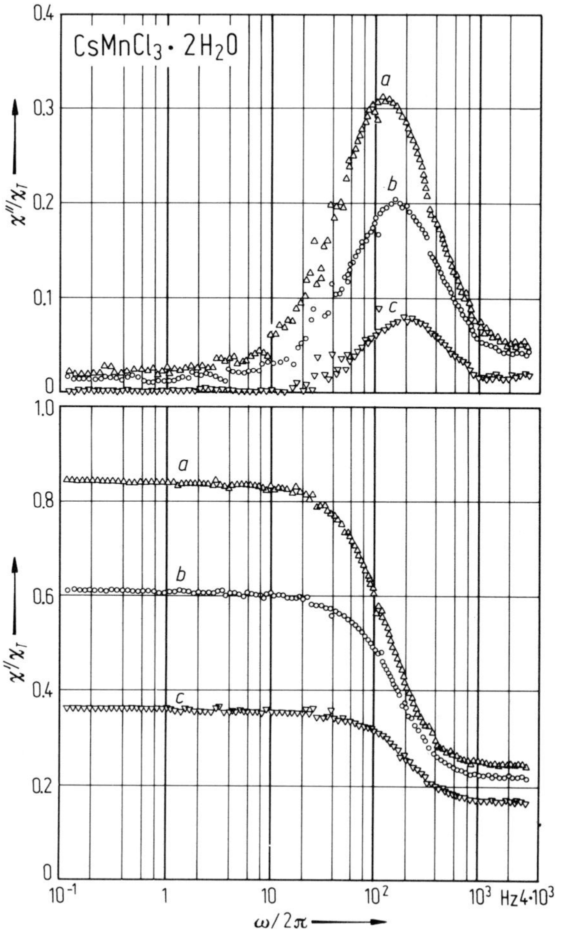

Fig. 438. $CsMnCl_3 \cdot 2H_2O$. Effect of small magnetic field changes on the frequency dependence of χ' and χ'' in the AF–SF intermediate state at $T = 1.955$ K. Curve a: $H = 1.4338 \cdot 10^6$ A/m, $\tau^{-1} = 7.85 \cdot 10^2\,s^{-1}$, b: $H = 1.4334 \cdot 10^6$ A/m, $\tau^{-1} = 10.0 \cdot 10^2\,s^{-1}$, c: $H = 1.4331 \cdot 10^6$ A/m, $\tau^{-1} = 13.5 \cdot 10^2\,s^{-1}$ [85C1]. τ: relaxation time of domain wall relaxation.

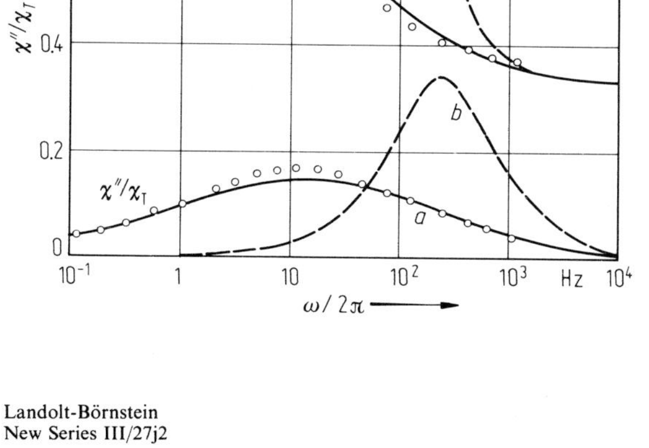

Fig. 439. $CsMnCl_3 \cdot 2H_2O$. Comparison between the experimental data and theoretical calculations (curves a) of $\chi'(\omega)$, $\chi''(\omega)$ at $H = 1.4624 \cdot 10^6$ A/m, $T = 2.358$ K. The expected intrinsic relaxation behaviour is represented by the curves b for $\tau = 0.616 \cdot 10^{-3}$s [85C1].

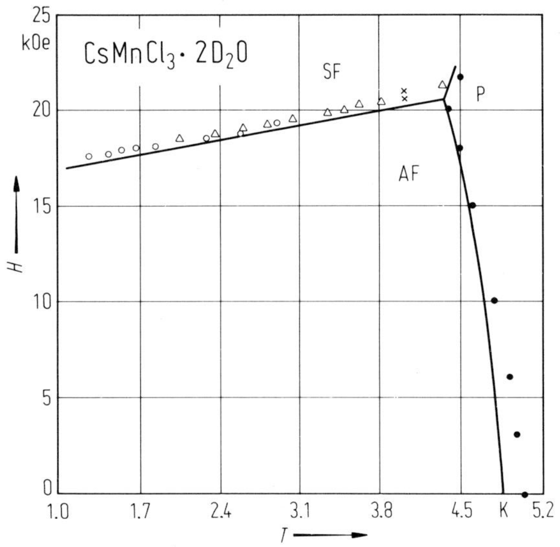

Fig. 440. $CsMnCl_3 \cdot 2D_2O$. Magnetic phase diagram. The solid lines represent the phase boundaries for $CsMnCl_3 \cdot 2H_2O$. All field measurements were made in the *bc* plane [72F1]. Measuring points observed by: open circles, triangles, crosses: isentropes, full circles: heat capacity.

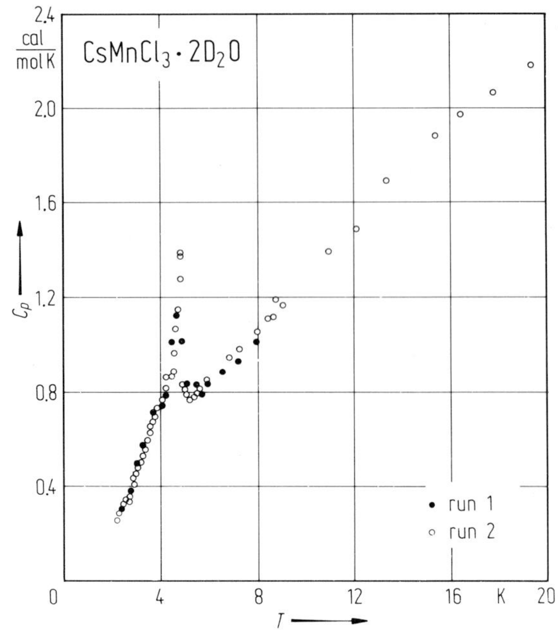

Fig. 441. $CsMnCl_3 \cdot 2D_2O$. Temperature dependence of heat capacity C_p [72F1].

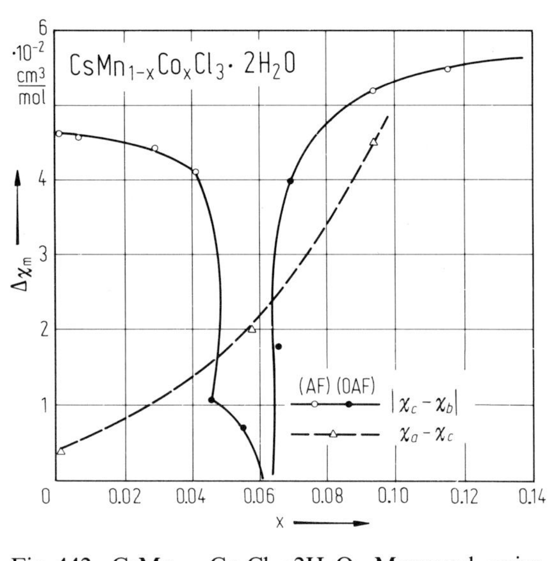

Fig. 442. $CsMn_{1-x}Co_xCl_3 \cdot 2H_2O$. Measured anisotropies in the susceptibility at 2.0 K [80Y1]. OAF: oblique antiferromagnetic (phase).

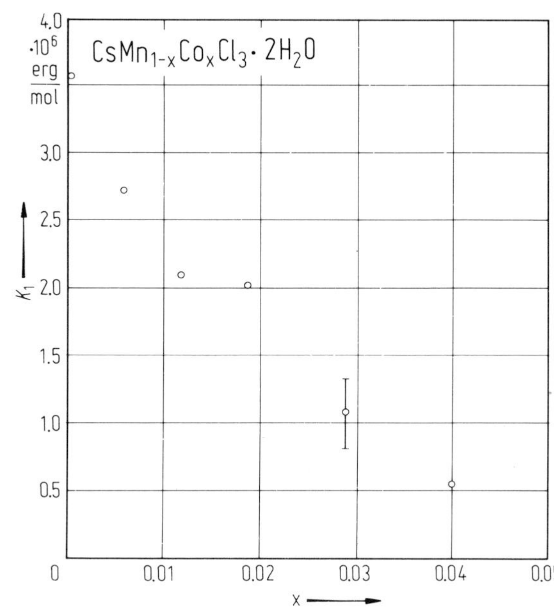

Fig. 443. $CsMn_{1-x}Co_xCl_3 \cdot 2H_2O$. Concentration dependence of the anisotropy constant K_1, the anisotropy energy in the *bc* plane at 4.2 K [80Y1].

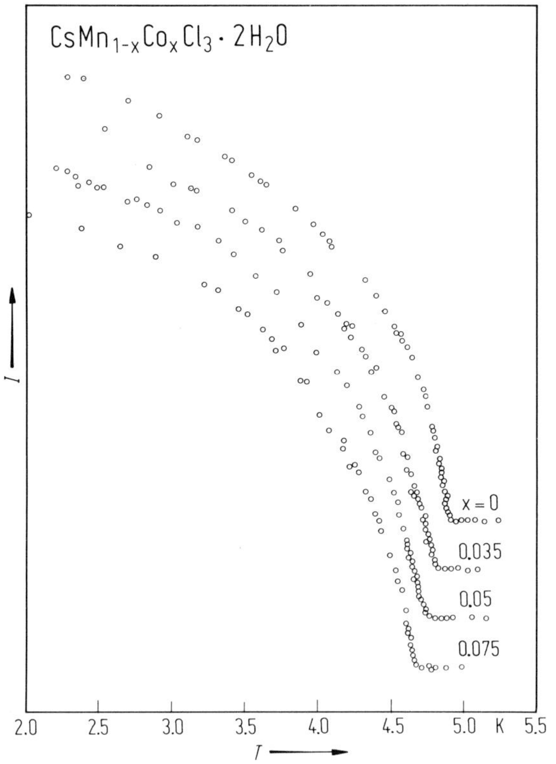

Fig. 444. $CsMn_{1-x}Co_xCl_3\cdot 2H_2O$. Temperature dependence of intensity for (1, 1/2, 1) magnetic reflection for x = 0, 0.035, 0.05 and 0.075 [82A1].

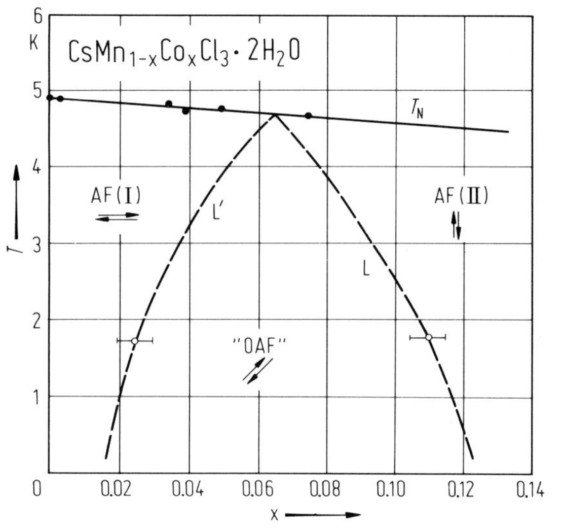

Fig. 445. $CsMn_{1-x}Co_xCl_3\cdot 2H_2O$. Phase diagram. The T_N are due to [82A1]. The boundary between the antiferromagnetic and OAF (oblique antiferromagnetic) phase is obtained from the concentration dependence of NMR spectra in zero applied field [84K2, 84K1]. The lines L and L′ represent transition temperatures of the $S_{\parallel}$ and $S_{\perp}$ domains, respectively, where $S_{\parallel}$: domain with moments parallel to the *b* axis, $S_{\perp}$: domain with moments parallel to the *c* axis. See also [85Y1, 89K1].

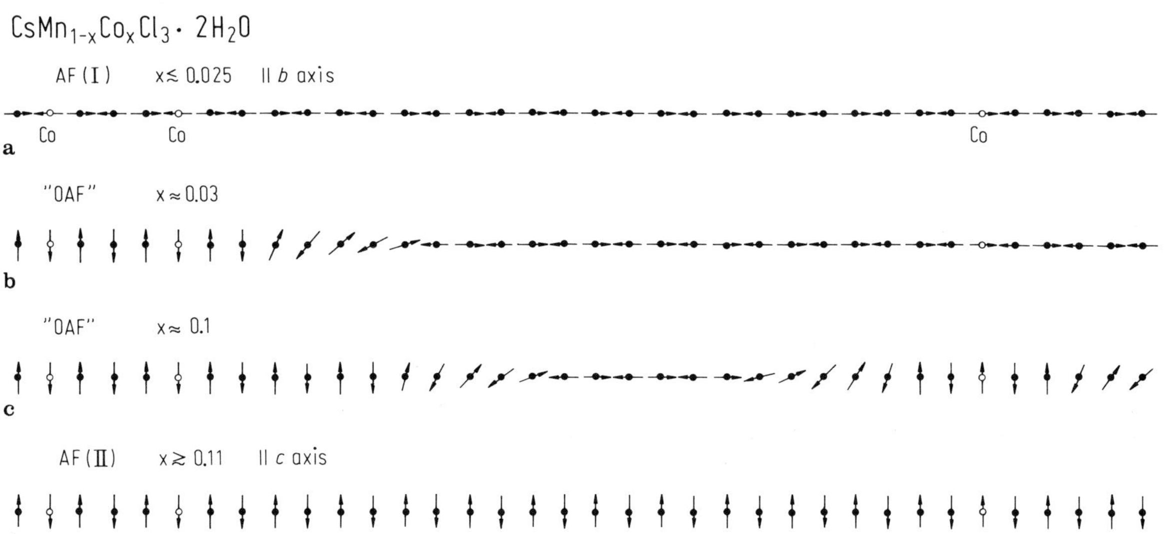

Fig. 446. $CsMn_{1-x}Co_xCl_3\cdot 2H_2O$. Proposed model of the spin structure at 1.7 K. For simplicity, the one-dimensional model i.e. the chain along *a* axis is assumed. The transient region will spread more than 10 lattice points [84K1, 84K2].

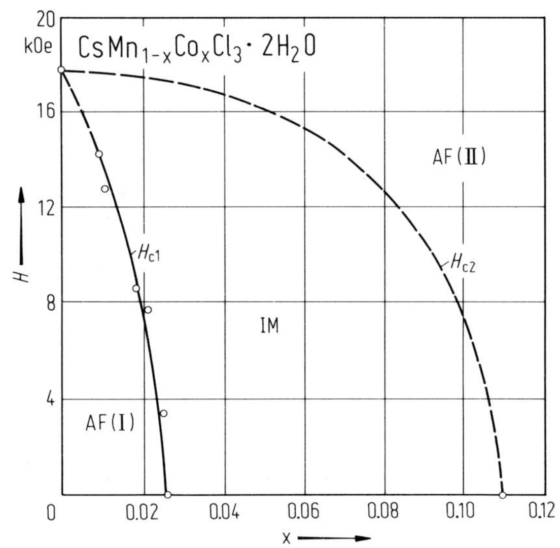

Fig. 447. $CsMn_{1-x}Co_xCl_3 \cdot 2H_2O$. Phase diagram at 1.7 K in the external field parallel to the *b* axis. IM: intermediate phase [84K3, 86K1]. cf. Fig. 449.

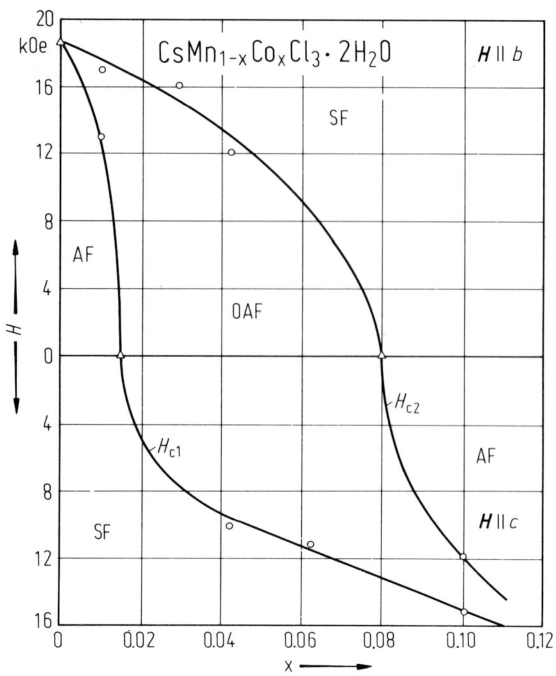

Fig. 449. $CsMn_{1-x}Co_xCl_3 \cdot 2H_2O$. H–x phase diagram of random antiferromagnetic mixture obtained from magnetization (circles) and magnetic resonance (triangles) measurements. [86N1, 90I1].

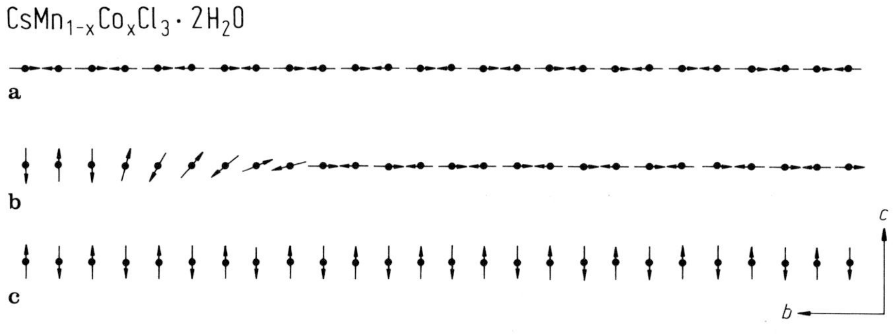

Fig. 448. $CsMn_{1-x}Co_xCl_3 \cdot 2H_2O$. Spin structure. (a) The antiferromagnetic phase I at $H \leq H_{c1}$, (b) the intermediate phase at $H_{c1} \leq H \leq H_{c2}$, (c) the antiferromagnetic phase II at $H \geq H_{c2}$. See also Fig. 447 [84K2, 84K3].

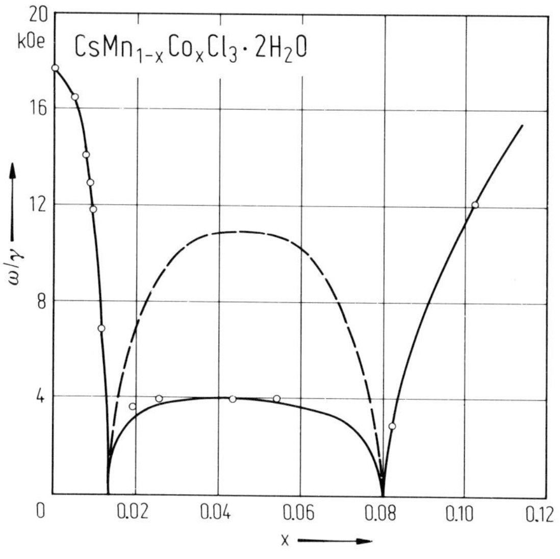

Fig. 450. $CsMn_{1-x}Co_xCl_3 \cdot 2H_2O$. Concentration dependence of zero-field resonance frequencies. The dashed curve shows the result of the calculation for a two-sublattice model with anisotropy terms K_1 (x) and L (x) determined from H–x phase diagram of Fig. 449 [86N1, 90I1].

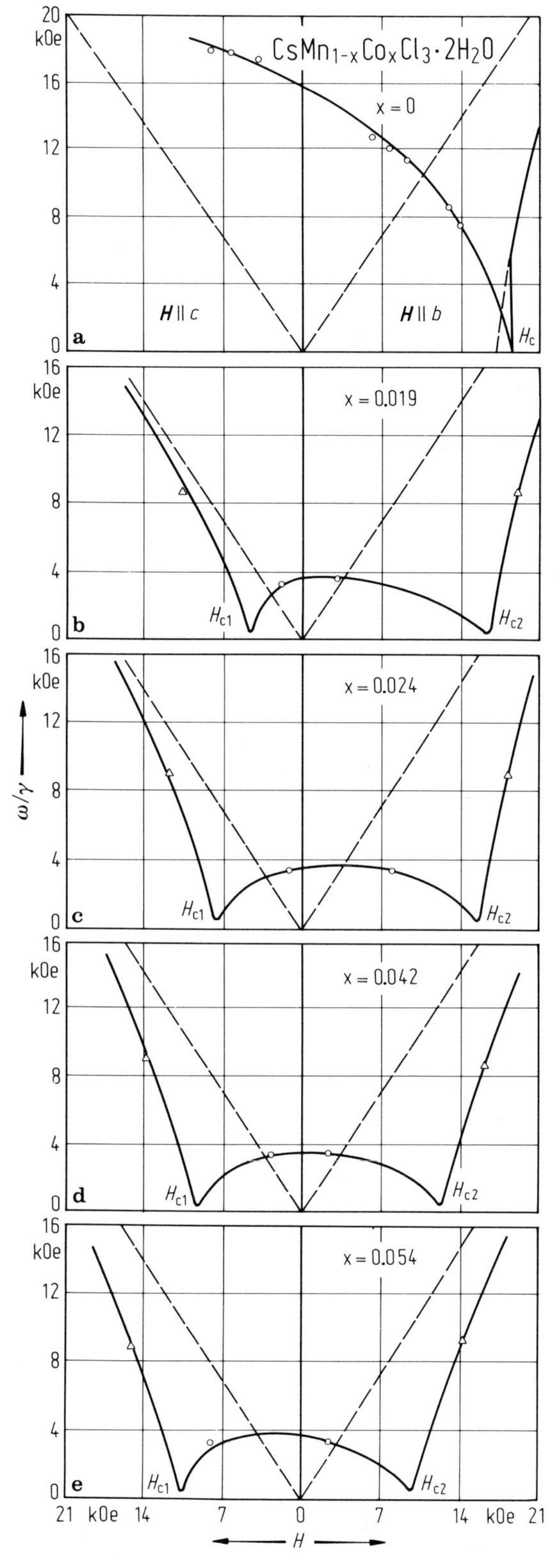

Fig. 451. $CsMn_{1-x}Co_xCl_3 \cdot 2H_2O$. ω/γ–H diagrams. Dashed lines: paramagnetic. Solid lines based on theoretical results. Open triangles estimated by extrapolation from angular dependence of resonance fields [86N1].

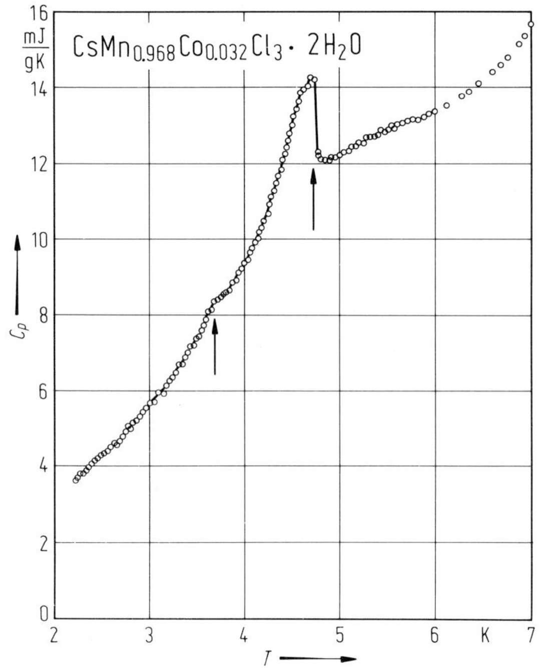

Fig. 452. $CsMn_{0.968}Co_{0.032}Cl_3 \cdot 2H_2O$. Temperature dependence of the specific heat capacity including contribution from the lattice. Arrows show the transition temperatures [85Y1].

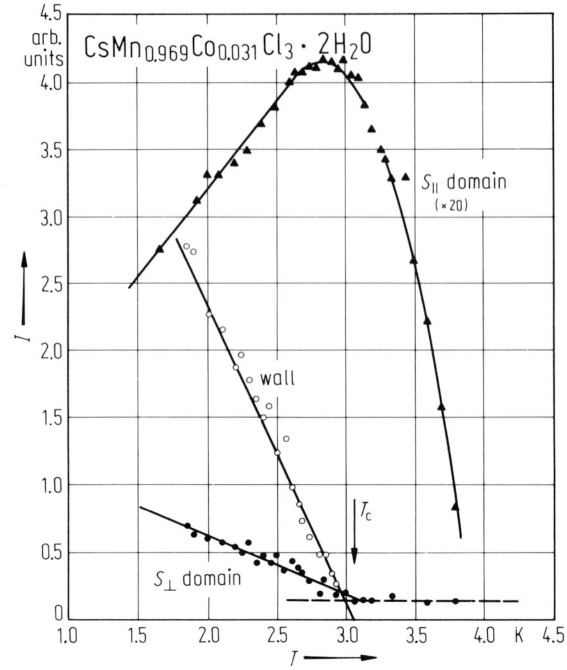

Fig. 454. $CsMn_{0.969}Co_{0.031}Cl_3 \cdot 2H_2O$. Temperature dependence of NMR intensities of the lines due to $S_{||}$ and $S_{\perp}$ domains and the wall (see Fig. 455). Solid and dashed lines are guide to the eye [89Kl].

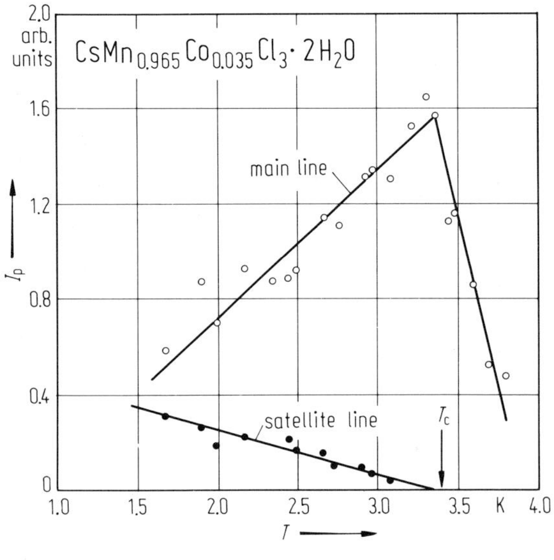

Fig. 453. $CsMn_{0.965}Co_{0.035}Cl_3 \cdot 2H_2O$. Temperature dependence of NMR peak intensity in zero applied field. The transition from the AF (I) to IM phase occurs at T_c (see Fig. 445) [89K1].

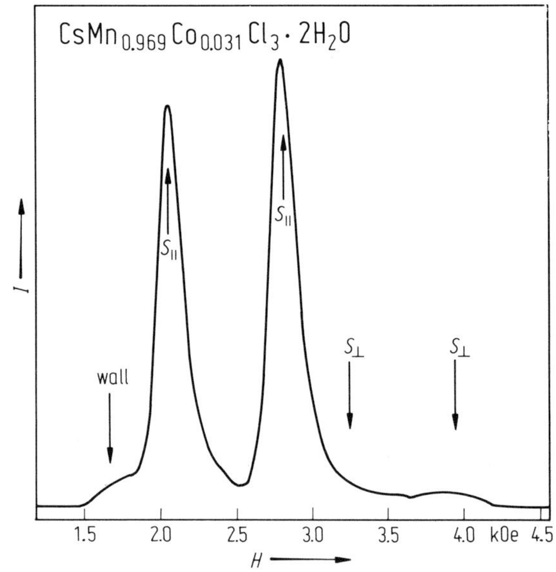

Fig. 455. $CsMn_{0.969}Co_{0.031}Cl_3 \cdot 2H_2O$. External magnetic field dependence of the proton NMR lineshape at 1.7 K and frequency of 20 MHz in a field parallel to the c axis [89Kl].

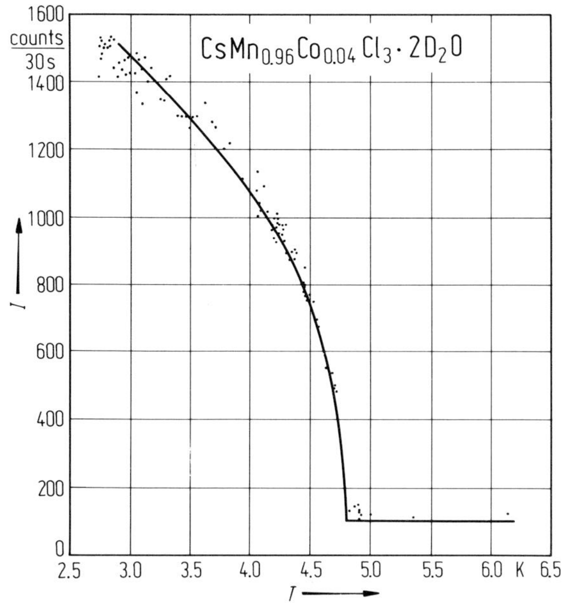

Fig. 456. $CsMn_{0.96}Co_{0.04}Cl_3 \cdot 2D_2O$. Temperature dependence of intensity of magnetic (1, $\frac{1}{2}$, 1) reflection around T_N (= 4.80 K). Solid line is represented by $I = A\,(1 - T/T_N)^{2\beta}$ with $\beta = 0.23$ [81A1].

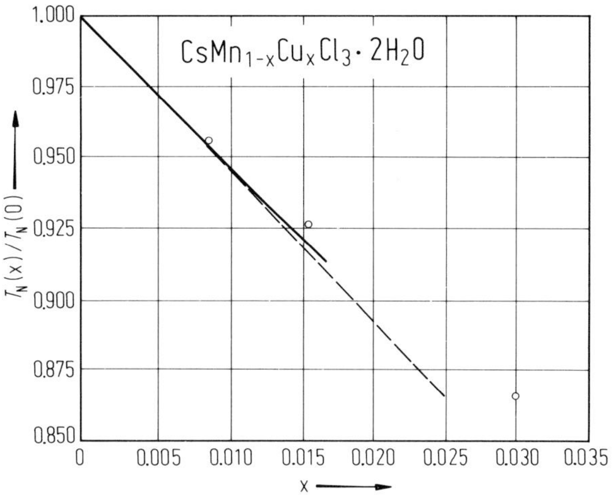

Fig. 457. $CsMn_{1-x}Cu_xCl_3 \cdot 2H_2O$. Concentration dependence of $T_N(x)/T_N(0)$. Dashed line represents theoretical prediction [80S1].

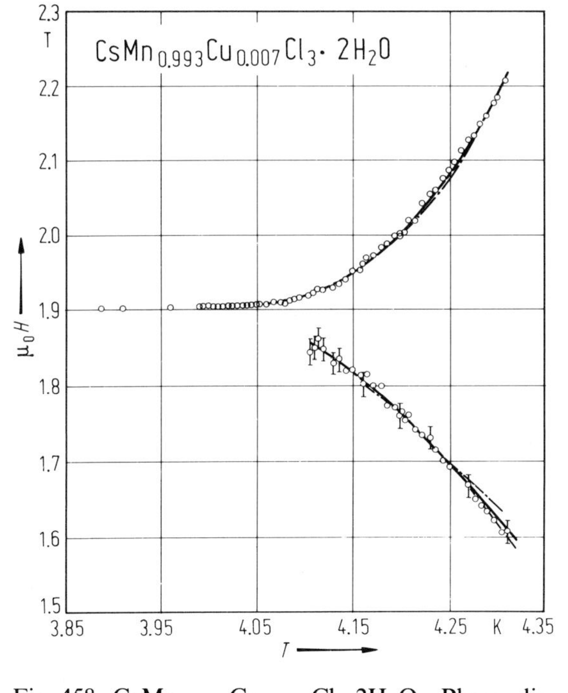

Fig. 458. $CsMn_{0.993}Cu_{0.007}Cl_3 \cdot 2H_2O$. Phase diagram near the bicritical point. Three lines represent different fits [83V1].

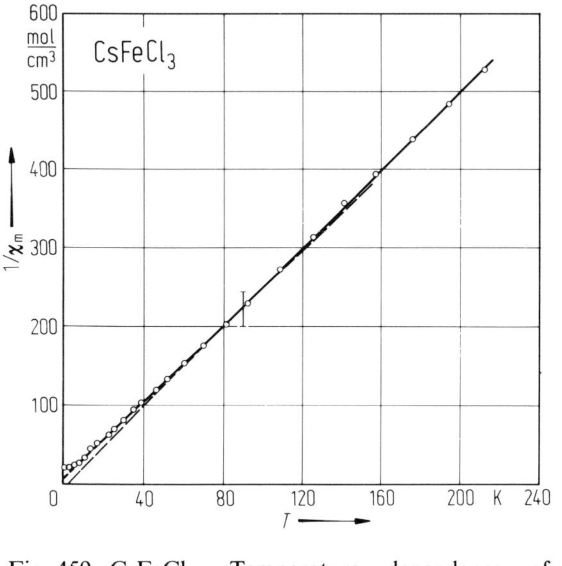

Fig. 459. $CsFeCl_3$. Temperature dependence of χ_m^{-1} [73W1].

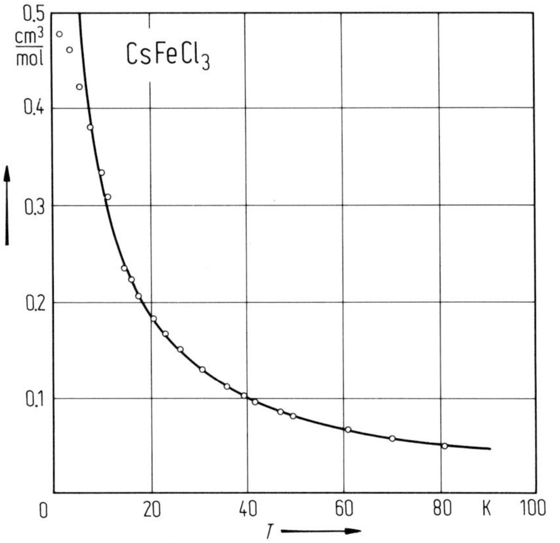

Fig. 460. $CsFeCl_3$. Temperature dependence of χ_m. Solid line: theoretical fit [73W1].

For Fig. 461, see next page.

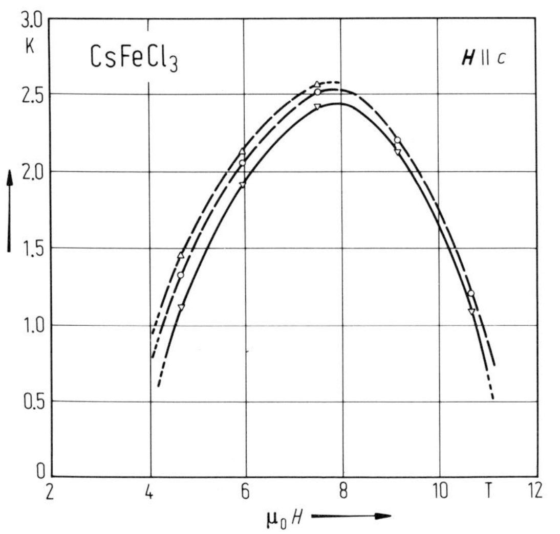

Fig. 463. $CsFeCl_3$. T–H phase boundary [81H1]. Curves correspond to maxima of heat capacity vs. H plots of Fig. 462.

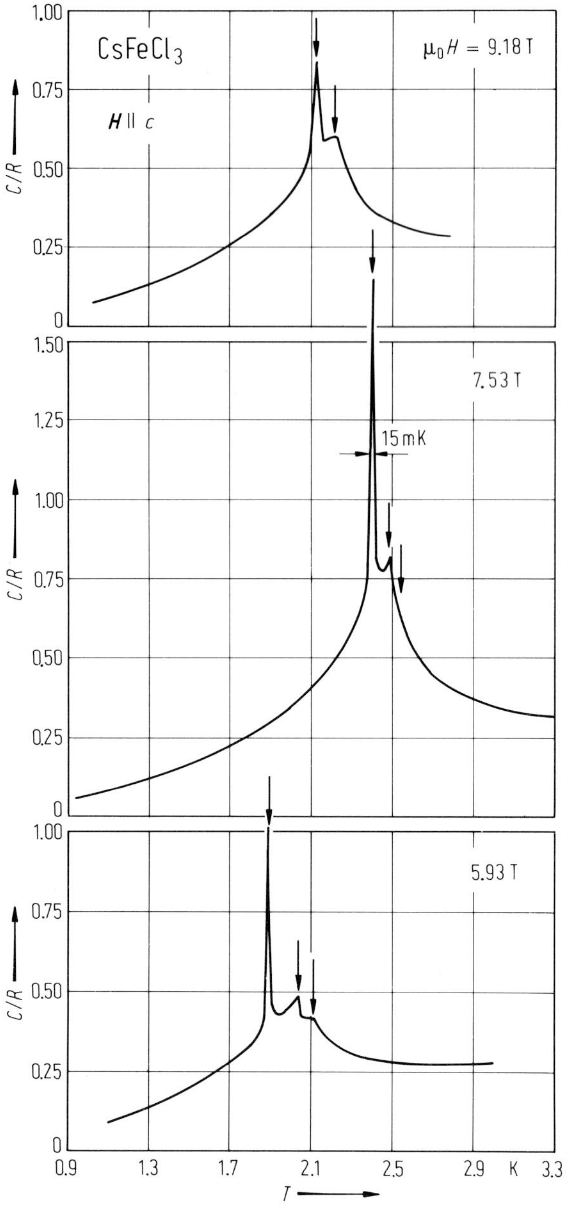

Fig. 462. $CsFeCl_3$. Heat capacity under several magnetic fields for $\boldsymbol{H}$ parallel to the c axis [81H1]. Arrows point to maxima.

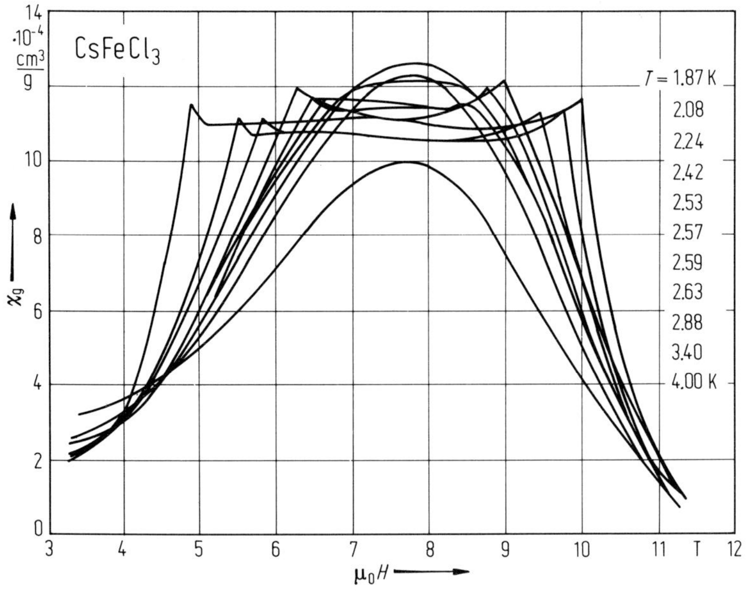

Fig. 461. $CsFeCl_3$. Magnetic field dependence of ac susceptibility χ_g under the magnetic field parallel to the c axis. At higher temperatures the broad peak appears just at the field value $H_{cross} = D/g\mu_B$ where one of the excited doublet levels makes cross to the ground singlet [81H1].

For Fig. 461, see next page.

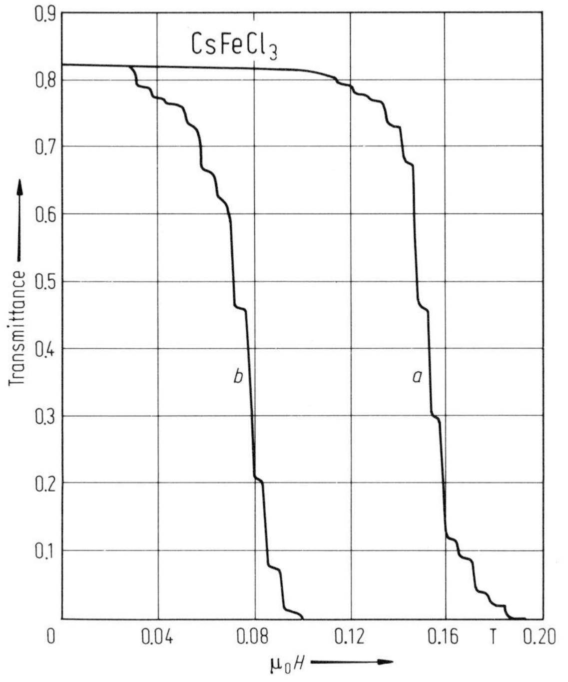

Fig. 464. $CsFeCl_3$. Magnetic field dependence of the transmittance at 21K. The field was applied perpendicular to the c axis. The light of 540 nm wavelength was propagated along H. Curve a was obtained in an increasing field, curve b in a decreasing field [85T2].

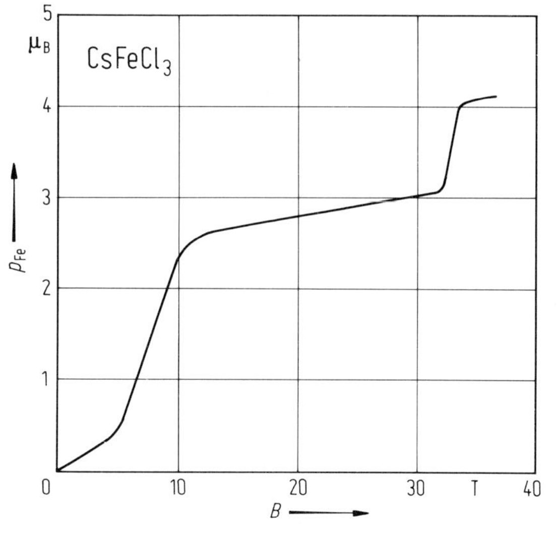

Fig. 465. $CsFeCl_3$. Magnetic field dependence of p_{Fe} at 4.2K measured by a pulsed magnet for $\boldsymbol{B} \parallel c$ [87C1].

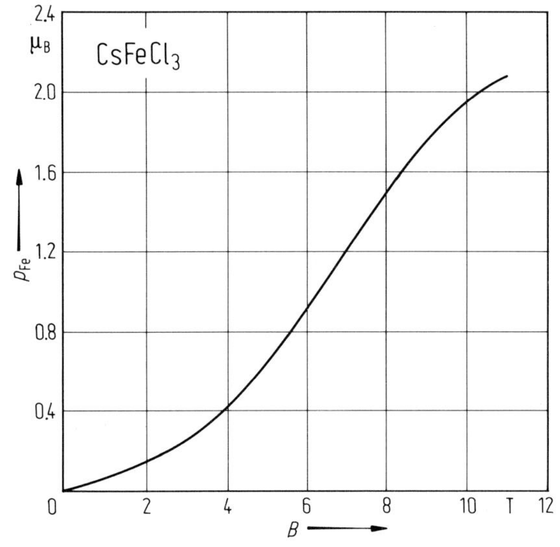

Fig. 466. $CsFeCl_3$. Magnetic field dependence of p_{Fe} at 4.2 K using a superconducting selenoid [87C1].

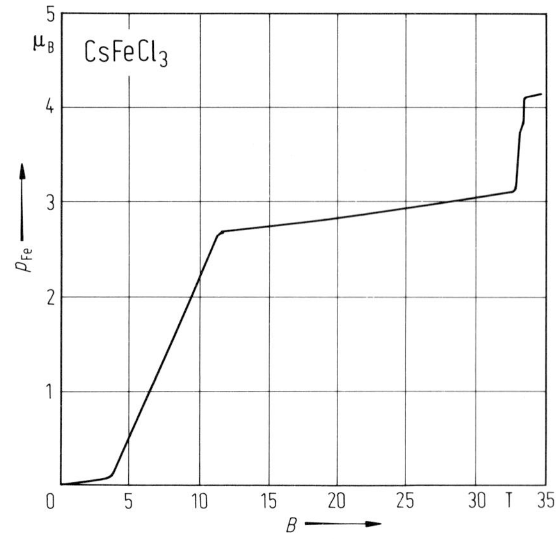

Fig. 467. $CsFeCl_3$. Magnetic field dependence of p_{Fe} for $\boldsymbol{B} \parallel c$, at 1.3 K observed by a pulsed magnet [87C1].

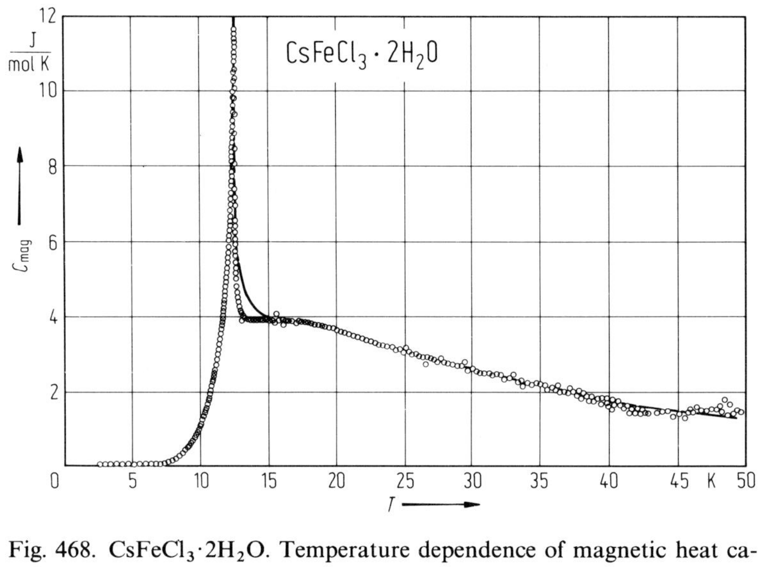

Fig. 468. $CsFeCl_3 \cdot 2H_2O$. Temperature dependence of magnetic heat capacity. Circles: experimental data minus calculated lattice contribution. Full curve is calculated [85K1].

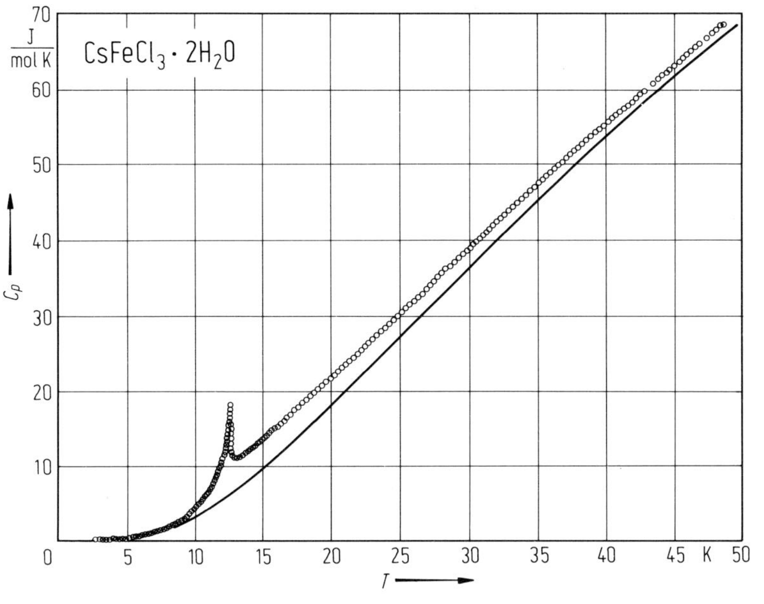

Fig. 469. $CsFeCl_3 \cdot 2H_2O$. Temperature dependence of C_p. Solid line: lattice contribution [85K1].

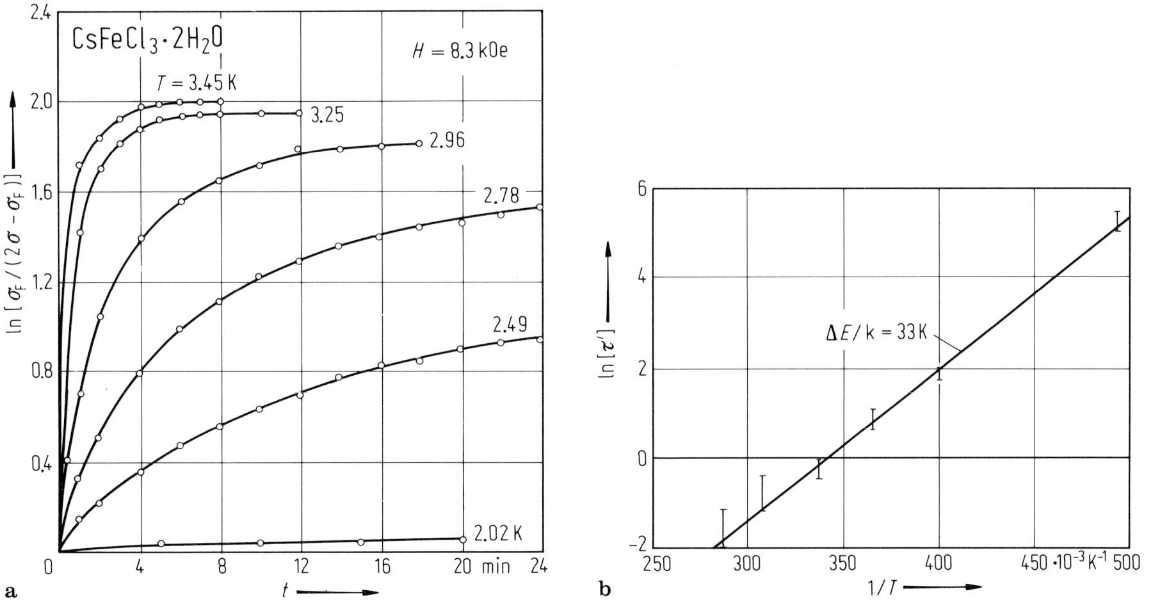

Fig. 470. $CsFeCl_3 \cdot 2H_2O$. (a) Variation of the magnetization with time for different temperatures, after a rapid field change starting in the F phase. (b) The initial relaxation time $\tau'(t = 0)$ (in min) plotted against T^{-1}. The drawn line corresponds to $\tau'(t = 0) \propto \exp(\Delta E/kT)$ [85S2].

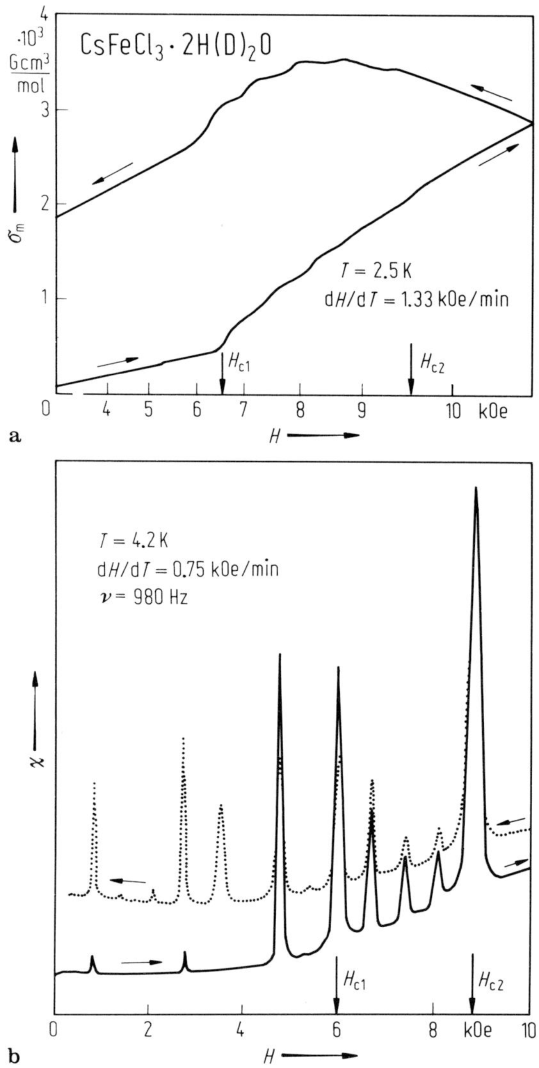

Fig. 471. $CsFeCl_3 \cdot 2H(D)_2O$. Magnetic field dependence of σ_m (a) and χ (in arb. units) (b) with $\boldsymbol{H} \parallel c$ at low *T*. The field values at which an instability occurs in σ_m are not affected by the direction of the field sweep. Note the shift in the vertical χ axis [85S2].

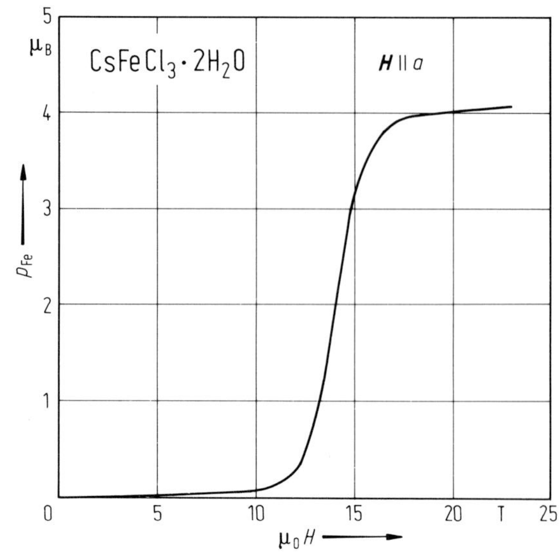

Fig. 472. $CsFeCl_3 \cdot 2H_2O$. Magnetic field dependence of p_{Fe} along *a* axis at 4.2K [89T1].

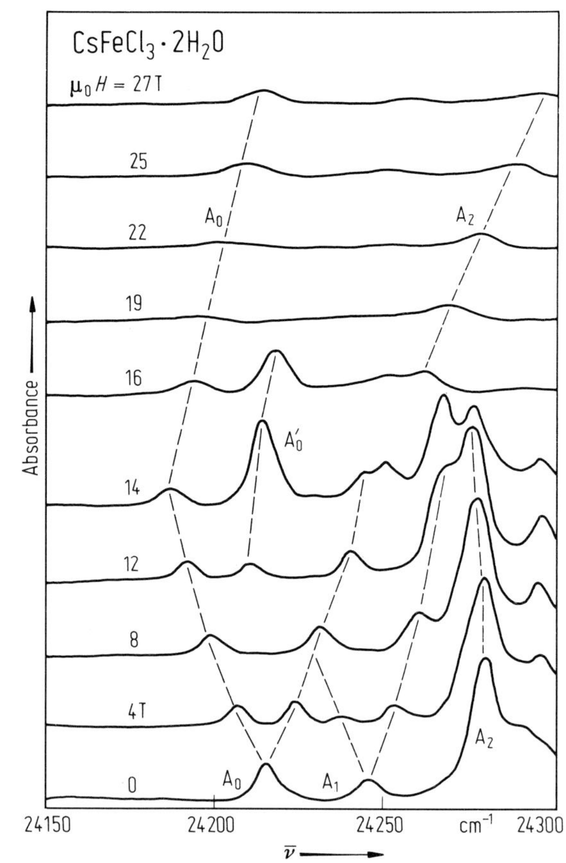

Fig. 473. $CsFeCl_3 \cdot 2H_2O$. Absorption spectra at various magnetic fields along the *a* axis at 4.2K, measured in Voigt configuration with a polarization $\tilde{\boldsymbol{E}} \parallel a$; $H \parallel a$. Broken lines are guide for the eye [89T1].

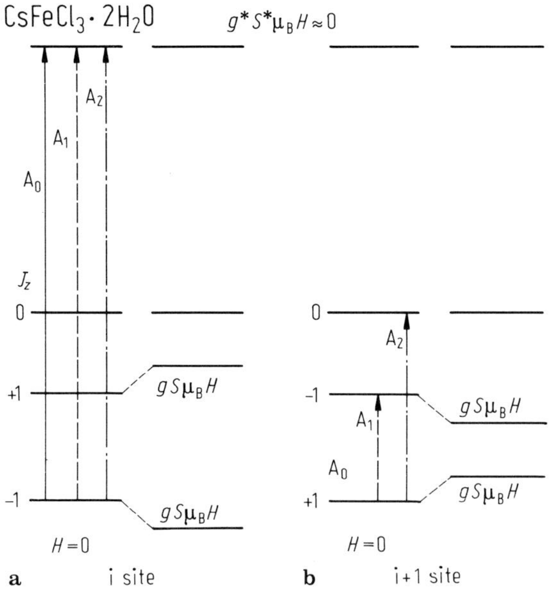

Fig. 474. $CsFeCl_3 \cdot 2H_2O$. Schematic energy diagram (a) at site i, (b) the nearest neighbor site i + 1 in an antiferromagnetic chain. A_0: exciton line, A_1: magnon sideband, A_2: another magnon sideband (see Fig. 473) [89T1].

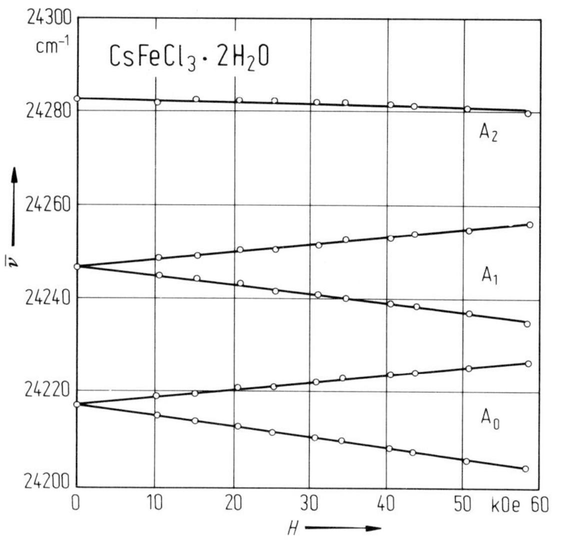

Fig. 475. $CsFeCl_3 \cdot 2H_2O$. Magnetic field dependence of energy shift of A_0, A_1 and A_2 lines with electric and magnetic field of incident light $\tilde{\boldsymbol{E}} \parallel a$, $\tilde{\boldsymbol{H}} \parallel b$ and $\boldsymbol{H} \parallel a$ at 1.7K [90O1]. Solid lines are guide to the eye.

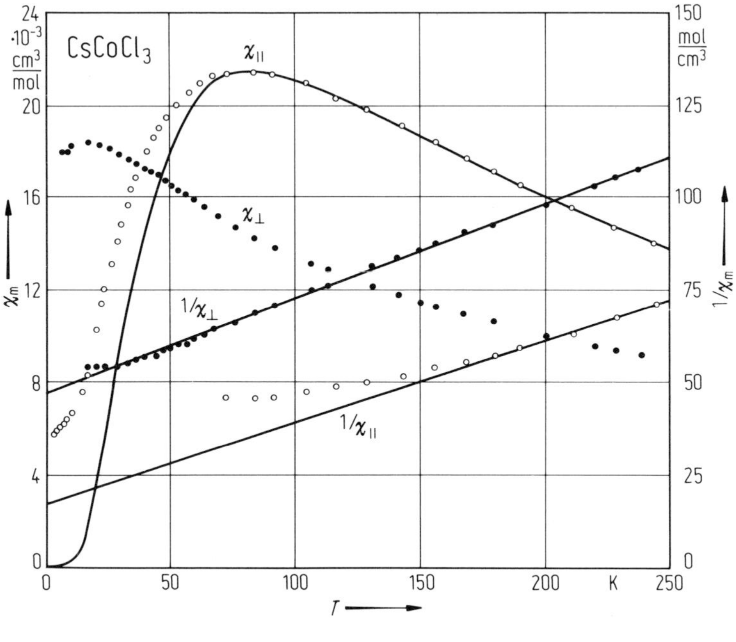

Fig. 476. $CsCoCl_3$. Temperature dependence of χ_m and χ_m^{-1} of a single crystal. The solid curve is calculated for parallel χ (along [0001]) for linear antiferromagnetic chains with Ising spin $S = 1/2$ [69A1]. $\chi_{||}$: along [0001], $\chi_{\perp}$: along [11$\bar{2}$0].

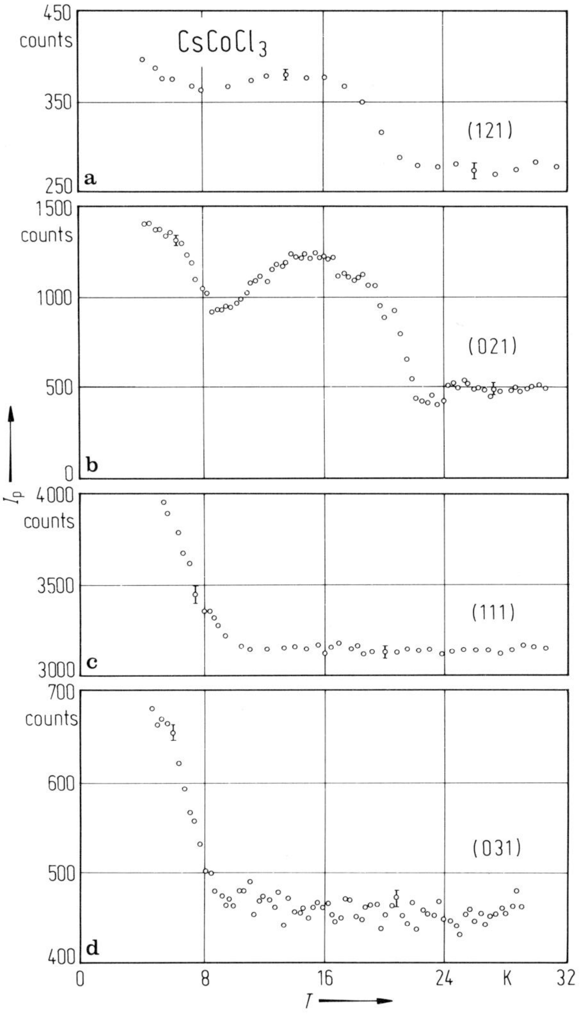

Fig. 477. $CsCoCl_3$. Temperature dependence of the neutron diffraction ($\lambda = 1.02$Å) for the magnetic reflection (a) powder (1,2,1), (b) single crystal (0,2,1), (c) single crystal (1,1,1), (d) single crystal (0,3,1) [74M2].

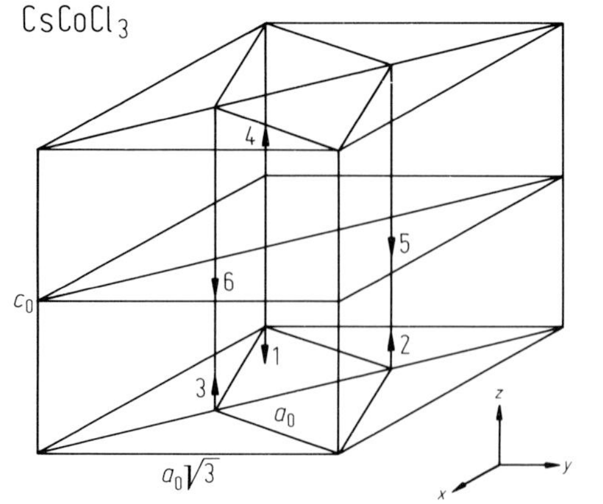

Fig. 478. $CsCoCl_3$. Magnetic unit cell (H-cell). The spins of the Co^{2+} ions are shown [74M2]. a_0: unit cell parameter according to space group $P6_3/mmc$.

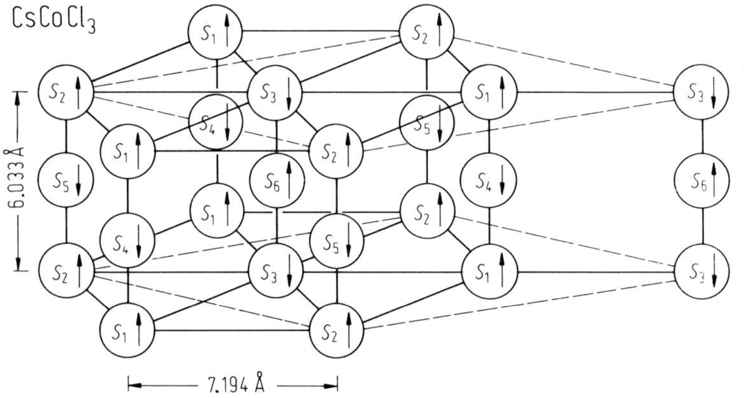

Fig. 479 $CsCoCl_3$. Magnetic structure. Dashed lines show magnetic unit cell. In high temperature phase, spins of sublattices 1 and 4 are uncorrelated to those of other sublattices [78M1].

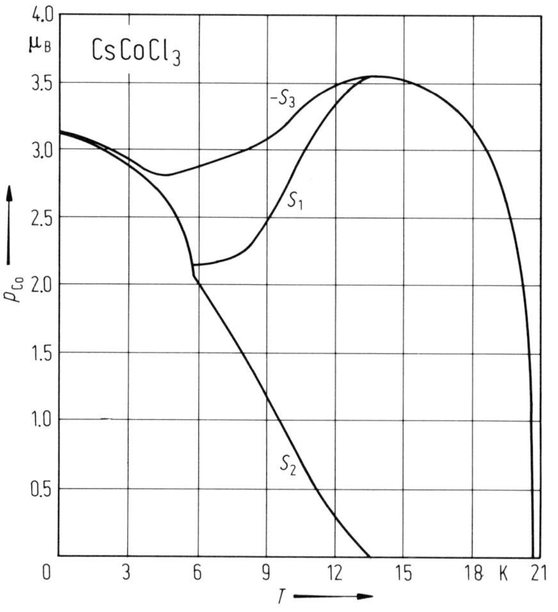

Fig. 480. $CsCoCl_3$. Probable temperature dependence of sublattice magnetization [78M1].

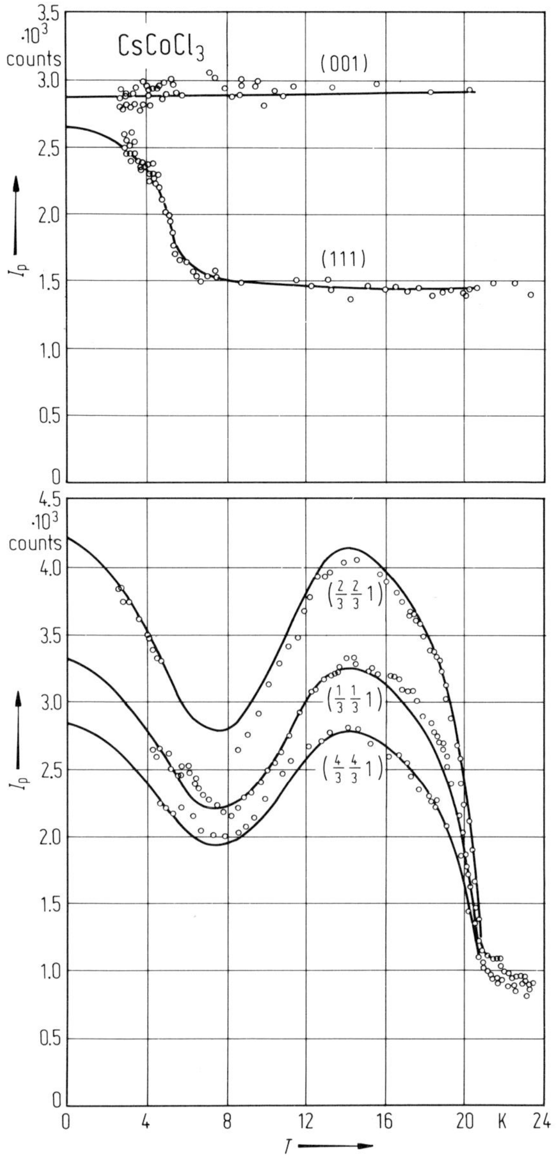

Fig. 481. $CsCoCl_3$. Temperature dependence of neutron peak intensity of typical magnetic reflections [78M1]. Solid curves calculated for sublattice magnetization given in Fig. 480.

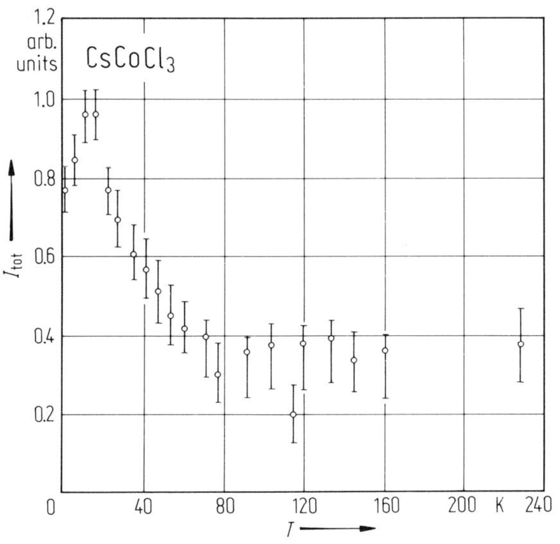

Fig. 482. $CsCoCl_3$. Temperature dependence of the total intensity of the magnon-band from Raman scattering [79B2].

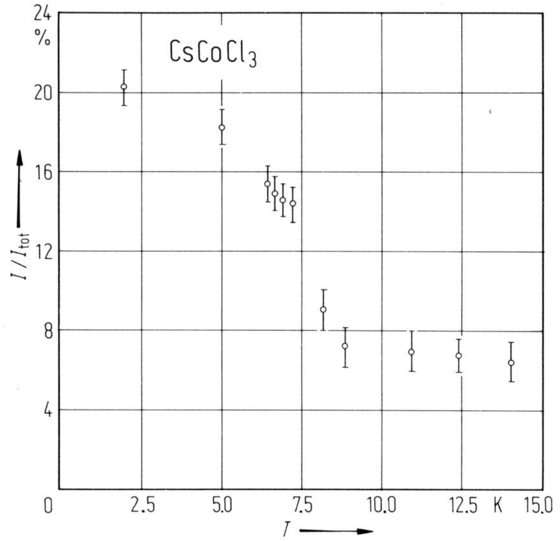

Fig. 483. $CsCoCl_3$. Temperature dependence of the magnon-line at 90.5 cm^{-1} [79B2].

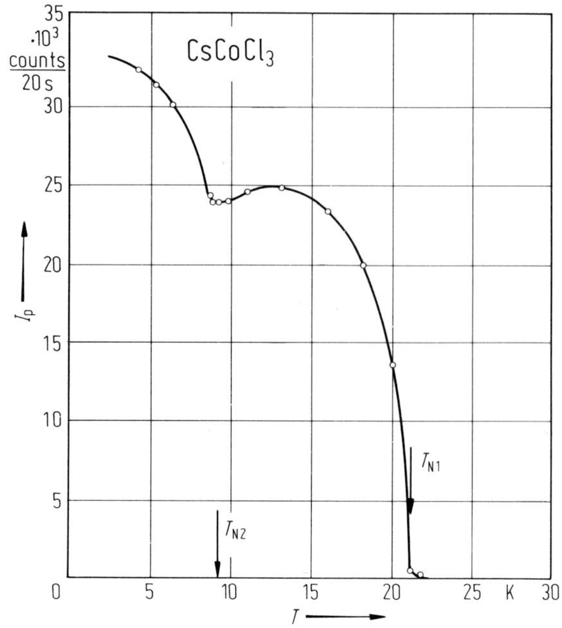

Fig. 484. $CsCoCl_3$. Temperature dependence of the Bragg neutron scattering intensity of the ($\bar{2}$,0,1) reflection [79Y3].

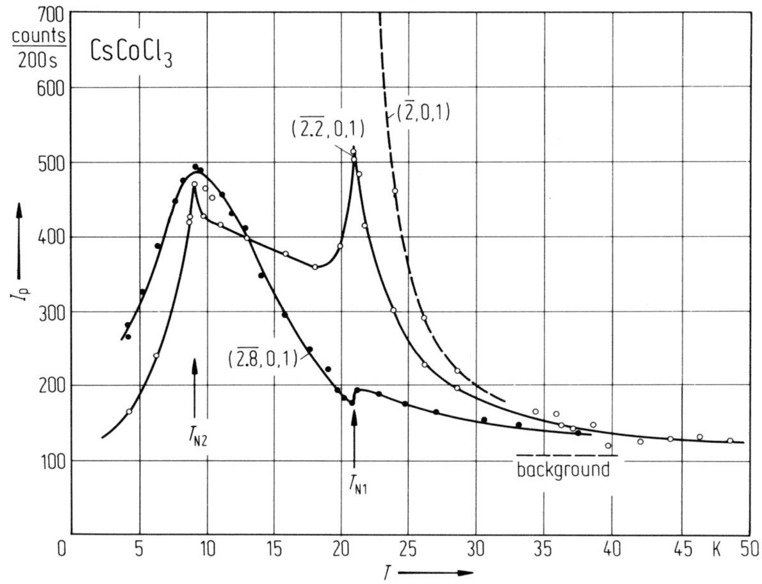

Fig. 485. $CsCoCl_3$. Temperature dependence of neutron peak intensity, at three different points of the reciprocal space [79Y3].

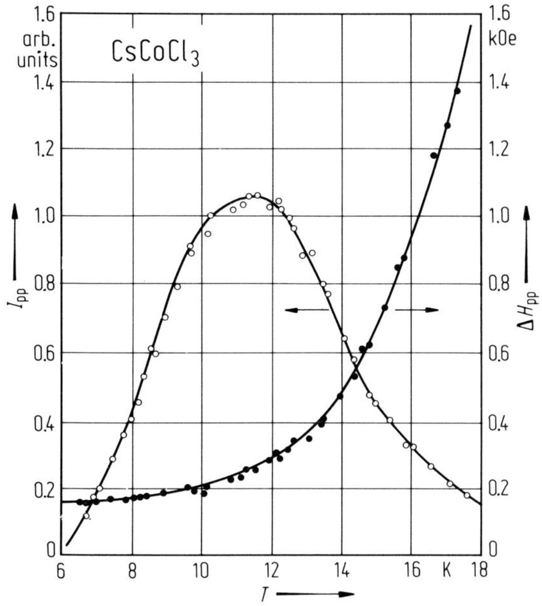

Fig. 486. $CsCoCl_3$. Temperature dependence of EPR peak-to-peak height I_{pp} and peak-to-peak linewidth ΔH_{pp} [81A2].

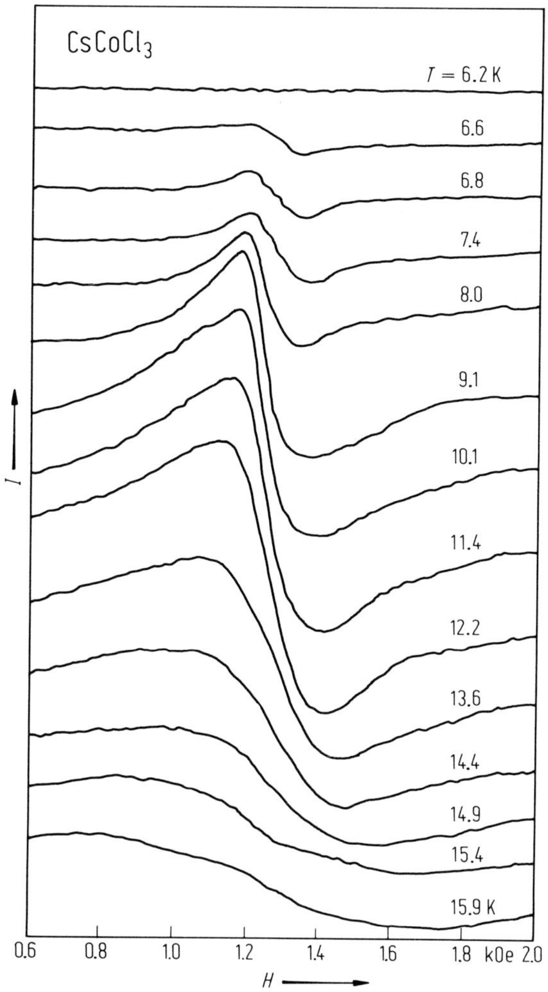

Fig. 488. $CsCoCl_3$. EPR resonance spectra at various temperatures. Magnetic field is applied along c axis [81A2].

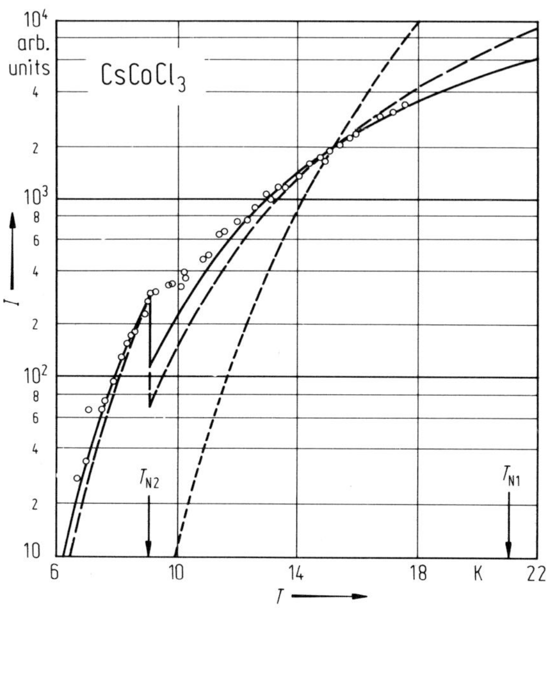

Fig. 487. $CsCoCl_3$. Temperature dependence of ESR absorption intensity. Solid, long and short dashed lines are theoretical [81A2].

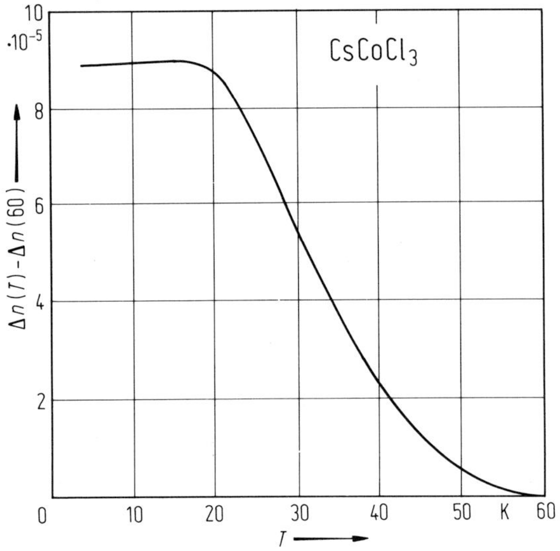

Fig. 489. $CsCoCl_3$. Temperature dependence of birefringer.ce $\Delta n = n_c - n_a$. One cannot see any meaningful anomaly at the successive transition points $T_{N1} = 21.5K$ and $T_{N2} = 9.2K$. A slight decrease of Δn below about 15K may be attributed to a trivial lattice natural birefringence [86I1].

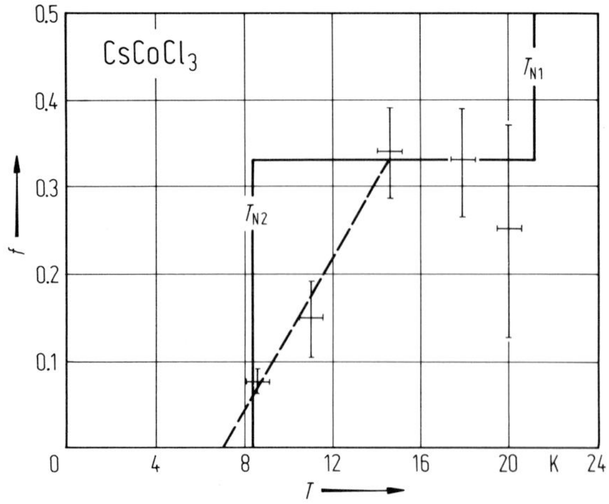

Fig. 490. $CsCoCl_3$. Temperature dependence of the fraction of Co^{2+} chains that are one-dimensionally ordered, deduced from Mössbauer experiment data [87W1].

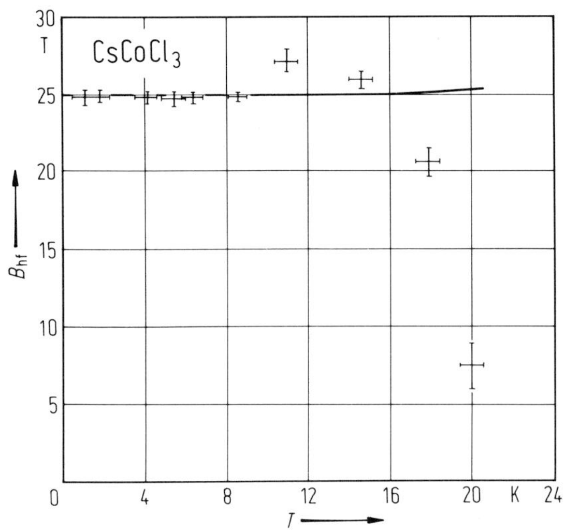

Fig. 491. $CsCoCl_3$. Temperature dependence of the ^{57}Fe hyperfine magnetic field B_{hf} experimental values deduced from Mössbauer experiment [87W1].

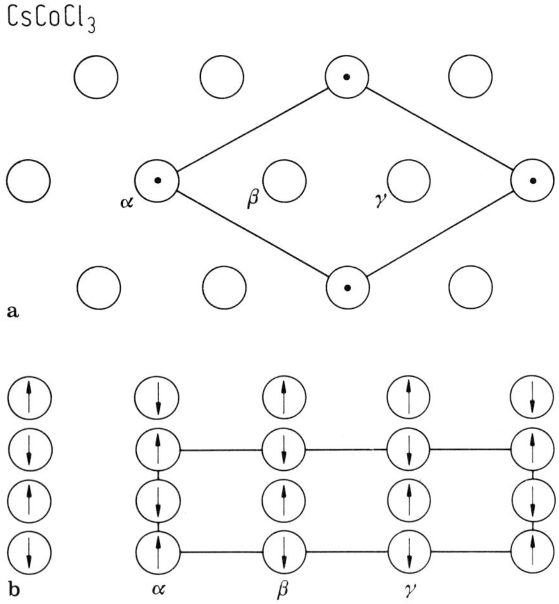

Fig. 492. $CsCoCl_3$. The spatial arrangement of Co^{2+} ions (a) in the crystallographic *ab* plane (b) in plane containing *c* axis and the three chains α, β and γ. Only Co^{2+} ions are shown. Full line indicates magnetic unit cell [87W1].

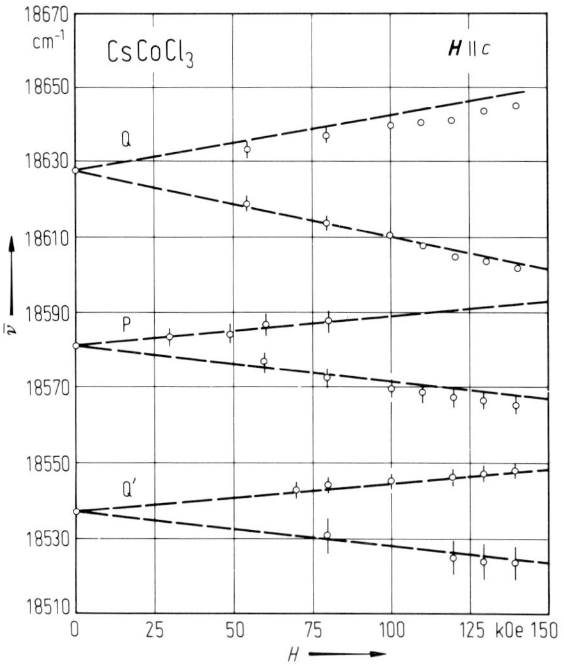

Fig. 493. $CsCoCl_3$. Magnetic field dependence of energy shifts of optical absorption lines of $\boldsymbol{H} \parallel c$. P and Q band measured at 4.2K, and Q′ at 22.2K. Dashed line for Q band is calculated for sublattice splitting of cold magnon sideband. Dashed lines for P and Q′ bands are for sublattice splitting of exciton line [89M1].

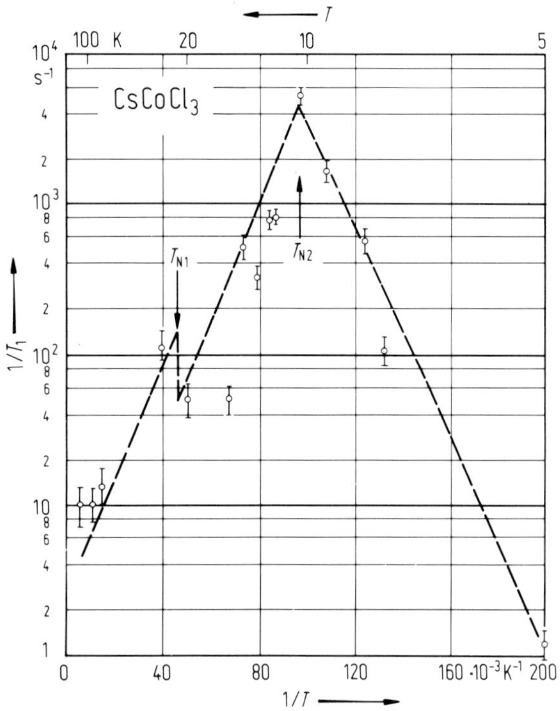

Fig. 494. $CsCoCl_3$. Inverse temperature dependence of nuclear relaxation rate T_1^{-1} of ^{133}Cs. Dashed line: theoretical predictions [90M3].

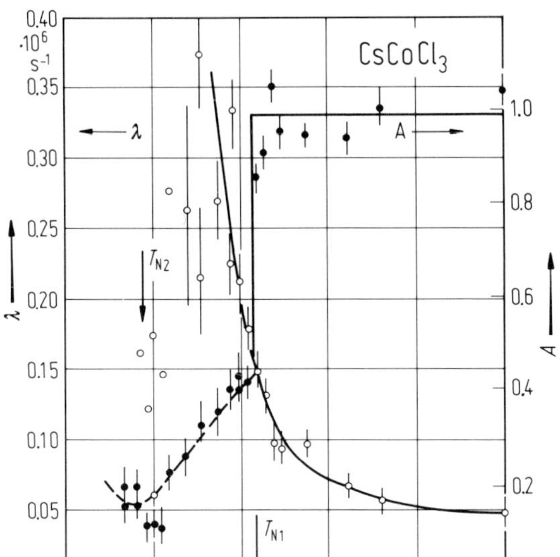

Fig. 495. $CsCoCl_3$. Temperature dependence of relaxation rate λ and amplitude A of the longitudinal polarization of μ^+ spins. A: initial amplitude of the relaxation component, normalized to 1 at room temperature [90M3].

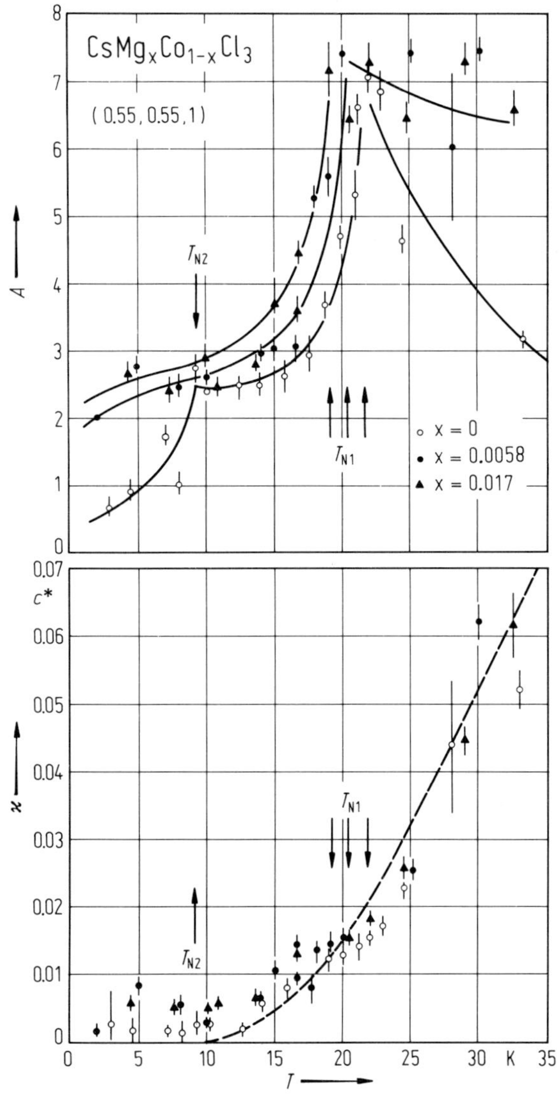

Fig. 496. CsMg$_x$Co$_{1-x}$Cl$_3$. Temperature dependence of the width κ and amplitude A of diffuse neutron scattering along the c axis around (0.55, 0.55, 1) obtained by fitting the observed profiles to Lorentzian curves. Dashed line: solution for an Ising spin chain [90M4].

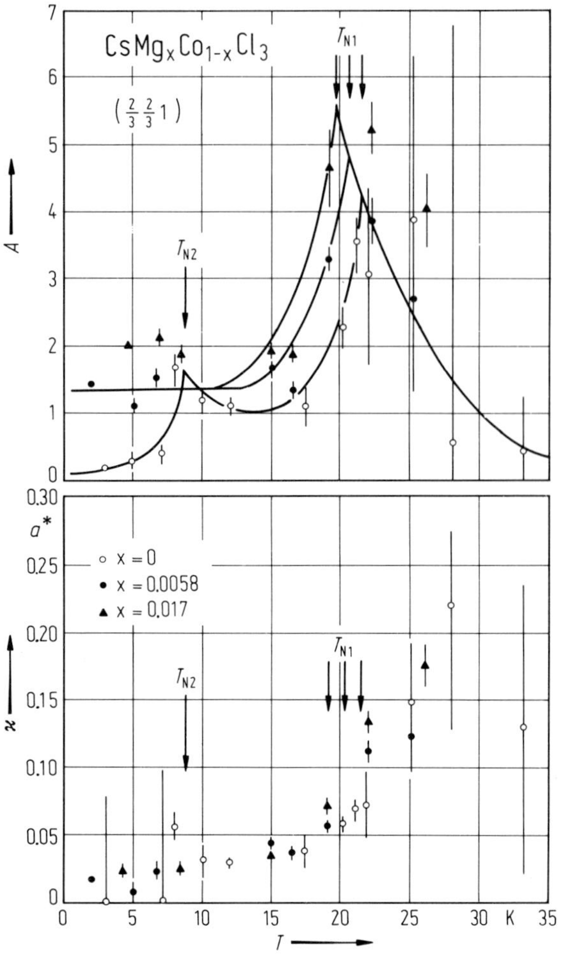

Fig. 497. CsMg$_x$Co$_{1-x}$Cl$_3$. Temperature dependence of the width κ and amplitude A of diffuse neutron scattering along the c axis around (2/3, 2/3, 1) [90M4]. See Fig. 496.

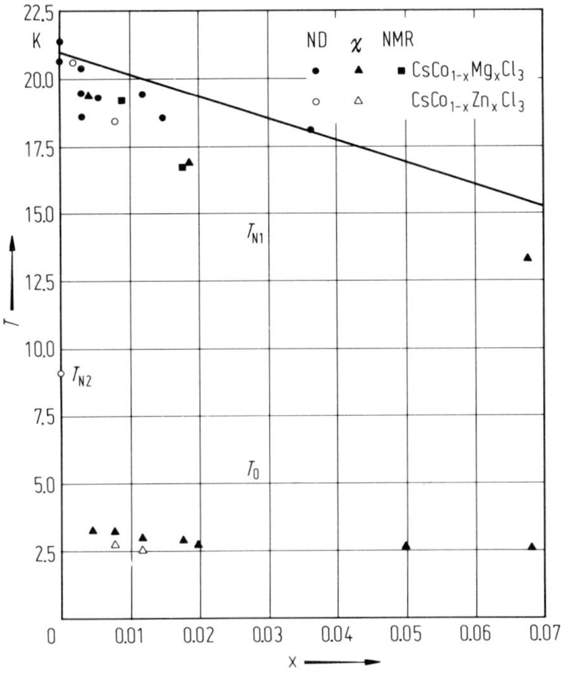

Fig. 498. $CsCo_{1-x}Mg_xCl_3$, $CsCo_{1-x}Zn_xCl_3$. Magnetic transition temperatures vs. concentration. Solid curve calculated. T_0: see Fig. 406, could not be found by neutron scattering [87M3].

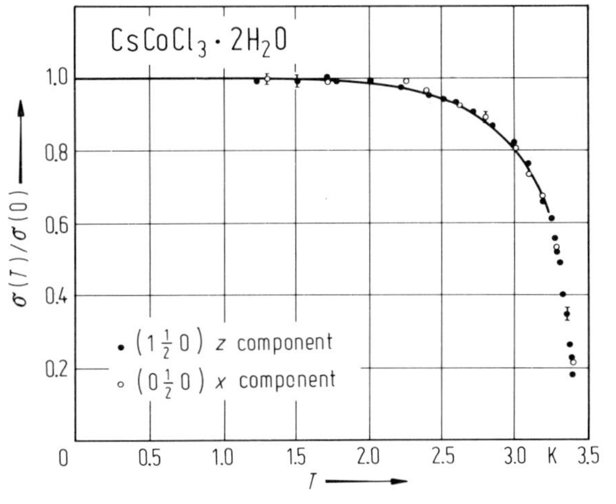

Fig. 499. $CsCoCl_3 \cdot 2H_2O$. Temperature dependence of reduced sublattice magnetization $\sigma(T)/\sigma(0)$ from neutron diffraction. The drawn line represents the NMR results [72B1].

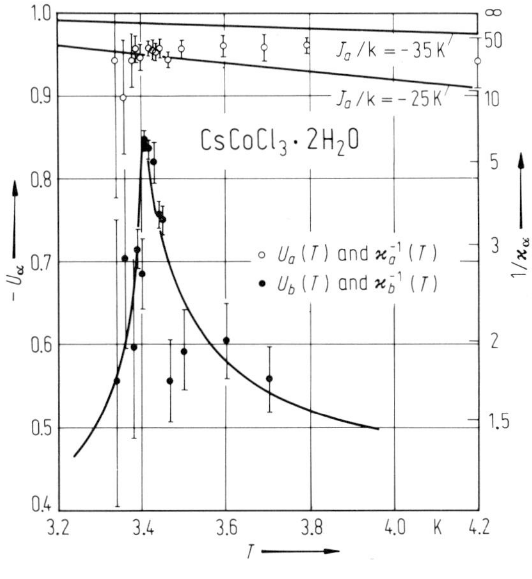

Fig. 500. $CsCoCl_3 \cdot 2H_2O$. Temperature dependence of the correlation parameter U_α and the correlation length $\kappa_\alpha^{-1}(T)$ $(=(-\ln|U_\alpha|)^{-1})$ in units of interatomic distances along a and b. Drawn lines for U_α are calculated for an Ising linear chain [72B1].

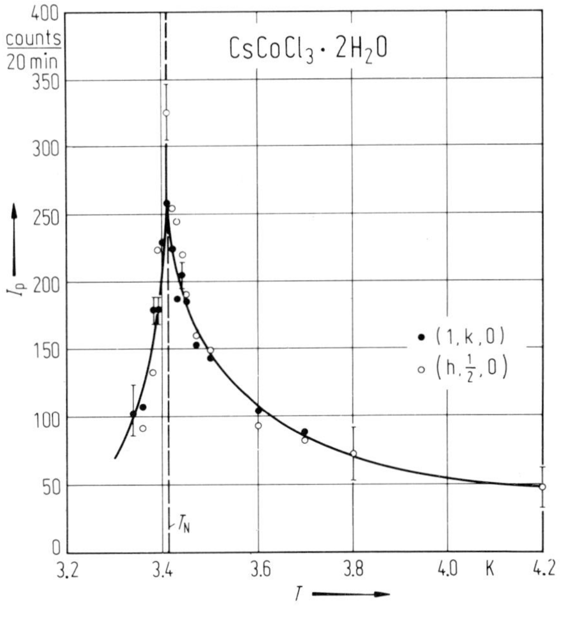

Fig. 501. $CsCoCl_3 \cdot 2H_2O$. Temperature dependence of the peak intensity of the Lorentz profiles from the $(h, \frac{1}{2}, 0)$ and $(1,k,0)$ scans [72B1].

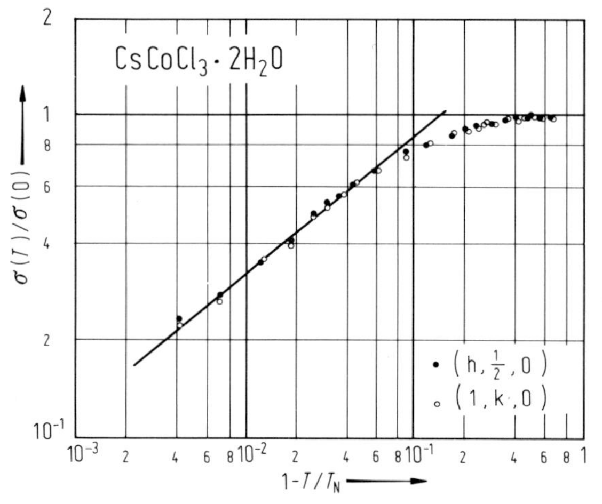

Fig. 502. $CsCoCl_3 \cdot 2H_2O$. Reduced sublattice magnetization determined from neutron reflection near T_N. Solid line represents the best fit for $3.20 \leq T \leq 3.39K$ to $\sigma(T)/\sigma(0) = A(1 - T/T_N)^{\beta}$, where $T_N = 3.404(3)K$, $\beta = 0.44(3)$ and $A = 2.34(18)$ [72B1].

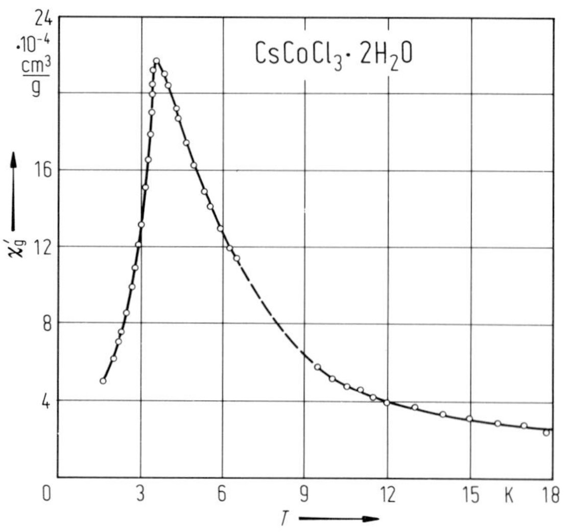

Fig. 503. $CsCoCl_3 \cdot 2H_2O$. Low temperature dependence of dynamic χ'_g at 1200 Hz with ac field parallel to the *a* axis [72H1].

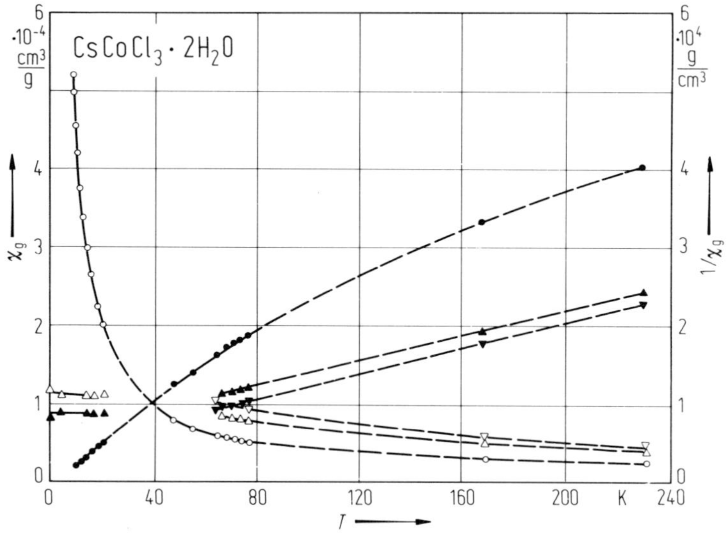

Fig. 504. $CsCoCl_3 \cdot 2H_2O$. Temperature dependence of χ_g (open symbols) and χ_g^{-1} (full symbols), (circles: $\boldsymbol{H} \parallel a$; △, ▲: $\boldsymbol{H} \parallel b$; ▽, ▼: $\boldsymbol{H} \parallel c$) [72H1].

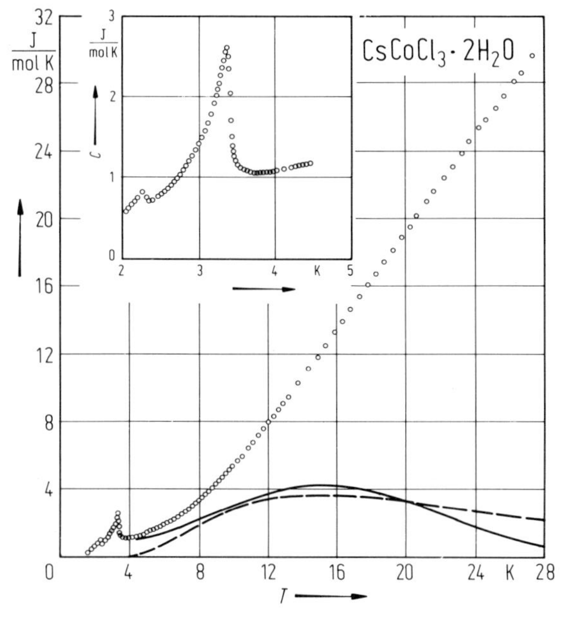

Fig. 505. $CsCoCl_3 \cdot 2H_2O$. Temperature dependence of heat capacity. Solid line: magnetic part after subtraction of lattice term, dashed line: Ising calculation with $J/k = 72K$ [72H1].

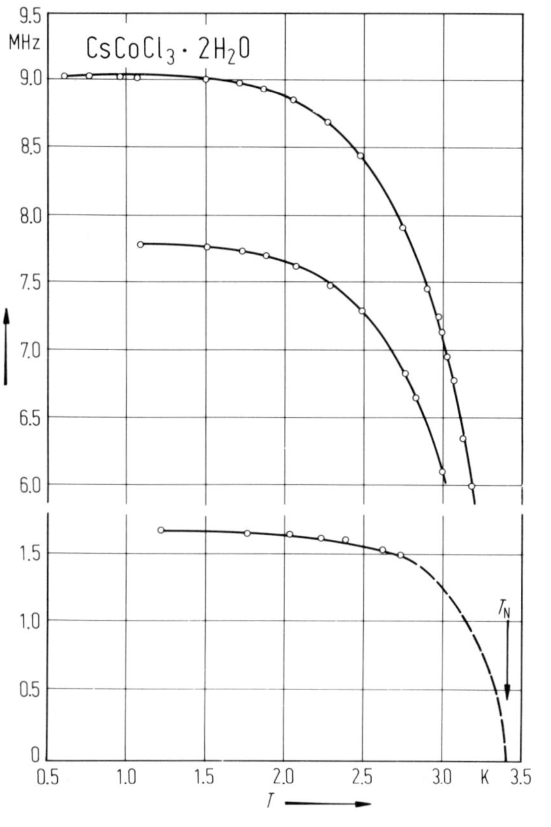

Fig. 506. $CsCoCl_3 \cdot 2H_2O$. Temperature dependence of proton (upper two curves) and Cs (lower curve) NMR frequency [72H1]. See also [71J1].

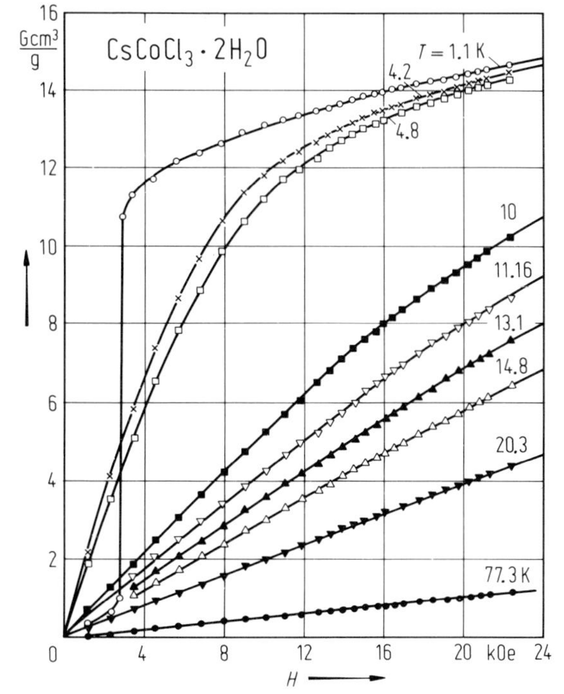

Fig. 507. $CsCoCl_3 \cdot 2H_2O$. Magnetic field dependence of σ for $\boldsymbol{H} \parallel a$ at various temperatures [72H1].

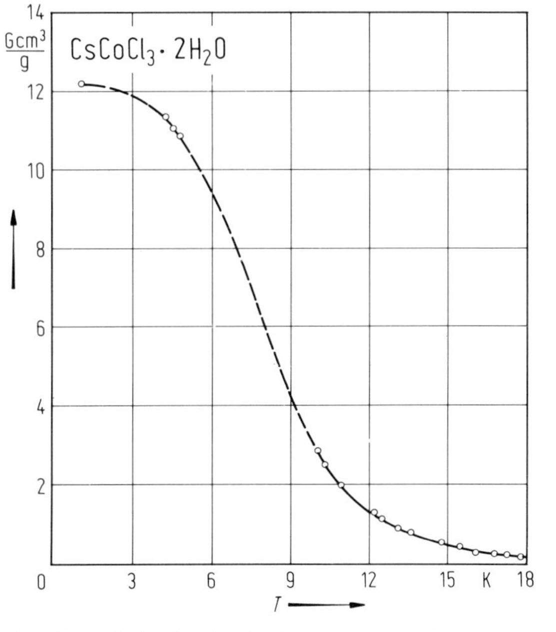

Fig. 508. $CsCoCl_3 \cdot 2H_2O$. Temperature dependence of induced ferromagnetic moment σ extrapolated from 20 kOe ($\boldsymbol{H} \parallel a$) [72H1].

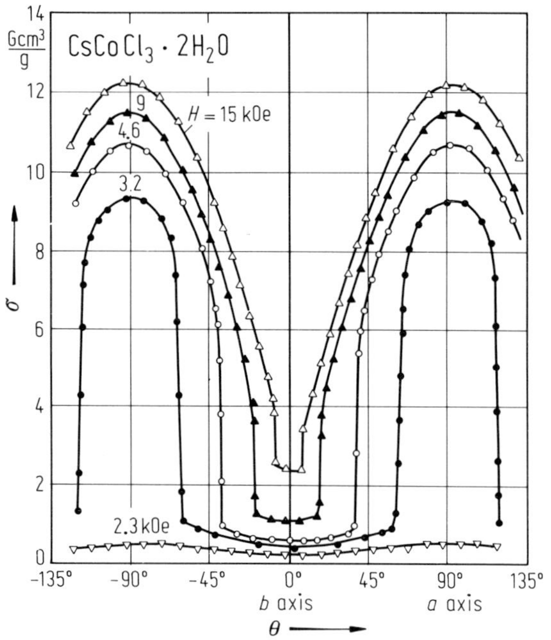

Fig. 509. $CsCoCl_3 \cdot 2H_2O$. Magnetization in the *ab* plane vs. angle between ***H*** and *b* axis at 1.1K for various magnetic fields [72H1].

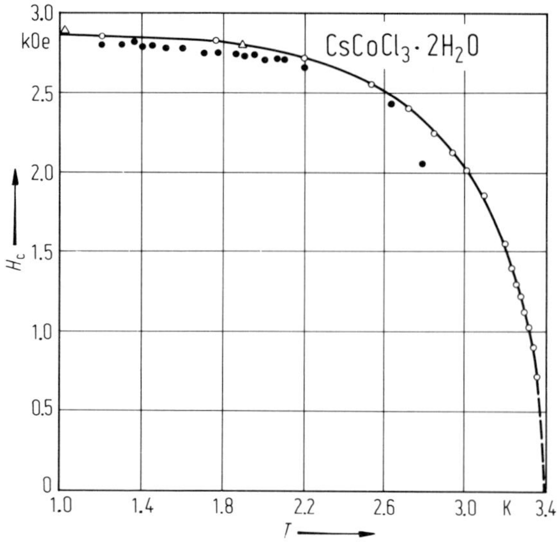

Fig. 510. $CsCoCl_3 \cdot 2H_2O$. Temperature dependence of the critical field, Open circles: NMR, full circles: AFMR, triangles: magnetization [72H1].

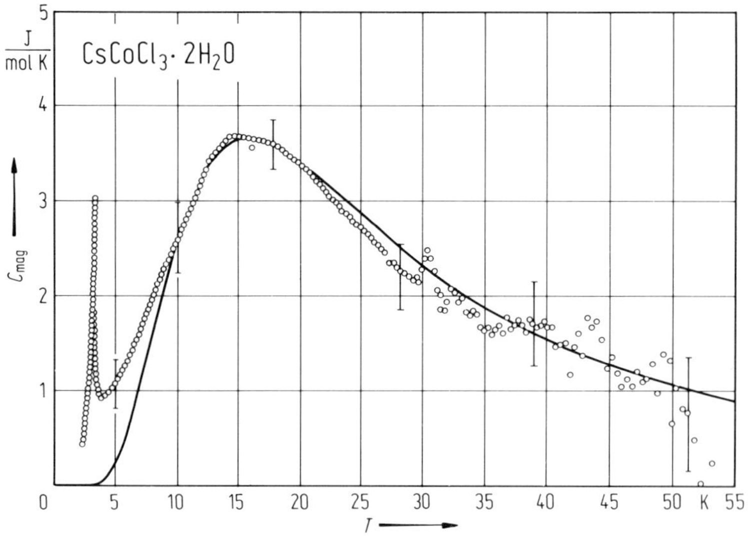

Fig. 511. $CsCoCl_3 \cdot 2H_2O$. Temperature dependence of magnetic heat capacity C_{mag}. Open circles are the experimental points corrected for the lattice contribution. The solid line is theoretical for Ising linear chain with $S = 1/2$ and $|J/k| = 38.6$ K [77K1].

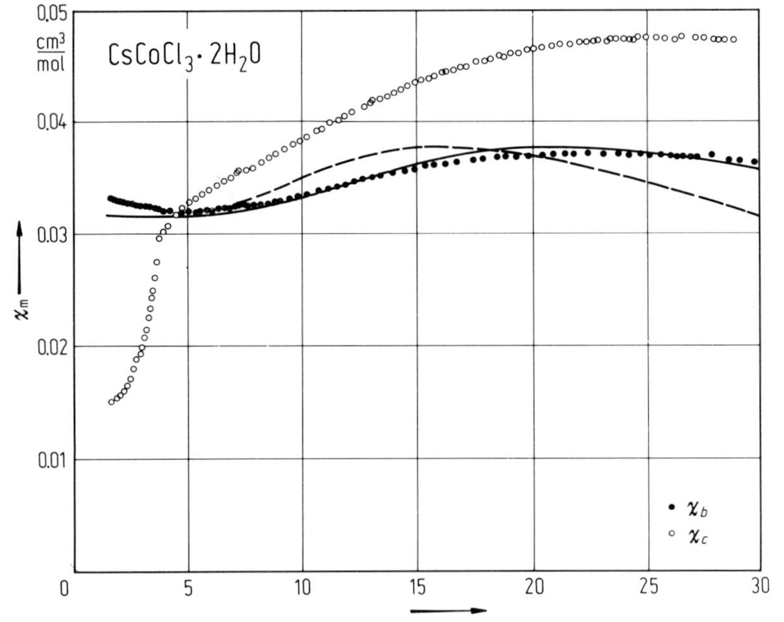

Fig. 512. $CsCoCl_3 \cdot 2H_2O$. Temperature dependence of χ_m along c and b axes. The solid curve corresponds to the Ising linear chain model with $J/k = -52K$, and the dashed line for $J/k = -38.6K$ [77M1].

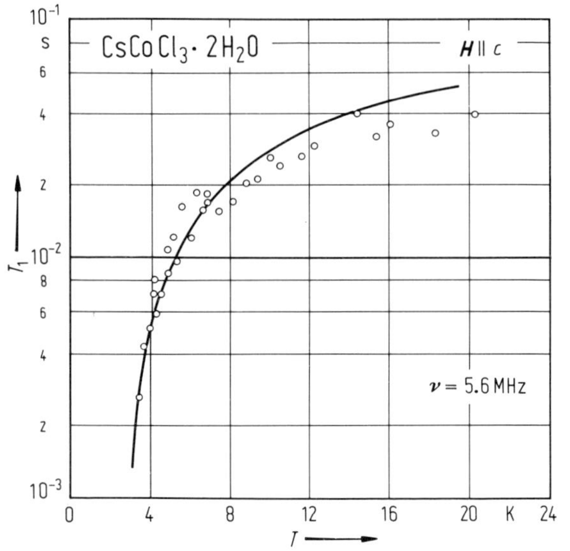

Fig. 513. $CsCoCl_3 \cdot 2H_2O$. Temperature dependence of NMR ^{133}Cs relaxation time T_1 for $\boldsymbol{H} \parallel c$ at 5.6 MHz. Solid line is the best fit for $|J/k| = 12K$ [81G1].

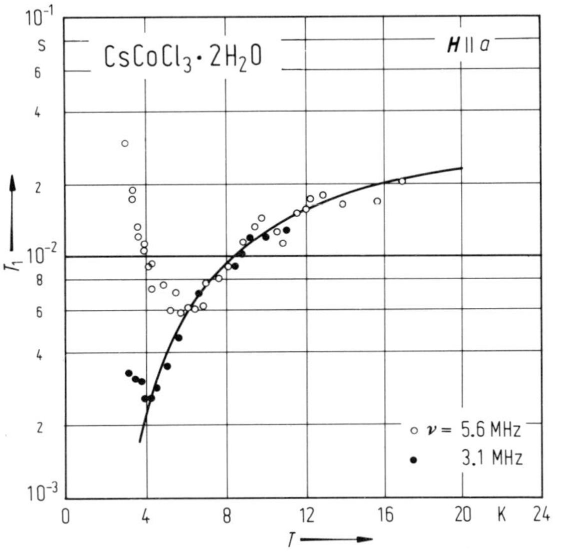

Fig. 514. $CsCoCl_3 \cdot 2H_2O$. Temperature dependence of NMR ^{133}Cs relaxation time T_1 for $\boldsymbol{H} \parallel a$ at 5.6 and 3.1 MHz. Solid line is the best fit for $|J/k| = 12K$ [81G1].

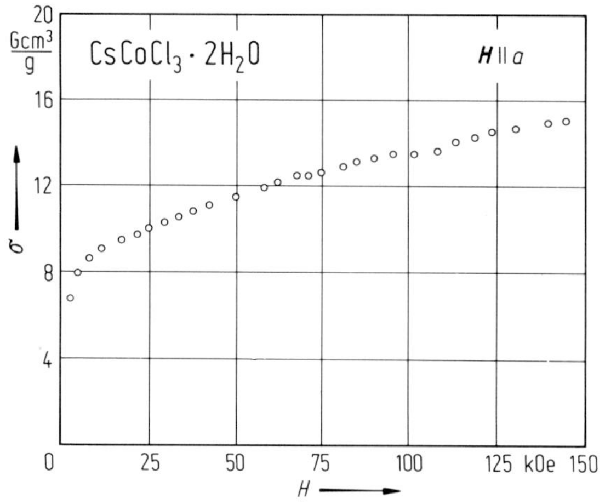

Fig. 515. $CsCoCl_3 \cdot 2H_2O$. Magnetic field dependence of σ at 1.3 K for $\boldsymbol{H} \parallel a$ [86M2].

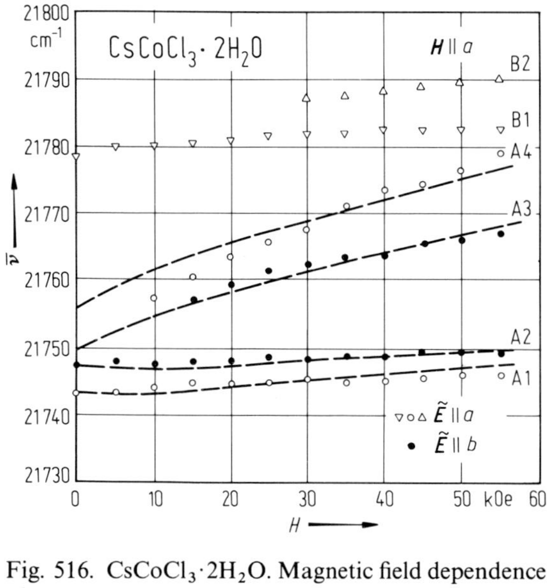

Fig. 516. $CsCoCl_3 \cdot 2H_2O$. Magnetic field dependence of energy shift of the A-exciton lines and their magnon sidebands B with $\boldsymbol{H} \parallel a$ at 1.7 K from optical absorption measurements [86M2].

For Fig. 517, see next page.

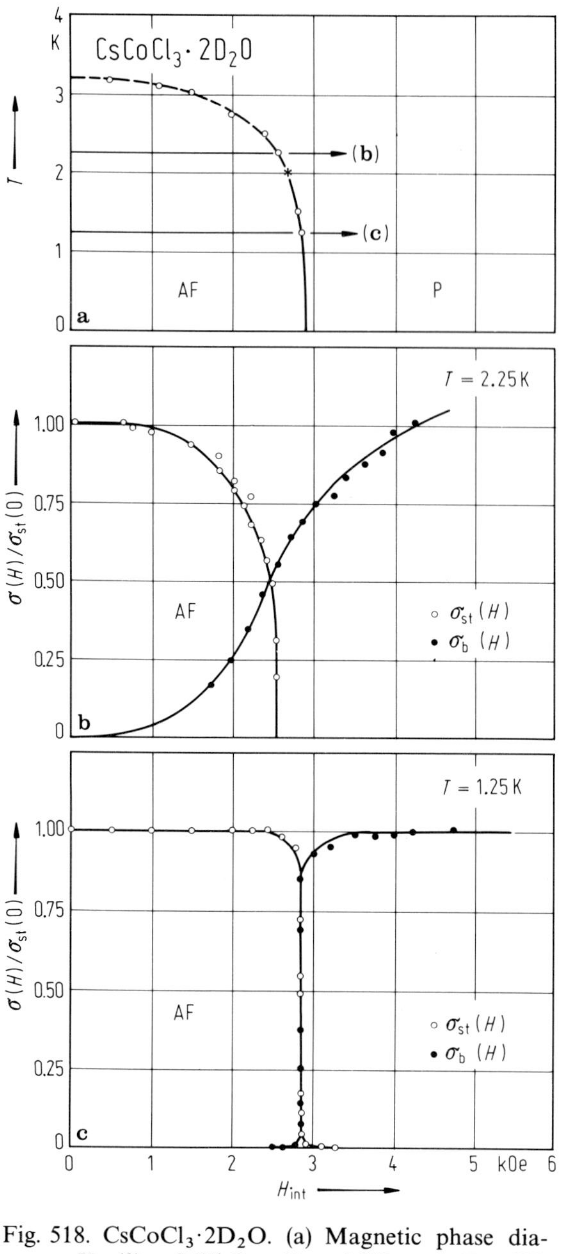

Fig. 518. $CsCoCl_3 \cdot 2D_2O$. (a) Magnetic phase diagram, H_c (0) = 2.85kOe, T_N = 3.2K and T_t = 2K. (b), (c) Magnetic field dependence of σ_{st} and σ_b at 2.25 and 1.25K. σ_{st}: staggered magnetization obtained by measuring the square root of the Bragg intensity of the magnetic superlattice reflection $(1,\frac{1}{2},0)$ and $(2,\frac{1}{2},0)$. σ_b: uniform bulk magnetization of the (1,0,0) magnetic reflection [74B2].

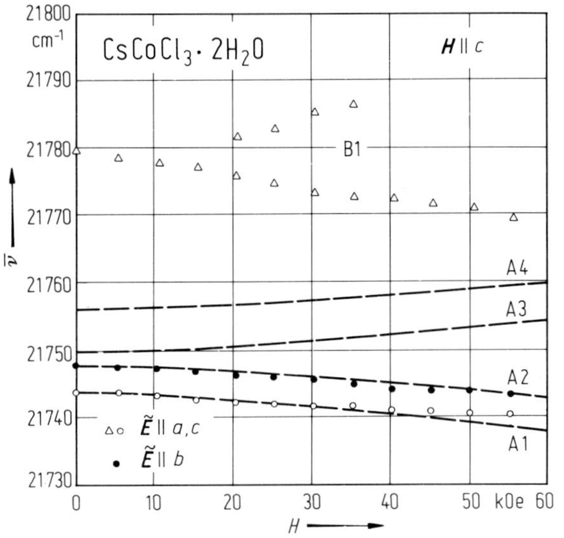

Fig. 517. $CsCoCl_3 \cdot 2H_2O$. Magnetic field dependence of energy shift of the A- exciton lines and their magnon sideband B1 with $\boldsymbol{H} \parallel c$ at 1.7K [86M2].

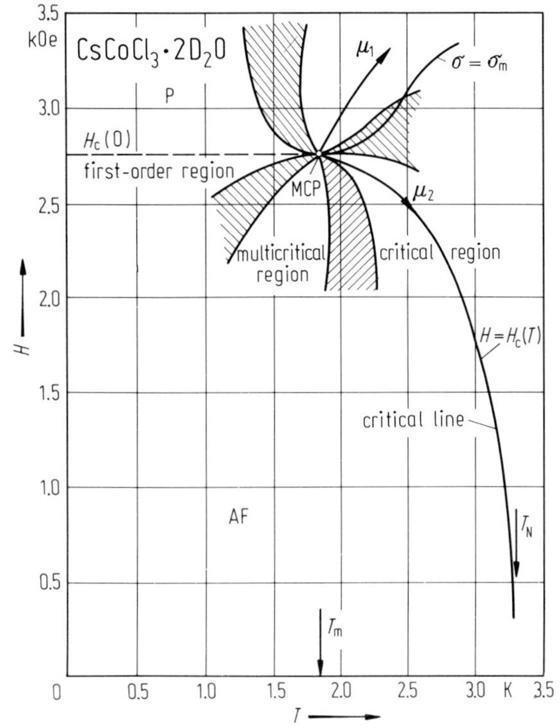

Fig. 520. $CsCoCl_3 \cdot 2D_2O$. Magnetic phase diagram as a function of magnetic field and temperature T; $H_c(0) = 2.8$kOe, and $T_N = 3.3$K. The multicritical point (MCP) $T_m = 1.85$K and $H_m = 2.75$kOe, μ_1 and μ_2 are scaling fields. Shaded regions correspond to the 25% crossover regions [76B1, 77B1]. ($T_m = T_t$ of Fig. 521).

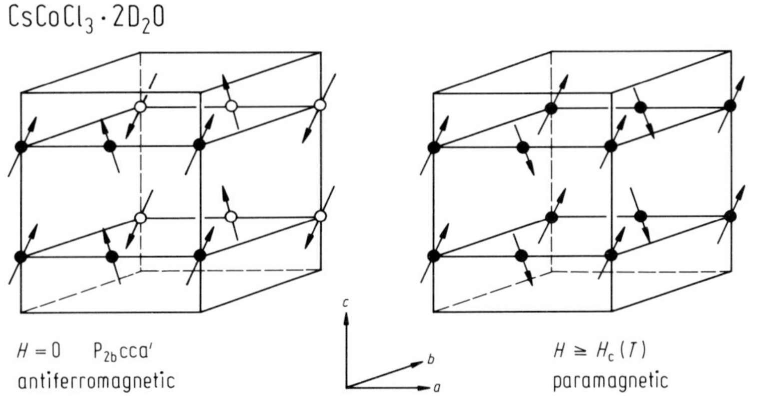

Fig. 519. $CsCoCl_3 \cdot 2D_2O$. Magnetic structure for $H = 0$ and $H > H_c$ (0). All magnetic moments are in the ac plane. For $H > H_c$ (0) the moments in alternate ac planes are reversed [72B1, 72H1, 76B1, 77B1, 86M2].

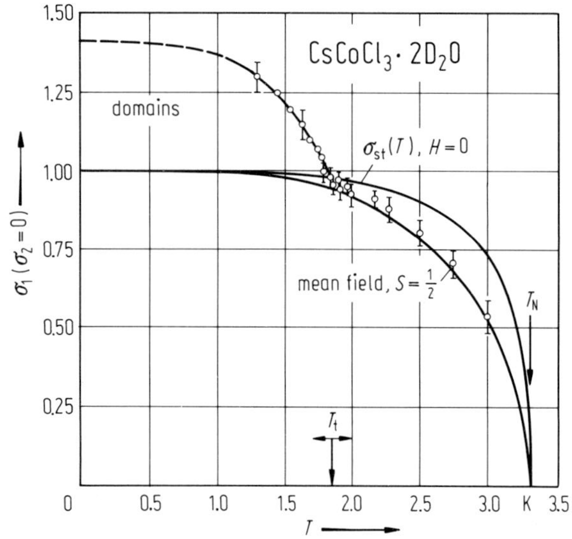

Fig. 521. $CsCoCl_3 \cdot 2D_2O$. Temperature dependence of magnetization of sublattice 1 σ_1 (if $\sigma_2 = 0$) determined from the intersection of the I vs. H curves of the (100) and $(1\,\frac{1}{2}\,0)$ reflection, i.e. $I_{1\frac{1}{2}0}(H,T) = I_{100}(H,T)$. For $T > 1.9$K the replacement of the molecular field of a sublattice by the internal magnetic field in the sample induces mean field behavior. Below 1.8K presence of domains in the sample is suggested [77B1]. (σ_1, σ_2 from square root of Bragg reflection reduced to 1 at $T = 0$K).

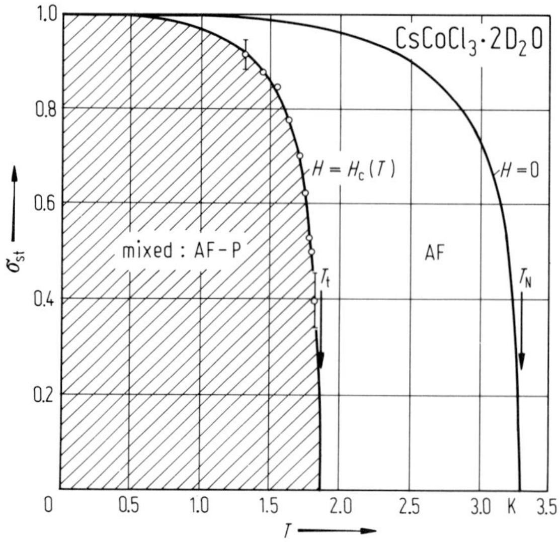

Fig. 522. $CsCoCl_3 \cdot 2D_2O$. Temperature dependence of the staggered magnetization for $H = H_c\,(T)$ and $H = 0$. The full line through the data points is obtained from the zero-field data with $T_t = 1.85$ K [77B1].

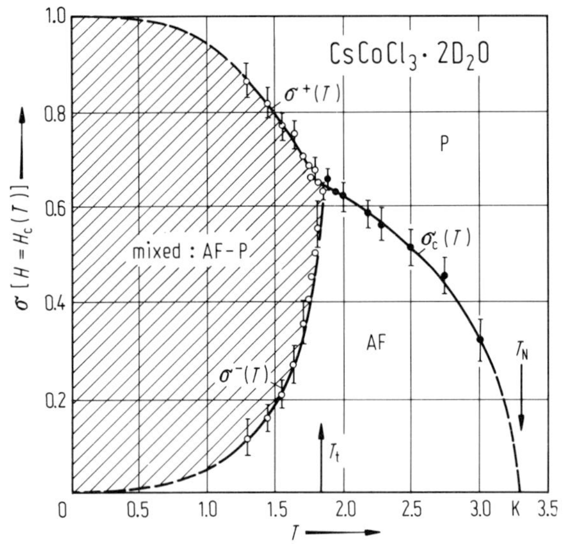

Fig. 523. $CsCoCl_3 \cdot 2D_2O$. Temperature dependence of the induced magnetization along the phase border $H = H_c\,(T)$. For $T < 1.85$ K there is a mixed, antiferromagnetic-paramagnetic state caused by the discontinuity in magnetization [77B1].

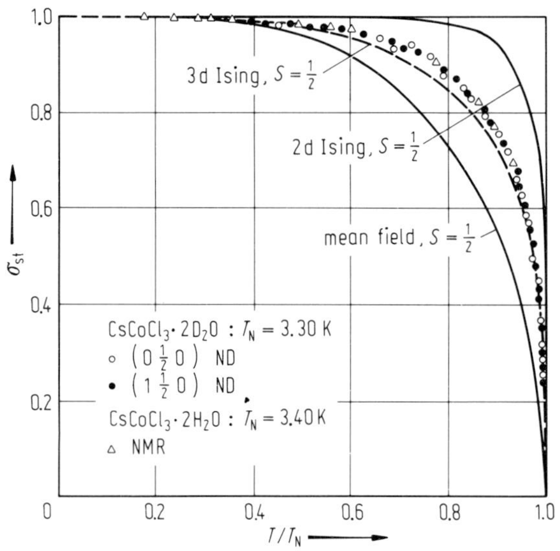

Fig. 524. $CsCoCl_3 \cdot 2D_2O$ ($\cdot 2H_2O$). Reduced temperature dependence of the zero-field staggered magnetization determined by NMR and neutron diffraction [77B1].

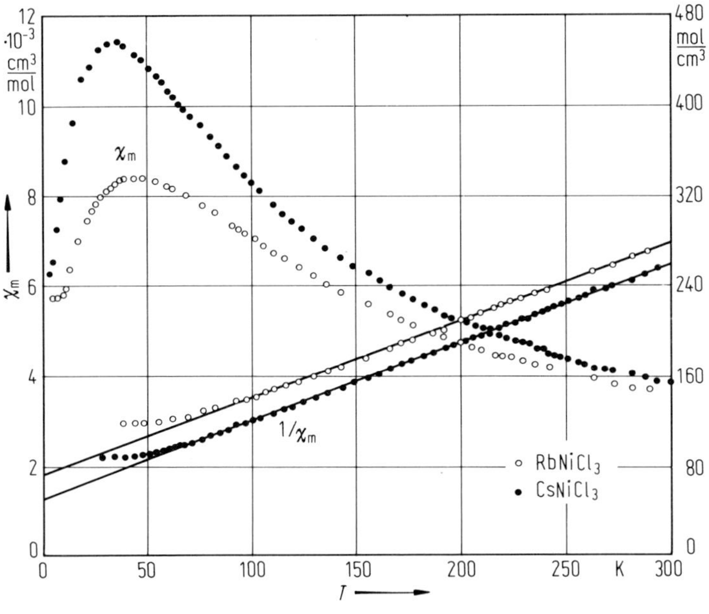

Fig. 525. $RbNiCl_3$, $CsNiCl_3$. Temperature dependence of χ_m and χ_m^{-1} for polycrystals. The correction of inner core orbital diamagnetism and Van Vleck paramagnetism were applied [69A1] See also [74W1].

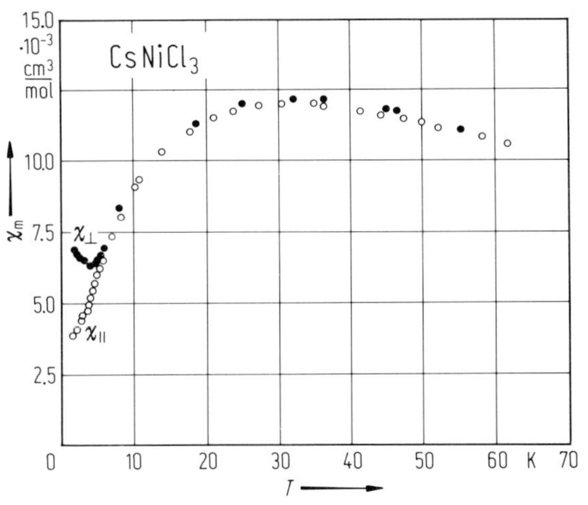

Fig. 526. $CsNiCl_3$. Temperature dependence of single crystal χ_m perpendicular to [11$\bar{2}$0] and parallel to [0001] axis [69A1].

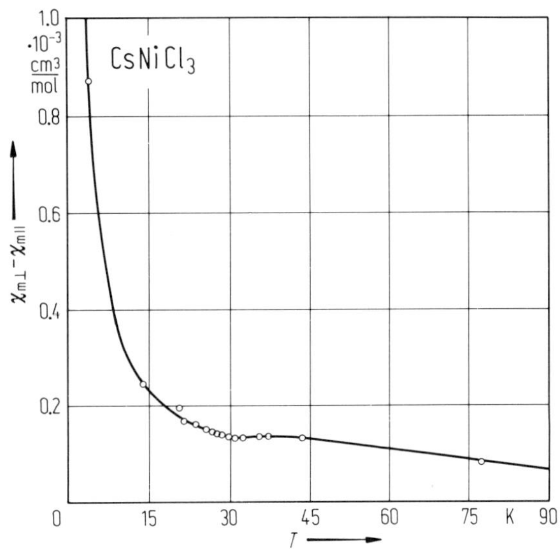

Fig. 527. $CsNiCl_3$. Temperature dependence of $(\chi_{m\perp} - \chi_{m\parallel})$ (∥ and ⊥ along [0001] and [11$\bar{2}$0] axis, respectively) obtained from magnetic torque measurement [69A1].

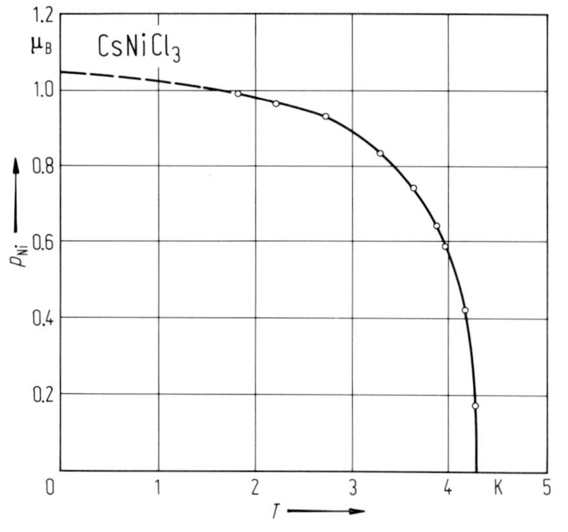

Fig. 528. $CsNiCl_3$. Temperature dependence of $p_{Ni^{2+}}$ moment from neutron diffraction [71C1].

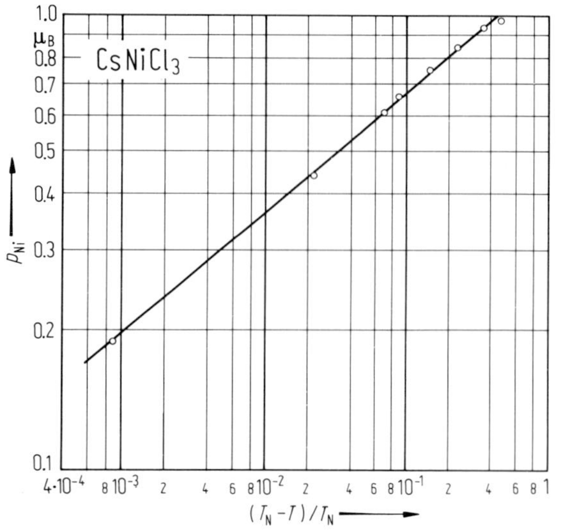

Fig. 529. $CsNiCl_3$. Temperature dependence of $p_{Ni^{2+}}$ moment near T_N [71C1].

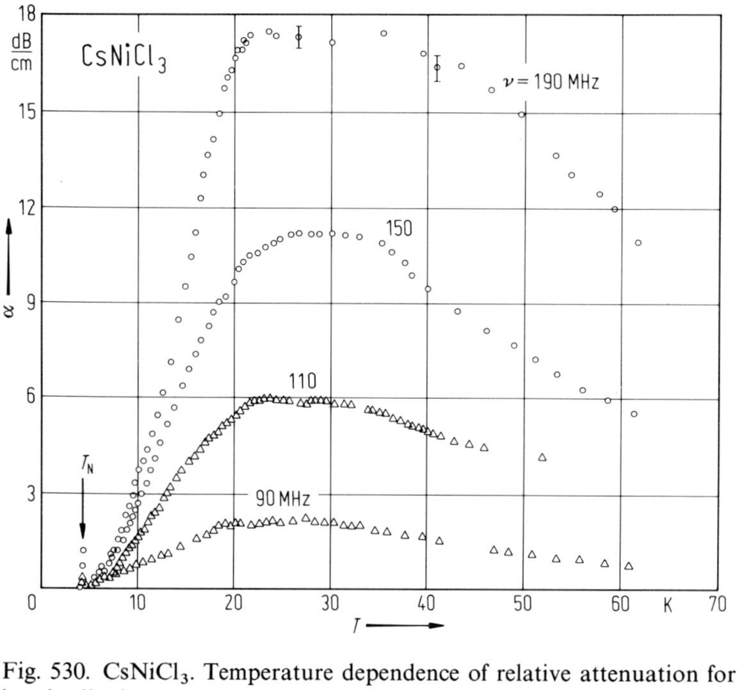

Fig. 530. $CsNiCl_3$. Temperature dependence of relative attenuation for longitudinal waves propagating along [0001] [72A1]. See also [77A1].

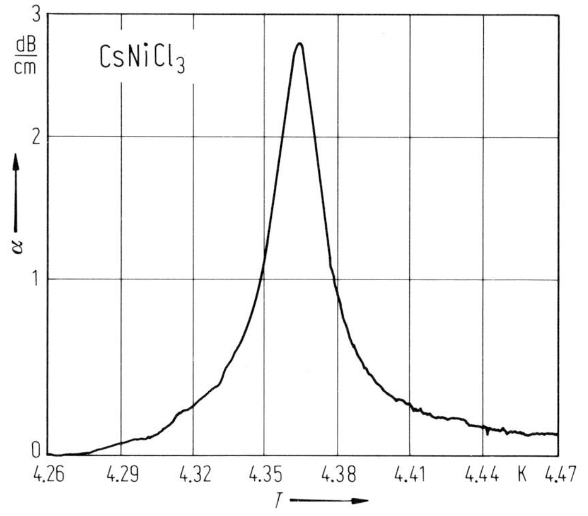

Fig. 531. $CsNiCl_3$. Temperature dependence of recording tracing data of relative attenuation (see Fig. 530) taken at 330 MHz near transition temperature [72A1].

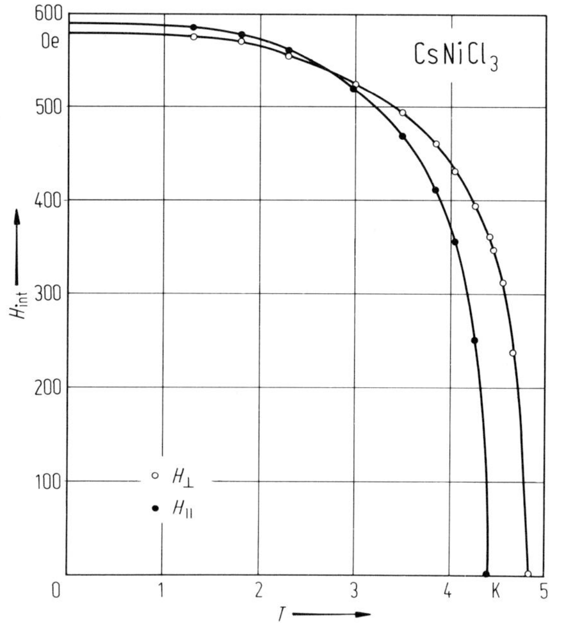

Fig. 532. $CsNiCl_3$. Temperature dependence of internal magnetic field $H_{int}^{\perp}$ and $H_{int}^{\parallel}$ (to c axis) at the Cs site with $\nu = 4.700$ MHz and $H \approx 8.4$ kOe from NMR [72C1].

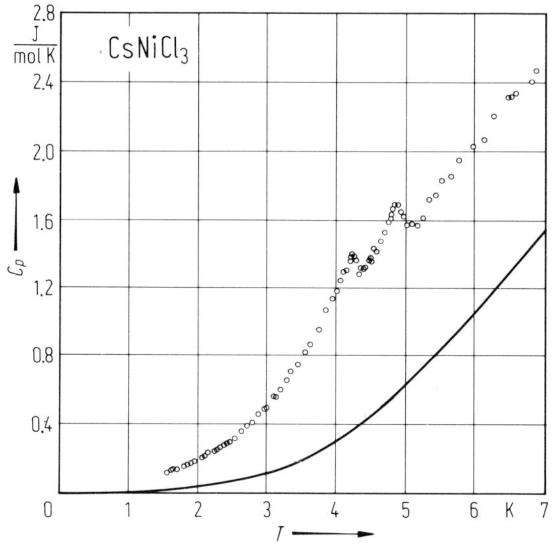

Fig. 533. $CsNiCl_3$. Temperature dependence of heat capacity C_p (circles). Data for $CsMgCl_3$ (solid curve) are also shown [73A1].

Fig. 534. $CsNiCl_3$. Temperature dependence of the strongest magnetic peaks from neutron diffraction [73Y1].

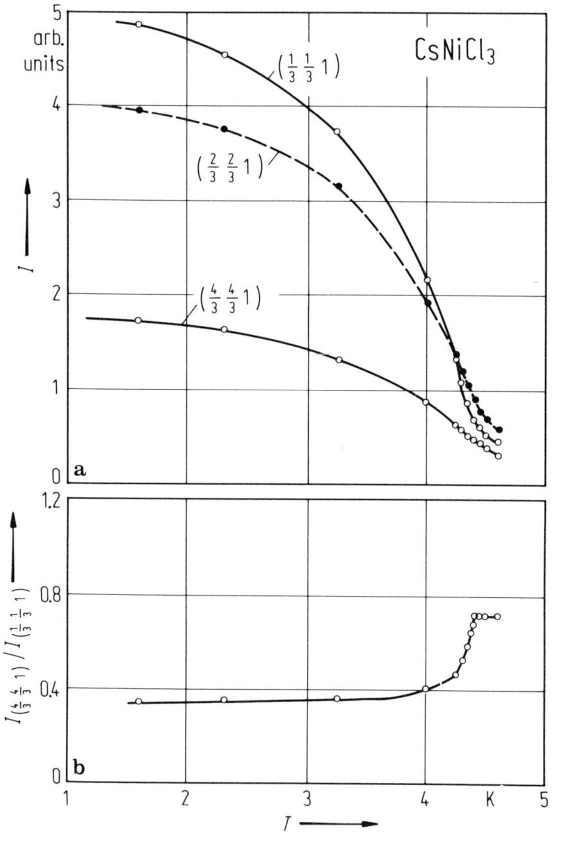

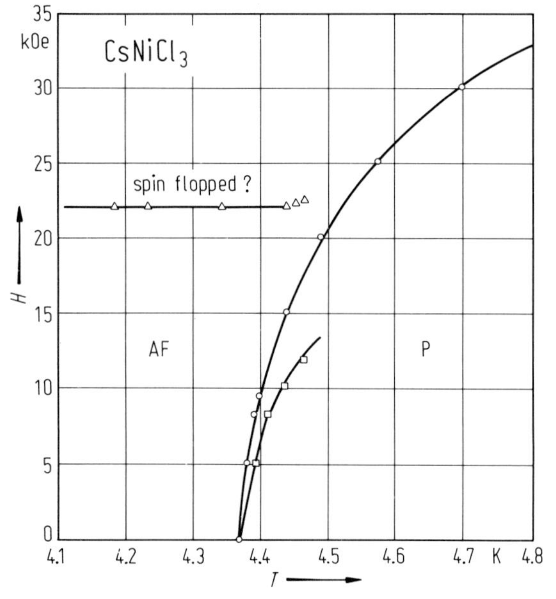

Fig. 535. $CsNiCl_3$. Phase diagram. Circles and squares obtained from temperature sweeps in constant magnetic field ∥ and ⊥ to [0001]. Triangles obtained from magnetic sweeps ∥ to [0001] axis at constant temperature as observed by ultrasonic attenuation [75A1, 77A1].

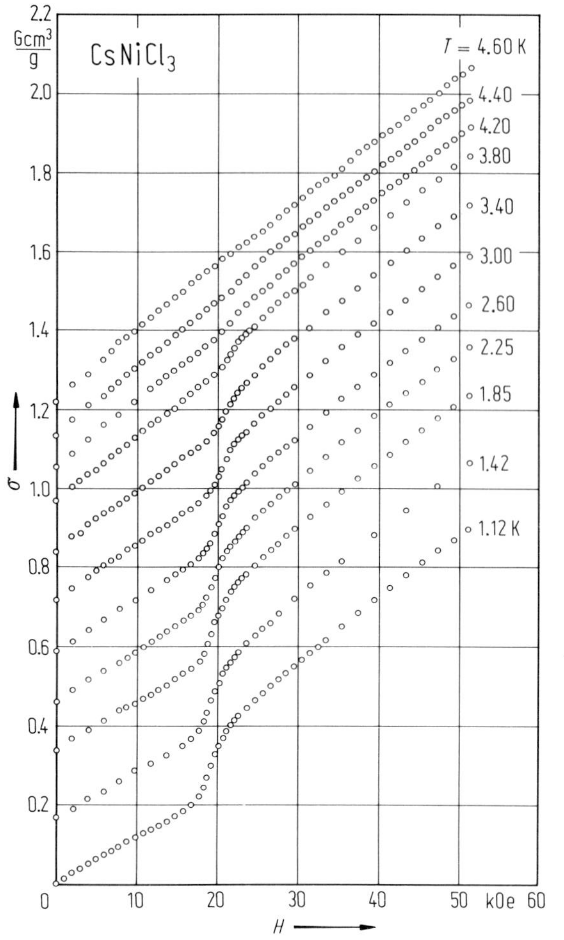

Fig. 537. $CsNiCl_3$. Magnetic field dependence of σ at different temperatures for $\boldsymbol{H} \parallel c$. Curves displaced vertically for clarity [79J1].

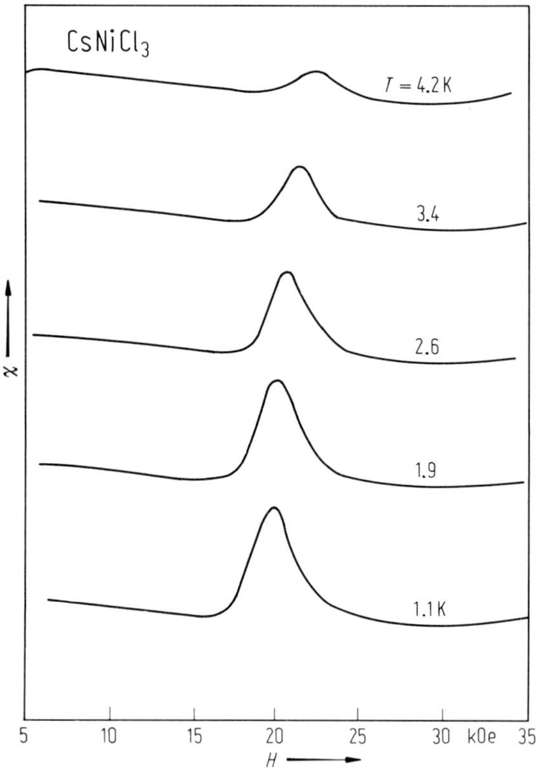

Fig. 536. $CsNiCl_3$. Magnetic field dependence of χ at different temperatures for $\boldsymbol{H} \parallel c$ showing spin flop [79J1].

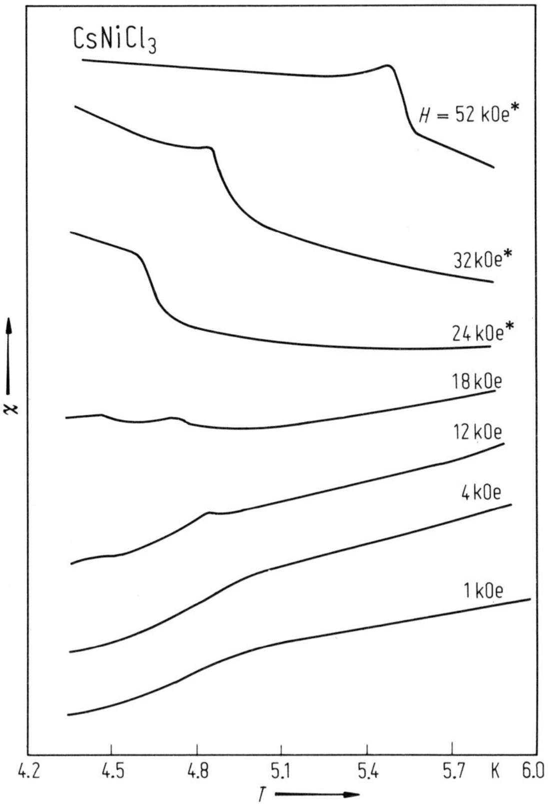

Fig. 538. $CsNiCl_3$. Temperature dependence of χ for different $\boldsymbol{H} \parallel c$. Curves marked by asterisk are with gain reduced by factor of 20 [79J1].

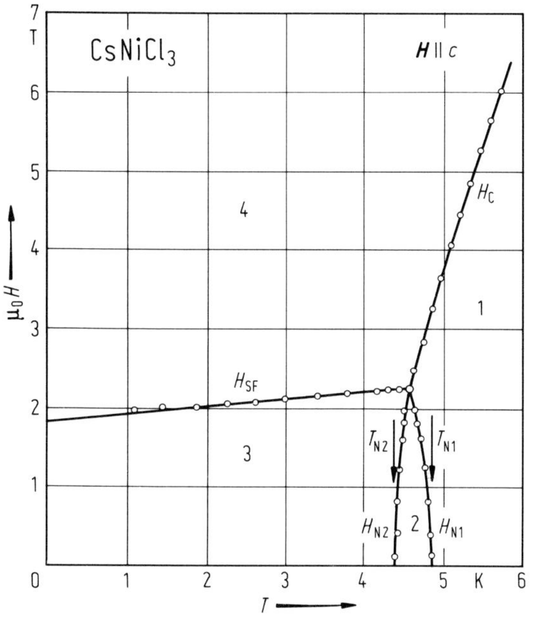

Fig. 539. $CsNiCl_3$. Magnetic phase diagram for $\boldsymbol{H} \parallel c$. Open circles from [79J1]. Solid line: theoretical fit [79J1, 88P2]. For discussion of phase 1–4 see [88P2].

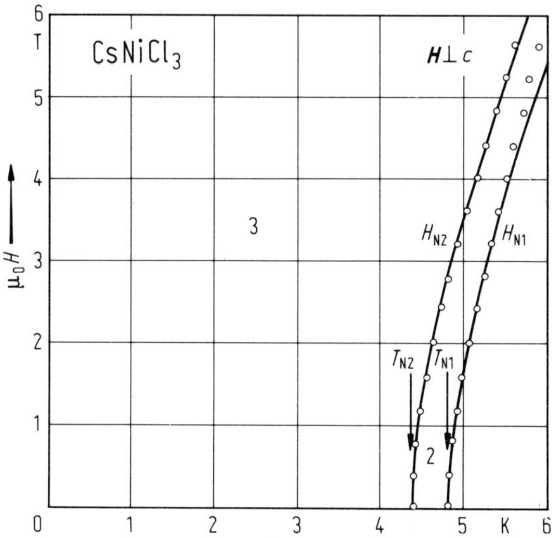

Fig. 540. $CsNiCl_3$. Phase diagram for $\boldsymbol{H} \perp c$. Circles from [79J1]. Solid line: theoretical fit [79J1, 88P2].

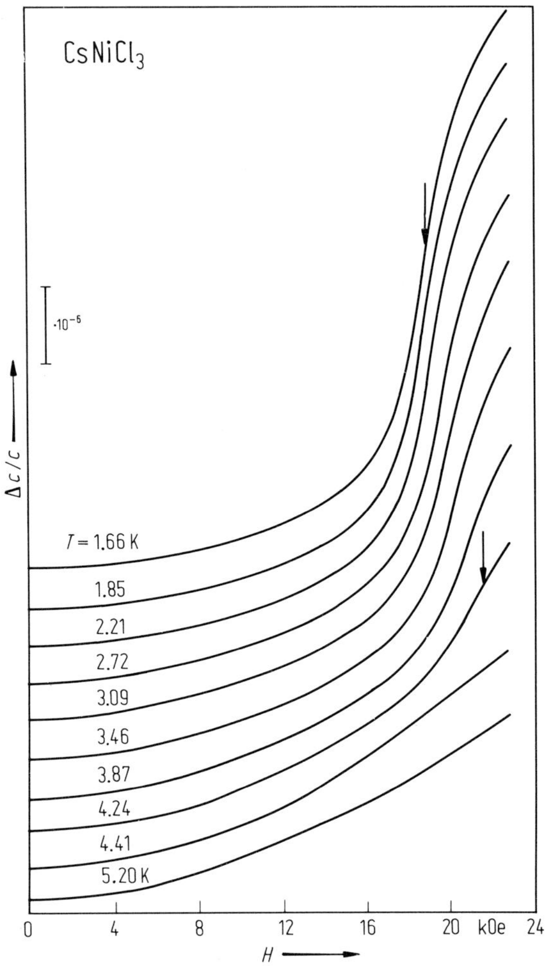

Fig. 541. $CsNiCl_3$. Magnetic field dependence of magnetostriction for $\boldsymbol{H} \parallel c$. Arrows correspond to spin-flop field at 4.24 K and 1.6 K, respectively [83 R1, 84 R1].

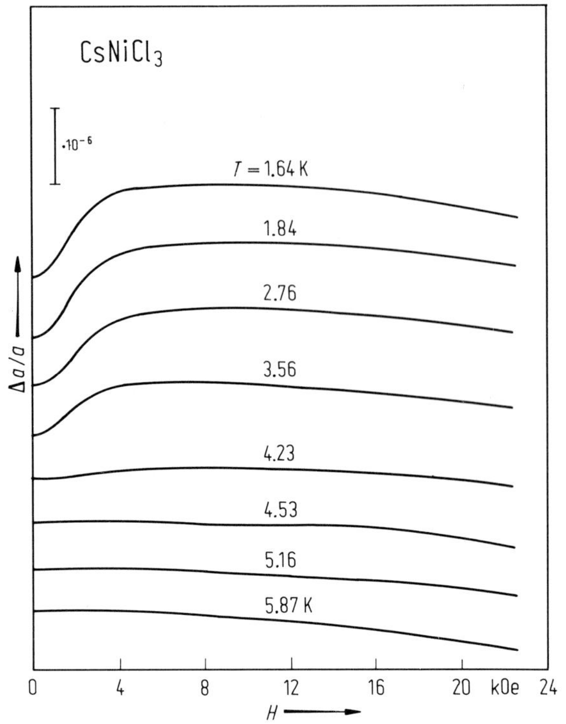

Fig. 542. $CsNiCl_3$. Magnetic field dependence of magnetostriction for $\boldsymbol{H} \perp c$. Note negative effect in paramagnetic phase above $T_N = 4.85$ K [83 R1, 84 R1].

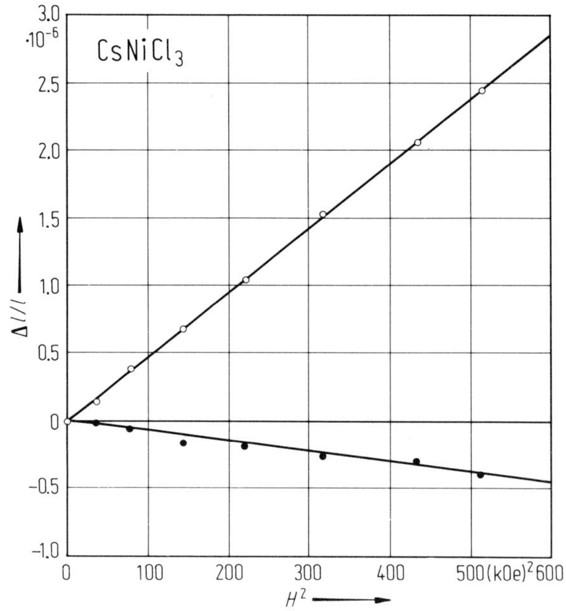

Fig. 543. $CsNiCl_3$. Magnetostriction $\Delta l/l$ vs. H^2 at 5.2 K; open circles: for $\boldsymbol{H} \parallel c$, full circles: $\boldsymbol{H} \perp c$ [83 R1, 84 R1].

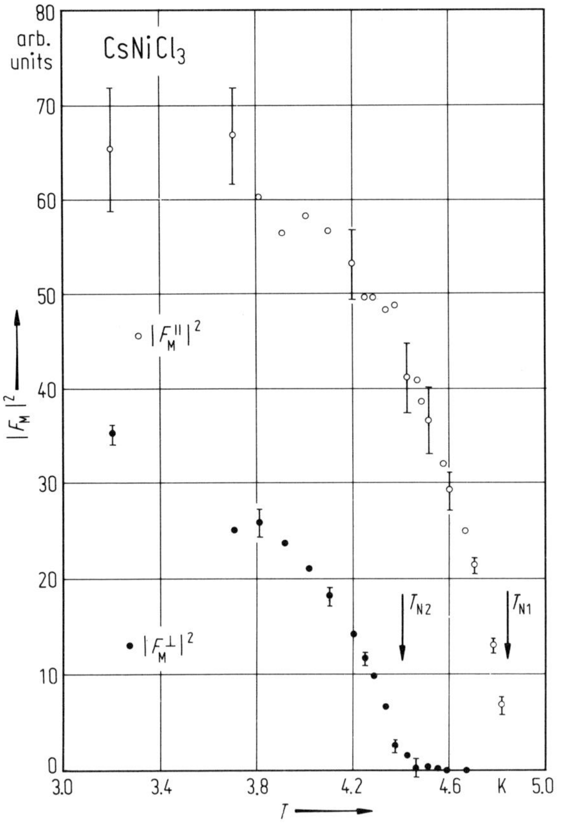

Fig. 544. $CsNiCl_3$. Temperature dependence of squares of structure factors $|F_M^{\parallel}|$ and $|F_M^{\perp}|$, which are proportional to the sublattice magnetizations parallel and perpendicular to the c axis, respectively [87K1].

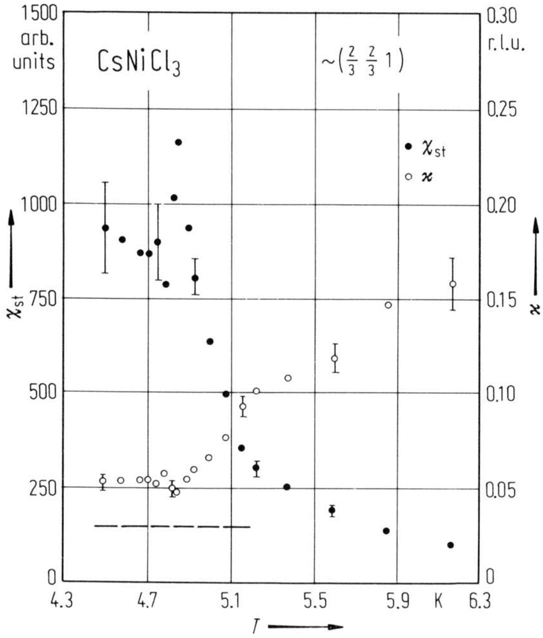

Fig. 545. $CsNiCl_3$. Temperature dependence of staggered susceptibility, χ_{st}, determined by fitting the critical neutron scattering profile around the $(\frac{2}{3}, \frac{2}{3}, 1)$ Bragg point. κ is the inverse correlation length in reciprocal lattice units 0.59 $Å^{-1}$. Dashed line is the resolution width [87K1].

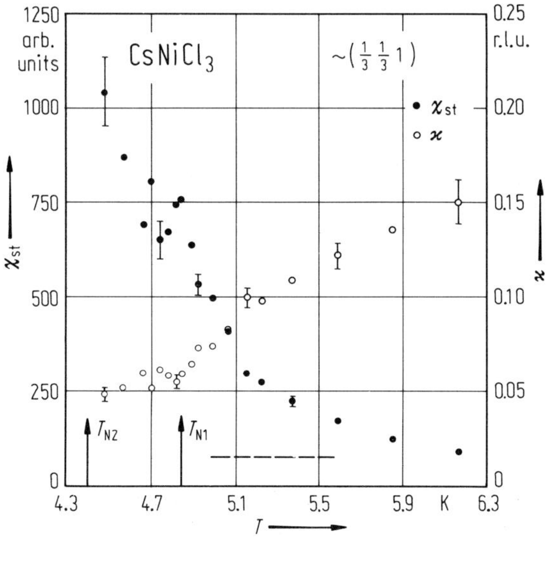

Fig. 546. $CsNiCl_3$. Temperature dependence of staggered susceptibility, χ_{st}, determined by fitting the critical neutron scattering profile around the $(\frac{1}{3}, \frac{1}{3}, 1)$ Bragg point. Dashed line and κ: see Fig. 545 [87K1].

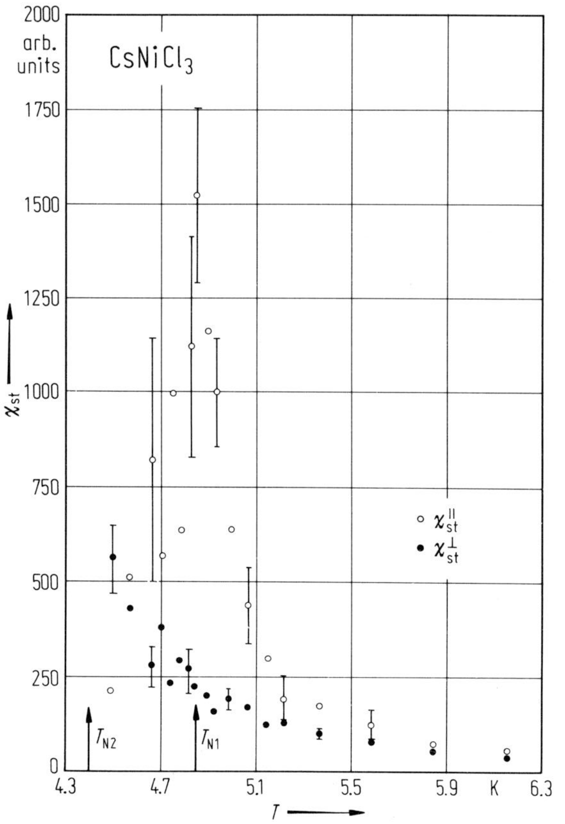

Fig. 547. $CsNiCl_3$. Temperature dependence of the staggered susceptibilities along $\chi_{st}^{\parallel}$ and perpendicular $\chi_{st}^{\perp}$ to the c axis [87 K1].

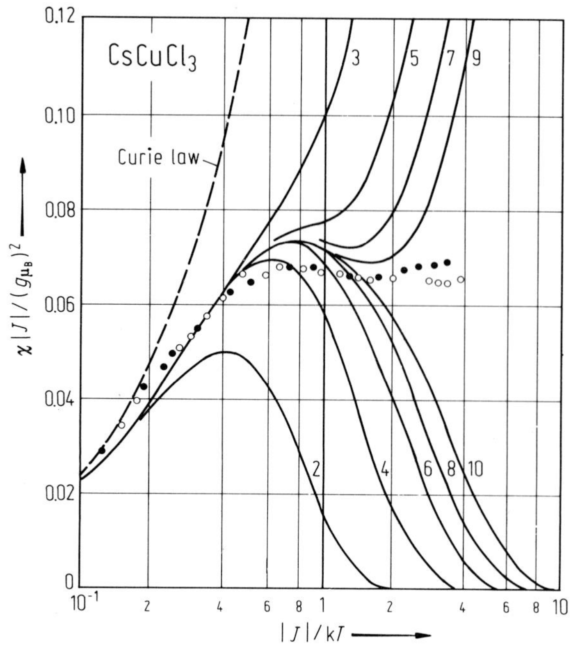

Fig. 549. $CsCuCl_3$. Normalized χ of a single crystal vs. T^{-1}. Solid curves are calculated for finite Heisenberg rings by [62G1] with antiferromagnetic coupling. Open and full circles are for $\chi_{\parallel}$ and $\chi_{\perp}$. Exchange interaction energies were obtained as $J_{\parallel}/k = -7.0$ K, $J_{\perp}/k = -6$ K [69A1].

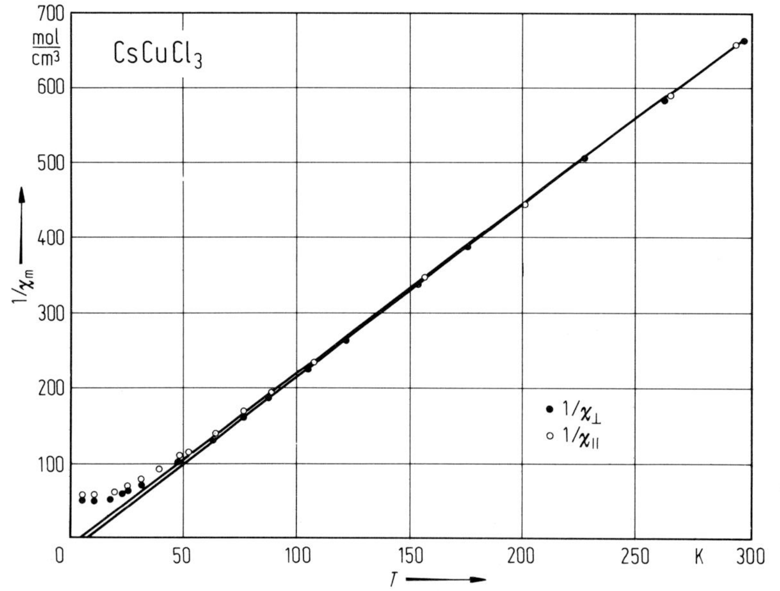

Fig. 548. $CsCuCl_3$. Temperature dependence of χ_m^{-1} for a single crystal. The corrections of core orbital diamagnetism and Van Vleck paramagnetism were applied. $\chi_{\parallel}$: along [0001], $\chi_{\perp}$: along [11$\bar{2}$0] [69A1].

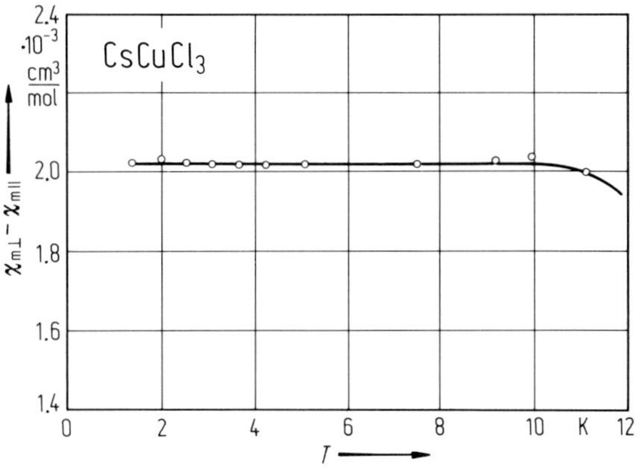

Fig. 550. $CsCuCl_3$. Temperature dependence of ($\chi_{m\perp}-\chi_{m\parallel}$) (∥ along [0001] and ⊥ along [11$\bar{2}$0], respectively) obtained from torque measurement [69A1].

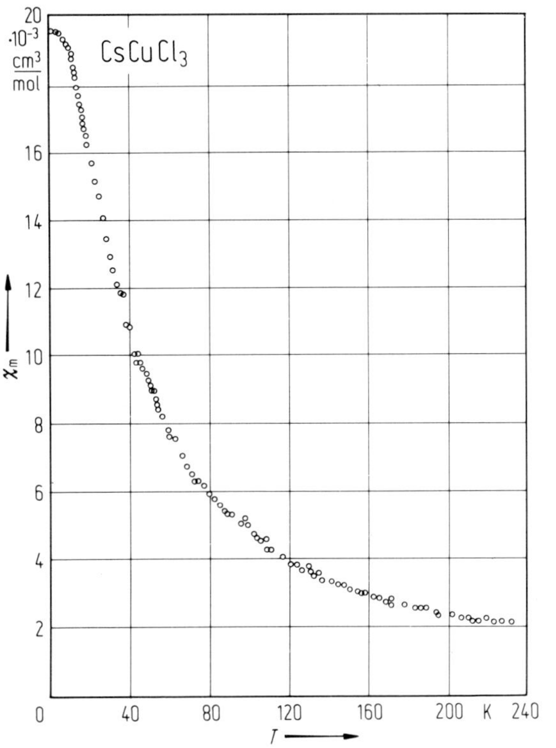

Fig. 551. $CsCuCl_3$. Temperature dependence of χ_m for powder sample [69R1].

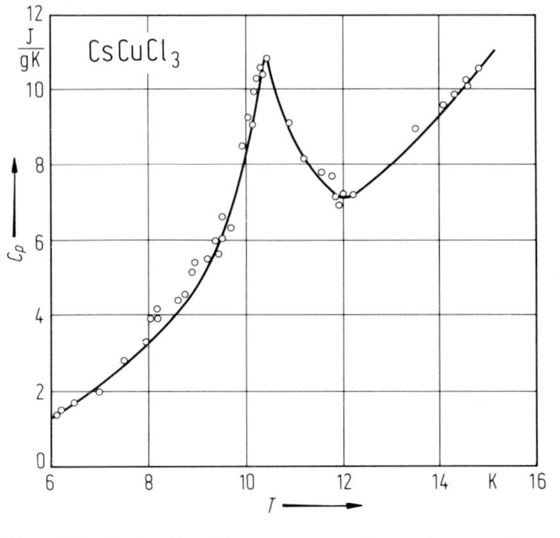

Fig. 552. $CsCuCl_3$. Temperature dependence of C_p for powder sample [69R1].

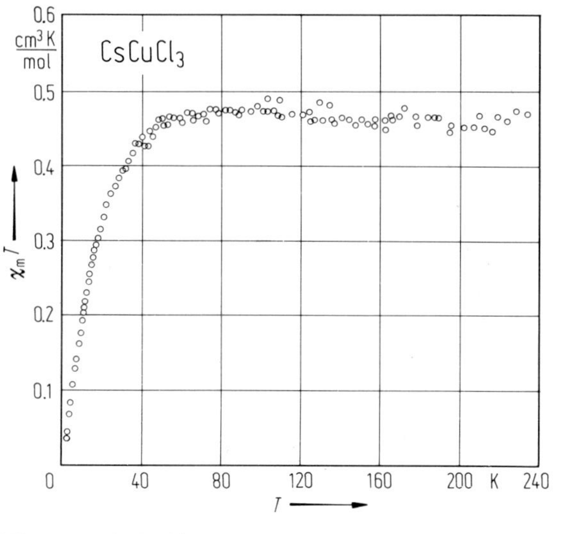

Fig. 553. $CsCuCl_3$. Temperature dependence of $\chi_m T$ for powder sample [69R1].

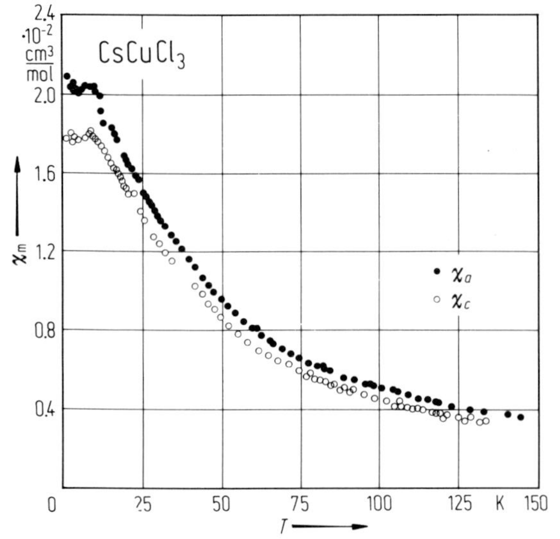

Fig. 554. $CsCuCl_3$. Temperature dependence of χ_m for a single crystal [70R1]. See also [81T1], Figs. 560, 562.

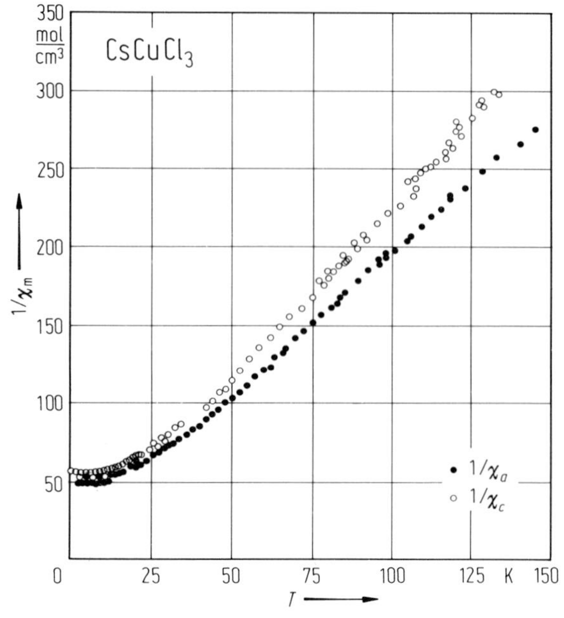

Fig. 555. $CsCuCl_3$. Temperature dependence of χ_m^{-1} for a single crystal [70R1].

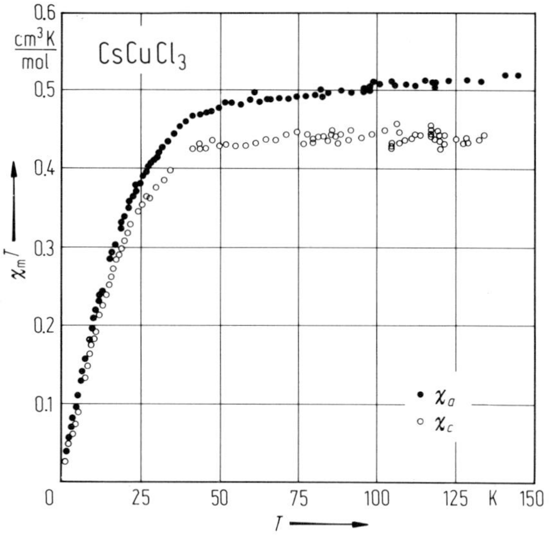

Fig. 556. $CsCuCl_3$. Temperature dependence of $\chi_m T$ for a single crystal [70R1].

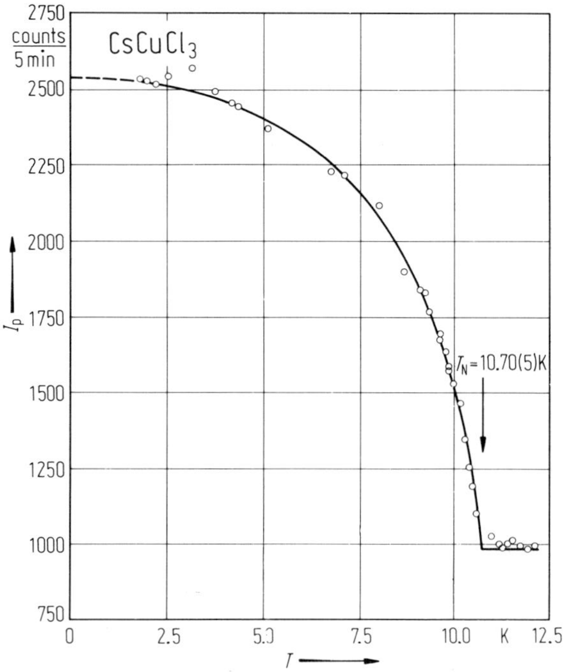

Fig. 557. $CsCuCl_3$. Temperature dependence of neutron diffraction $(\frac{1}{3},\frac{1}{3},6^+)$ peak intensity [80A1].

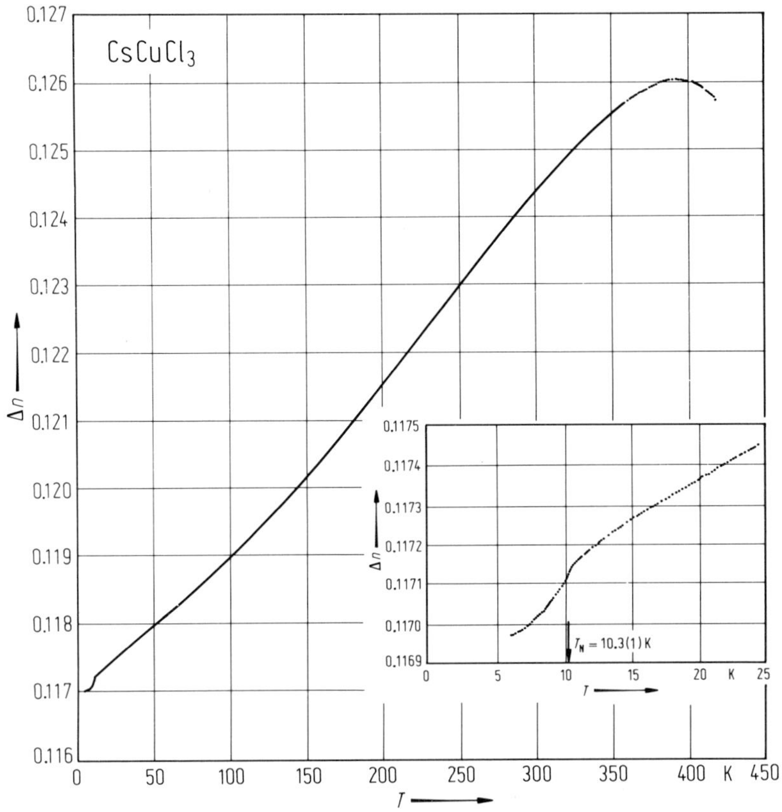

Fig. 558. $CsCuCl_3$. Temperature dependence of the birefringence $\Delta n = n_c - n_a$ for $\lambda = 0.6328$ μm [81H2].

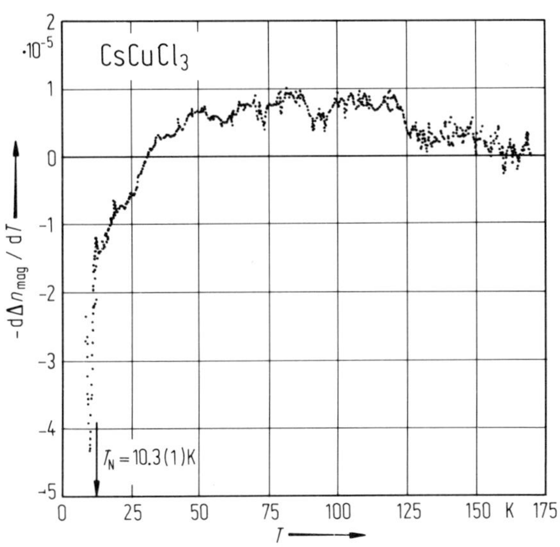

Fig. 559. $CsCuCl_3$. Calculated temperature dependence of the temperature derivative of the magnetic part of $-\mathrm{d}\Delta n_{mag}/\mathrm{d}T$ ($\lambda = 0.6328$ μm) obtained by numerical differentiation. The inverse λ-shaped spike appearing at 10.3(1) K indicates the three-dimensional antiferromagnetic phase transition point T_N [81H2].

For Fig. 560, see next page.

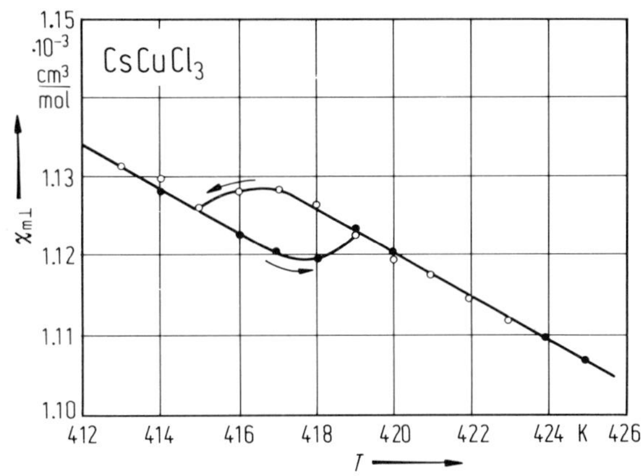

Fig. 561. $CsCuCl_3$. Temperature dependence of $\chi_{m\perp}$ around the phase transition temperature T_t (see Fig. 560) [81T1].

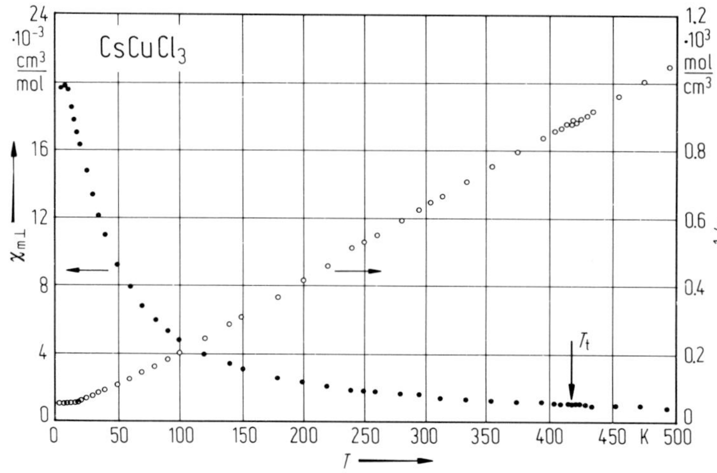

Fig. 560. $CsCuCl_3$. Temperature dependence of $\chi_{m\perp}$ and $\chi_{m\perp}^{-1}$ at 10 kOe measured perpendicular to the c axis [81T1].

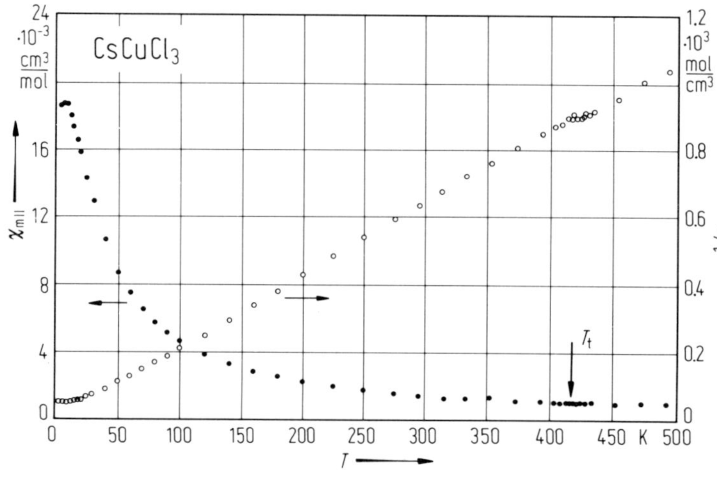

Fig. 562. $CsCuCl_3$. Temperature dependence of $\chi_{m\parallel}$ and $\chi_{m\parallel}^{-1}$ measured along the c axis [81T1].

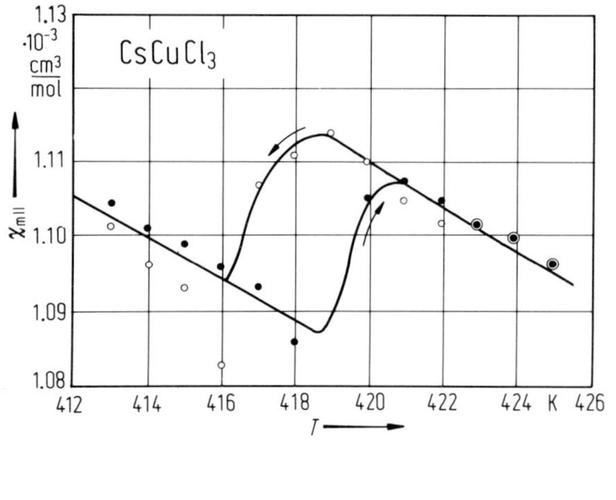

Fig. 563. $CsCuCl_3$. Temperature dependence of $\chi_{m\parallel}$ around the phase transition temperature T_t (see Fig. 562) [81T1].

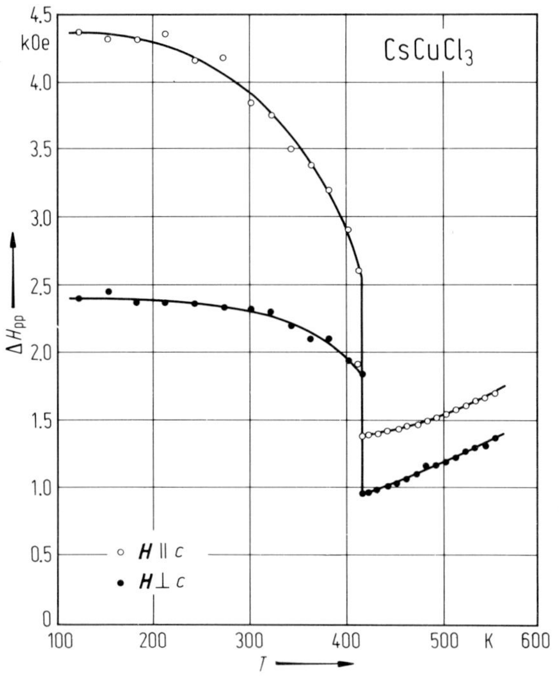

Fig. 564. $CsCuCl_3$. Temperature dependence of the ESR peak-to-peak derivative linewidth ΔH_{pp} at X-band [83T1].

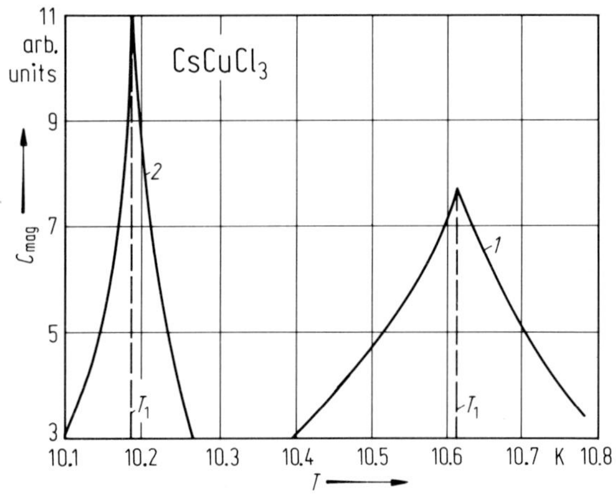

Fig. 565. $CsCuCl_3$. Temperature dependence of magnetic contribution to the heat capacity C_{mag} at 60 kOe (curve *1*) and 71.6 kOe (*2*) [85F1].

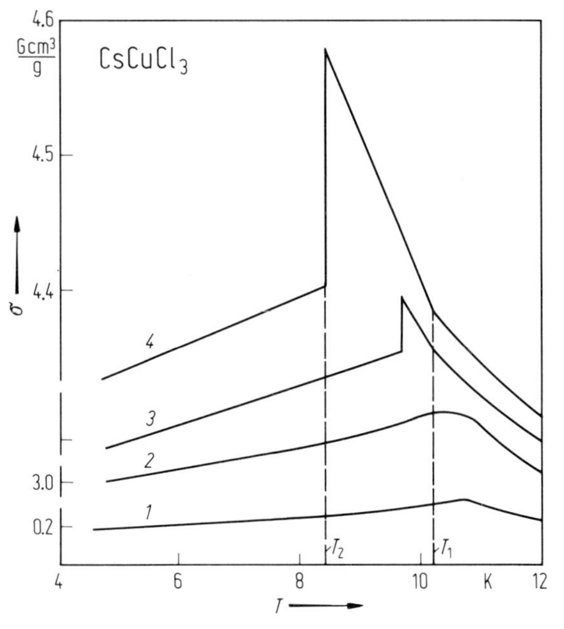

Fig. 566. $CsCuCl_3$. Temperature dependence of σ for $\boldsymbol{H} \parallel c$ axis for *1*: H = 3.3 kOe; *2*: 50 kOe; *3*: 60 kOe; *4*: 71.6 kOe [85F1].

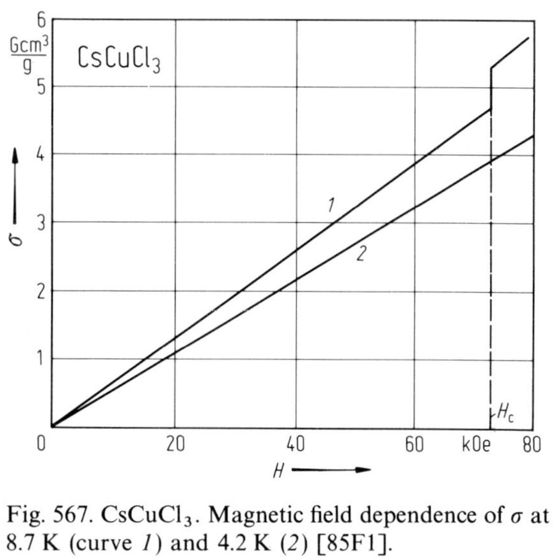

Fig. 567. $CsCuCl_3$. Magnetic field dependence of σ at 8.7 K (curve *1*) and 4.2 K (*2*) [85F1].

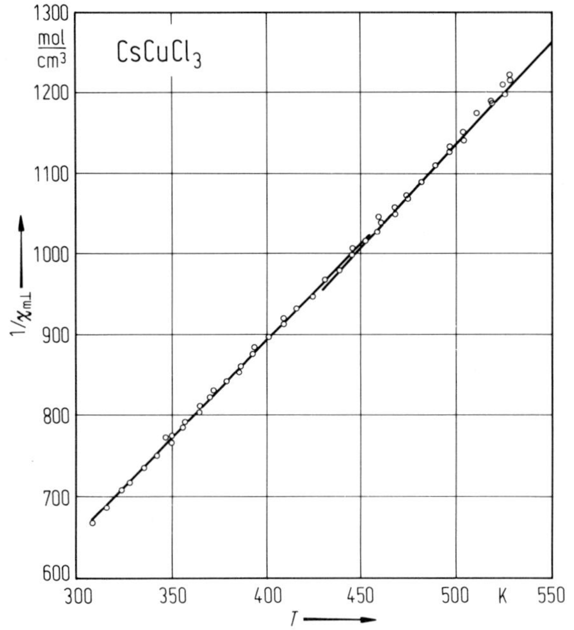

Fig. 568. $CsCuCl_3$. Temperature dependence of χ_m^{-1} for $\boldsymbol{H} \perp c$. The full line represents the Curie-Weiss fit [86H2].

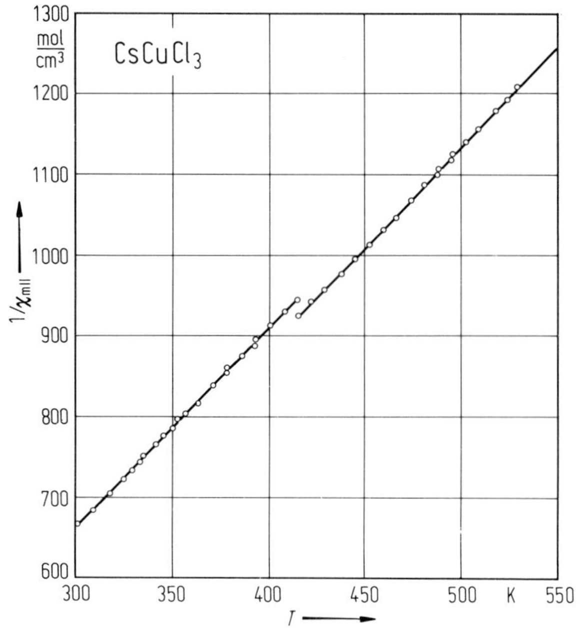

Fig. 569. $CsCuCl_3$. Temperature dependence of χ_m^{-1} for $\boldsymbol{H} \parallel c$. The full line represents the Curie-Weiss fit [86H2].

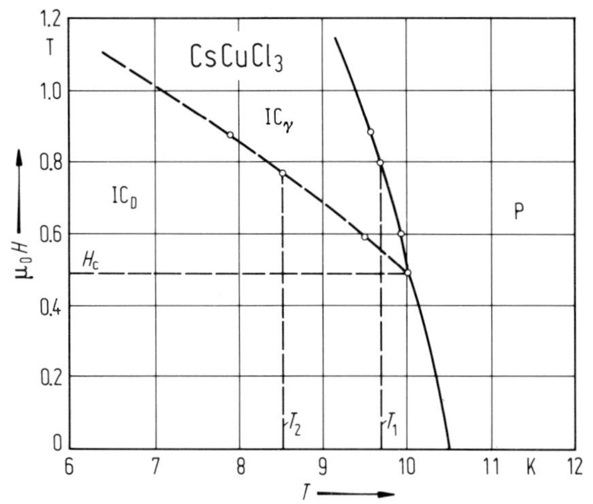

Fig. 570. $CsCuCl_3$. H–T phase diagram. Solid line indicates second-order transition. Circles: experimental points. IC_γ: sinusoidal incommensurate structure in the basal plane, IC_D: spiral structure modulated along c axis of the hexagonal lattice [89G2]. See also [85F1].

For Fig. 571, see next page.

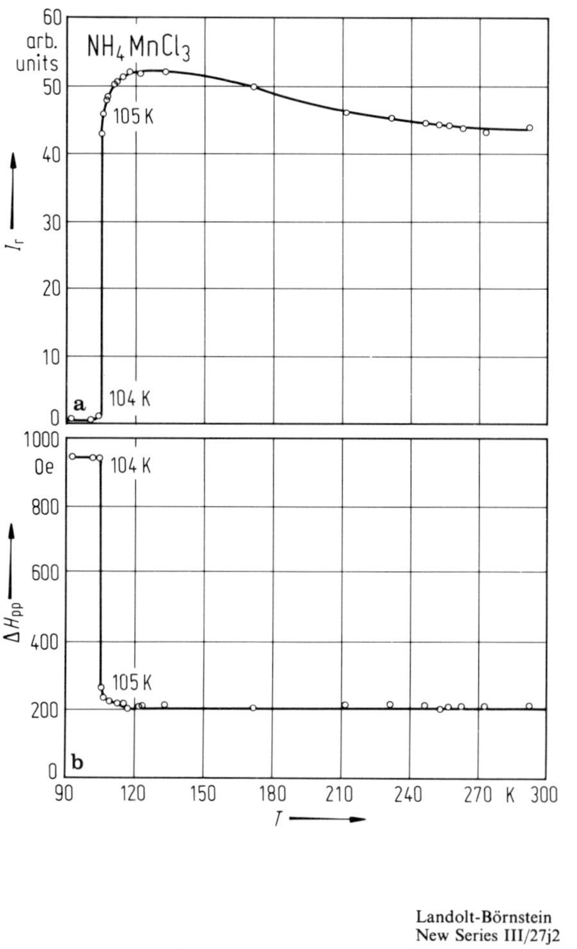

Fig. 572. NH_4MnCl_3. Temperature dependence of (a) relative resonance absorption intensity I_r and (b) width of the EPR line near T_N [75T2].

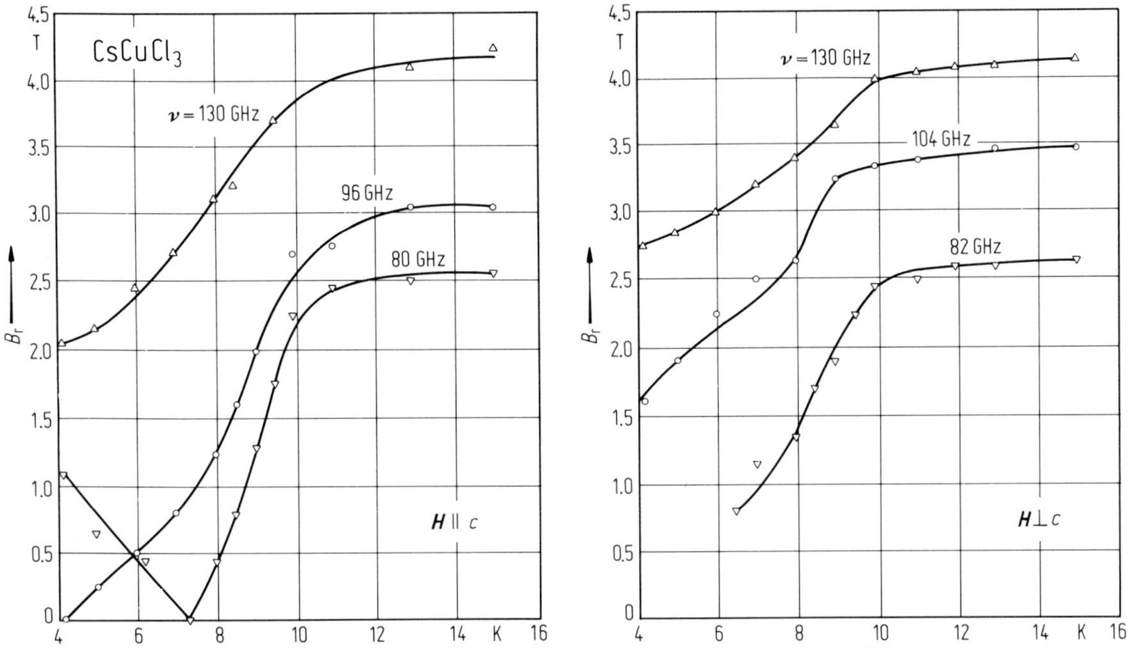

a b

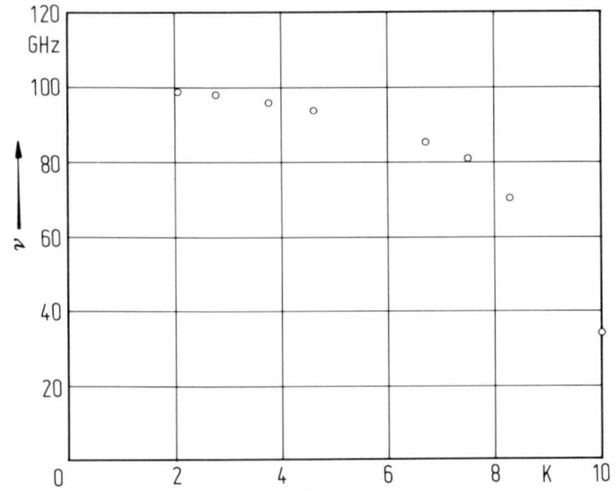

c

Fig. 571. $CsCuCl_3$. Temperature dependence of the resonance field for (a) $\boldsymbol{H} \parallel c$, (b) $\boldsymbol{H} \perp c$; (c) zero field resonance frequency [90P2].

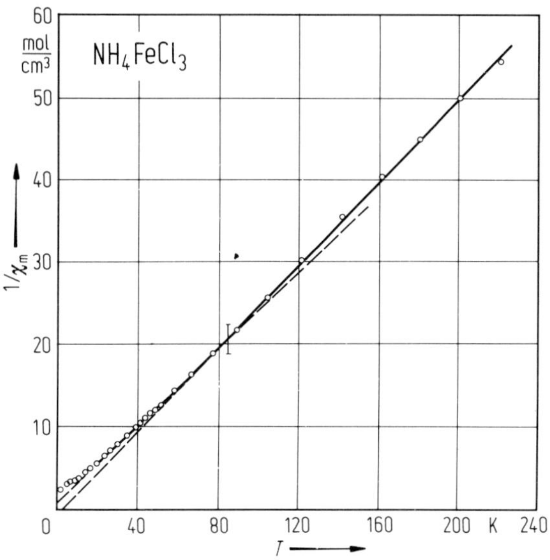

Fig. 573. NH_4FeCl_3. Temperature dependence of χ_m^{-1} [73W1].

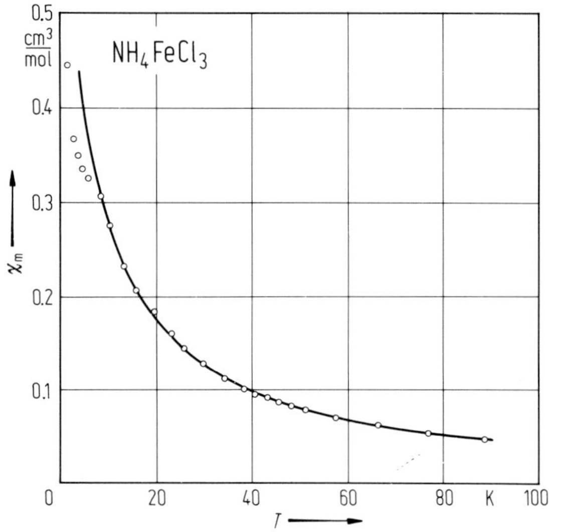

Fig. 574. NH_4FeCl_3. Temperature dependence of χ_m. Solid line: theoretical fit [73W1].

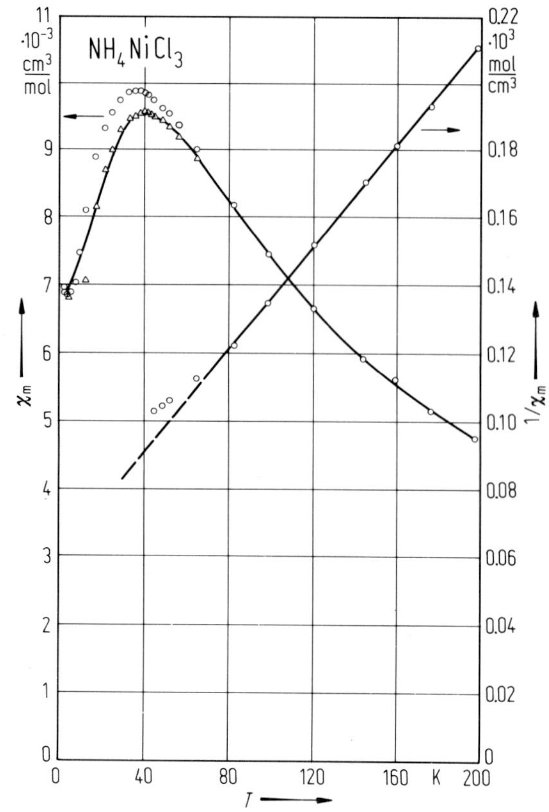

Fig. 575. NH_4NiCl_3. Temperature dependence of χ_m and χ_m^{-1}. Triangles: paramagnetic impurity corrected experimental points. Full curve is the theoretical fit for χ_m [74W1].

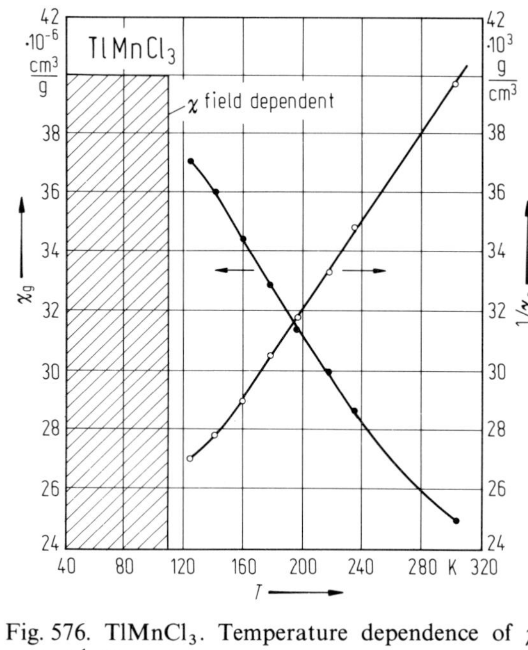

Fig. 576. $TlMnCl_3$. Temperature dependence of χ_g and χ_g^{-1} [73V1].

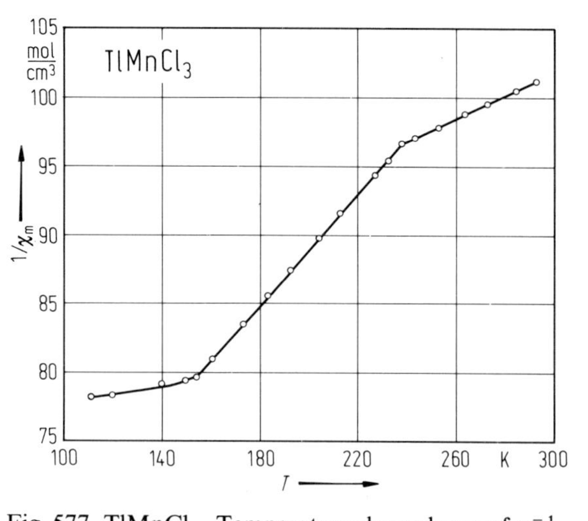

Fig. 577. $TlMnCl_3$. Temperature dependence of χ_m^{-1}. Susceptibility was found to be independent of the $\boldsymbol{H}$ orientation [77F1].

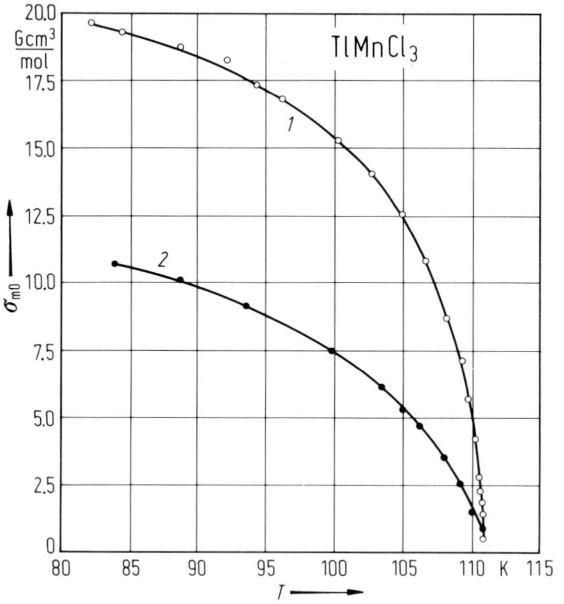

Fig. 578. $TlMnCl_3$. Temperature dependence of the spontaneous molar magnetization σ_{m0} in the [0 0 1] direction of the single crystal (curve *1*) and of the powder sample (*2*) [77F1].

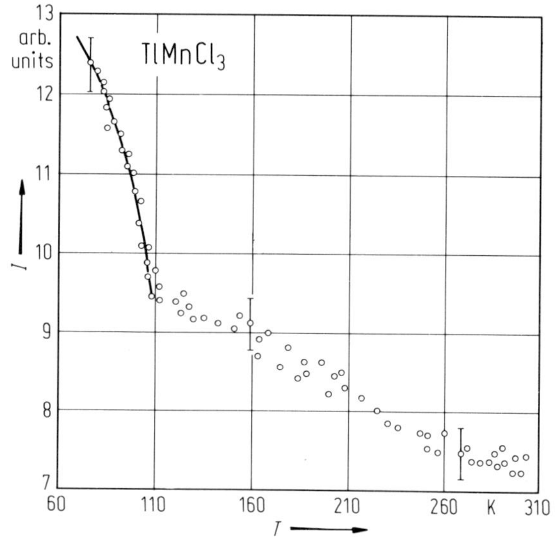

Fig. 579. $TlMnCl_3$. Temperature dependence of neutron diffraction intensity of the (2, 1, 1) reflection. The line is the Brillouin function for $S = 5/2$ [77F1].

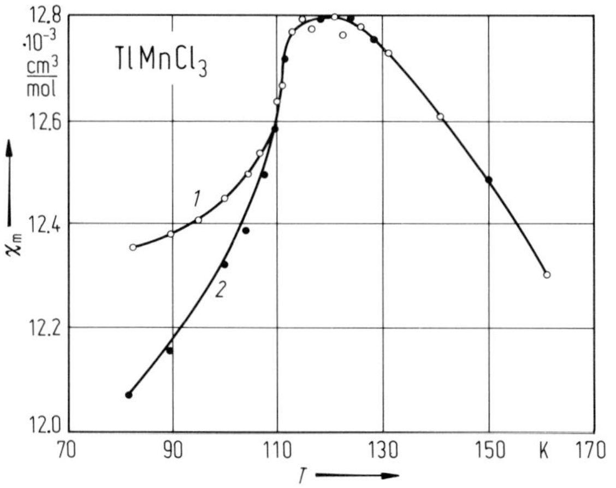

Fig. 580. $TlMnCl_3$. Temperature dependence of χ_m of a single-domain crystal. Curve *1*: $\boldsymbol{H}\|[1 1 0]$, *2*: $\boldsymbol{H}\|[0 0 1]$ or $[1 \bar{1} 0]$ [77F1].

→

Fig. 581. α-$TlMnCl_3$. Temperature dependence of χ_m in the direction of [1 1 0] at $H = 0$ (curve *1*); 50 (*2*) and 100 kOe (*3*) [87V1].

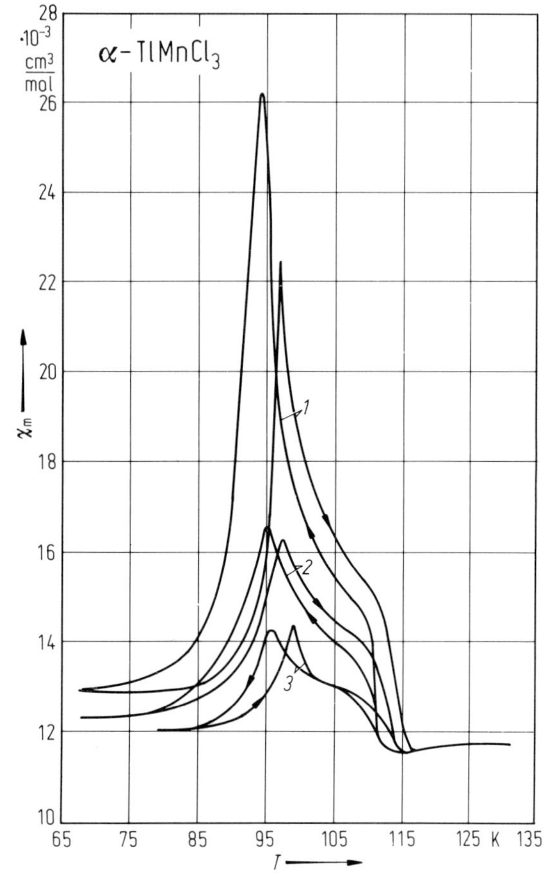

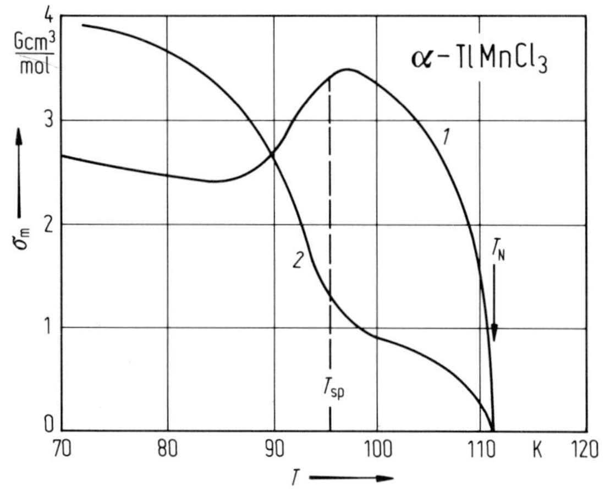

Fig. 582. α-$TlMnCl_3$. Temperature dependence of σ_m at $H = 0$ in the direction of [1 1 0] (*1*) and [0 0 1] (*2*) [87V1]. T_{sp}: magnetic moment reorientation temperature.

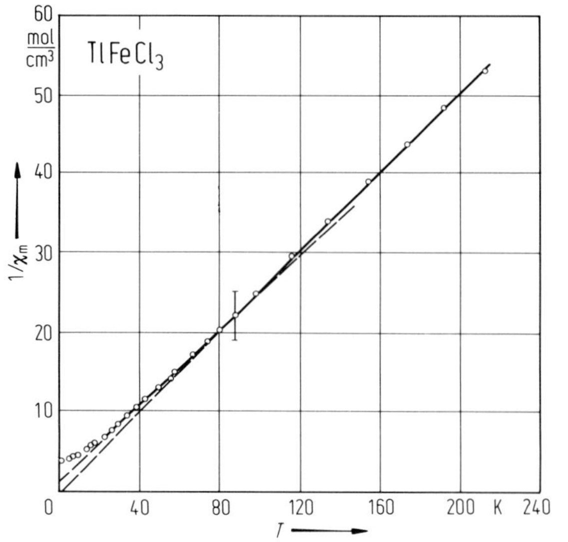

Fig. 583. $TlFeCl_3$. Temperature dependence of χ_m^{-1} [73W1].

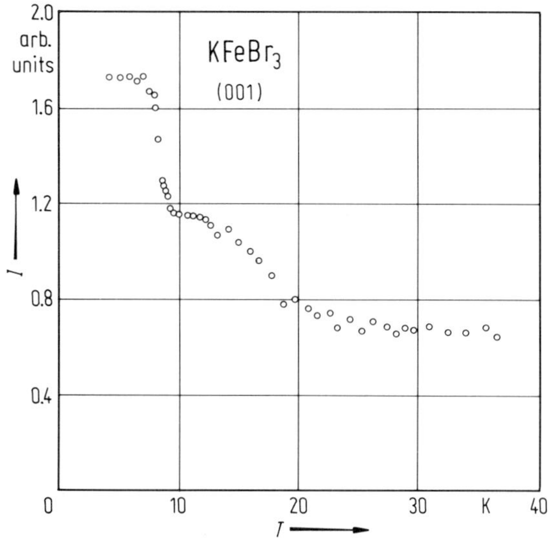

Fig. 584. $KFeBr_3$. Temperature dependence of neutron magnetic (0, 0, 1) reflection [82G3].

For Fig. 585, see next page.

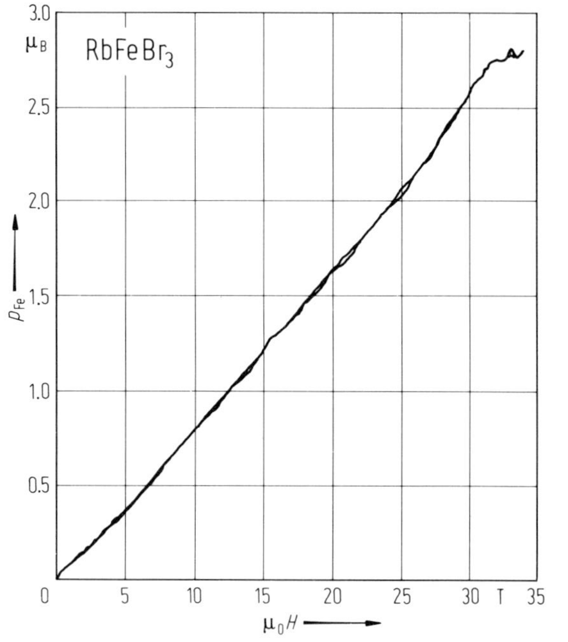

Fig. 586. $RbFeBr_3$. Magnetic field dependence of magnetic moment p_{Fe}; $\boldsymbol{H} \parallel c$ and $T = 1.8$ K [88A2].

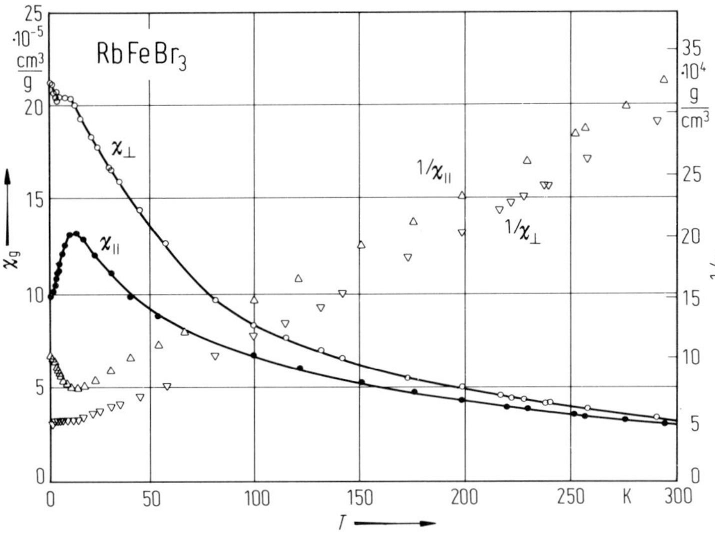

Fig. 585. $RbFeBr_3$. Temperature dependence of χ_g and χ_g^{-1} at 15.3 kOe for single crystal measured parallel and perpendicular to the c axis [75L1].

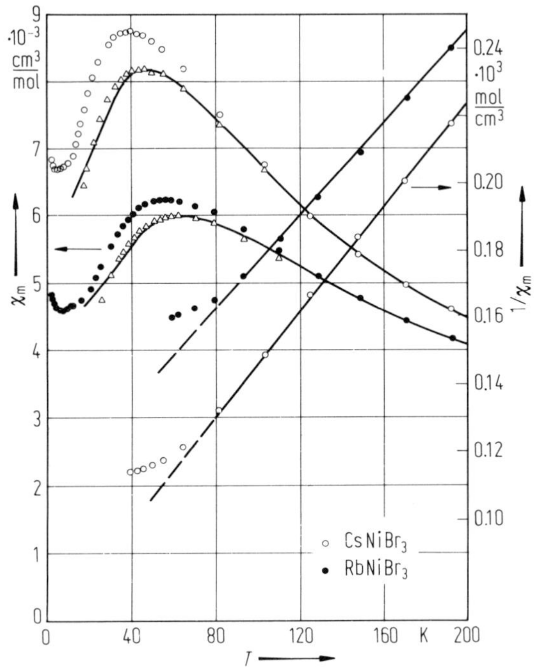

Fig. 587. $CsNiBr_3$, $RbNiBr_3$. Temperature dependence of χ_m and χ_m^{-1}. Triangles: paramagnetic impurity corrected experimental points. The full curves are the theoretical fits for χ_m [74W1].

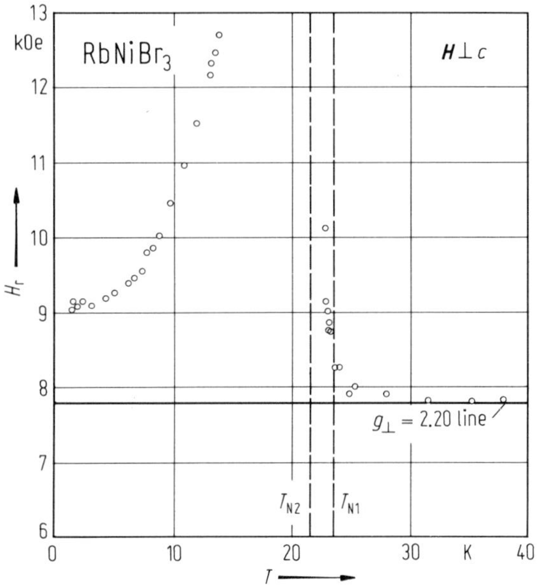

Fig. 588. $RbNiBr_3$. Temperature dependence of ESR field H_r at 24.02 GHz and $\boldsymbol{H} \perp c$ axis [90T1]. $T_{N1} = 23.50$ K, $T_{N2} = 21.47$ K.

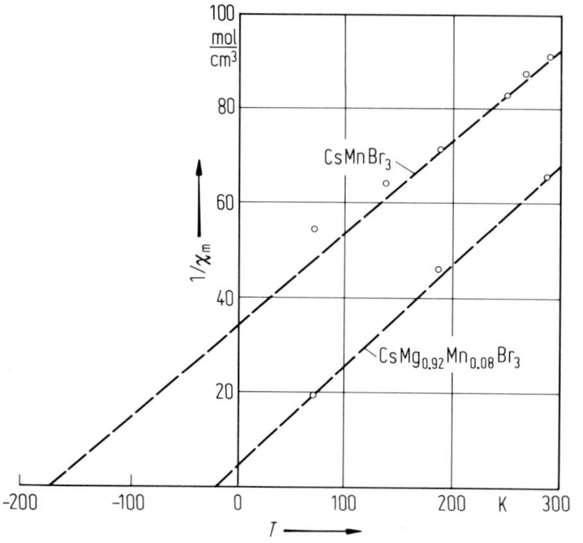

Fig. 589. $CsMnBr_3$, $CsMg_{0.92}Mn_{0.08}Br_3$. Temperature dependence of χ_m^{-1} [74M3].

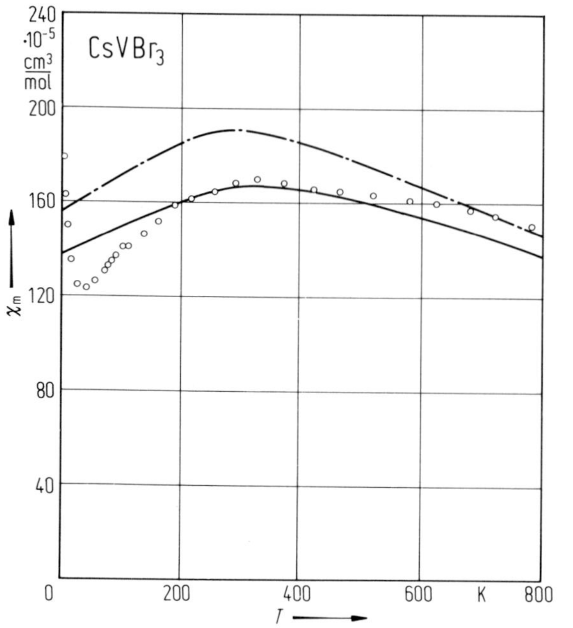

Fig. 590. $CsVBr_3$. Temperature dependence of χ_m. Solid and dashed lines are theoretical fits for $J/k = -91$ K and -80 K [77N1].

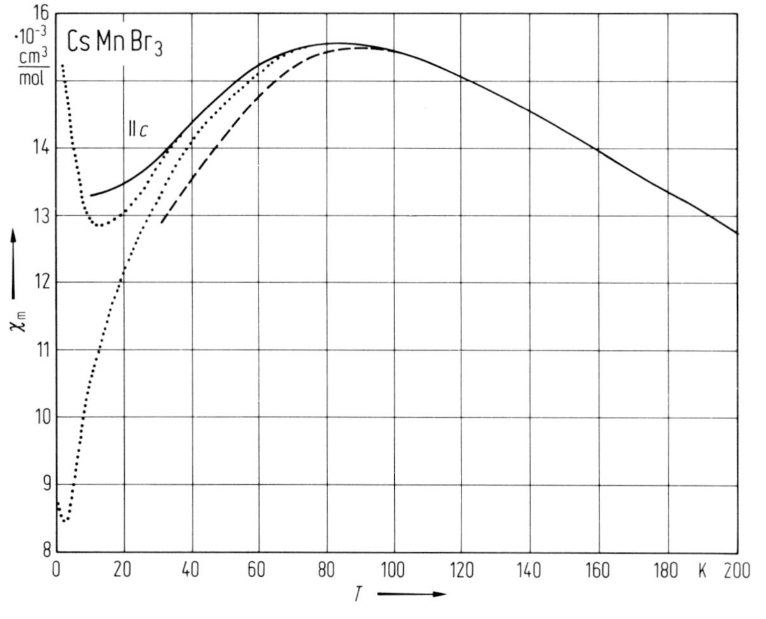

Fig. 591. $CsMnBr_3$. Temperature dependence of χ_m for $\boldsymbol{H} \parallel c$ and $\boldsymbol{H} \perp c$ at 3.5 kOe. The full curve is the fit by Fisher's [64B1] model with $J/k = -9.64$ K. The broken curve represents the fit with $J/k = 9.15$ according to Weng's [68W1] extrapolation scheme [82F1].

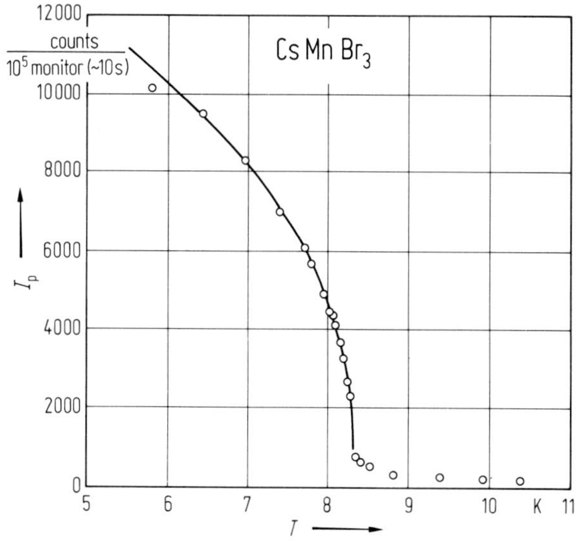

Fig. 592. $CsMnBr_3$. Temperature dependence of intensity for the $(\frac{1}{3}, \frac{1}{3}, 1)$ magnetic peak. The line is calculated for critical exponent $\beta = 0.21$, $T_N = 8.3$ K [87M2].

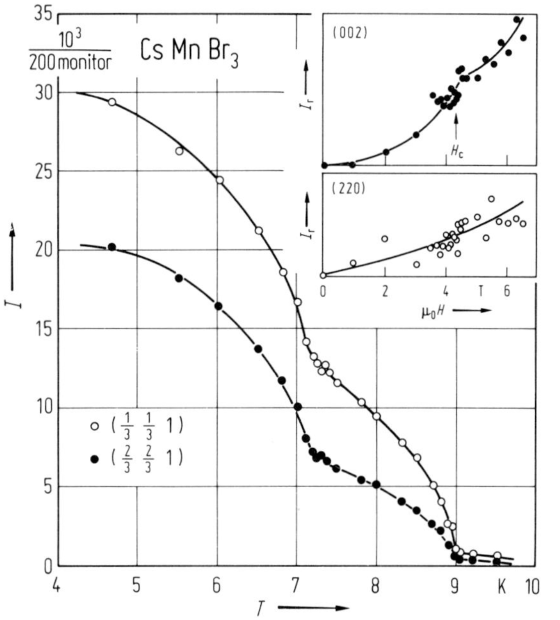

Fig. 593. $CsMnBr_3$. Temperature dependence of $(\frac{1}{3}, \frac{1}{3}, 1)$ and $(\frac{2}{3}, \frac{2}{3}, 1)$ magnetic Bragg intensities in a magnetic field of 4.2 T in the [1 0 0] direction. $T_{II} = 9.00$ (10) K and $T_I = 7.15$ (10) K are successive phase transitions from paramagnetic to the spin-flopped phase and to the triangular phase respectively (see Fig. 598). Inset: field dependence of (2, 2, 0) and (0, 0, 2) Bragg peaks [89G1]. I_r: relative intensity. See also [87G1].

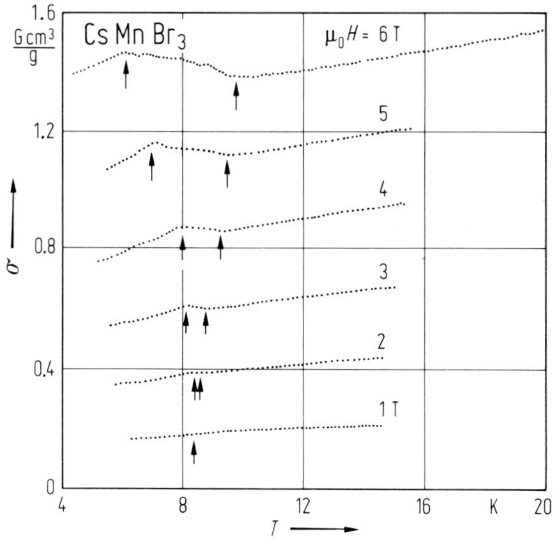

Fig. 594. $CsMnBr_3$. Temperature dependence of σ for different H in the c plane. Transition temperatures are indicated by arrows [90A1].

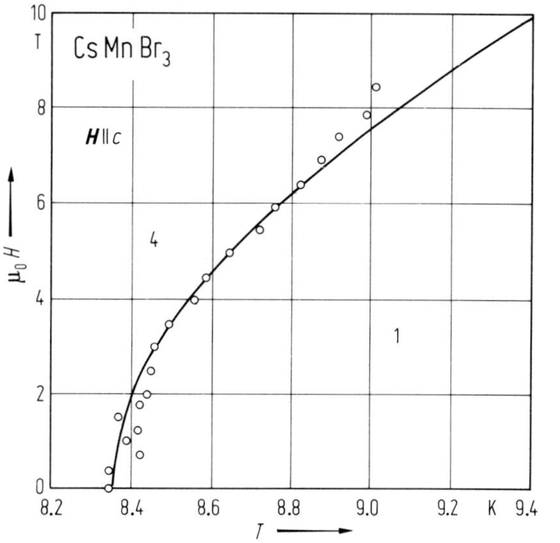

Fig. 595. $CsMnBr_3$. H–T phase diagram for $\boldsymbol{H} \parallel c$. Full line is theoretical. 1: paramagnetic phase, 4: helically polarized magnetic phase [90C1].

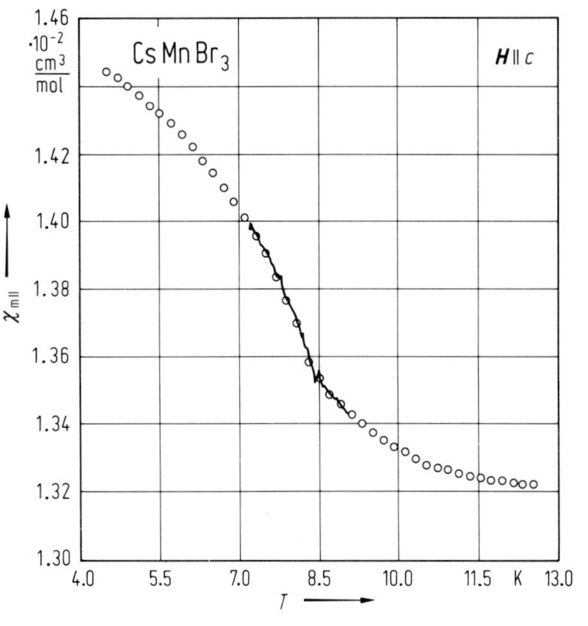

Fig. 596. $CsMnBr_3$. Temperature dependence of χ_m at 0.5 kOe for $\boldsymbol{H} \parallel c$. The line shows data for $H = 0.1$ kOe [90M2].

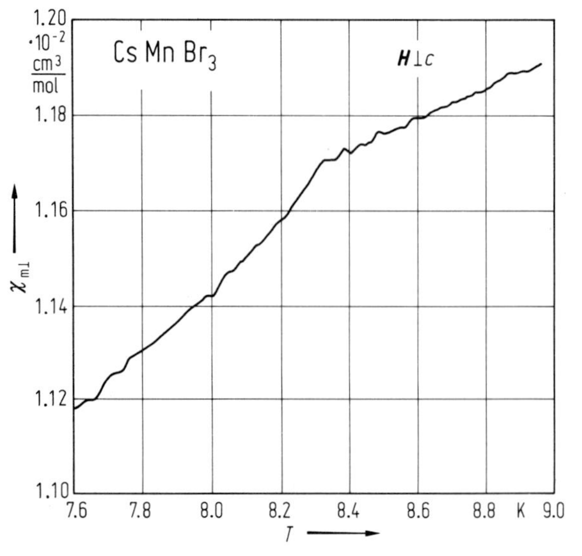

Fig. 597. $CsMnBr_3$. Temperature dependence of χ_m for $\boldsymbol{H} \perp c$ at 0.5 kOe [90M2].

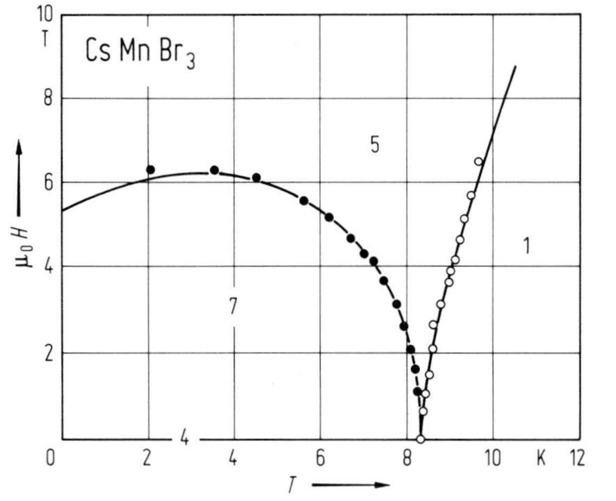

Fig. 598. $CsMnBr_3$. Magnetic phase diagram for $\boldsymbol{H} \perp c$. Phases 1, 4, 5 and 7 refer to paramagnetic, helical, linear (spin-flop) and elliptical (triangular) phases, respectively. The lines are results of theoretical fitting [90P1]. See also [89G1, 90A1, 90P3].

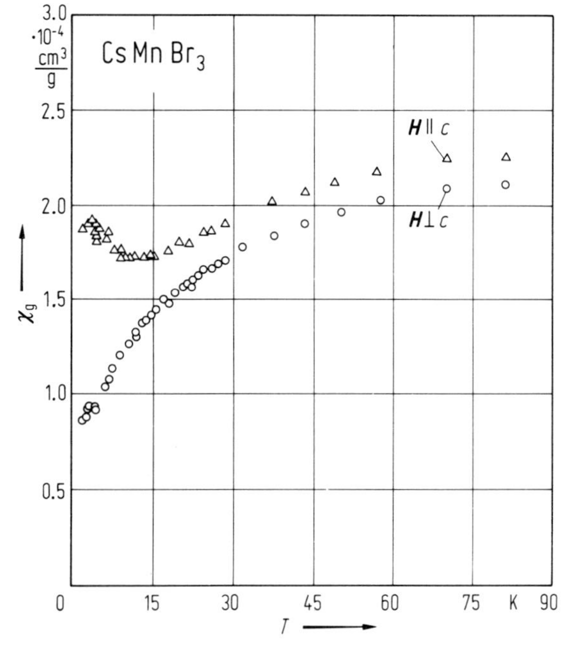

Fig. 599. $CsMnBr_3$. Temperature dependence of χ_g for $\boldsymbol{H} \parallel c$ and $\boldsymbol{H} \perp c$ at 22.5 kOe [91K1].

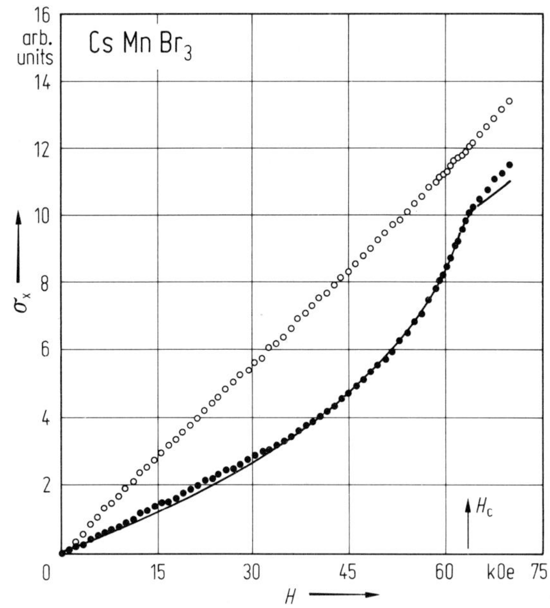

Fig. 600. $CsMnBr_3$. Magnetic field dependence of $\sigma_{x\|}$ (open circles) and $\sigma_{x\perp}$ (full circles) at 1.7 K. $\sigma_{x\|}$ is for $\boldsymbol{H} \| c$ and $\sigma_{x\perp}$ for $\boldsymbol{H} \perp c$. Full line calculated using [88C1] theory [91K1].

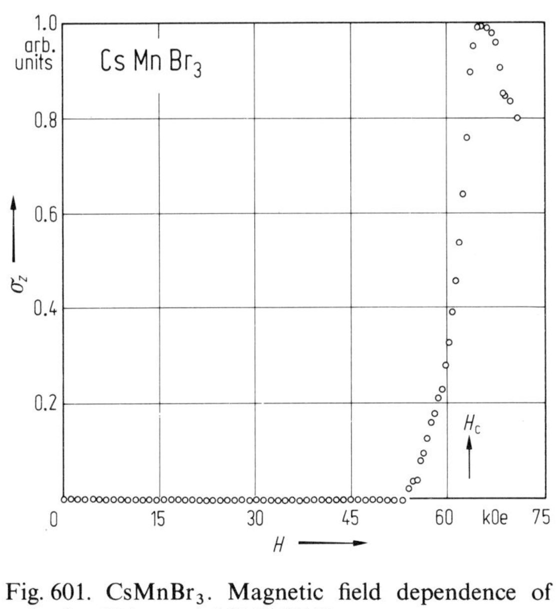

Fig. 601. $CsMnBr_3$. Magnetic field dependence of $\sigma_z \| c$, for $\boldsymbol{H} \perp c$ at 1.7 K [91K1].

For Fig. 602, see next page.

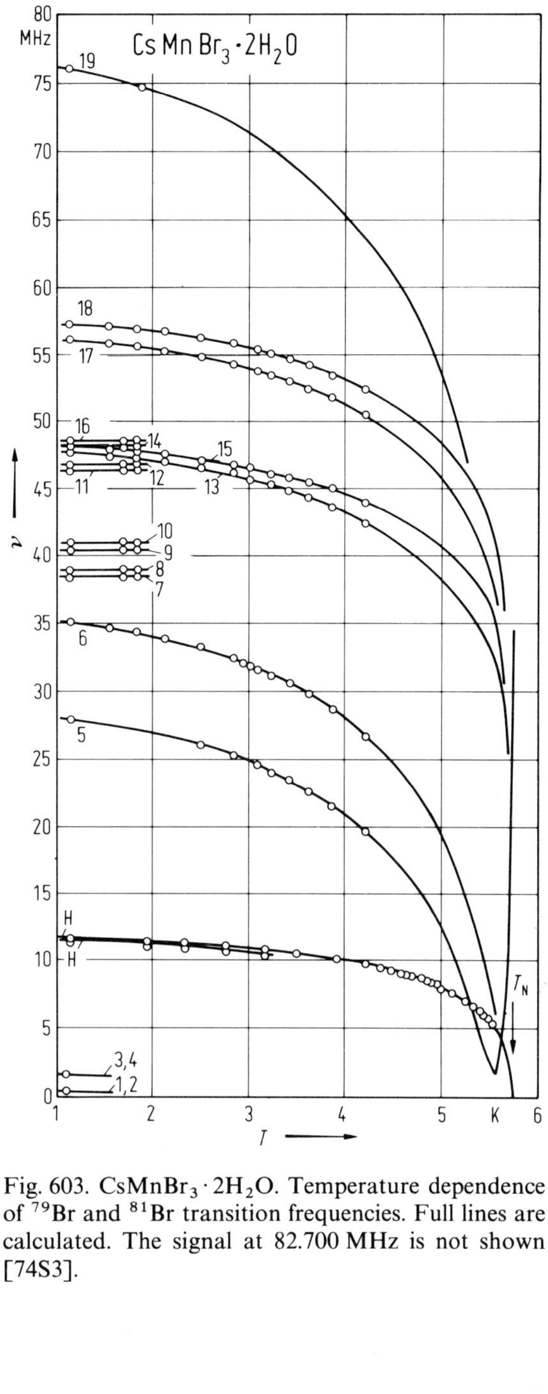

Fig. 603. $CsMnBr_3 \cdot 2H_2O$. Temperature dependence of ^{79}Br and ^{81}Br transition frequencies. Full lines are calculated. The signal at 82.700 MHz is not shown [74S3].

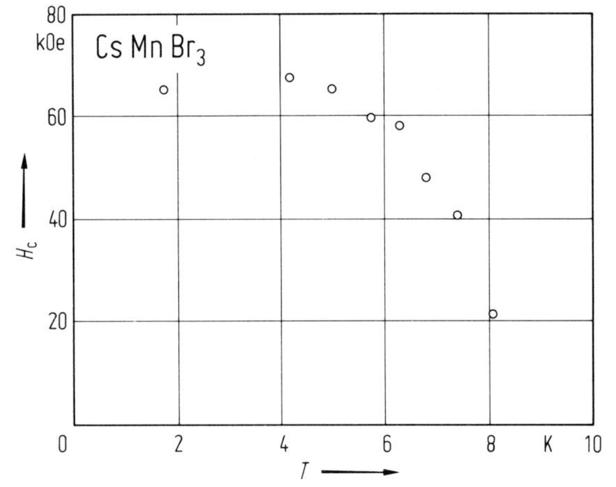

Fig. 602. $CsMnBr_3$. Temperature dependence of critical magnetic field H_c, obtained from the $\sigma_z(H)$ curve [91K1].

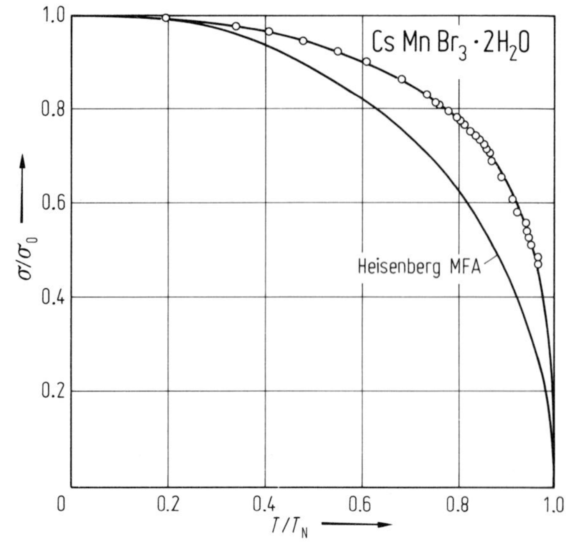

Fig. 604. $CsMnBr_3 \cdot 2H_2O$. Reduced temperature dependence of the proton-resonance lines. Solid line represents the predicted sublattice magnetization in Heisenberg MFA for $S = 5/2$ [74S3].

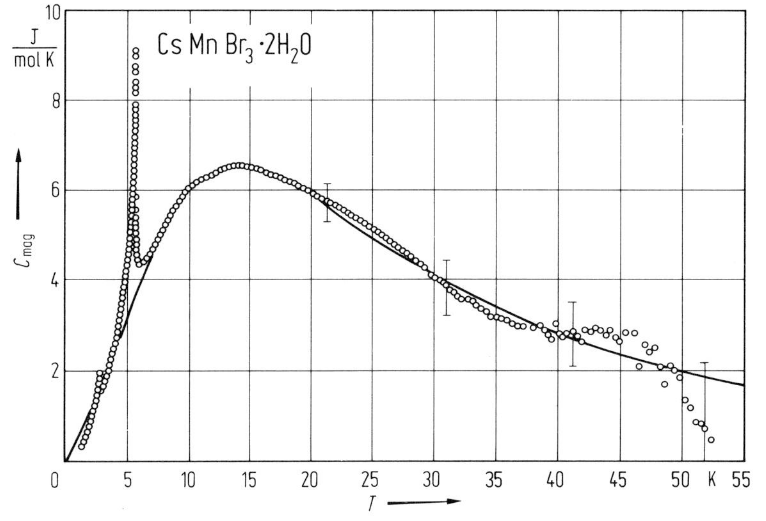

Fig. 605. $CsMnBr_3 \cdot 2H_2O$. Temperature dependence of the magnetic heat capacity. The anomaly at 2.8 K is due to a small fraction of $Cs_2MnBr_4 \cdot 2H_2O$. The solid line is drawn with $J/k = -2.6$ K. For explanation see Fig. 434 [77K3].

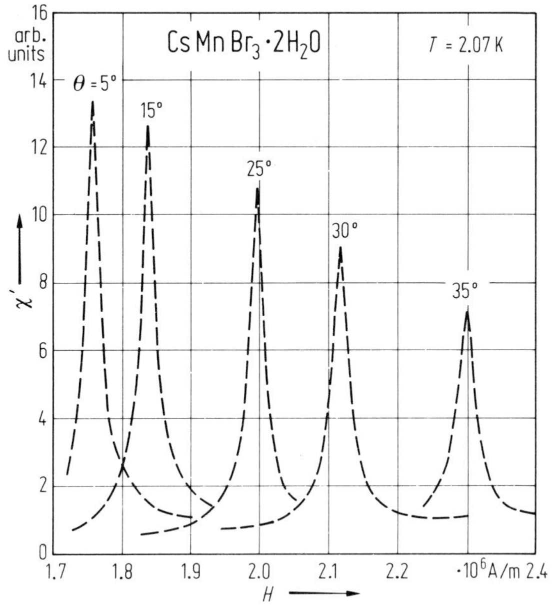

Fig. 606. $CsMnBr_3 \cdot 2H_2O$. Magnetic field dependence of χ' near the spin-flop transition measured in the *ab* plane (orthorhombic crystal structure) at $\nu = 1.08$ kHz and $H_p \approx 170$ A/m. Curves for $\theta = 0°$, 10° and 20° omitted for clarity (see Fig. 607) [83C1].

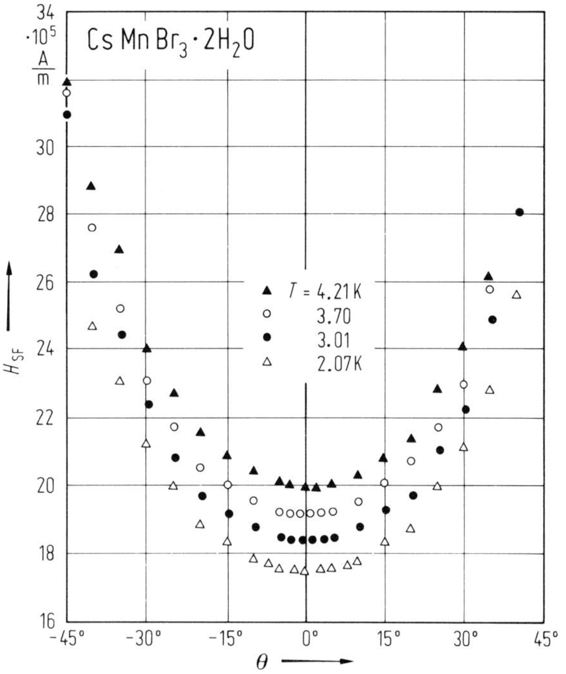

Fig. 607. $CsMnBr_3 \cdot 2H_2O$. Some critical hyperbolae of the spin-flop field H_{SF} as a function of the direction θ of the applied field in the *ab* plane relative to the *b* axis for different temperatures [83C1].

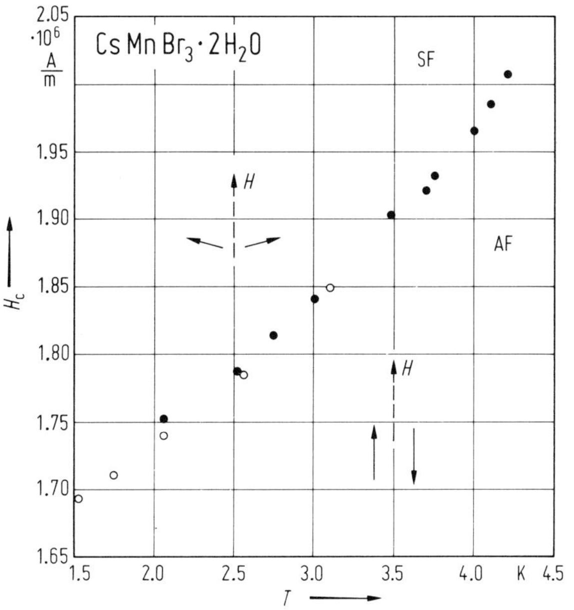

Fig. 608. $CsMnBr_3 \cdot 2H_2O$. Temperature dependence of the critical spin-flop field H_c. $H_c = H_{SF}$ ($\theta = 0$). Open circles: $\nu = 12.7$ Hz, discrete field sweeps; full circles: $\nu = 1.08$ kHz, continuous field sweeps [83C1].

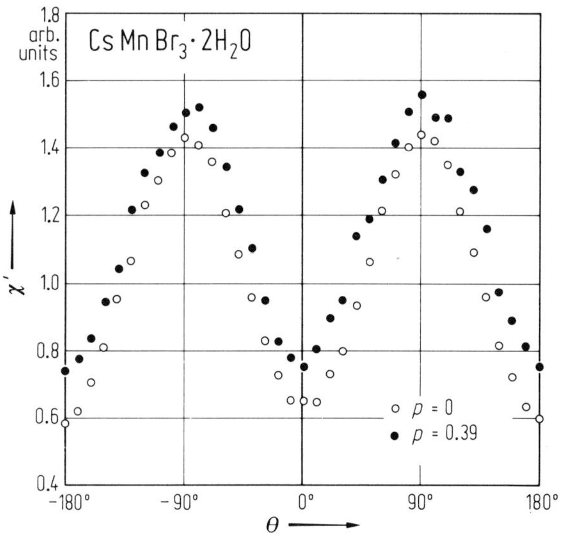

Fig. 609. $CsMnBr_3 \cdot 2H_2O$. Susceptibility rotation diagram in *ab* plane at 4.00 K, $\nu = 1.08$ kHz for $p = H/H_c \ll 1$ (see Fig. 607) [83C1].

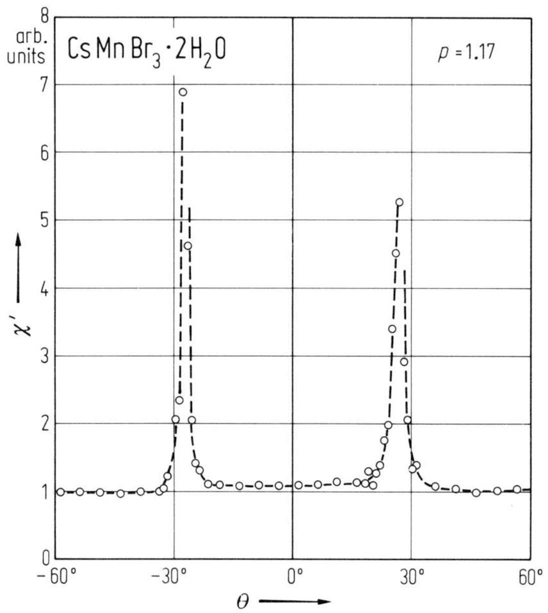

Fig. 610. $CsMnBr_3 \cdot 2H_2O$. Susceptibility rotation diagram in *ab* plane at 4.00 K, $\nu = 1.08$ kHz and $p = H/H_c = 1.17$ (see Fig. 607) [83C1].

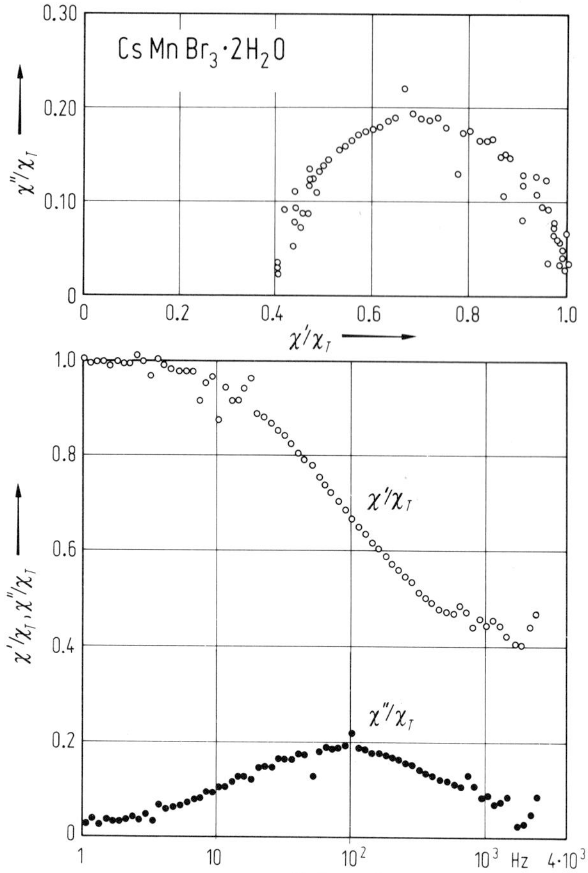

Fig. 611. $CsMnBr_3 \cdot 2H_2O$. Frequency dependence of χ'/χ_T and χ''/χ_T (χ_T: isothermal value in the low frequency region) at the spin-flop transition, $H_{SF} = 1.793 \cdot 10^6$ A/m, $T_{SF} = 2.557$ K. Argand diagram (χ'' vs. χ') is also shown [83C1].

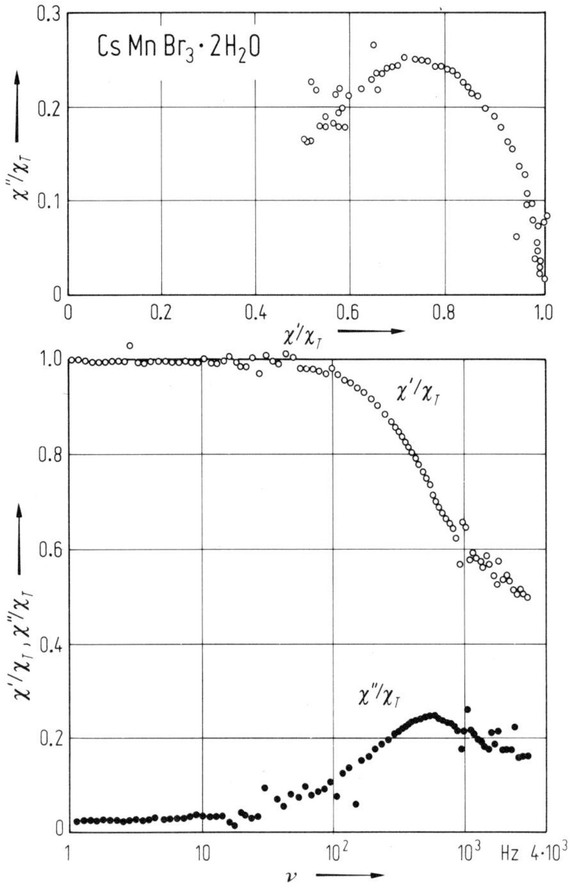

Fig. 612. $CsMnBr_3 \cdot 2H_2O$. Frequency dependence of χ'/χ_T and χ''/χ_T at the spin-flop transition, $H_{SF} = 1.7387 \cdot 10^6$ A/m, $T_{SF} = 2.044$ K (see Fig. 611) [83C1].

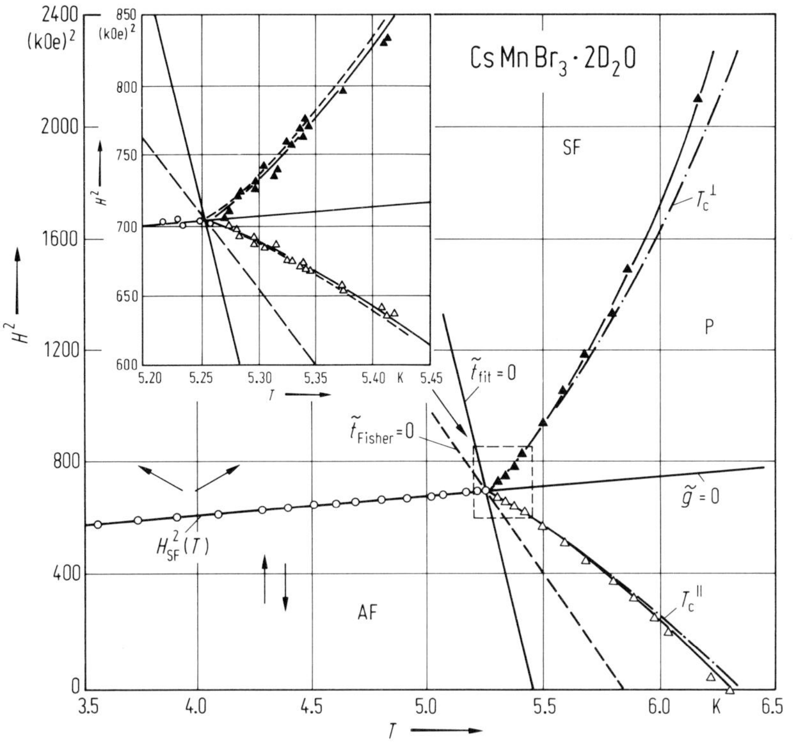

Fig. 613. $CsMnBr_3 \cdot 2D_2O$. Magnetic phase diagram. Solid and dash-dotted curves represent best fit to the data. $\tilde{g}$ and $\tilde{t}$ are optimum scaling axes. The theoretical (dashed) and fitted (solid) $\tilde{t} = 0$ axes are shown [78B3].

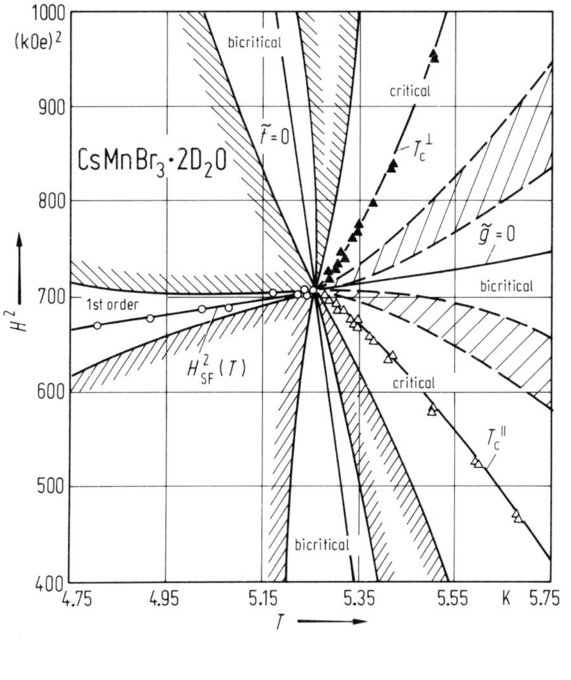

For Fig. 614, see next page.

Fig. 615. $CsMnBr_3 \cdot 2D_2O$. Part of the (H^2, T) phase diagram with the location of the $\tilde{g}$ and $\tilde{t}$ scaling axes. The shaded edges enclose the experimentally determined 25% crossover regions. For completeness, the crossover regions in the paramagnetic phase are also sketched [79B1].

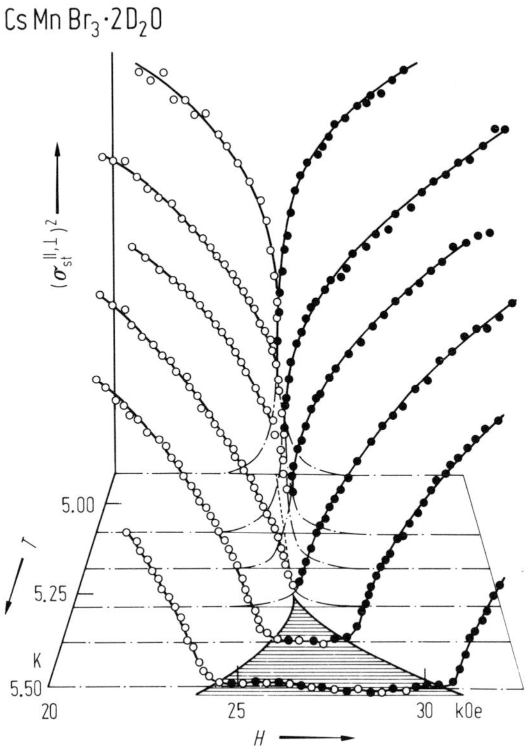

Fig. 614. $CsMnBr_3 \cdot 2D_2O$. Magnetic field dependence of $(\sigma_{st}^{\parallel})^2$ (open circles) and $(\sigma_{st}^{\perp})^2$ (closed circles) close to the bicritical point [79B1].

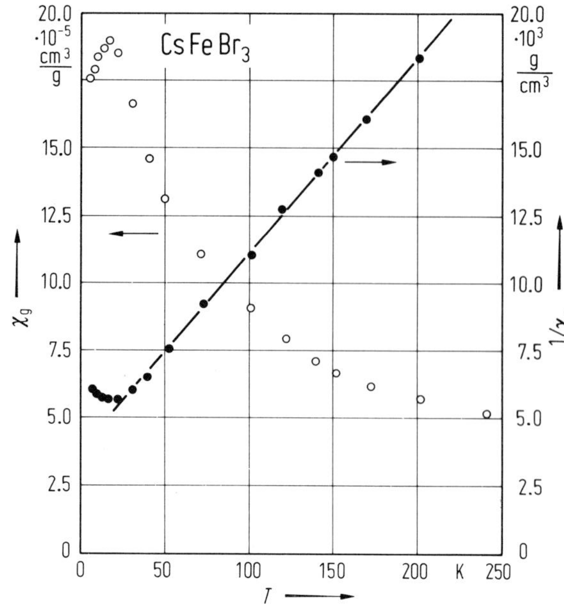

Fig. 616. $CsFeBr_3$. Temperature dependence of χ_g and χ_g^{-1} [74T1].

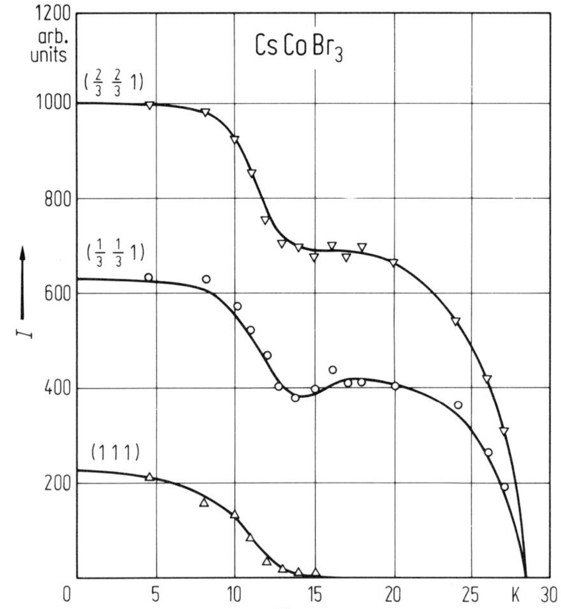

Fig. 617. $CsCoBr_3$. Temperature dependence of some of the strong neutron reflections [75Y1].

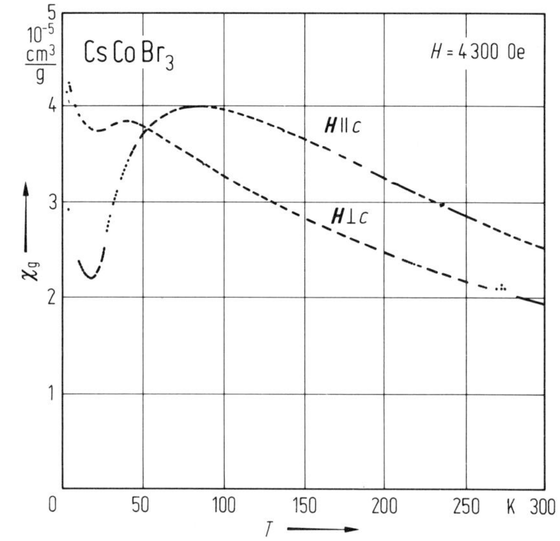

Fig. 618. $CsCoBr_3$. Temperature dependence of χ_g of a single crystal parallel and perpendicular to the c axis [75Y1].

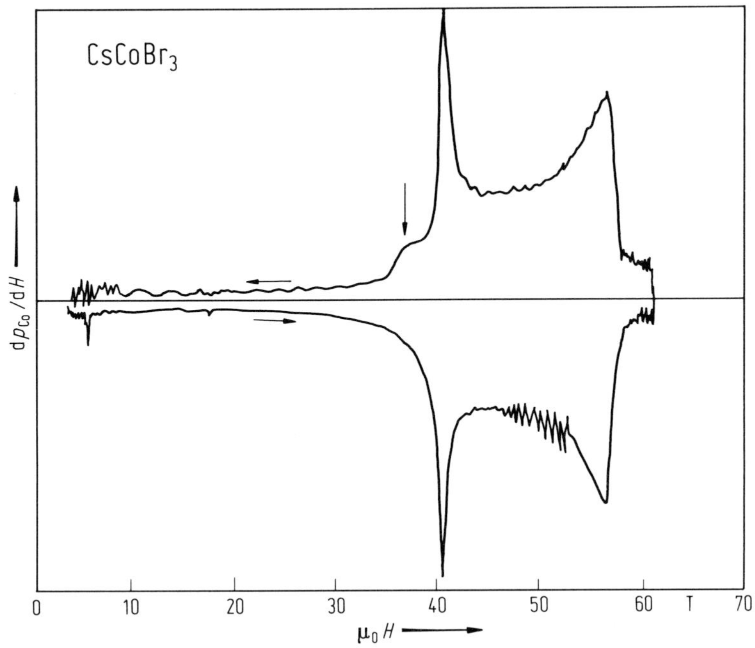

Fig. 619 $CsCoBr_3$.$dp_{Co^{2+}}/dH$ curves in increasing and decreasing mode of the pulsed magnetic field at 4.2 K and $\boldsymbol{H}\parallel c$ [90H2].

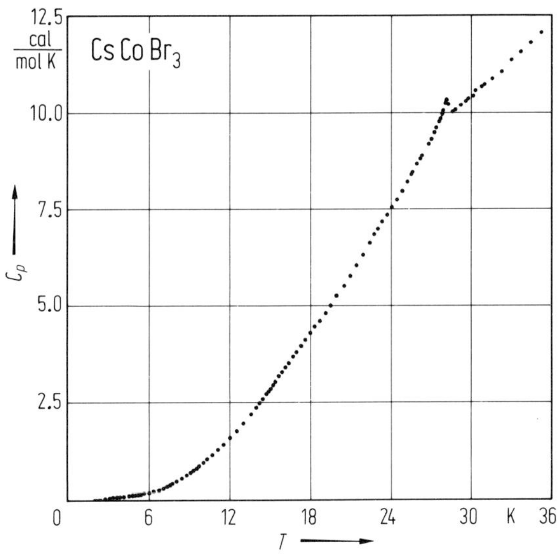

Fig. 620. $CsCoBr_3$. Temperature dependence of the heat capacity C_p. $T_N = 28$ K [75Y1].

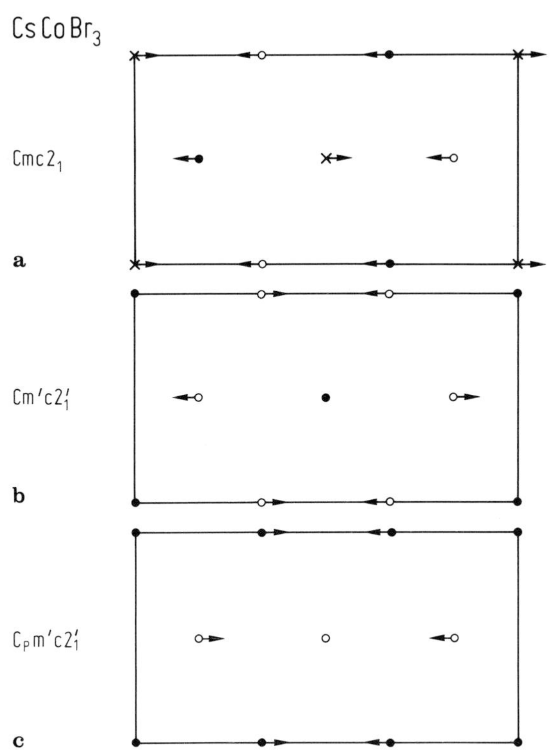

Fig. 621. $CsCoBr_3$. Proposed orthorhombic models for magnetic structure. Arrows indicate in-plane components, solid and open circles moments into and out of the plane respectively, crosses disordered axial components of the moment. (a): high temperature structure, $Cmc2_1$, (b): low temperature structure observed in single crystals, $Cm'c2'_1$, (c): low temperature structure from powder m′ type peaks, $C_pm'c2'_1$ [75Y1].

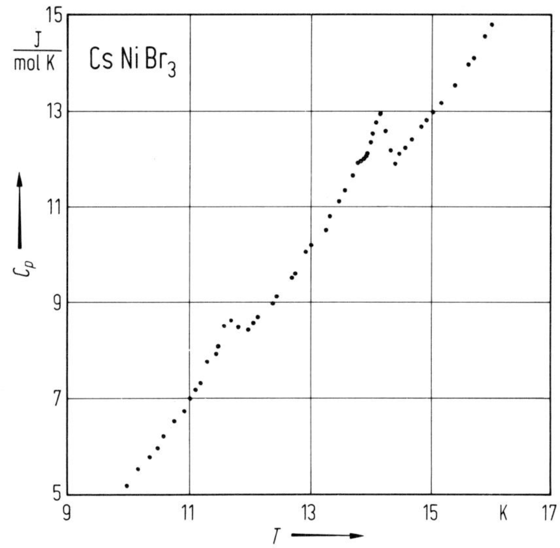

Fig. 622. $CsNiBr_3$. Temperature dependence of the heat capacity C_p [75B1, 77B4].

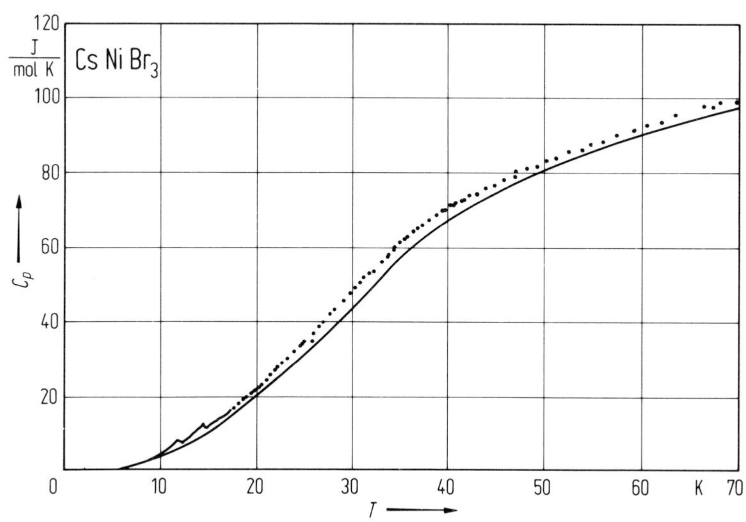

Fig. 623. $CsNiBr_3$. Temperature dependence of heat capacity C_p at $H = 0$. Solid line denotes estimated lattice contribution [75B1, 77B4].

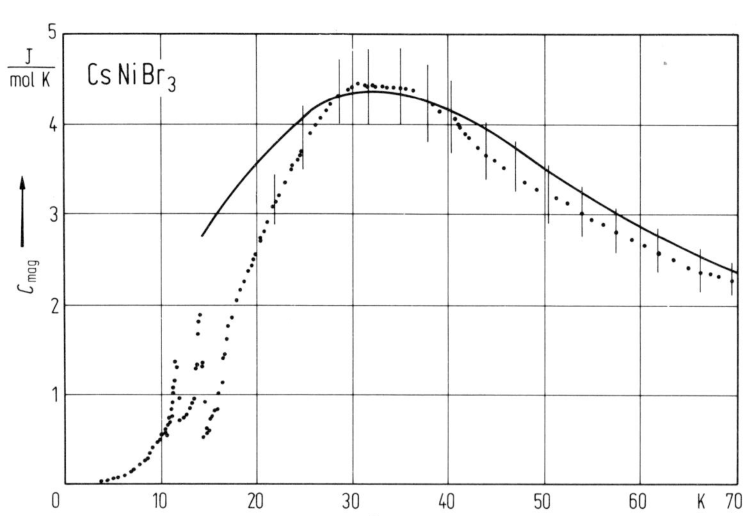

Fig. 624. $CsNiBr_3$. Temperature dependence of magnetic heat capacity. Experimental points are corrected for lattice contribution. The solid line is a fit according to Weng's theory with $J/k = -17.5$ K [77B4].

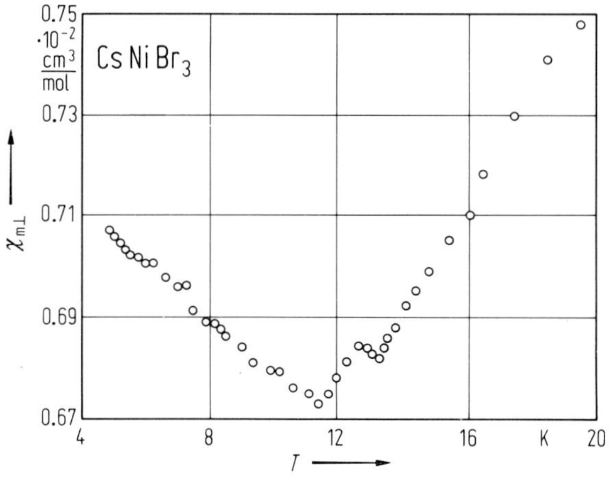

Fig. 625. $CsNiBr_3$. Temperature dependence of $\chi_{m\perp}$ near the magnetic transition point [77B4].

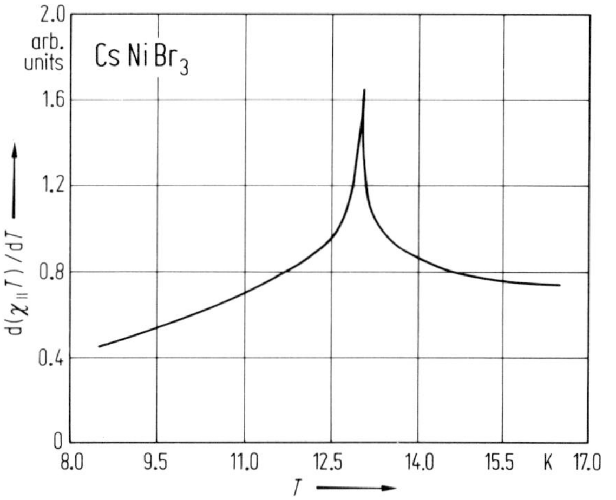

Fig. 626. $CsNiBr_3$. Temperature dependence of $d(\chi_{\parallel}T)/dT$ in the transition region [77B4].

For Fig. 627, see next page.

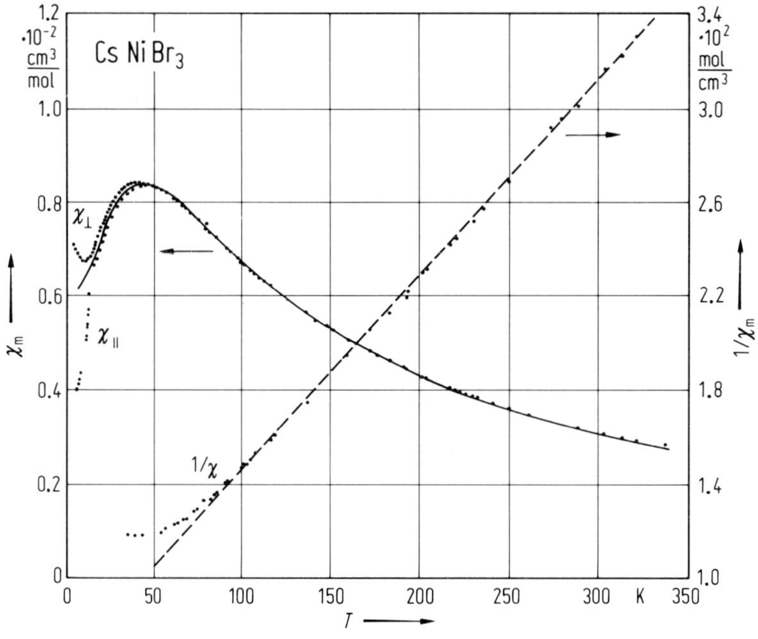

Fig. 628. $CsNiBr_3$. Temperature dependence of χ_m and χ_m^{-1} of single crystal at 10 kOe. Solid line is a fit according to Weng's theory with $J/k = -17.0$ K, $g = 2.09$ [77B4].

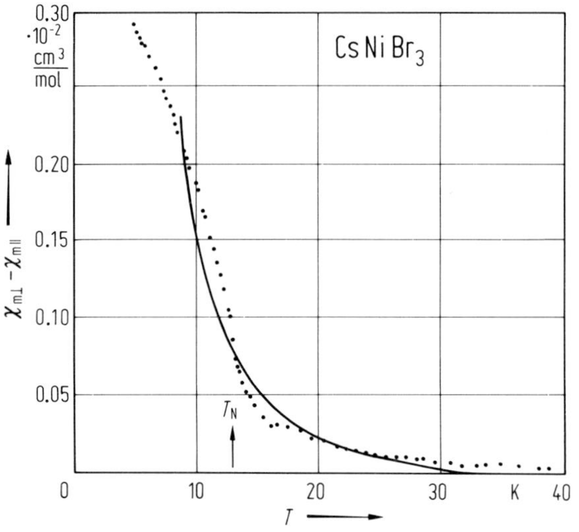

Fig. 627. $CsNiBr_3$. Temperature dependence of $\chi_{m\perp} - \chi_{m\parallel}$. Solid line is the best fit with $J/k = -17.0$ K, $D/k = -1.5$ K, $g = 2.09$ [77B4].

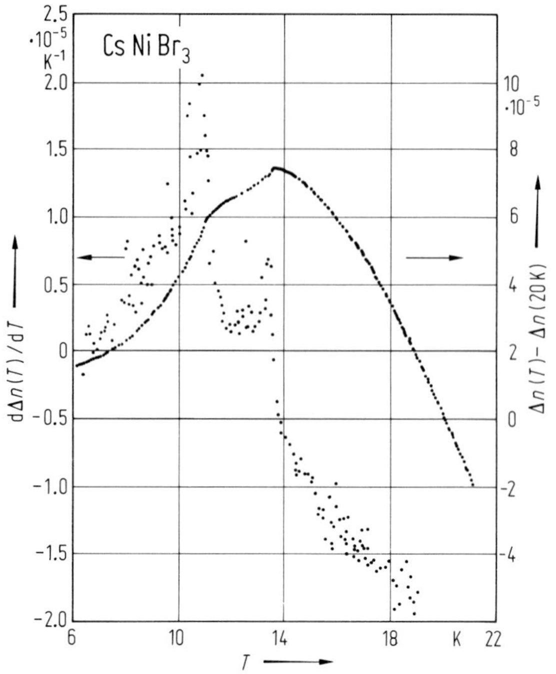

Fig. 629. $CsNiBr_3$. Temperature dependence of birefringence $\Delta n = n_c - n_a$ and that of the temperature derivative of Δn at a wavelength of 0.6328 μm. The Δn has the singularities at the phase transition points $T_{N1} = 13.46$ K and $T_{N2} = 11.07$ K [86I1].

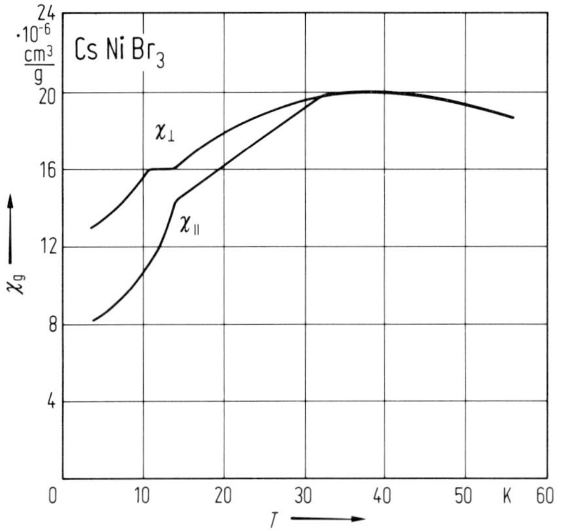

Fig. 630. $CsNiBr_3$. Temperature dependence of χ_g at 1.2 kOe for $\boldsymbol{H} \parallel c$ and $\boldsymbol{H} \perp c$ [89P1].

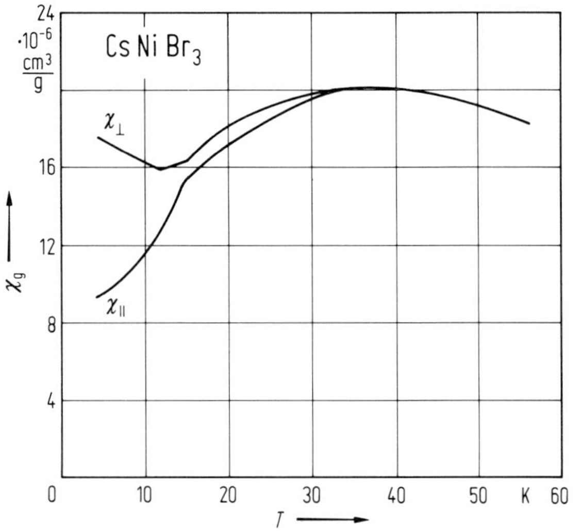

Fig. 631. $CsNiBr_3$. Temperature dependence of χ_g at 60 kOe for $\boldsymbol{H} \parallel c$ and $\boldsymbol{H} \perp c$ [89P1].

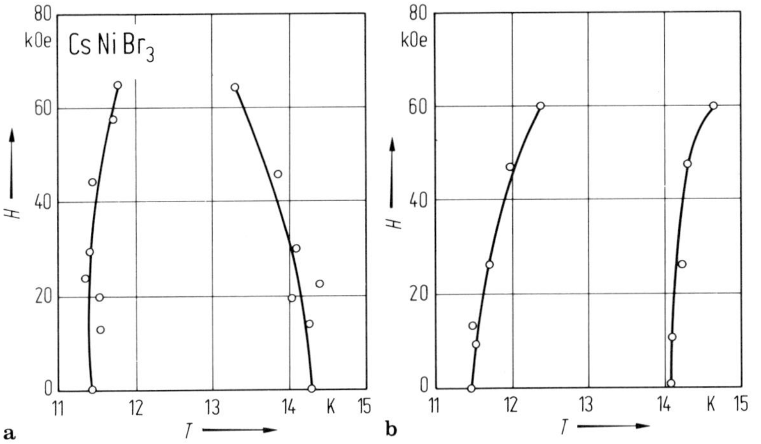

Fig. 632. CsNiBr$_3$. H–T phase diagram with magnetic field $\boldsymbol{H} \parallel c$ (a) and $\boldsymbol{H} \perp c$ (b) [89P1].

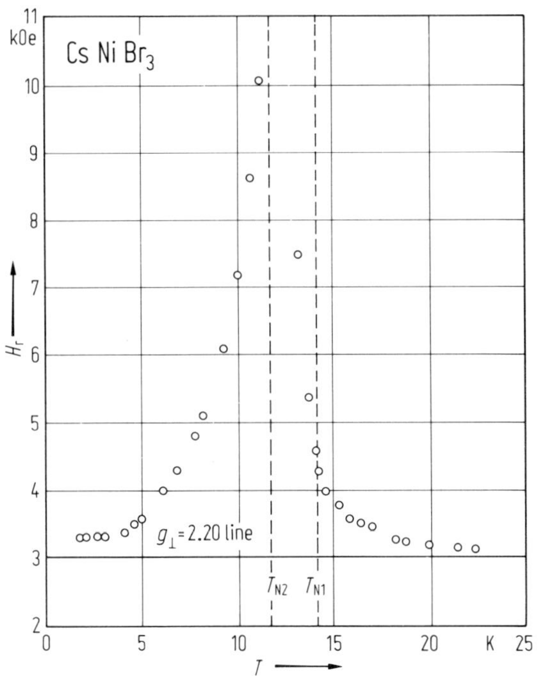

Fig. 633. CsNiBr$_3$. Temperature dependence of ESR resonance field H_r at 9.30 GHz and $\boldsymbol{H} \perp c$ [90T1]. T_{N1} = 14.25 K, T_{N2} = 11.75 K.

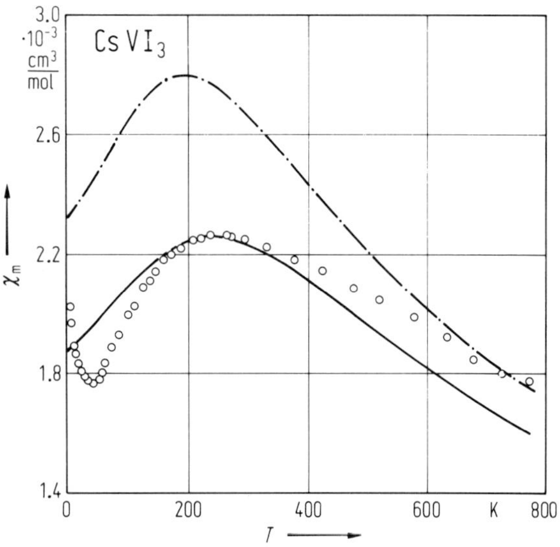

Fig. 634. CsVI$_3$. Temperature dependence of χ_m, reduced to spin contribution. Solid line is theoretical fit according to Smith and Friedberg [68S4] with $J/k = -67$ K, and dashed line with $J/k = -54$ K [77N1, 77N2].

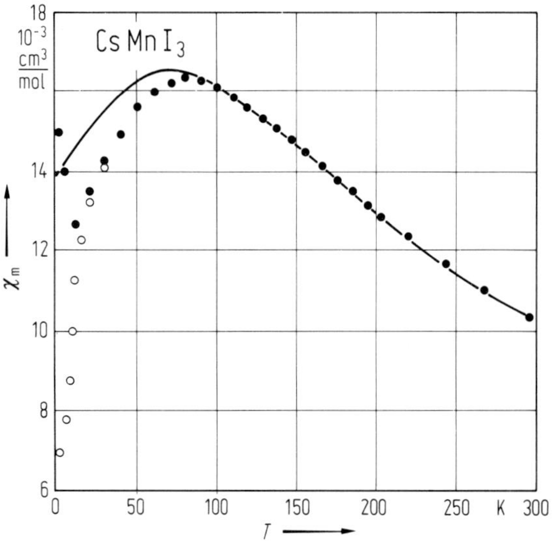

Fig. 635. $CsMnI_3$. Temperature dependence of χ_m at 11.2 kOe along *c* axis (open circles) and *a* axis (full circles). Solid curve is the best fit with Fisher's approximation [80Z2].

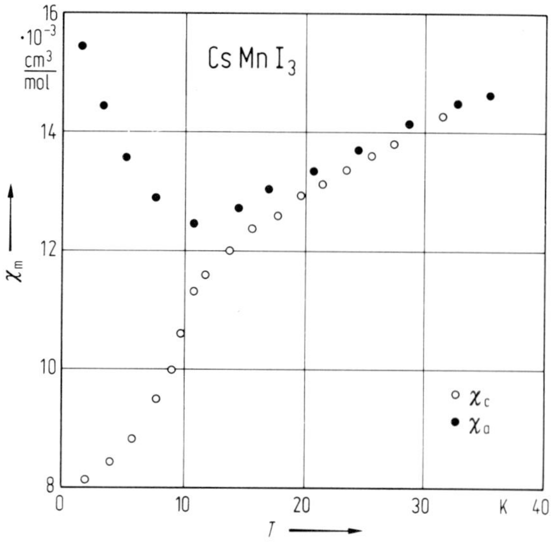

Fig. 636. $CsMnI_3$. Temperature dependence of χ_m at low temperature (see Fig. 635) [80Z2].

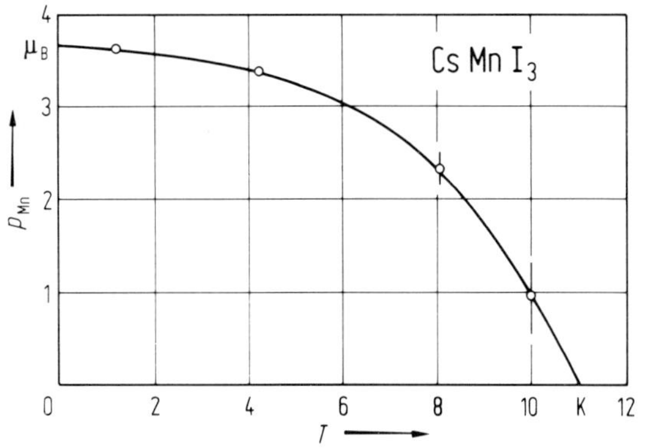

Fig. 637. $CsMnI_3$. Temperature dependence of p_{Mn} from neutron diffraction [80Z2].

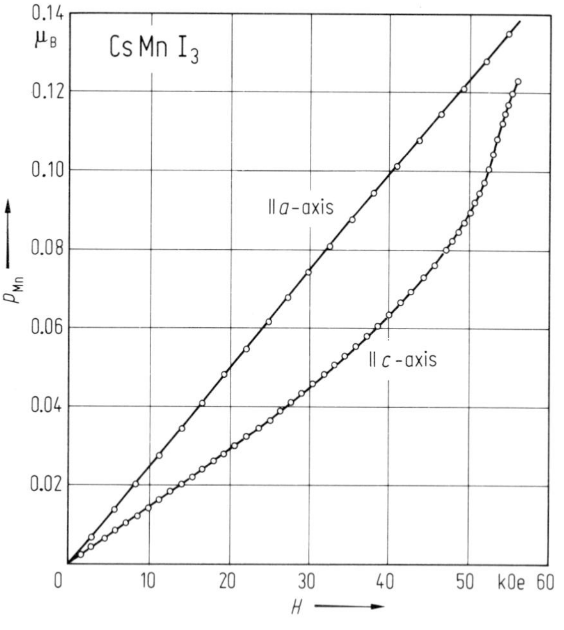

Fig. 638. $CsMnI_3$. Magnetic field dependence of p_{Mn} at 2 K [80Z2].

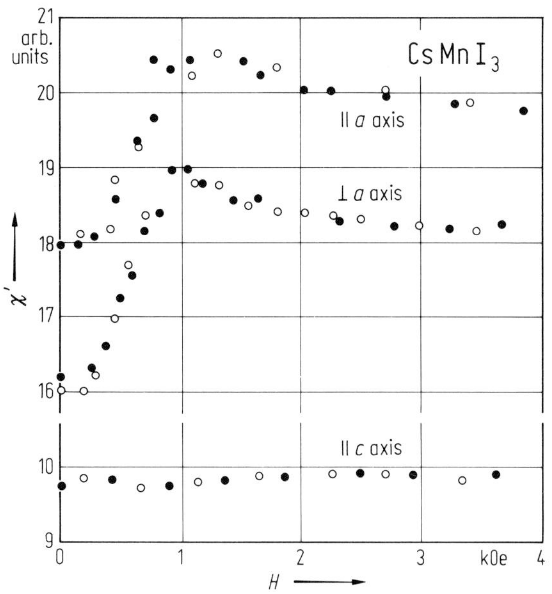

Fig. 639. $CsMnI_3$. Magnetic field dependence of ac χ' (in arbitrary units) measured with increasing (solid circles), and with decreasing (open circles) magnetic field [80Z2].

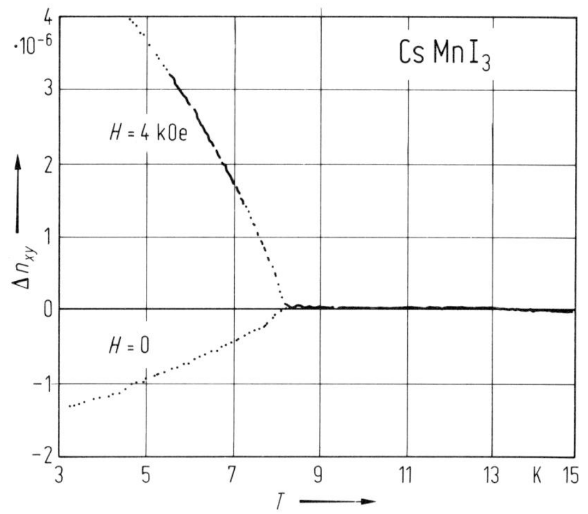

Fig. 640. $CsMnI_3$. Temperature dependence of in-plane birefringence $\Delta n_{xy}(T) = n_x(T) - n_y(T)$ around $T_{N2} = 8.20$ K, for light propagating along c axis ($\lambda = 632.8$ nm). Δn_{xy} appears indicating the establishment of an optically biaxial property. x, y and z are axes $\perp$ and $\parallel$ to [1 1 0] and the c axis, respectively [88I1].

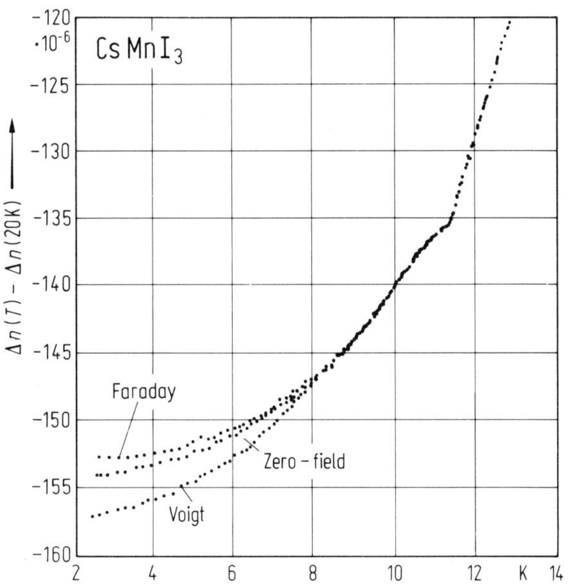

Fig. 641. $CsMnI_3$. Temperature dependence of $\Delta n(T) - \Delta n(20\ \text{K})$ for light perpendicular to the cleavage (110) face ($\lambda = 632.8$ nm) around $T_{N1} = 11.42$ K under the action of external magnetic field in the basal plane. Voigt and Faraday configurations denote the $\boldsymbol{H}\perp$ and $\boldsymbol{H}\parallel$ to the wave vector of light, respectively [88I1].

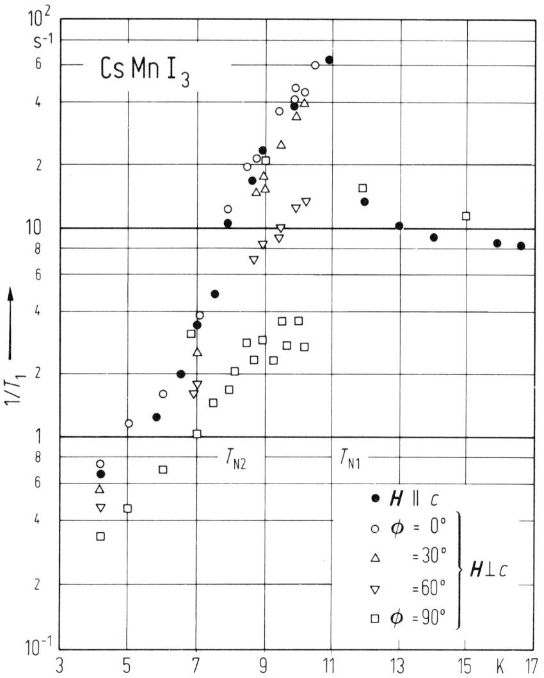

Fig. 642. $CsMnI_3$. Temperature dependence of NMR relaxation rates of ^{133}Cs in single crystals, at 4 MHz. 3 peaks observed by sweeping field parallel to the a axis correspond to angles $\phi = 30°$, $90°$ and $150°$ ($30°$). Four peaks for $\boldsymbol{H}\perp a$ axis in the c plane correspond to $\phi = 0°$, $60°$, $120°$ ($60°$) and $180°$ ($0°$). ϕ is angle between external and internal fields [90M5].

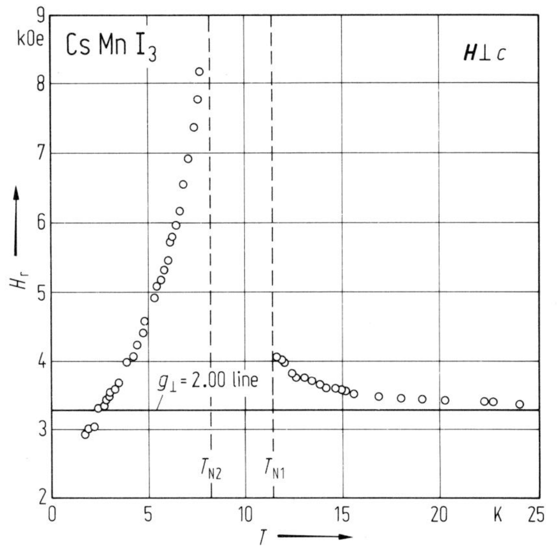

Fig. 643. $CsMnI_3$. Temperature dependence of ESR field H_r at 9.26 GHz and $H \perp c$ axis [90T1]. $T_{N1} = 11.42$ K, $T_{N2} = 8.20$ K.

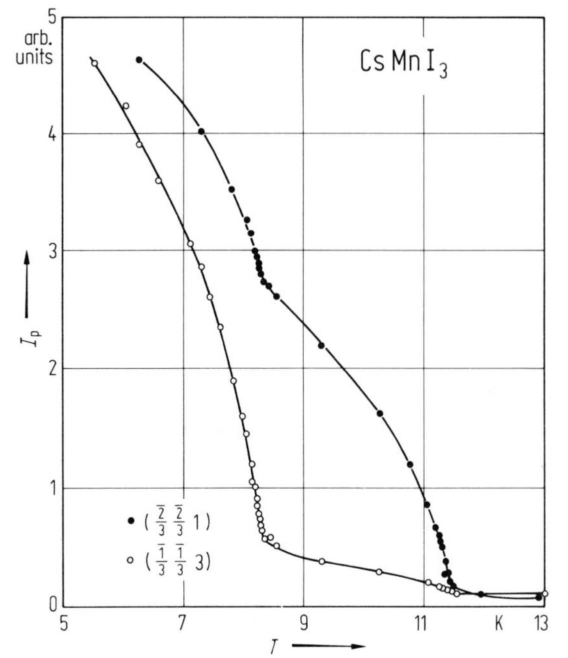

Fig. 644. $CsMnI_3$. Temperature dependence of neutron diffraction $(\frac{\bar{2}}{3}, \frac{\bar{2}}{3}, 1)$ and $(\frac{\bar{1}}{3}, \frac{\bar{1}}{3}, 3)$ Bragg peak intensity at $H = 0$ [91H2].

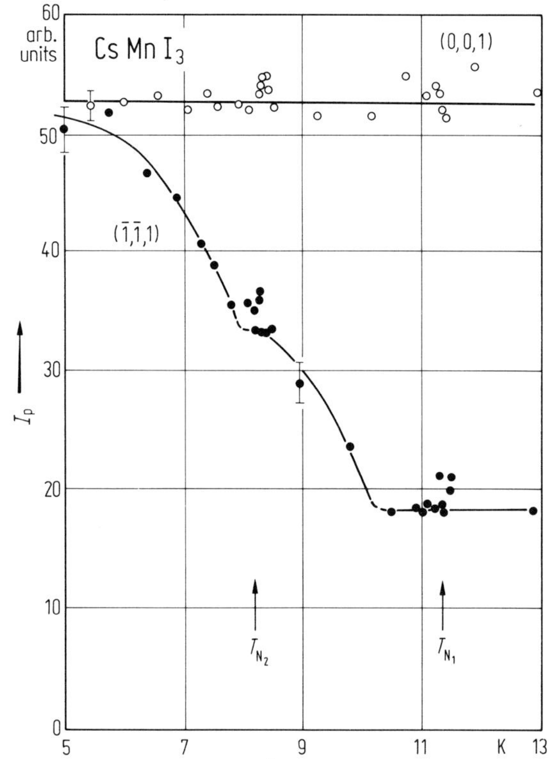

Fig. 645. $CsMnI_3$. Temperature dependence of neutron diffraction (0, 0, 1) and $(\bar{1}, \bar{1}, 1)$ Bragg peak intensity for $H = 0$ [91H2].

9.6.3 References for 9.6 and 9.9

60H1 Hirakawa, K., Hirakawa, K., Hashimoto, T.: J. Phys. Soc. Jpn. **15** (1960) 2063.
60S1 Shulman, R. G., Knox, K.: Phys. Rev. **119** (1960) 94.
61B1 Beckman, O., Knox, K.: Phys. Rev. **121** (1961) 376.
61H1 Heeger, A. J., Beckman, O., Portis, A. M.: Phys. Rev. **123** (1961) 1652.
61H2 Heeger, A. J., Portis, A. M., Teaney, D. T., Witt, G.: Rhys. Rev. Lett. **7** (1961) 307.
61S1 Scatturin, V., Corliss, L., Elliott, N., Hastings, J.: Acta Crstallogr. **14** (1961) 19.
62G1 Griffith, R. B.: Ph.D. Thesis 1962, Stanford University.
62J1 Jensen, S. J., Andersen, P., Rasmunsen, S. E.: Acta Chem. Scand. **16** (1962) 1310.
62L1 Lea, K. R., Leask, M. J. M., Wolf, W. P. J.: Phys. Chem. Solids **23** (1962) 1381.
62T1 Taney, D. T., Freiser, M. J.: Phys. Rev. Lett. **9** (1962) 212.
62T2 Taney, D. T., Blackburn, J. S., Stevenson, R. W. H.: Bull. Am. Phys. Soc. **7** (1962) 201.
63F1 Freiser, M. J., Seiden, P. E., Teaney, D. T.: Phys. Rev. Lett. **10** (1963) 293.
63H1 Hashimoto, T.: J. Phys. Soc. Jpn. **18** (1963) 1140.
63K1 Knox, K., Shulman, R. G., Sugano, S.: Phys. Rev. **130** (1963) 512.
63L1 Lee, K., Portis, A. M., Witt, G. L.: Phys. Rev. **132** (1963) 144.
63R1 Rüdorf, W., Lincke, G., Babel, D.: Z. Anorg. Allg. Chem. **320** (1963) 150.
63S1 Shulman, R. G., Sugano, S.: Phys. Rev. **130** (1963) 506.
64B1 Bonner, J. C., Fisher, M. E.: Phys. Rev. A**135** (1964) 640.
64D1 Denison, A. B., James, L. W., Currin, J. D., Tanittila, W. H.: Phys. Rev. Lett. **12** (1964) 244.
64H1 Hirakawa, K.: J. Phys. Soc. Jpn. **19** (1964) 1678.
64H2 Heeger, A. J., Teaney, D. T.: J. Appl. Phys. **33** (1964) 846.
64H3 Hutchings, M. T.: Solid State Phys. **16** (1964) 227.
64N1 Nakamura, A., Minkiewicz, V., Portis, A. M.: J. Appl. Phys. **33** (1964) 842.
64P1 Pickart, S. J., Alperin, H. A., Nathans, R.: J. Phys. (Paris) **25** (1964) 565.
64S1 Sumune, Y., Ikawa, H.: J. Phys. Soc. Jpn. **19** (1964) 1686.
64S2 Sumune, Y.: J. Phys. Soc. Jpn. **19** (1964) 2234.
65F1 Ferguson, J., Guggenheim, H. J., Tanabe, Y.: J. Appl. Phys. **36** (1965) 1046.
65H1 Holloway jr., W. W., Kestigian, M.: Phys. Rev. Lett. **15** (1965) 17.
65K1 Kedzie, R. W., Shane, J. R., Kestigian, M., Croft, W. J.: J. Appl. Phys. **36** (1965) 1197.
65P1 Pisarev, R. V.: Zh. Eksp. Teor. Fiz. (SSSR) **49** (1965) 1445.
65P2 Prohorofsky, E. W., Holloway jr., W. W., Kestigian, M.: J. Appl. Phys. **36** (1965) 1041.
65P3 Petrov, M. P., Smolensky, G. A.: Fiz. Tverd. Tela (SSSR) **7** (1965) 2156.
65S1 Shulman, R. G., Knox, K.: J. Chem. Phys. **42** (1965) 813.
65T1 Tsuchida, A., Nakagawa, I.: J. Phys. Soc. Jpn. **20** (1965) 1726.
66B1 Boyd, E. L.: Phys. Rev. **145** (1966) 174.
66C1 Cooper, M. J., Nathans, R.: J. Appl. Phys. **37** (1966) 1041.
66C2 Cole, P. H., Ince, W. J.: Phys. Rev. **150** (1966) 377.
66E1 Eastman, D. E., Joenk, R. J., Teaney, D. T.: Phys. Rev. Lett. **17** (1966) 300.
66F1 Ferguson, J., Guggenheim, H. J., Tanabe, Y.: J. Phys. Soc. Jpn. **21** (1966) 692.
66F2 Ferguson, J., Guggenheim, H. J., Tanabe, Y.: J. Chem. Phys. **45** (1966) 1134.
66H1 Horai, K., Saiki, K.: J. Phys. Soc. Jpn. **21** (1966) 397.
66I1 Ince, W.: J. Appl. Phys. **37** (1966) 1132.
66L1 Lvov, V. S., Petrov, M. P.: Phys. Status Solidi **13** (1966) K65.
66P1 Petrov, M. P., Smolensky, G. A.: Zh. Eksp. Teor. Fiz. (SSSR) **50** (1966) 871.
66P2 Pickart, S. J., Collins, M. F., Windsor, C. G.: J. Appl. Phys. **37** (1966) 1054.
66S1 Smolensky, G. A., Ludin, V. M., Syrnikov, P. P., Sherman, A. B.: Fiz. Tverd. Tela (SSSR) **8** (1966) 2965.
66S2 Shane, J. R., Kedzie, R. W., Kestigian, M.: J. Appl. Phys. **37** (1966) 1134.

66T1 Teaney, D. T., Moruzzi, V. L., Argyle, B. E.: J. Appl. Phys. **37** (1966) 1122.
66T2 Tsuchida, A.: J. Phys. Soc. Jpn. **21** (1966) 2497.
66W1 Wang, F. Y., Kestigian, M.: J. Appl. Phys. **37** (1966) 975.
66Y1 Yoneyama, S., Hirakawa, K.: J. Phys. Soc. Jpn. **21** (1966) 183.
66Z1 Zhogolev, D. A.: Fiz. Tverd. Tela (SSSR) **8** (1966) 2798.
67A1 Aoyagi, K.: J. Phys. Soc. Jpn. **22** (1967) 1516.
67C1 Cole, P. H., Courtney, W. E.: J. Appl. Phys. **38** (1967) 1278.
67E1 Eastman, D. E., Shafer, M. W., Figat, R. A.: J. Appl. Phys. **38** (1967) 5209.
67E2 Eastman, D. E., Shafer, M. W.: J. Appl. Phys. **38** (1967) 1274.
67E3 Eastman, D. E.: Phys. Rev. **156** (1967) 645.
67F1 Fatehally, R., Shenoy, G. K., Sastry, N. P., Nagarajan, R.: Phys. Lett. **25A** (1967) 453.
67G1 Ganiel, U., Kestigian, M., Shtrikman, S.: Phys. Lett. **24A** (1967) 577.
67H1 Hirakawa, K., Kadota, S.: J. Phys. Soc. Jpn. **23** (1967) 757.
67H2 Holloway jr., W. W., Kestigian, M.: J. Appl. Phys. **38** (1967) 1480.
67H3 Huang, Nai-Li., Orbach, R.: Phys. Rev. **154** (1967) 487.
67J1 Judin, V. M., Sherman, A. B.: Phys. Status Solidi **20** (1967) 759.
67K1 Kohn, K., Fukuda, R., Iida, S.: J. Phys. Soc. Jpn. **22** (1967) 333.
67K2 Kadota, S., Yamada, I., Yoneyama, S., Hirakawa. K.: J. Phys. Soc. Jpn. **23** (1967) 751.
67K3 Kharchenko, N. F., Eremenko, V. V.: Fiz.. Tverd. Tela (SSSR) **9** (6) (1967) 1655.
67L1 Lines, M. E.: Phys. Rev. **164** (1967) 736.
67M1 Maass, G. J., Gerstein, B. C., Willett, R. D.: J. Chem. Phys. **46** (1967) 401.
67M2 Mehra, A., Venkateswarlu, P.: J. Chem. Phys. **47** (1967) 2334.
67N1 Nestorova, N. N., Pisarev, R. V.: Fiz. Tverd. Tela (SSSR) **9** (1967) 1550.
67N2 Nestorova, N. N., Siny, I. G., Pisarev, R. V., Syrnikov, P. P.: Fiz. Tverd. Tela (SSSR) **9** (1967) 21.
67P1 Pisarev, R. V., Siny, I. G., Smolensky, G. A.: Fiz. Tverd. Tela (SSSR) **9** (1967) 3149.
67P2 Perry, C. H., Young, E. F.: J. Appl. Phys. **38** (1967) 4616.
67P3 Pisarev, R. V., Prokhorova, S. D.: Fiz. Tverd. Tela (SSSR) **9** (1967) 1879.
67S1 Sakamoto, N., Yamaguchi, Y.: J. Phys. Soc. Jpn. **22** (1967) 885.
67S2 Shafer, M. W., McGuire, T. R., Argyle, B. E., Fan, G. J.: Appl. Phys. Lett. **10** (1967) 202.
67S3 Smolensky, G. A., Petrov, M. P., Pisarev, R. V.: J. Appl. Phys. **38** (1967) 1269.
67S4 Shane, J. R., Lyons, D. H., Kestigian, M.: J. Appl. Phys. **38** (1967) 1280.
67T1 Testardi, L. R., Levinstein, H. J., Guggenheim, H. J.: Phys. Rev. Lett. **19** (1967) 503.
67W1 Wertheim, G. K., Guggenheim, H. J., Williams, H. J., Buchanan, D. N. E.: Phys. Rev. **158** (1967) 446.
67W2 Welsh, L. B.: Phys. Rev. **156** (1967) 370.
67Y1 Yeremenko, V. V., Popkov, Yu. A., Novikov, V. P., Belyaeva, A. I.: Zh. Eksp. Teor. Fiz. (SSSR) **52** (1967) 463.
68B1 Belyaeva, A. I., Yeremenko, V. V.: Zh. Eksp. Teor. Fiz. (SSSR) **54** (1968) 1303.
68E1 Epstein, A., Makovsky, J., Melamud, M., Shaked, H.: Phys. Rev. **174** (1968) 560.
68F1 Fleury, P. A.: Phys. Rev. Lett. **21** (1968) 151.
68F2 Fatehally, R., Sastry, N. P., Nagarajan, R.: Phys. Status Solidi **26** (1968) 91.
68F3 Forstat, H., McElearney, J. N., Bailey, P. T.: Phys. Lett. **27A** (1968) 549.
68G1 Gurievich, A. G., Golovenchits, E. J., Sanina, V. A.: J. Appl. Phys. **39** (1968) 1023.
68G2 Ganiel, U., Shtrikman, S.: J. Appl. Phys. **39** (1968) 1254.
68G3 Golding, B.: Phys. Rev. Lett. **20** (1968) 5.
68H1 Hinderks, L. W., Richards, P. M.: J. Appl. Phys. **39** (1968) 824.
68I1 Ince, W. J., Platzker, A.: Phys. Rev. **175** (1968) 650.

68M1 McGuire, T. R., Shafer, M. W.: J. Appl. Phys. **39** (1968) 1130.
68M2 Meltzer, R. S., Chen, M. Y., McClure, D. S., Lowe-Pariseau, M.: Phys. Rev. Lett. **21** (1968) 913.
68M3 Melcher, R. L., Bolef, D. I.: Phys. Rev. Lett. **20** (1968) 1338.
68N1 Nathans, R., Menzinger, F.: J. Appl. Phys. **39** (1968) 1237.
68P1 Pisarev, R. V., Siny, I. G., Nesterova, N. N., Smolensky, G. A., Syrkov, P. P.: Phys. Status Solidi **30** (1968) 367.
68P2 Petrov, M. P., Nedlin, G. M.: J. Appl. Phys. **39** (1968) 1012.
68P3 Pisarev, R. V., Prokhorova, S. D.: Fiz. Tverd. Tela (SSSR) **10** (1968) 2129.
68S1 Smith, T., Friedberg, S.A.: Phys. Rev. **176** (1968) 660.
68S2 Sintani, K., Tomono, Y.: J. Phys. Soc. Jpn. **25** (1968) 99.
68W1 Wigomore, J. K., Rosenberg, H. M., Garrod, D. K.: J. Appl. Phys. **39** (1968) 682.
68Y1 Yeremenko, V. V., Novikov, V. P., Popkov, Yu. A.: Zh. Eksp. Teor. Fiz. (SSSR) **54** (1968) 1043.
69A1 Achiwa, N.: J. Phys. Soc. Jpn. **27** (1969) 561.
69B1 Botterman, A. C., de Jonge, W. J. M., de Leeuw, P.: Phys. Lett. **30**A (1969) 150.
69C1 Chinn, S. R., Zeiger, H. J., O'Connor, J. R.: J. Appl. Phys. **40** (1969) 1603.
69C2 Cape, J. A., White, R. L., Feigelson, R. S.: J. Appl. Phys. **40** (1969) 5001.
69F1 Fleury, P. A., Worlock, J. M., Guggenheim, H. J.: Phys. Rev. **185** (1969) 738.
69G1 Gyorgy, E. M., Levinstein, H. J., Dillon jr., J. F., Guggenheim, H. J.: J. Appl. Phys. **40** (1969) 1599.
69G2 Gooen, K., DiBartolo, B., Alam, M.: Phys. Rev. **177** (1969) 615.
69G3 Golovenchitz, E., Sanina, V. A., Gurievich, A. G.: Fiz. Tverd. Tela **11** (1969) 642; Sov. Phys. Solid State (English Transl.) **11** (1969) 516.
69H1 Hutchings, M. T., Samuelsen, E. J., Shirane, G., Hirakawa, K.: Phys. Rev. **188** (1969) 919.
69I1 Ince, W. J.: J. Appl. Phys. **40** (1969) 1595.
69K1 Krebs, J. J.: J. Appl. Phys. **40** (1969) 1137.
69L1 Longo, J. M., Kafalas, J. A.: J. Appl. Phys. **40** (1969) 1601.
69M1 Melcher, R. L., Bolef, D. I.: Phys. Rev. **178** (1969) 864.
69M2 Melcher, R. L., Bolef, D. I.: Phys. Rev. **186** (1969) 491.
69M3 Miyashita, T., Kondo, H.: J. Phys. Soc. Jpn. **27** (1969) 256.
69P1 Petrov, M. P., Korneev, V. R., Sirnikov, P. P.: Phys. Lett. **28**A (1969) 618.
69R1 Rioux, F. J., Gerstein, B. C.: J. Chem. Phys. **50** (1969) 758.
69S1 Spence, R. D., de Jonge, W. J. M., Rama Rao, K. V. S.: J. Chem. Phys. **51** (1969) 4694.
69S2 Shapiro, M. M., Stevenson, R.: J. Appl. Phys. **40** (1969) 989.
69S3 Seavey, M. H.: J. Appl. Phys. **40** (1969) 1597.
69S4 Svensson, E. C., Buyers, W. J. L., Holden, T. M., Cowley, R. A., Stevenson, R. W. H.: Can. J. Phys. **47** (1969) 1983.
69S5 Smolensky, G. A., Pietrov, M. P.: Postepy Fiz **XX** (1969) 18.
69T1 Testardi, L. R., Levinstein, H. J., Gyorgy, E. M., Guggenheim, H. J.: Solid State Commun. **7** (1969) 241.
69Y1 Yamaguchi, Y., Sakamoto, N.: J. Phys. Soc. Jpn. **27** (1969) 144.
70B1 Breed, D. J., Gilyamse, K., Storkenberg, J. W. E., Mediema, A. R.: J. Appl. Phys. **41** (1970) 1267.
70B2 Borovik-Romanov, A. S., Kotyuzhansky, B. Ya., Prozorova, L. A.: Zh. Eksp. Teor. Fiz. (SSSR) **58** (1970) 1911.

70B3 Baltes, H. P., Tosi, M., Kneubuhl, F. K.: J. Phys. Chem. Solids **31** (1970) 321.
70F1 Friedman, Z., Melamud, M., Makovsky, J., Shaked, H.: Phys. Rev. B**2** (1970) 179.
70L1 Longo, J. M., Kafalas, J. A., O'Connor, J. R., Goodenough, J. B.: J. Appl. Phys. **41** (1970) 935.
70M1 Minkiewicz, V. J., Fujii, Y., Yamada, Y.: J. Phys. Soc. Jpn. **28** (1970) 443.
70M2 Minkiewicz, V. J., Cox, D. E., Shirane, G.: Solid State Commun. **8** (1970) 1001.
70M3 McGuire, T. R., Moruzzi, V. L., Shafer, M. W.: J. Appl. Phys. **41** (1970) 956.
70P1 Petrov, M. P., Moskalev, V. V., Smolensky, G. A.: Solid State Commun. **8** (1970) 157.
70R1 Rioux, F. J., Gerstein, B.C.: J. Chem. Phys. **53** (1970) 1789.
70S1 Saito, F.: Solid State Commun. **8** (1970) 969.
70S2 Skalyo jr., J., Shirane, G., Friedberg, S. A., Kobayashi, H.: Phys. Rev. B**2** (1970) 1310.
70S3 Sinii, I. G., Pisarev, P. V.: Fiz. Tverd. Tela (SSSR) **12** (1970) 114.
71C1 Cox, D. E., Minkiewicz, V. J.: Phys. Rev. B**4** (1971) 2209.
71E1 Epstein, A., Makovsky, J., Shaked, H.: Solid State Commun. **9** (1971) 249.
71F1 Forstat, H., Bailey, P. T., Riks, J. R.: J. Appl. Phys. **42** (1971) 1559.
71I1 Ikebe, M., Date, M.: J. Phys. Soc. Jpn. **30** (1971) 93.
71J1 de Jonge, W. J. M., Rama Rao, K. V. S., Swüste, C. H. W.: Physica **51** (1971) 620.
71K1 Karraker, D. G.: J. Chem. Phys. **55** (1971) 1084.
71L1 Longo, J. M., Kafalas, J. A., Menyuk, N., Dwight, K.: J. Appl. Phys. **42** (1971) 1561.
71M1 Minkiewicz, V. J., Cox, D. E., Shirane, G.: J. Phys. (Paris) C**32** (1971) 892.
71M2 Melamud, M., Makovsky, J., Shaked, H.: Phys. Rev. B**3** (1971) 821.
71M3 Melamud, M., Makovsky, J., Shaked, H.: Phys. Rev. B**3** (1971) 3873.
71M4 Melemud, M., Pinto, H., Shachar, G., Makovsky, J., Shaked, H.: Phys. Rev. B**3** (1971) 2344.
71P1 Popkov, Yu. A., Eremienko, V. V., Fomin, V. I.: Fiz. Tverd. Tela (SSSR) **13** (1971) 2028.
71S1 Steiner, M.: Z. Angew. Phys. **32** (1971) 116.
71S2 Steiner, M., Krüger, W.: Solid State Commun. **9** (1971) 227.
71S3 Shachar, G., Makovsky, J., Shaked, H.: Solid State Commun. **9** (1971) 493.
71T1 Toyota, E., Hirakawa, K.: J. Phys. Soc. Jpn. **30** (1971) 692.
71W1 Wang, F. F. Y., Cox, D. E., Kestigian, M.: Phys. Rev. B**3** (1971) 3946.
72A1 Almond, D. P., Rayne, J. A.: Phys. Lett. **25**A (1972) 295.
72B1 Bongaarts, A. L. M., van Laar, B.: Phys. Rev. B**6** (1972) 2669.
72B2 Bragg, E. E., Seehra, M. S.: Phys. Lett. **39**A (1972) 29.
72C1 Clark, R. H., Moulton, W. G.: Phys. Rev. B**5** (1972) 788.
72F1 Forstat, H., Rudnik, D. M.: Phys. Lett. **42**A (1972) 125.
72G1 Gulley, J., Jaccarino, V.: Phys. Rev. B**6** (1972) 58.
72G2 Gupta, R. R., Seehra, M. S., Vehse, W. E.: Phys. Rev. B**5** (1972) 92.
72H1 Herweijer, A., de Jonge, W. J. M., Botterman, A. L. M., Bongaarts, A. L. M., Cowen, J. A.: Phys. Rev. B**5** (1972) 4618.
72M1 Mattheiss, L. F.: Phys. Rev. B**6** (1972) 4718.
72M2 Maartense, I., Searle, C. W.: Phys. Rev. B**6** (1972) 894.
72M3 Montano, P. A., Cohen, E., Shechter, H.: Phys. Rev. B**6** (1972) 1053.
72N1 Nagata, K., Tazuke, Y.: J. Phys. Soc. Jpn. **32** (1972) 337.
72S1 Steiner, M.: Solid State Commun. **11** (1972) 73.
72S2 Saiki, K.: J. Phys. Soc. Jpn. **33** (1972) 1284.
72S3 Srivastava, V. C., Stevenson, R.: Solid State Commun. **11** (1972) 41.
72W1 Witt, G. L.: Physica **61** (1972) 476.
72Y1 Yelon, W. B., Cox, D. E.: Phys. Rev. B**6** (1972) 204.
72Y2 Yoshioka, H., Saiki, K.: J. Phys. Soc. Jpn. **33** (1972) 1566.

72Z1 Zimmerman, N. J., Bastmeijer, J. D., Van den Handel, J.: Phys. Lett. **40**A (1972) 259.
73A1 Adachi, K., Mekata, M.: J. Phys. Soc. Jpn. **34** (1973) 269.
73B1 Breed, D. J., Gilijamse, K., Sterkenburg, J. W. E., Miediema, A. R.: Physica **68** (1973) 303.
73B2 Butterworth, G. J., Woollam, J. A., Aron, D.: Physica **70** (1973) 547.
73H1 Hirakawa, K., Miike, H.: Phys. Lett. **45**A (1973) 79.
73I1 Ikeda, H., Hirakawa, K.: J. Phys. Soc. Jpn. **35** (1973) 722.
73L1 Lebesque, J. V., Snel, J., Smit, J. J.: Solid State Commun. **13** (1973) 371.
73M1 Montano, P. A., Cohen, E., Shechter, H.: Phys. Rev. **B7** (1973) 1180.
73M2 Maartense, I.: Solid State Commun. **12** (1973) 1133.
73M3 Metselaar, J. W., de Klerk, D.: Physica **69** (1973) 499.
73P1 Pisarev, R. V., Moch, P., Dugautier, C.: Phys. Rev. **B7** (1973) 4185.
73R1 Ryan, J. F., Scott, J. F., Nouet, J.: Solid State Commun. **13** (1973) 793.
73S1 Saiki, K., Horai, K., Yoshioka, H.: J. Phys. Soc. Jpn. **35** (1973) 1016.
73T1 Toms, D. J., Ryan, J. F., Scott, J. F., Nouet, J.: Phys. Lett. **44**A (1973) 187.
73V1 Vijayaraghavan, R., Karkhanavala, M. D., Damle, S. D., Gupta, L. C., Rao, U. R. K.: Pramãna **1** (1973) 155.
73W1 Witteveen, H. T., van Veen, J. A. R.: J. Chem. Phys. **58** (1973) 186.
73Y1 Yelon, W. B., Cox, D. E.: Phys. Rev. **B7** (1973) 2024.
73Y2 Yamaguchi, Y., Sakuraba, T.: Phys. Lett. **43**A (1973) 185.
74B1 Bucher, E., Guggenheim, H. J., Andres, K., Hull jr., G. W., Cooper, A. S.: Phys. Rev. **B10** (1974) 2945.
74B2 Bongaarts, A. L. M.: Phys. Lett. **49**A (1974) 211.
74E1 Ebara, K., Tanabe, Y.: J. Phys. Soc. Jpn. **36** (1974) 93.
74G1 Gurewitz, E., Makovsky, J., Shaked, H.: Phys. Rev. **B9** (1974) 1071.
74H1 Hoehn, M. V., Karraker, D. G.: J. Chem. Phys. **60** (1974) 393.
74H2 Hendricks, M. E., Jones jr., E. R.: J. Chem. Phys. **60** (1974) 2095.
74I1 Ikeda, H.: J. Phys. Soc. Jpn. **37** (1974) 660.
74J1 de Jongh, L. J., Breed, D. J.: Solid State Commun. **15** (1974) 1061.
74M1 Montano, P. A., Shechter, H., Cohen, E., Makovsky, J.: Phys. Rev. **B9** (1974) 1066.
74M2 Melamud, M., Pinto, H., Makovsky, J., Shaked, H.: Phys. Status Solidi (b) **63** (1974) 699.
74M3 McPherson, G. L., Aldrich, H. S., Chang, I. R.: J. Chem. Phys. **60** (1974) 534.
74P1 Pisarev, R. V., Ferre, J., Petit, R. H., Krichevtsov, B. B., Syrnikov, P. P.: J. Phys. C **7** (1974) 4143.
74P2 Petrouleas, V.: Solid State Commun. **15** (1974) 1097.
74S1 Saiki, K., Yoshioka, H.: Solid State Commun. **15** (1974) 1067.
74S2 Steiner, M., Dachs, H.: Solid State Commun. **14** (1974) 841.
74S3 Swüste, C. H. W., de Jonge, W. J. M., Van Meijel, J. A. G. W.: Physica **76** (1974) 21.
74T1 Takeda, Y., Shimada, M., Kanamaru, F.: J. Phys. Soc. Jpn. **37** (1974) 276.
74W1 Witteveen, H. T., van Veen, J. A. R.: J. Phys. Chem. Solids **35** (1974) 337.
75A1 Almond, D. P., Rayne, J. A.: Phys. Lett. **55**A (1975) 137.
75B1 Brener, R., Moses, D., Shechter, H., Ehrenfreund, E., Makovski, J., Wei, T.: Phys. Status Solidi (a) **31** (1975) K31.
75D1 Dunlap, B. D., Shenoy, G. K.: Phys. Rev. **B12** (1975) 2716.
75D2 Dupas, C., Renard, J. P.: Proc. 14th Int. Conf. of Low Temp. Phys. Vol. 5 **1975** Amsterdam: North Holland Publ. Co.
75E1 Eibschütz, M., Lines, M. E., Sherwood, R. C.: Phys. Rev. **B11** (1975) 4595.
75F1 Fleury, P. A., Hayes, W., Guggenheim, H. J.: J. Phys. C **8** (1975) 2183.

75F2 Flokstra, J., Gerritsma, G. J., Van den Brandt, B., Van der Marel, L. C.: Phys. Lett. **53**A (1975) 159.
75F3 Fedoseyeva, N. V., Beznosikov, B. V.: Zh. Eksp. Teor. Fiz. Pisma (SSSR) **21** (1975) 108.
75H1 Hidaka, M.: J. Phys. Soc. Jpn. **39** (1975) 103.
75H2 Hidaka, M., Ohama, N., Okazaki, A., Sakashiba, H.: Solid State Commun. **16** (1975) 1121.
75I1 Ito, A., Morimoto, S.: J. Phys. Soc. Jpn. **39** (1975) 884.
75I2 Iwashita, T., Uryû, N.: J. Phys. Soc. Jpn. **39** (1975) 1226.
75K1 Kopinga, K., de Neef, T., de Jonge, M.: Phys. Rev. **B11** (1975) 2364.
75L1 Lines, M. E., Eibschütz, M.: Phys. Rev. **B11** (1975) 4583.
75L2 Loveluck, J. M., Lovesey, S. W., Aubury, S.: J. Phys. C **8** (1975) 3841.
75M1 Miike, H., Hirakawa, K.: J. Phys. Soc. Jpn. **38** (1975) 92.
75N1 de Neef, T.: Eindhoven University of Technology, Report THE/VVS/71275, Ph.D. Thesis 1975, Eindhoven, Netherlands
75P1 Petrouleas, V., Simopoulos, A., Kostikas, A.: Phys. Rev. **B12** (1975) 4675.
75S1 Saiki, K.: J. Phys. Soc. Jpn. **38** (1975) 373.
75T1 Tucker, J. W.: J. Phys. C **8** (1975) 353.
75T2 Tornero, J. D., Lopez, F. J., Cabrera, J. M.: Solid State Commun. **16** (1975) 53.
75Y1 Yelon, W. B., Cox, D. E., Eibschütz, M.: Phys. Rev. **B12** (1975) 5007.
75Y2 Yamaguchi, Y., Sakuraba, T.: J. Phys. Soc. Jpn. **38** (1975) 1011.
76B1 Bongaarts, A. L. M., de Jonge, W. J. M., van der Leeden, P.: Phys. Rev. Lett. **37** (1976) 1007.
76B2 Bernard, D. J., Walker, W. C.: Rev. Sci. Instrum. **47** (1976) 122.
76B3 Belyaeva, A. I., Kotlyarskii, M. M.: Phys. Status solidi (b) **76** (1976) 419.
76J1 de Jonge, W. J. M., Kopinga, K., Swüste, C. H. W.: Phys. Rev. **B14** (1976) 2137.
76J2 Jakubowski, P., Kraska-Skrzypek, D., Ratuszna, A., Chełkowski, A.: Acta Phys. Pol. A**49** (1976) 285.
76K1 Kubo, H., Tanaka, F., Kaneshima, N., Hirakawa, K.: Solid State Commun. **18** (1976) 79.
76K2 Kubo, H., Yahara, I., Hirakawa, K.: J. Phys. Soc. Jpn. **41** (1976) 442.
76K3 Karraker, D. G., Dunlap, B. D.: J. Chem. Phys. **65** (1976) 2032.
76K4 Katsumata, K., Kikuchi, Y.: J. Phys. Soc. Jpn. **41** (1976) 449.
76K5 Kubo, H., Yahara, I., Hirakawa, K.: J. Phys. Soc. Jpn. **40** (1976) 591.
76L1 Lebesque, J. V., Huyboom, N. F.: Commun. Phys. (GB) **1** (1976) 33.
76S1 Steiner, M., Axmann, A.: Solid State Commun. **19** (1976) 115.
76S2 Steiner, M., Villain, J., Windsor, C. G.: Adv. Phys. **25** (1976) 87.
77A1 Almond, D. P.: Physica **86–88**B (1977) 651.
77B1 Bongaarts, A. L. M., de Jonge, W. J. M.: Phys. Rev. **B15** (1977) 3424.
77B2 Barmatz, M., Testardi, L. R., Eibschütz, M., Guggenheim, H. J.: Phys. Rev. **B15** (1977) 4370.
77B3 Bartolomé, J., Navarro, R., González, D., de Jongh, L. J.: Physica **92B** (1977) 45.
77B4 Brener, E., Ehrenfreund, E., Shechter, H., Makovsky, J.: J. Phys. Chem. Solids **38** (1977) 1023.
77B5 Breitling, W., Lehmann, W., Weber, R., Lehner, N., Wagner, U.: J. Magn. Magn. Mater. **6** (1977) 113.
77D1 Dupas, C., Renard, J-P.: J. Phys. C. **10** (1977) 5057.
77D2 Le Dang, K., Veillet, P., Renard, J-P.: Solid State Commun. **24** (1977) 313.
77F1 Fedoseyeva, N. V., Spevakova, I. P., Beznosikov, B. V., Naiden, E. P., Khokhlov, V. A.: Phys. Status Solidi (a) **44** (1977) 429.
77G1 Grill, R. J., Dürr, U., Weber, R.: Physica **86–88**B (1977) 673.
77H1 Hartmann, J. B.: Phys. Rev. **B15** (1977) 273.

77I1 Ikeda, H.: J. Phys. C **10** (1977) L469.
77K1 Kopinga, K., Van Vlimmeren, Q. A. G., Bongaarts, A. L. M., de Jonge, W. J. M.: Physica **86–88**B (1977) 671.
77K2 Krausz, E., Viney, S., Day, P.: J. Phys. C **10** (1977) 2685.
77K3 Kopinga, K.: Phys. Rev. B**16** (1977) 427.
77K4 Khatamian, D., Collins, M. F.: Can. J. Phys. **55** (1977) 773.
77M1 McElearney, J. N.: Solid State Commun. **24** (1977) 863.
77N1 Niel, M., Cros, C., le Flem, G., Pouchard, M., Hagenmuller, P.: Physica **86–88**B (1977) 702.
77N2 Niel, M., Cros, C., Pouchard, M., Chamminade, J-P.: J. Solid State Chem. **20** (1977) 1.
77R1 Rosinski, Ch., Elschner, B.: J. Magn. Magn. Mater. **4** (1977) 193.
77S1 Struge, M. D., Cohen, E.: Bull. Am. Phys. Soc. **22** (1977) 375.
77S2 Steiner, M., Kjems, J. K.: J. Phys. C **10** (1977) 2665.
77S3 Scherer, C., Barjhoux, Y.: Phys. Status Solidi (b) **80** (1977) 313.
77S4 Shapira, Y., Becerra, C. C.: Phys. Rev. Lett. **38** (1977) 358.
77S5 Smolensky, G. A., Pisarev, R. V., Markovin, P. A., Krichevzov, B. B.: Physica **86–88**B (1977) 1205.
77S6 Suzuki, N., Isu, T., Motizuki, K.: Solid State Commun. **23** (1977) 319.
77Y1 Yamaguchi, Y., Sakuraba, T.: J. Phys. Chem. Solids **38** (1977) 957.
78A1 Anistratov, A. T., Beznosikov, B. V., Gusar, V. A.: Sov. Phys. Solid State **20** (12) (1978) 2138.
78B1 Bazhan, A. N., Fedoseyeva, N. V., Spevakova, I. P.: Zh. Eksp. Teor. Fiz. (SSSR) **75** (1978) 577.
78B2 Becerra, C. C., Shapira, Y., Oliveira jr., N. F.: Phys. Rev. B**18** (1978) 5060.
78B3 Basten, J. A. J., Frikkee, E., de Jonge, W. J. M.: Phys. Lett. **68**A (1978) 385.
78B4 Basten, J. A. J., Van Vlimmeren, Q. A. G., de Jonge, W. J. M.: Phys. Rev. B**18** (1978) 2179.
78B5 Bazhan, A. N., Fedoseyeva, N. V., Pietrov, S. V., Beznosikov, B. V.: Zh. Eksp. Teor. Fiz. (SSSR) **74** (1978) 1698.
78E1 Ehrenfreund, E., Moses, D., Brener, R.: Phys. Lett. **66**A (1978) 317.
78I1 Iio, K., Hyodo, H., Nagata, K., Yamada.: J. Phys. Soc. Jpn. **44** (1978) 1393.
78J2 Jakubowski, P.: Acta Phys. Pol. A**54** (1978) 397.
78K1 Kim, H., Lange, J.: Phys. Rev. B**17** (1978) 4207.
78M1 Mekala, M., Adachi, K.: J. Phys. Soc. Jpn. **44** (1978) 806.
78N1 Nagata, K., Nishino, T., Hirosawa, T., Komatsubara, T.: J. Phys. Soc. Jpn. **44** (1978) 813.
78P1 Plesko, S., Kind, R., Roos, J.: J. Phys. Soc. Jpn. **45** (1978) 553.
78S1 Shapira, Y., Oliveira jr., N. F.: Phys. Rev. B**17** (1978) 4432.
78S2 Shapira, Y., Oliveira jr., N. F.: J. Appl. Phys. **49** (1978) 1374.
78T1 Tsuda, T., Yasuoka, H., Miyauchi, T.: J. Phys. Soc. Jpn. **45** (1978) 1551.
78Y1 Yamaguchi, Y., Sakuraba, T.: J. Phys. Soc. Jpn. **44** (1978) 1389.
78Y2 Yamamoto, I.: J. Phys. Soc. Jpn. **45** (1978) 1545.
79A1 Arakawa, M., Ebisu, H.: J. Phys. Soc. Jpn. **46** (1979) 1571.
79B1 Basten, J. A. J., Frikkee, E., de Jonge, W. J. M.: Phys. Rev. Lett. **42** (1979) 897.
79B2 Bretling, B., Lehmann, W., Weber, R.: J. Magn. Magn. Mater. **10** (1979) 25.
79D1 Day, P., Gregson, A. K., Leech, D. H., Hutchings, M. T., Rainford, B. D.: J. Magn. Magn. Mater. **14** (1979) 166.

79D2 Dance, J-M., Tressaud, A.: Mater. Res. Bull. **14** (1979) 37.
79E1 Engelsberg, M., Rezende, S. M., Soares, E. A.: J. Appl. Phys. **50** (1979) 1929.
79E2 Engelsberg, M.: Phys. Rev. **B19** (1979) 5942.
79F1 Fedoseyeva, N. V., Valikanova, T. A., Zvegintsev, A. G.: Phys. Status Solidi (a) **51** (1979) K93.
79G1 Guriewitz, E., Horowitz, A., Shaked. H.: Phys. Rev. **B20** (1979) 4544.
79J1 Johnson, P. B., Rayne, J. A.., Friedberg, S. A.: J. Appl. Phys. **50** (1979) 1853.
79L1 Lockwood, D. J., Coombs, G. J., Cowley, R. A.: J. Phys. C **12** (1979) 4611.
79N1 Nevald, R., Voss, F. W., Nielsen, O. N., Amberger, H-D., Fischer, R. D.: Solid State Commun. **32** (1979) 1223.
79O1 Oliveira jr., N. F., Shapira, Y.: J. Appl. Phys. **50** (1979) 1790.
79P1 Prater, J. T., Egami, T.: J. Appl. Phys. **50** (1979) 1732.
79S1 Shapira, Y., Oliveira jr., N. F., Chang, T. S.: Phys. Rev. Lett. **42** (1979) 1292.
79S2 Smit, J. J., Van Wijk, H. J., de Jongh, L. J.: Chem. Phys. Lett. **62** (1979) 158.
79V1 Van Vlimmeren, Q. A. G., de Jonge, W. J. M.: Phys. Rev. **B19** (1979) 1503.
79Y1 Yamazaki, H., Soares, E., Paneucci, H., Morishige, Y.: J. Phys. Soc. Jpn. **47** (1979) 1464.
79Y2 Yosida, T., Aoki, H., Takeuchi, H., Arakawa, M., Horai, K.: J. Phys. Soc. Jpn. **46** (1979) 1661.
79Y3 Yoshizawa, H., Hirakawa, K.: J. Phys. Soc. Jpn. **46** (1979) 448.
80A1 Adachi, K., Achiwa, N., Mekata, M.: J. Phys. Soc. Jpn. **49** (1980) 545.
80A2 Andrienko, A. V., Prozorova, L. A.: Zh. Eksp. Teor. Fiz. (SSSR) **78** (1980) 2411.
80B1 Basten, A. J., Frikkee, E., de Jonge, W. J. M.: Phys. Rev. **B22** (1980) 1429.
80B2 Belyaeva, A. I., Pietrov, S. V., Stelmakhov, Yu. N.: Zh. Eksp. Teor. Fiz. (SSSR) **79** (1980) 2252.
80C1 Clark, G. F., Tanner, B. K.: Phys. Status Solidi (a) **59** (1980) 241.
80D1 Dubicki, L., Ferguson, J., van Oosterhout, B.: J. Phys. C **13** (1980) 2791.
80F1 Fedoseyeva, N., Spevakova, I., Petrakovskii, G., Chuev, V., Petrov, S.: J. Magn. Magn. Mater. **15–18** (1980) 539.
80G1 Gusmão, M. V., Scherer, C.: Phys. Status Solidi (b) **100** (1980) K63.
80I1 Jio, K., Hyodo, H., Nagata, K.: J. Phys. Soc. Jpn. **49** (1980) 1336.
80K1 Krejov, K., Bidikov, J., Brankov, J., Sidjimov, B.: J. Phys. C **13** (1980) 5413.
80P1 Perez-Mato, J. M., Mañes, J. L., Tello, M. J.: J. Phys. C **13** (1980) 2667.
80S1 Schouten, J. C., Boersma, F., Kopinga, K.: J. Magn. Magn. Mater. **15–18** (1980) 1047.
80S2 Shapira, Y., Oliveira jr., N. F.: Phys. Rev. **B21** (1980) 1271.
80S3 Steiner, M.: Phase Trans. **1** (1980) 269.
80U1 Urland, W.: Z. Naturforsch. **35**a (1980) 403.
80V1 Velu, E., Megy, R.: J. Magn. Magn. Mater. **15–18** (1980) 1017.
80V2 Van Vlimmeren, Q. A. G., Swüste, C. H. W., de Jonge, W. J. M., van der Steeg, M. J. H., Stoelinga, J. H. M., Wyder, P.: Phys. Rev. **B21** (1980) 3005.
80Y1 Yamamoto, I.: J. Phys. Soc. Jpn. **49** (1980) 74.
80Z1 Zandbergen, H. W., Ijdo, D. J.: J. Solid State Chem. **34** (1980) 65.
80Z2 Zandbergen, H. W.: J. Solid State Chem. **35** (1980) 367.
81A1 Ajiro, Y., Adachi, K., Meketa, M.: Solid State Commun. **37** (1981) 449.
81A2 Adachi, K.: J. Phys. Soc. Jpn. **50** (1981) 3904.
81G1 Goto, T., Kawai, T.: J. Phys. Soc. Jpn. **50** (1981) 3531.
81G2 Ganot, F., Dugautier, S., Moch, P., Nouet, J.: J. Appl. Phys. **52** (1981) 2289.
81H1 Haseda, T., Wada, N., Hata, M., Amaya, K.: Physica **108B** (1981) 841.
81H2 Hyiodo, H., Iio, K., Nagata, K.: J. Phys. Soc. Jpn. **50** (1981) 1545.

81M1 Motokawa, M., Suzuki, N., Motizuki, K.: J. Phys. Soc. Jpn. **50** (1981) 3588.
81S1 Someya, Y., Ito, A.: J. Phys. Soc. Jpn. **50** (1981) 1891.
81S2 Someya, Y., Ito, A., Morimoto, S.: J. Phys. Soc. Jpn. **50** (1981) 1883.
81T1 Tazuke, Y., Tanaka, H., Iio, K., Nagata, K.: J. Phys. Soc. Jpn. **50** (1981) 3919.
81Y1 Yoshizawa, H., Axe, J. D., Shirane, G.: Solid State Commun. **38** (1981) 241.
82A1 Ajiro, Y., Steiner, M., Dachs.: Solid State Commun. **41** (1982) 203.
82A2 D'Ariano, G.: Phys. Rev. **B26** (1982) 6215.
82B1 Bontemps, N., Grisolia, C., Nerozzi, M., Briat, B.: J. Appl. Phys. **53** (1982) 2710.
82C1 Chaddha, G., Seehra, M. S.: Solid State Commun. **44** (1982) 1097.
82F1 Fitzgerald, W. J., Visser, D., Ziebeck, K. R. A.: J. Phys. C **15** (1982) 795.
82F2 Ferré, J., Jamet, J. P.: Solid State Commun. **44** (1982) 485.
82G1 Gurewitz, E., Melamud, M., Horowitz, A.: Phys. Rev. **B25** (1982) 5220.
82G2 Ganot, F., Dugautier, C., Moch, P., Nouet, J.: J. Phys. C **15** (1982) 801.
82G3 Gurewitz, B., Shaked, H.: Acta Crytallogr. **B38** (1982) 2771.
82H1 Hirakawa, K., Yoshizawa, H., Ubukoshi, K.: J. Phys. Soc. Jpn. **51** (1982) 1119.
82J1 Johnstone, I. W., Jones, G. D., Lockwood, D. J.: J. Phys. C **15** (1982) 2043.
82L1 Lockwood, D. J., Johnstone, I. W.: J. Appl. Phys. **53** (1982) 8169.
82N1 Novikov, W. P., Eremienko, W. W., Kachir, I. S.: Zh. Eksp. Teor. Fiz. (SSSR) **82** (1982) 346.
82N2 Nagler, S. E., Buyers, J. L., Amstrong, R. L., Briat, B: Phys. Rev. Lett. **49** (1982) 590.
82S1 Suzuki, H., Miyamoto, M., Masuda, Y., Ohtsuka, T.: J. Low Temp. Phys. **48** (1982) 297.
82V1 Veenendaal, E. J., Brom, H. B., Ihringer, J.: Physica **114B** (1982) 31.
82W1 Waada, N., Ubukoshi, K., Hirakawa, K.: J. Phys. Soc. Jpn. **51** (1982) 2833.
83A1 Ajiro, Y., Adachi, K., Mekata, M.: J. Magn. Magn. Mater. **31–34** (1983) 1141.
83B1 Bartolomé, J., Rojo, J. A., Navarro, R., González, D., Ibarra, M. R., del Moral, A.: J. Magn. Magn. Mater. **31–34** (1983) 1052.
83B2 Bartolomé, R., Burriel, R., Palacio, F., González, D.: Physica **115B** (1983) 190.
83B3 Baines, J. A., Johnson, C. E., Thomas, M. F.: J. Phys. C **16** (1983) 3579.
83C1 Chirwa, M., Top, J., Flokstra, J.: Physica **123B** + C (1983) 53.
83F1 Ferré, J.: J. Phys. C **16** (1983) 3971.
83G1 Goto, T., Yamaguchi, Y.: J. Magn. Magn. Mater. **31–34** (1983) 1211.
83H1 Hauser, A., Falk, U., Fischer, P., Furrer, A., Güdel, H. U.: J. Magn. Magn. Mater. **31–34** (1983) 1139.
83H2 Hirakawa, K., Ikeda, H., Kadowaki, H., Ubukoshi, K.: J. Phys. Soc. Jpn. **52** (1983) 2882.
83I1 Iio, K., Hoshino, T., Nagata. K.: J. Magn. Magn. Mater. **31–34** (1983) 587.
83L1 Lockwood, D. J., Johnstone, I. W., Labbe, H. J., Briat, B.: J. Phys. C **16** (1983) 6451.
83N2 Nordblad, P., Belanger, D. P., King, A. R., Jaccarino, V.: J. Magn. Magn. Mater. **31–34** (1983) 1093.
83R1 Rayne, J. A., Collins, J. G., White, G. K.: Solid State Commun. **45** (1983) 681.
83S1 Suzuki, N.: J. Phys. Soc. Jpn. **52** (1983) 3199.
83S2 Shiba, H., Suzuki, N.: J. Phys. Soc. Jpn. **52** (1983) 1382.
83S3 Smeets, J. P. M., Maessen, K. M. H., Frikkee, E., Kopinga, K.: J. Magn. Magn. Mater. **31–34** (1983) 1163.
83T1 Tanaka, H., Iio, K., Nagata, K.: J. Magn. Magn. Mater. **31–34** (1983) 641.
83V1 Vélu, E., Mégy, R., Seiden, J., Renard, J. R.: J. Magn. Magn. Mater. **31–34** (1983) 1069.
83V2 Veenendaal, E. J., Brom, H. B., Huiskamp, W. J.: Physica **121B** (1983) 1.
83V3 Velikanova, T. A., Fedoseyeva, N. V., Balaev, A. D.: Fiz. Tverd. Tela (SSSR) **25** (1983) 2917.
83W1 Wada, N., Sumiyoshi, K., Watanabe, T., Amaya, K.: J. Phys. Soc. Jpn. **52** (1983) 1893.

83W2 Wada, N., Amaya, K., Tokuda, S., Tomikawa, T.: J. Magn. Magn. Mater. **31–34** (1983) 721.

84B1 Belyaeva, A. J., Kotlyarskii, M. M., Miloslavskaya, O. V., Stelmakhov, Yu. N., Yurev, V. P.: Phys. Status Solidi (a) **83** (1984) 245.

84C1 Celiński, Z., Skrzypek, D.: Acta Phys. Pol. **A65** (1984) 149.

84F1 Fedoseyeva, N. V., Velikanova, T. A., Balaev, A. D.: Solid State Commun. **52** (1984) 241.

84I1 Imppu, E., Laiho, R., Levola, T.: Phys. Rev. **B30** (1984) 232.

84K1 Kubo, H., Uryû, N., Nishihara, H.: J. Phys. Soc. Jpn. **53** (1984) 2099.

84K2 Kubo, H., Uryû, N.: J. Phys. Soc. Jpn. **53** (1984) 499.

84K3 Kubo, H., Uryû, N., Nishihara, H., Yasuoka, H.: J. Phys. Soc. Jpn. **53** (1984) 3747.

84K4 Krichevtsov, B. B., Markovin, P. A., Petrov, S. V., Pisarev, R. V.: Zh. Eksp. Teor. Fiz. (SSSR) **86** (1984) 2262; Sov. Phys JETP (English Transl. **59** (1984) 1316.

84R1 Rayne, J. A., Collins, J. G., White, G. K.: J. Appl. Phys. **55** (1984) 2404.

84S1 Safonov, V. L.: Phys. Status Solidi (b) **126** (1984) 197.

84T1 Tazuke, Y., Tanaka, H., Iio, K., Nagata, K.: J. Phys. Soc. Jpn. **53** (1984) 3191.

85C1 Chirawa, M., Top, J. L., Flokstra, J.: Physica **128B** + C (1985) 13.

85C2 Causa, M. T., Passeggi, M. C. G.: Phys. Rev. **B32** (1985) 3229.

85F1 Fedoseyeva, N. V., Gekht, R. S., Velikanova, T. A., Balaev, A. D.: Zh. Eksp. Teor. Fiz. Pisma (SSSR) **41** (1985) 332.

85H1 Hauser, A., Falk, U., Fischer, P., Güdel, H. U.: J. Solid State Chem. **56** (1985) 343.

85K1 Kopinga, K., Steiner, M., de Jonge, W. J. M.: J. Phys. C **18** (1985) 3511.

85K2 Kato, H., Tomikawa, T., Amaya, K., Wada, N.: J. Phys. Soc. Jpn. **54** (1985) 3942.

85L1 Levola, T., Kleemann, W.: J. Phys. C **18** (1985) 1513.

85P1 Popov, E. A., Kotlyarskii, M. M., Edelman, I. S.: Phys. Status Solidi (b) **132** (1985) 465.

85P2 Pezzoni, R., Rigamonti, A., Torre, S.: Solid State Commun. **55** (1985) 899.

85R1 Ramirez, A. P., Wolf, W. P.: Phys. Rev. **B32** (1985) 1639.

85S1 Smit, H. H. A., de Groot, H. J. M., Thiel, R. C., de Jongh, L. J.: Solid State Commun. **53** (1985) 573.

85S2 Smeets, J. P. M., Frikkee, E.: Phys. Rev. **B31** (1985) 7323.

85T1 Tsuda, T., Itoh, M., Yasuoka, H.: J. Phys. Soc. Jpn. **54** (1985) 1599.

85T2 Tsuboi, T., Laiho, R.: Phys. Status Solidi (b) **132** (1985) K33.

85T3 Tsuboi, T., Laiho, R.: Phys. Rev. **B32** (1985) 1673.

85T4 Tsuboi, T.: Phys. Rev. **B32** (1985) 3164.

85Y1 Yayama, H., Kubo, H., Tomokiyo, A.: J. Phys. Soc. Jpn. **54** (1985) 4096.

86C1 Caillé, A., Poirier, M.: Solid State Commun. **60** (1986) 945.

86H1 Harrison, A., Vissert, D., Day, P., Knop, W., Steiner, M.: J. Phys. C **19** (1986) 6811.

86H2 Haije, W. G., Maaskant, W. J. A.: J. Phys. C **19** (1986) 6943.

86H3 Hidaka, M., Fujii, H., Maeda, S.: Phase Trans. **6** (1986) 101.

86I1 Iio, K., Nagata, K.: J. Magn. Magn. Mater. **54–57** (1986) 1393.

86K1 Kubo, H., Uryû, N., Hamasaki, T., Hidaka, M., Nishihara, H., Yasuoka, H.: J. Magn. Magn. Mater. **54–57** (1986) 35.

86L1 Laiho, R., Tsuboi, T.: J. Magn. Magn. Mater. **54–57** (1986) 1265.

86M1 Makata, M., Ajiro, Y., Adachi, K.: J. Magn. Magn. Mater. **54–57** (1986) 1267.

86M2 Mogi, I., Kojima, N., Ban, T., Tsujikawa, I.: J. Phys. Soc. Jpn. **55** (1986) 3664.

86M3 Mogi, I., Okamoto, T., Kojima, N., Ban, T., Tsujikawa, I.: J. Phys. Soc. Jpn. **55** (1986) 987.

86N1 Nagata, K., Iio, K., Saito, Y., Igarashi, M., Tazuke, Y.: J. Magn. Magn. Mater. **54–57** (1986) 33.

86S1 Suzuki, H., Masuda, Y., Miyamoto, M., Ohtsuka, T.: J. Low Temp. Phys. **62** (1986) 497.

86T1 Tazuke, Y., Kinouchi, S-i., Tanaka, H., Iio, K., Nagata, K.: J. Phys. Soc. Jpn. **55** (1986) 4020.

87C1 Chiba, M., Tsuboi, T., Hori, H., Shiozaki, I., Date, M.: Solid State Commun. **63** (1987) 427.

87F1 Falk, U., Furrer, A., Güdel, H. U., Kjems, J. K.: Phys. Rev. **B35** (1987) 4888.

87F2 Falk, U., Furrer, A., Furrer, N., Güdel, H. U., Kjems, J. K.: Phys. Rev. **B35** (1987) 4893.

87G1 Gaullin, B. D., Collins, M. F., Buyers, W. J. L.: J. Appl. Phys. **61** (1987) 3409.

87K1 Kadowaki, H., Ubukoshi, K., Hirakawa, K.: J. Phys. Soc. Jpn. **56** (1987) 751.

87M1 Mukhopadhyay. S., Chatterjee, I.: J. Magn. Magn. Mater. **67** (1987) 221.

87M2 Mason, T. E., Collins, M. F., Gaulin, B. D.: J. Phys. C **20** (1987) L945.

87M3 Mekata, M., Tatsumi, T. Nakashima, T., Adachi, K., Ajiro, Y.: J. Phys. Soc. Jpn. **56** (1987) 4544.

87M4 Mogi, I., Kojima, N., Ajiro, Y., Kikuchi, H., Ban, T., Tsujikawa, I.: J. Phys. Soc. Jpn. **56** (1987) 4592.

87S1 Simpson, A. M., Caillé, A., Jericho, M. H.: Solid State Commun. **64** (1987) 1117.

87V1 Velikanova, T. A., Balaev, A. D., Fedoseyeva, N. V.: Fiz. Tverd. Tela (SSSR) **29** (1987) 283.

87W1 Ward, J. B., McCann, V. H., Pankhurst, Q. A., Hassett, W. L., Price, D. C.: J. Phys. C **20** (1987) 1689.

88A1 Ajiro, Y., Nakashima, T., Unno, Y., Kadowaki, H., Makata, M., Achiwa, N.: J. Phys. Soc. Jpn. **57** (1988) 2648.

88A2 Amaya, K., Ishizuka, M., Nakagawa, T., Saratani, S., Sakakibara T., Takeyama, S., Nakao, K., Goto, T., Miura, N., Unno, Y., Ajiro, Y.: J. Phys. Soc. Jpn. **57** (1988) 38.

88B1 Bocquet, S., Ward, J. B., McCann, V. H.: J. Phys. C **21** (1988) 367.

88C1 Chubukov, A. V.: J. Phys. C **21** (1988) L441.

88I1 Iio, K., Sano, M., Masuda, H., Tanaka, H., Nagata, K.: J. Phys. Soc. Jpn. **57** (1988) 50.

88K1 Kadowaki, H., Shapiro, S. M., Inami, T., Ajiro, Y.: J. Phys. Soc. Jpn. **57** (1988) 2640.

88M1 Maegawa, S., Goto, T., Ajiro, Y.: J. Phys. Soc. Jpn. **57** (1988) 1402.

88M2 Meidenbauer, K., Gliemann, G., Oomen, E. W. J. L., Blasse, G.: J. Phys. C **21** (1988) 4703.

88P1 Pawłowska, M., Bronowska, W., Wojciechowski, W.: Phys. Status Solidi (b) **146** (1988) K51.

88P2 Plumer, M. L., Hood, K., Caillé, A.: Phys. Rev. Lett. **60** (1988) 45.

88R1 Ratuszna, A., Glazer, A. M.: Phase Trans. **12** (1988) 347.

88S1 Suzuki, H., Masuda, Y., Authur, J.: J. Low Temp. Phys. **70** (1988) 301.

88S2 Skrzypek, D., Białas-Borgieł, K., Celiński, Z.: Phys. Status Solidi (a) **105** (1988) 567.

88T1 Tanaka, H., Teraoka, S., Kakehashi, E., Iio, K., Nagata, K.: J. Phys. Soc. Jpn. **57** (1988) 3979.

89F1 Fedoseyeva, N. V., Velikanova, T. A.: Fiz. Tverd. Tela (SSSR) **31** (1989) 224.

89G1 Gaulin, B. D., Mason, T. E., Collins, M. F.: Phys. Rev. Lett. **62** (1989) 1380.

89G2 Gekht, R. S., Fedoseyeva, N. V., Dolina, V. A., Balaev, A. D.: Phys. Status Solidi (b) **155** (1989) 639.

89G3 Gaulin, B. D., Collins, M. F., Mason, T. E.: Physica **B156–157** (1989) 244.

89H1 Harrison, A., Visser, D.: J. Phys. Condens. Matter **1** (1989) 733.

89K1 Kubo, H., Hamasaki, T., Uryû, N.: Phys. Rev. **B40** (1989) 9098.

89K2 Kikuchi, H., Ajiro, Y., Goto, T.: J. Phys. Soc. Jpn. **58** (1989) 43.

89M1 Mogi, I., Takeda, M., Kido, G., Nakagawa, Y., Kikuchi, H., Ajiro, Y.: J. Phys. Soc. Jpn. **58** (1989) 2188.

89P1 Petrakovski, G. A., Fedoseyeva, N. V., Aplesnin, S. S., Korolev, V. K.: Fiz. Tverd. Tela (SSSR) **31** (1989) 169.
89T1 Takeda, M., Kido, G., Mogi, I., Nakagawa, Y., Okada, H., Kojima, N.: J. Phys. Soc. Jpn. **58** (1989) 3418.
90A1 Ajiro, Y., Inami, T., Goto, T.: J. Magn. Magn. Mater. **90–91** (1990) 273.
90C1 Caillé, A., Plumer, M. L., Poirier, M., Gaulin, B. D.: Physica **B165–166** (1990) 169.
90F1 Fargin, E., Lestienne, B., Dance, J. M.: Solid State Commun. **75** (1990) 769.
90H1 Harrison, A., Visser, D.: J. Phys. Condens. Matter **2** (1990) 10487.
90H2 Hori, H., Mikami, H., Date, M., Amaya, K.: Physica **B165–166** (1990) 237.
90H3 Hori, H., Amaya, K., Mikami, H., Date, M.: J. Magn. Magn. Mater **90–91** (1990) 269.
90I1 Igarashi, M., Nagata, K.: J. Magn. Magn. Mater. **90–91** (1990) 363.
90M1 Merkle, L. D.: Phys. Rev. B **42** (1990) 3783.
90M2 Mason, T. E., Stager, C. V., Gaulin, B. D., Collins, M. F.: Phys. Rev. **B42** (1990) 2715.
90M3 Mekata, M.: J. Magn. Magn. Mater. **90–91** (1990) 247.
90M4 Mekata, M., Okamoto, S., Onoe, S., Matsuda, S., Yoshizawa, H.: J. Magn. Magn. Mater. **90–91** (1990) 267.
90M5 Maegawa, S., Fujiwara, N., Hohmoto, T., Goto, T.: J. Magn. Magn. Mater. **90–91** (1990) 271.
90O1 Okada, H., Kojima, N., Tsujikawa, I.: Phys. Rev. **B42** (1990) 11619.
90P1 Plumer, M. L., Caillé, A.: Phys. Rev. **B41** (1990) 2543.
90P2 Palme, W., Mertens, F., Born, O., Lüthi, B., Schotte, U.: Solid State Commun. **76** (1990) 873.
90P3 Poirier, M., Castonguay, M., Caillé, A., Plumer, M. L., Gaulin, B. D.: Physica **B165–166** (1990) 171.
90R1 Ratuszna, A., Majewska, K.: Powder Diffr. **5** (1990) 41.
90S1 Strager, R., Moran, G., Krausz, E., Güdel, H., Furrer N.: Mol. Phys. **69** (1990) 11.
90S2 Skrzypek, D.: Phys. Status Solidi (b) **157** (1990) 695.
90S3 Suzuki, N.: J. Magn. Magn. Mater. **90–91** (1990) 69.
90T1 Tanaka, H.: J. Magn. Magn. Mater. **90–91** (1990) 251.
90T2 Takeda, M., Mogi, I., Kido, G., Nakagawa, Y., Okada, H., Kojima. N.: J. Magn. Magn. Mater. **90–91** (1990) 244.
91C1 Campana, L. S., Caramico D'Auria, A., Esposito, F., Esposito, U., Kamieniarz, G.: Physica **B168** (1991) 153.
91H2 Harrison, A., Collins, M. F., Abu-Dayyeh, J., Stager, C. V.: Phys. Rev. **B43** (1991) 679.
91K1 Kotyuzhanskij, B. Ya., Nikiforov, D. V.: Phys. Cond Matter **3** (1991) 385.
91O1 Oohara, Y., Kadowaki, H., Iio, K.: J. Phys. Soc. Jpn. **60** (1991) 393.
91S1 Skrzypek, D.: J. Magn. Magn. Mater. **102** (1991) 175.

9.7 AMF_4 and $AMF_4 \cdot H_2O$ compounds (A = alkali element, Tl or NH_4, M = 3d element)

(For symbols and abbreviations, see Section 9.6.0.2)

9.7.1 AMF_4 compounds

AMF_4 compounds have the orthorhombic layered $TlAlF_4$-type crystal structure. $[MF_6]$ octahedra of six F ions with an M ion in the centre have four of their vertices in common and form thus layers parallel to the *ab* plane. These four vertices are either situated in a flat plane (A = K, Rb, Cs, Tl, NH_4; M = 3d, Al) (Fig. 1) or in a puckered plane (A = Na, M = V, Cr, Fe). A^+ ions are situated between these layers in such a way that the $[A^+F_6]$ octahedra form similar layers which alternate with the previous ones. Jahn-Teller effects results in distortions of the octahedra leading to a so-called antiferro-distortive order of elongated octahedra. Crystallographic parameters are given in Table 1. For exact coordinates of the structures of the various compounds is referred to the original literature. Generally these compounds are layer antiferromagnets: there exists antiferromagnetic coupling between the M ions, the intralayer coupling is by far the most prominent one and its magnitude is given by the exchange constant *J*. For a survey of the magnetic properties, see Tables 2–6.

Table 1. AMF_4. RT crystallographic parameters of the unit cell as derived from X-ray diffraction data. Crystal types: M: monoclinic, T: tetragonal, O: orthorhombic, H: hexagonal. Number of chemical formulae per unit cell: *Z*.

Compound	Type	Space group	*Z*	*β* [°]	*a* [Å]	*b* [Å]	*c* [Å]	Ref.
α-$NaTiF_4$	O	Pbcn	4		4.976 (3)	5.755 (9)	11.070 (5)	76O
$NaVF_4$	M	$P2_1/c$		102	7.90	5.33	7.55	76D
KVF_4	O				7.596	7.738	12.28	68B, 74B
	O		8		7.656	7.508	12.120	73C
α-$RbVF_4$	T				7.596		6.315	68B
					7.580		6.370	73C
β-$RbVF_4$	O		4		7.608	7.550	6.260	73C
$CsVF_4$	T		4		7.77		6.58	73C
	T				7.76		6.574	74B
$TlVF_4$	T		8		7.43		12.64	73C
	O				5.353	5.305	12.88	74B
$NaCrF_4$	T				7.438		6.442	68B
	M	$P2_1/c$	4	101.8 (2)	7.81 (1)	5.31 (1)	7.39 (1)	75K
$KCrF_4$	O	Pnma	24		15.76 (1)	7.43 (1)	18.38 (1)	77D1
$CsCrF_4$	H	$P\bar{6}2m$			9.650 (5)		3.857 (3)	78B
$TlCrF_4$	T				7.38		12.87	68K
$LiMnF_4$	M	$P2_1/c$	2	113.24 (2)	5.414 (1)	4.629 (1)	5.694 (1)	87W
$NaMnF_4$[a])	M	$P2_1/c$	2	108.07 (2)	5.736 (2)	4.892 (1)	5.748 (2)	91M
$KMnF_4$	M	$P2_1/a$	4	90.54 (3)	7.699 (2)	7.644 (2)	5.769 (1)	92M
	M	id.		90.625 (3)	7.7062 (4)	7.6571 (4)	5.7889 (3)	93M1
	M[f,h])	id.		90.432 (2)	7.7062 (2)	7.6568 (2)	5.7889 (1)	93M2

(*continued*)

Table 1 (*continued*)

Compound	Type	Space group	Z	β [°]	a [Å]	b [Å]	c [Å]	Ref.
$RbMnF_4$	M	$P2_1/a$	4	90.83 (4)	7.822 (5)	7.777 (3)	6.050 (2)	92M
	M[g])	id.		90.775 (3)	7.8136 (4)	7.7748 (4)	6.0466 (3)	93M1
	M	id.		90.443 (4)	7.8119 (4)	7.7761 (4)	6.0469 (3)	93M2
$CsMnF_4$	Me[b])	P4/n	4		7.9440 (6)		6.6376 (9)	92M
NH_4MnF_4	T		4		7.752 (6)		6.153 (12)	77B1
	Tric[c])				7.731 (1)	7.752 (2)	6.113 (2)	77M1
$TlMnF_4$	O				5.44	5.40	12.49	78K
	M	I2/a	4	90.19	5.397 (2)	5.441 (2)	12.484 (5)	92N
$NaFeF_4$[d])	M	$P2_1/c$	4	101.88	7.921 (5)	5.353 (5)	7.543 (5)	70T
$KFeF_4$	O				7.59	7.76	12.27	68B
	O	C_{2v}^8 or D_{2h}^9	8		7.76 (2)	7.59 (2)	12.30 (5)	70T
	O	Amma-D_{2h}^{17}	4		7.596	3.884	12.27	71H
α-$RbFeF_4$[e])	T				7.63		6.27	68B
	O	C_{2v}^5	4		7.65 (1)	6.24 (1)	7.62 (1)	69T, 70T
	O	D_{2h}^{11}-Pmab	4		7.6651 (13)	7.6316 (15)	6.2789 (6)	90M2
β-$RbFeF_4$[e])	T	P4/mmm-D_{2h}^{11}	1		3.85 (2)		6.29	69T, 70T
α-$CsFeF_4$[i])	O	C_{2v}^5	4		7.73 (1)	7.78 (1)	6.56 (1)	70T
	O	D_{2h}^{11}	4		7.796 (1)	7.801 (1)	6.5574 (1)	72E2
β-$CsFeF_4$	T	P4/nmm-D_{4h}^7	4		7.794		6.553	74B
NH_4FeF_4	O	C_{2v}^5	4		7.58 (1)	7.58 (1)	6.36 (1)	70T
$TlFeF_4$	O				5.364	5.308	12.88	74B
$LiCoF_4$	M	?	2	114.25 (5)	5.447 (3)	4.665 (1)	5.5540 (3)	82F
$NaCoF_4$	O		4		4.965 (1)	5.516 (1)	10.800 (2)	82F
$KCoF_4$	O	C_{2v}^5	4		7.526 (1)	7.584 (1)	5.792 (1)	82F
$RbCoF_4$	T	?			12.222 (1)		12.356 (2)	82F
$CsCoF_4$	T	P4/mnm	20		12.478 (1)		12.971 (1)	82F

[a]) Lattice parameters (in Å) from neutron diffraction data [91M]:

	70 K	4 K
a	5.755 (2)	5.755 (1)
b	4.892 (1)	4.889 (1)
c	5.760 (2)	5.755 (1)
β	108.62 (1)°	108.67 (1)°

[b]) Merohedral (110) twin sample [92M].

[c]) Triclinic $\alpha = 90.02$ (6)°, $\beta = 90.65$ (3)°, $\gamma = 90.46$ (6)° [77M1].

[d]) At 4.2K the symmetry is triclinic with $a = 7.88$Å, $b = 5.315$Å, $c = 7.51$Å; $\alpha = 90.4°$, $\beta = 101.7°$ and $\gamma = 91.1°$ [76D].

[e]) Allotropic phases, low-temperature phase α, high-temperature phase β stable above 650°C [69T].

[f]) Neutron diffraction data at RT.

[g]) Lattice parameters in Å from neutron diffraction data at 5.0K [93M1]: $a = 7.790$ (3), $b = 7.726$ (3), $c = 6.005$ (2) and $\beta = 90.67$ (1)°.

[h]) From neutron diffraction at 8.1K: $a = 7.667$ (3)Å, $b = 7.614$ (3)Å, $c = 5.731$ (2)Å and $\beta = 90.55$ (1)° [93M1].

[i]) In [82F] a third (tetragonal) phase obtained by fluorination of bromides is mentioned.

Table 2. $AMnF_4$. Survey of the magnetic properties. Curie-Weiss constants: Θ_p: paramagnetic Curie point, C_m: molar Curie-Weiss constant, p_{eff}: effective magnetic moment per FU. Properties of the magnetically ordered state: T_N, T_C: Néel and Curie temperature, respectively, intra-layer magnetic exchange energy J/k, magnetic moment $p_{Mn^{3+}}$ derived from neutron diffraction spectra and gyromagnetic factor g.

Compound	Figs.	Θ_p [K]	C_m [$cm^3K\ mol^{-1}$]	p_{eff} [μ_B/FU]	T_N [K]	J/k [K]	$p_{Mn^{3+}}$ [b] [μ_B]	g	Ref.
$NaMnF_4$	2···4	− 19	2.93	4.84	13.0 (5)	− 1.15	3.52 (5)	1.99	91M
$KMnF_4$	4···8	− 27		4.87					61H
		− 15 (4)	2.93 (6)	4.83 (5)	5.5	− 0.65	3.06 (3)	1.975	93M1
$RbMnF_4$	4, 8, 9	− 7		4.99 (5)					61H
		− 14 (2)			< 4				78K
		− 7			4.2	− 0.275	2.47 (4)	1.99	93M1
$CsMnF_4$	4, 10, 11	+ 27 (2)			T_C = 21 (2)		2.8 (1)[a]		78K
		+ 21			T_C = 23	+ 1.30			93M1
					T_C = 18.9 (5)		4.04 (2)		80M
NH_4MnF_4	12	− 12 (2)			T_C = 10 (1)		1.1 (2)[a]		78K
$TlMnF_4$	4, 13	− 7	2.87		4.2 (5)	− 0.45	3.13		92N

[a]) Magnetic moment per FU in a magnetic field of 18.735 kOe at 4.5K
[b]) See also Table 3.

Table 3. $AMnF_4$. Components of the magnetic moments of the Mn^{3+} ions along the three crystallographic axes, $p^x_{Mn^{3+}}$, $p^y_{Mn^{3+}}$ and $p^z_{Mn^{3+}}$ as derived from neutron powder diffraction patterns at low temperatures T.

Compound	T [K]	$p^x_{Mn^{3+}}$ [μ_B]	$p^y_{Mn^{3+}}$ [μ_B]	$p^z_{Mn^{3+}}$ [μ_B]	$p_{Mn^{3+}}$ [μ_B]	α_1 [a]) [°]	α_2 [b]) [°]	β [c]) [°]	Ref.
$NaMnF_4$	4	2.7	2.1	1.0	3.52(5)	13	16	0	91M
$KMnF_4$	1.5	2.55(4)	0.42 (7)	1.18 (7)	2.83 (2)			17	93M2
	1.6	2.99 (3)	0.49 (4)	0.45 (11)	3.06 (3)	≃9.3	≃8.5		93M1
$RbMnF_4$	1.5	1.97 (4)	1.97 (4)	1.03 (9)	2.97 (3)			0	93M2, 93M3
	1.5	1.80 (5)	1.42 (7)	0.95 (10)	2.47 (4)	38	23	0	93M1
$CsMnF_4$	2.2				4.04 (2)				80M
$TlMnF_4$	1.34				3.13	90	0		92N

[a]) Angle between magnetic moment and a axis.
[b]) Angle between magnetic moment and ab plane.
[c]) Canting between the magnetic moments of the two sublattices.

Table 4. $AFeF_4$. Survey of the magnetic properties. T_N: Néel temperature, J/k: intra-layer exchange energy. Quantities derived from magnetic susceptibility measurements (χ), from Mössbauer data (ME) and from neutron diffraction (ND).

Compound	Figs., Tabs.	T_N [K]	J/k [K]	Ref.
$NaFeF_4$	14; Tabs. 5⋯7	111.5 (3) (ME)	− 23 (χ)	76D
$KFeF_4$	15⋯19;	137 (1) (ME, ND)	− 26.5 (6) (χ)	71H
	Tabs. 4, 5	137 (2) (χ)	− 26.5 (5) (ME)	72E1
		148 (ND)		72H
		141.51 (ME)		72H
		135.79 (2) (ME)		83K
		136.50 (ME)		88C
α-$RbFeF_4$	15, 20⋯22;	190 (10) (χ)		70T
	Tabs. 5,6	133 (2) (χ)	− 22 (χ)	71E
		133.40 (5) (ME)		
		133.568 (5) (ME)		83K
α-$CsFeF_4$	21, 23, 24;	160 (2) (χ)	− 27.7 (9) (χ)	72E2
	Tabs. 5, 6	160.30 (5) (ME)	− 28.0 (5) (ME)	74E
NH_4FeF_4	25;	134.9 (5) (ME)	− 26 (1) (χ)	77D2
	Tabs. 5, 6			77M2

Table 5. $AFeF_4$. Static critical sublattice magnetisation parameters D, β and T_N as derived from Mössbauer data in the critical temperature region $t = 1 - T/T_N$ below the magnetic ordering temperature T_N. The parameters are defined by the equation $M(T) = M(0) \cdot D(1 - T/T_N)^\beta$. The magnetisation M is accepted to be proportional to the Mössbauer magnetic hyperfine field. Of most physical significance are the β values obtained by asymptotical extrapolation for temperatures close to T_N. For 2D and 3D Ising models of magnetic ordering the theoretical values β = 0.125 and β = 0.325 (1), have been calculated, respectively [77G].

Compound	t	T_N [K]	β	D	Ref.
$NaFeF_4$	$0 \cdots 4 \cdot 10^{-1}$	111.5 (3)	0.25 (1)	1.02	76D
$KFeF_4$	$2 \cdot 10^{-3} \cdots 3.6 \cdot 10^{-1}$	141.40 (2)	0.182 (2)	1.00	74E
	$2 \cdot 10^{-3} \cdots 4 \cdot 10^{-2}$	141.5 (5)	0.209 (8)	1.12	74E
	$10^{-2} \cdots 2.8 \cdot 10^{-1}$	137.2 (1)	0.185 (5)	0.99 (1)	71E
	$10^{-2} \cdots 2.7 \cdot 10^{-1}$	136.3 (1)	0.185 (1)	1.00 (1)	83K
	$2.4 \cdot 10^{-3} \cdots 1.2 \cdot 10^{-1}$	135.82 (2)	0.161 (3)	0.93 (2)	83K
	$3.8 \cdot 10^{-4} \cdots 5.7 \cdot 10^{-2}$	135.786 (6)	0.151 (3)	0.91 (2)	83K
α-$RbFeF_4$	$10^{-2} \cdots 6 \cdot 10^{-1}$	133.40 (5)	0.245 (5)	1.18 (1)	71E
	$10^{-2} \cdots 6 \cdot 10^{-1}$	133.08 (6)	0.249 (2)	1.05 (5)	83K
	$5 \cdot 10^{-4} \cdots 10^{-2}$	133.568 (5)	0.316 (3)	1.31 (6)	83K
α-$CsFeF_4$	$8 \cdot 10^{-3} \cdots 7 \cdot 10^{-1}$	160.33 (11)	0.278 (10)	1.14 (3)	74E
NH_4FeF_4	$10^{-2} \cdots 15 \cdot 10^{-2}$	134.9 (5)	0.26 (1)	1.06 (2)	77M2

Table 6. $AFeF_4$. Mössbauer parameters at various temperatures T. δ: isomer shift, Δ or 2ε: quadrupole splitting of the outermost Zeeman lines, H_{hf}: hyperfine field, η: asymmetry of the EFG, θ: angle between principal axis of EFG and H_{hf}, θ_γ: angle between principal axis of EFG and crystallographic c axis, Γ: linewidth at half height.

T [K]	δ[a]) [mms^{-1}]	Δ or 2ε [mms^{-1}]	H_{hf} [kOe]	η	θ [°]	θ_γ [°]	Γ [mms^{-1}]	Ref.
$NaFeF_4$, powder								
4.2	0.570 (1)	+ 0.680 (5)	562 (3)	≤ 0.10	< 10	< 10	0.28 (1)	76D
293	0.457 (5)	0.685 (5)[c])					0.27 (1)	76D
$KFeF_4$, powder								
4.2			540 (4)					72H
300	0.138 (2)[b])	1.460 (5)[c])			11 (2)			72H
$KFeF_4$, single crystal								
0[d])	0.560 (5)	− 1.47 (2)	540 (4)	0.00 (5)	14 (1)	11 (1)	0.202 (2)	83K
90	0.557 (2)	− 1.47 (2)	451 (2)	0.00 (5)	14.1 (5)	11.2 (5)	0.204 (2)	83K
296	0.455 (2)	− 1.444 (3)				13 (2)	0.199 (2)	83K
α-$RbFeF_4$, powder								
4.2	0.37 (4)[b])	− 1.60 (5)	530 (5)					71E
80	0.55 (1)	− 1.71 (2)	457 (2)		16.7 (4)			72K
α-$RbFeF_4$, single crystal								
4.2	0.54 (1)	− 1.69 (2)	534 (5)	0	16.5	16.5		76R
0[d])	0.551 (5)	− 1.70 (2)	537 (4)	0.00 (5)	16 (1)	15 (1)	0.202 (2)	83K
90	0.545 (2)	− 1.71 (2)	427 (2)	0.00 (5)	16.7 (5)	15.5 (5)	0.203 (2)	83K
296	0.447 (2)	− 1.676 (2)				17 (2)	0.201 (2)	83K
α-$CsFeF_4$								
4.2	0.55 (1)	− 1.54 (5)	540 (3)	0.3(3)	5 (4)			72E2
78	0.54 (1)	− 1.58 (7)	488 (3)	0.4 (3)	8 (4)			72E2
296	0.44 (1)	1.55 (1)[c])						72E2
NH_4FeF_4, powder								
4.2	0.58 (1)	− 1.4 (1)	542 (3)	0.0	16 (2)		0.31 (3)	77M2
78	0.56 (1)	− 1.4 (1)	460 (3)	0.0	16 (2)		0.30 (3)	77M2
293	0.44 (1)	1.34 (1)[c])					0.28 (1)	77M2

[a]) Relative to α-Fe at RT, unless otherwise stated.
[b]) Relative to ^{57}Co in Pd source, cooled to liquid N_2 temperature.
[c]) Only absolute value determined.
[d]) Extrapolated.

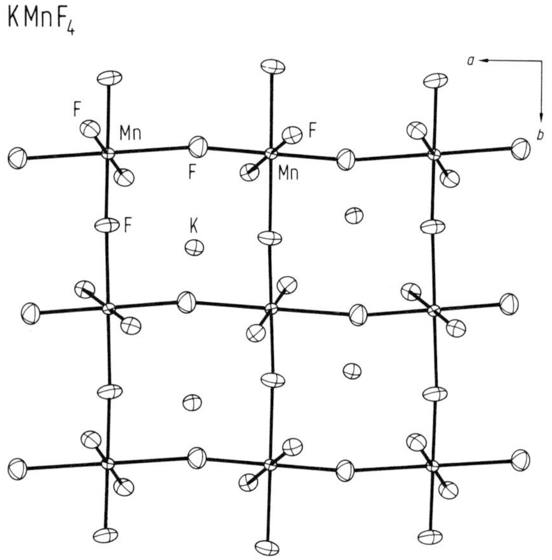

Fig. 1. $KMnF_4$. Arrangement of $[MnF_6]$ octahedra in a flat layer parallel to the *ab* plane [92M].

$NaMnF_4$

Magnetic data Fig. 2. Neutron diffraction reveals a 3D magnetic ordering below $T_N = 13.01\,(5)$ K. Magnetic structure Figs. 3 and 4, collinear AF. The intralayer magnetic coupling is antiparallel in both layer directions and also between spins in adjacent layers. Magnetic moment $p_{Mn^{3+}} = 3.52\,(5)\,\mu_B$, for its components see Table 3 [91M].

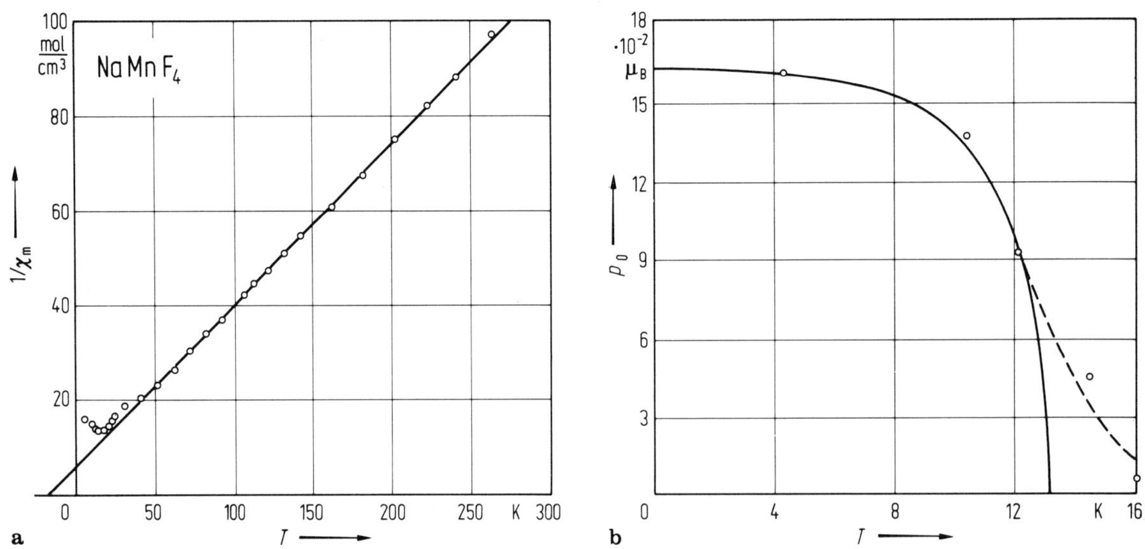

Fig. 2. $NaMnF_4$, powder. (a) Reciprocal molar magnetic susceptibility χ_m^{-1} and (b) spontaneous magnetic moment per FU, p_0 vs. temperature T. Full curve in (b) is an extrapolation to $T_N = 13$ K (derived from neutron diffraction spectra) [91M].

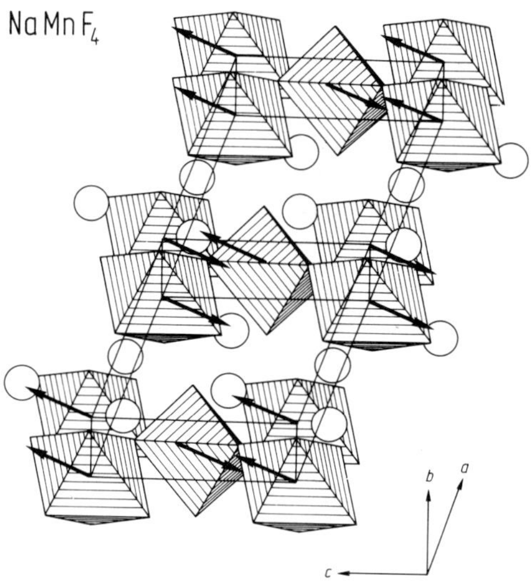

Fig. 3. $NaMnF_4$, single crystal. Magnetic structure as derived from neutron diffraction patterns below $T_N = 13$ K. Hatched $[MnF_6]$ octahedra with arrows representing the magnetic moments of the central Mn^{2+} ions. The circles represent the Na^+ ions [91M].

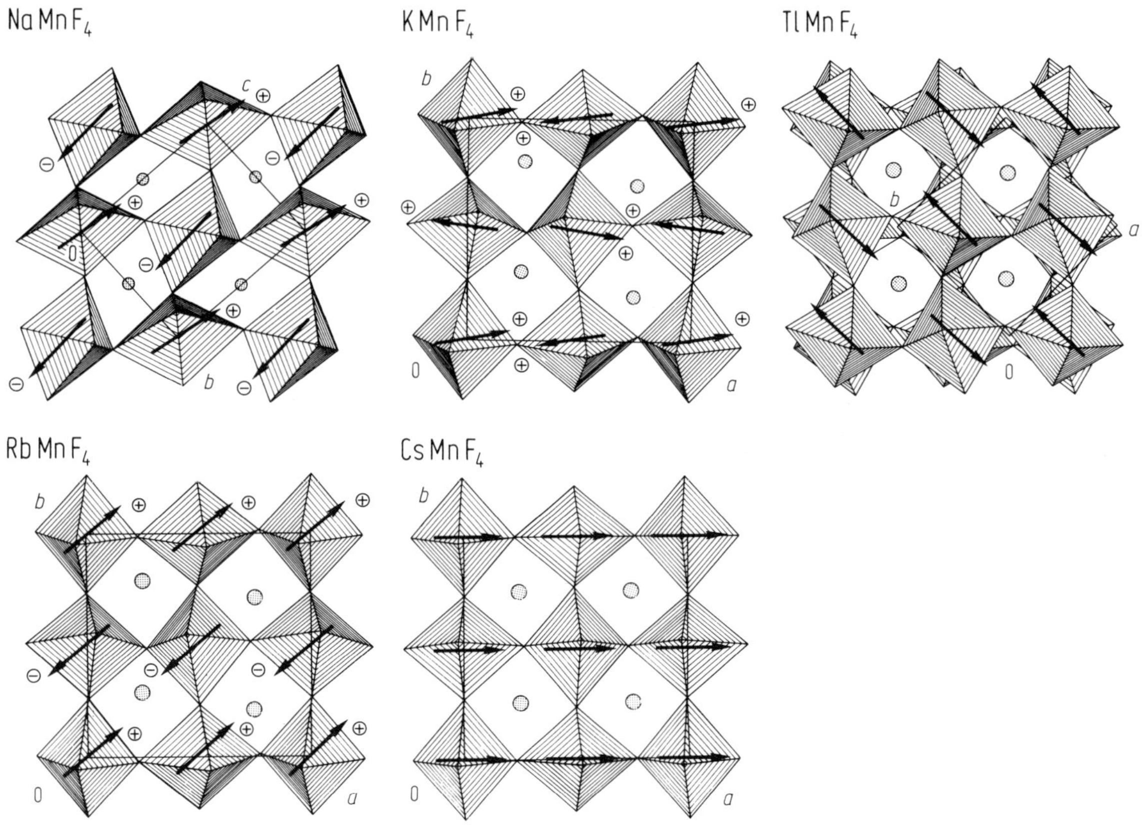

Fig. 4. $AMnF_4$. Magnetic structures in the crystallographic layers as derived from neutron diffraction experiments below the magnetic ordering temperature. *a*, *b* and *c* are the three crystallographic directions. ⊕ and ⊖ indicate the sign of the component of the magnetic moment perpendicular to the layer [93M1].

$KMnF_4$

Magnetic susceptibility Figs. 5 and 6. Magnetisation curves Fig. 7. Magnetic moment vs. temperature Fig. 8. Magnetic structure at low T from neutron diffraction Fig. 4: canted AF with resulting magnetic moment parallel to c axis. Within the layers the coupling is antiparallel in both directions, the interlayer coupling is ferromagnetic. It is assumed that the magnetic moments are oriented more or less parallel to the layers, making an angle of 8.5° with the layer plane and 9.3° with the a axis. The magnetic moments are 3.06 (3)μ_B/Mn^{3+} with $p_x = 2.99$ (3), $p_y = 0.49$ (4) and $p_z = 0.45$ (11) μ_B. Mössbauer parameters measured for $KMn_{0.99}{}^{57}Fe_{0.01}F_4$ are H_{hf} (1.6 K) = 481(3) kOe, Δ (> 10 K) = − 1.84 (2) mms^{-1}, $\delta = 0.420$ (8) mms^{-1} at 1.6K relative to iron metal. Polar angle $\theta = 61$ (3)° and $T_N = 5.5$ (2)K [93M1, 93M2].

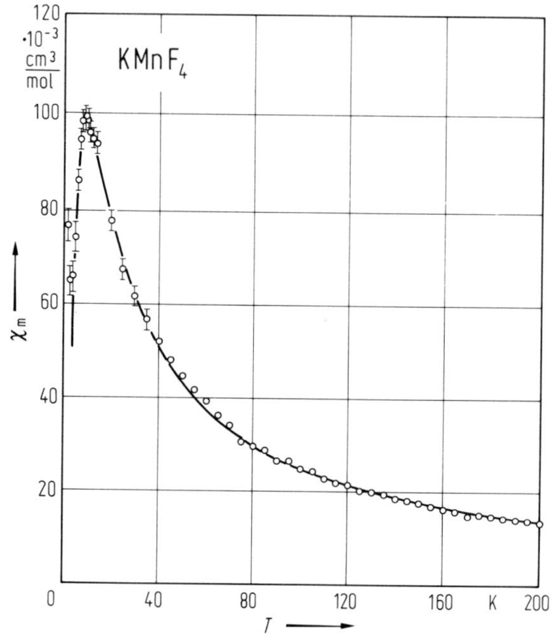

Fig. 5. $KMnF_4$, powder. Molar magnetic susceptibility χ_m vs. temperature T. The curve is calculated for quadratic-layer Heisenberg antiferromagnetism with $J/k = -0.65$ (2) K and $g = 1.975$ [93M1].

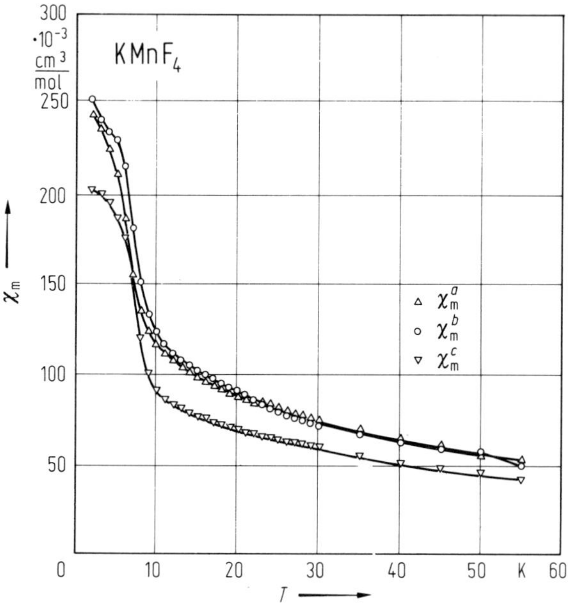

Fig. 6. $KMnF_4$, conglomerate of oriented small crystals. Molar magnetic susceptibility χ_m in the direction of the principal crystallographic axes as a function of temperature T [93M1].

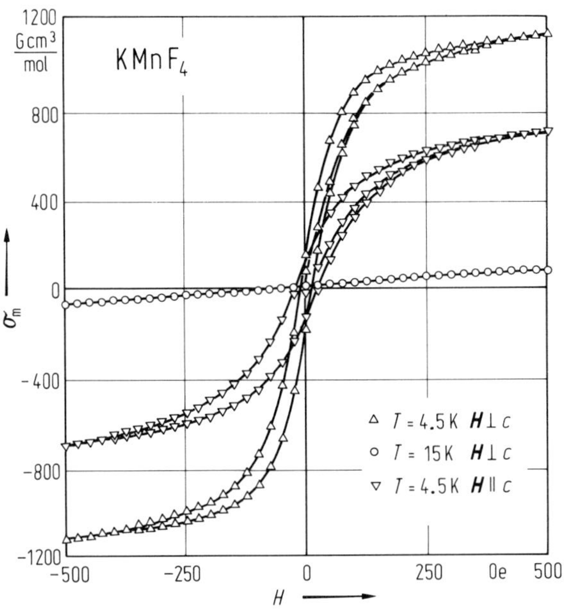

Fig. 7. $KMnF_4$, single crystal. Magnetisation curves in two directions at 4.5 K and perpendicular to the c axis at 15 K [93M1].

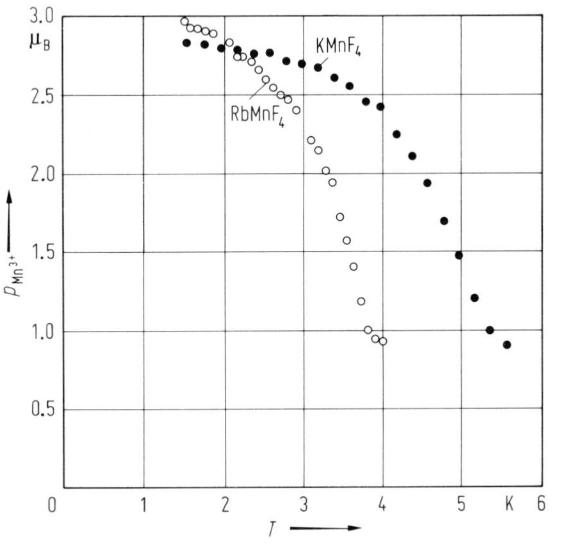

Fig. 8 $KMnF_4$, $RbMnF_4$, powder. Magnetic moment of the Mn^{3+} ion, $p_{Mn^{3+}}$ vs. temperature T as derived from neutron diffraction patterns [93M2].

$RbMnF_4$

Paramagnetic down to temperatures below 5 K, Fig. 9. Magnetic moment vs. temperature Fig. 8. Magnetic structure at low T from neutron diffraction, Fig. 4, canted AF. Within the layers the coupling is parallel along the a axis and antiparallel along the b axis, the inter-layer coupling is ferromagnetic. The spins make an angle $\approx 23°$ with the plane of the layers, the in-plane angle being $\approx 38°$. The magnetic moments of the Mn ions are: $p_x = 1.80$ (5), $p_y = 1.42$ (7) and $p_z = 0.95$ (10) μ_B with $p_{Mn^{3+}} = 2.47$ (4) μ_B [93M1, 93M2 and 93M3].

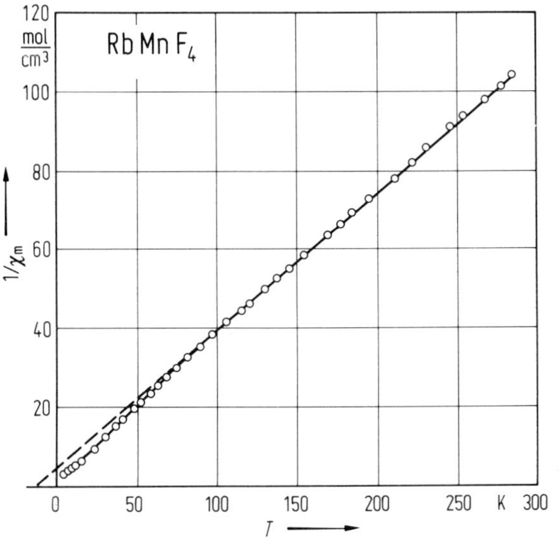

Fig. 9. $RbMnF_4$, powder. Reciprocal molar magnetic susceptibility χ_m^{-1} vs. temperature T [78K].

$CsMnF_4$

Magnetic data Figs. 10 and 11, showing a field-dependent magnetisation with $T_C = 21$ (2)K [78K, 80M]. Neutron diffraction patterns indicate $T_C = 18.9$ (5)K, within the magnetically-ordered state the magnetic moments are parallel to the *a* or *b* axis, Fig. 4, with a magnitude of $4.04\mu_B$. However, the *x* and *y* components are not equal [80M]. The inter-layer coupling is also ferromagnetic.

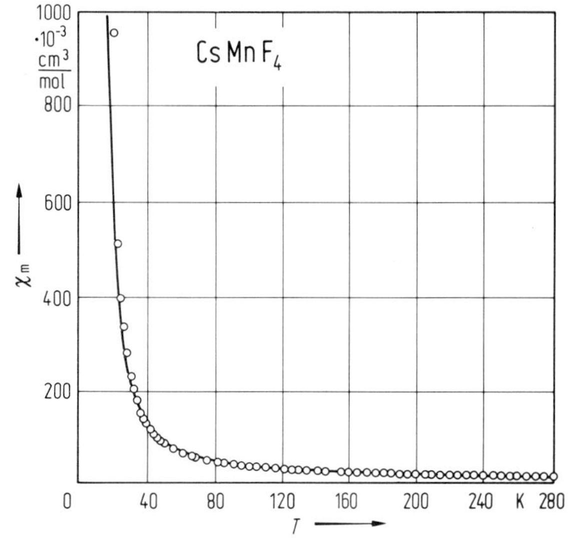

Fig. 10. $CsMnF_4$, powder. Molar magnetic susceptibility χ_m vs. temperature *T*. The curve is calculated for quadratic-layer Heisenberg antiferromagnetism with $J/k = 1.30$ K [93M1].

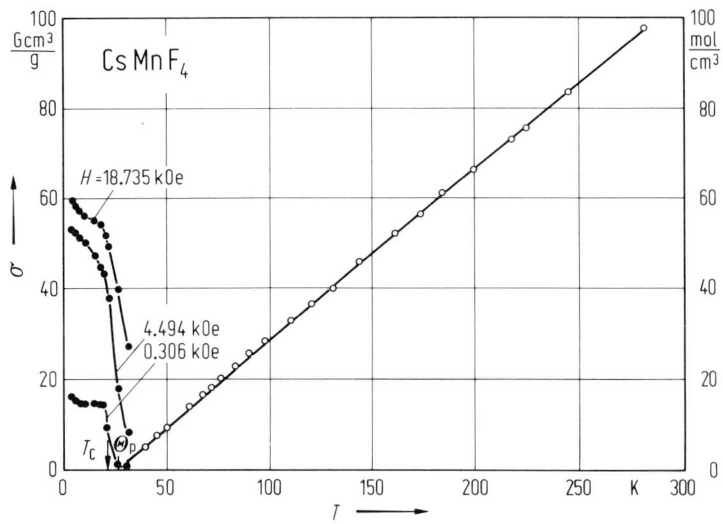

Fig. 11. $CsMnF_4$, powder. Reciprocal molar magnetic susceptibility χ_m^{-1} vs. temperature *T* and magnetisation σ in various magnetic fields at low temperatures [78K].

NH_4MnF_4

Similar to $CsMnF_4$ with T_C = 10 (1)K [78K]. See Fig. 12.

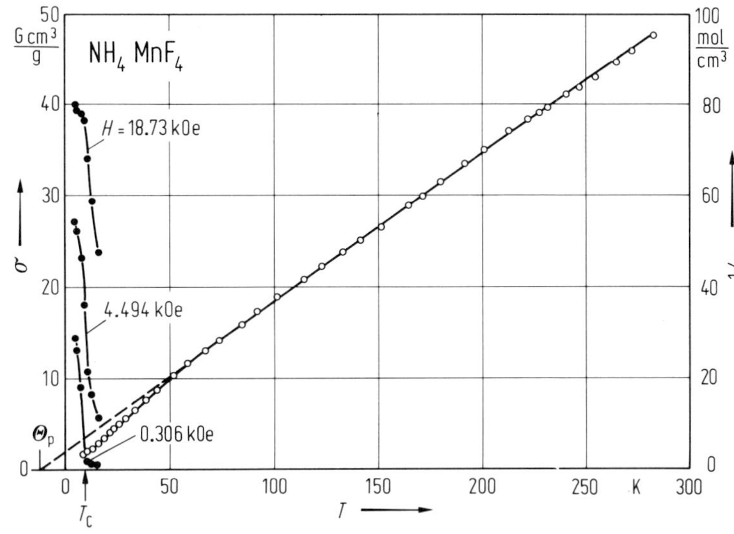

Fig. 12. NH_4MnF_4, powder. Similar to Fig. 11 [78K].

$TlMnF_4$

Susceptibility Fig. 13. Magnetic coupling AF in both layer directions. The inter-layer coupling is ferromagnetic Fig. 4 [92N].

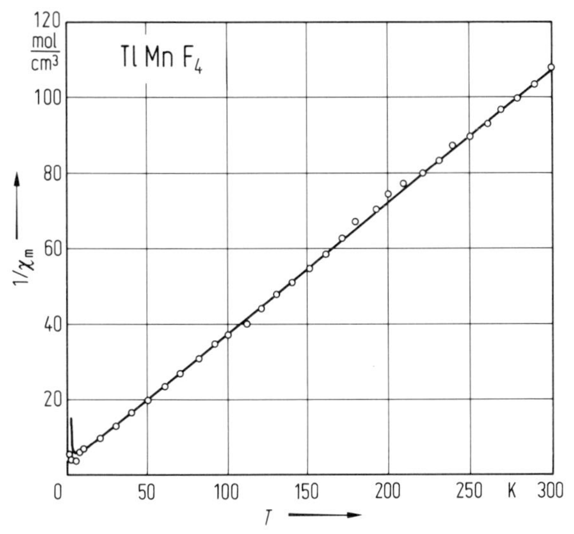

Fig. 13. $TlMnF_4$, powder. Inverse molar magnetic susceptibility χ_m^{-1} vs. temperature T. The solid curve is calculated for $J/k = -0.45$ K [92N].

$NaFeF_4$

Susceptibility vs. temperature curve Fig. 14. Above 450 K Curie-Weiss behavior with $\Theta_p = -325$ K, $C_m = 4.40$ cm^3 Kmol^{-1} and $p_{eff} = 5.93\mu_B$/FU. In the range 450–120 K the curve is characteristic for 2D magnetism still persisting above the ordering temperature T_N. Neutron diffraction spectra at 4.2 K show a magnetic cell doubled along the x axis as a consequence of 3D ordering, the crystallographic unit becoming triclinic. The iron moments are almost parallel to the b axis. They are mutually AF coupled with their four nearest neighbors in the same layer and also with their neighbors in adjacent layers [76D, 77D2]. For Mössbauer data see Tables 6, 7. The variation of the hyperfine field ($H_{hf}(T) = H_{hf}(0) \cdot D(1 - T/T_N)^\beta$), or the magnetisation, in the temperature range $0.6 \le T/T_N \le 1$ can be described with the parameters $D = 1.02$ (2), $T_N = 111.5$ (3) K and $\beta = 0.25$ (1). The direction of H_{hf} is perpendicular to the principal EFG axis.

Table 7. $NaFeF_4$. Hyperfine magnetic field H_{hf} at various temperatures T [76D].

$T \pm 0.2$ [K]	$H_{hf} \pm 3$ [kOe]	$T \pm 0.2$ [K]	$H_{hf} \pm 3$ [kOe]
4.2	562	105	284
70	451	106	266
80	420	107	251
85	397	108	243
90	379	109	225
95	354	110	194
100	327	110.5	171
102	312	111	145
104	295		

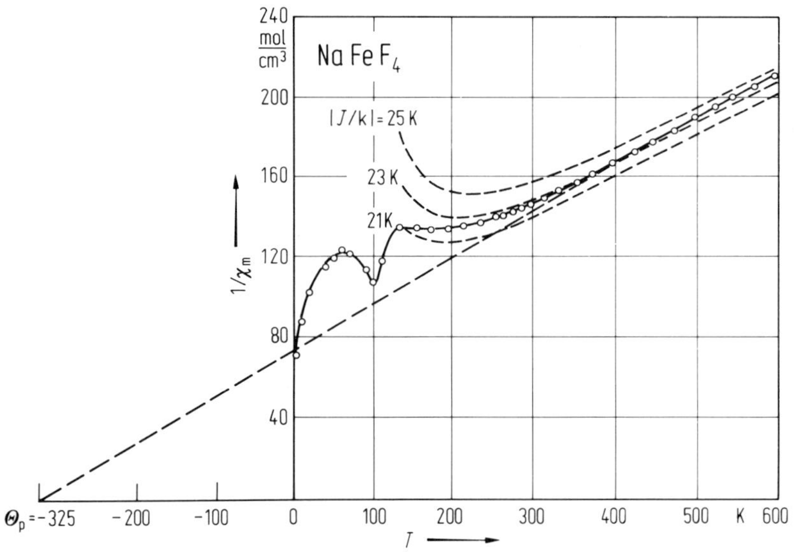

Fig. 14. $NaFeF_4$, powder. Inverse, molar magnetic susceptibility χ_m^{-1} vs. temperature T [76D]. Dashed curves are calculated for three different values of the intrachain exchange constant.

$KFeF_4$

[FeF_6] octahedra form sheets normal to the *c* axis, but puckered along the *a* axis with an angle of 11.65° [71H]. Adjacent magnetic layers are shifted by half a lattice constant, Fig. 15. Susceptibility Fig. 16. The compound represents a real planar antiferromagnet with magnetic moments oriented along the *c* axis. Mössbauer data at 300 K: $\delta = +0.138$ (2) mms^{-1} relative to Pd (^{57}Co) at liquid N_2 temperature, see also Fig. 17. Quadrupole splitting 1.460 (5) mms^{-1}, almost independent of temperature, see also Table 6 and Fig. 18. The hyperfine field $H_{hf}(0) =$ 540 (4) kOe. Principal axis of EFG tensor tilted 11 (2)° with respect to H_{hf} direction [72H, 93M1]. See also Fig. 19. In the asymptotical critical region below $T_N = 135.786$ (6) K the temperature dependence of H_{hf} can be described with the static critical exponent $\beta = 0.151$ (3) for $t = 1 - T/T_N$ values: $3.8 \cdot 10^{-4} < t < 5.7 \cdot 10^{-2}$ [83 K], characteristic for a 2D critical behaviour, and considerably different from the non-asymptotic values reported earlier [72H, 72E1]; see also Table 5. For critical spin fluctuations investigated by Mösssbauer experiments just above T_N, see [83K and 88C].

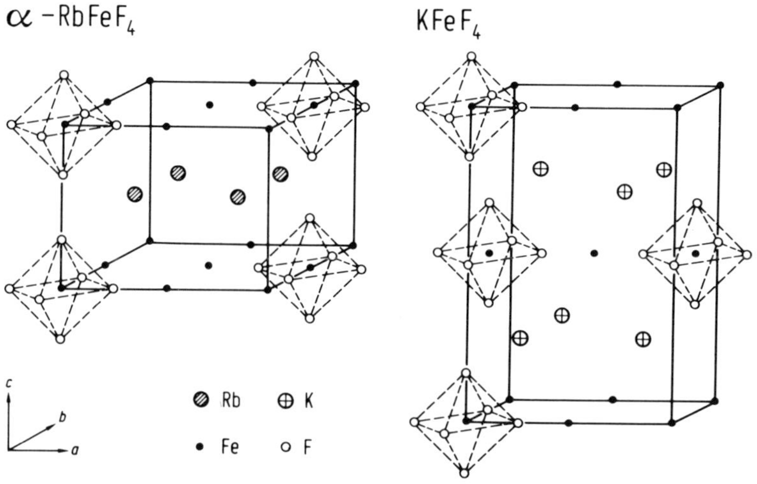

Fig. 15. $KFeF_4$, α-$RbFeF_4$. Idealised crystal structures. For clarity only a few [FeF_6] octahedra are shown, without distortion or tilting [83K].

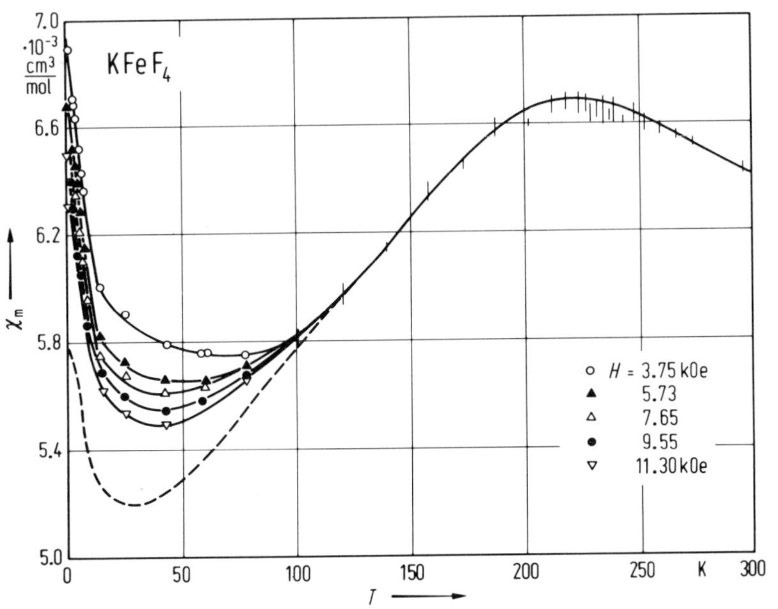

Fig. 16. $KFeF_4$, powder. Molar magnetic susceptibility χ_m vs. temperature T in various magnetic fields H. The field dependence will be caused by magnetic impurities, the dashed curve is an extrapolation avoiding their influence [71H].

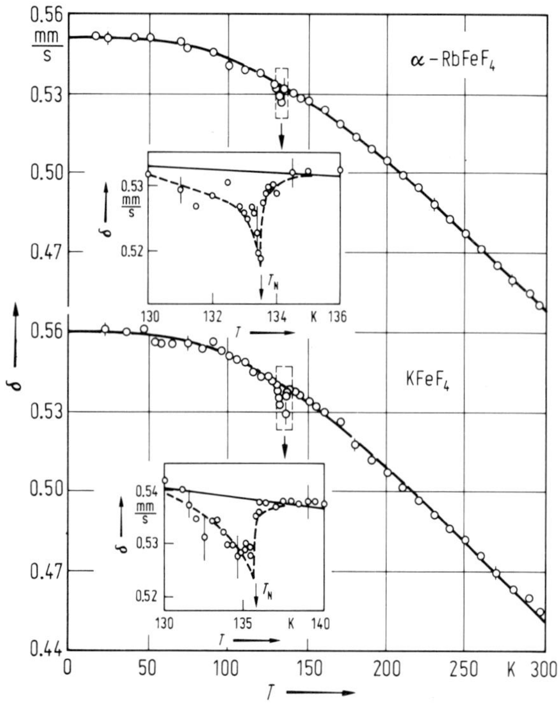

Fig. 17. $KFeF_4$, α-$RbFeF_4$, single crystals. Temperature dependence of the isomer shift δ (relative to metallic iron at RT). The solid lines correspond to Debye fits [83K]; for $KFeF_4$, see also [72H].

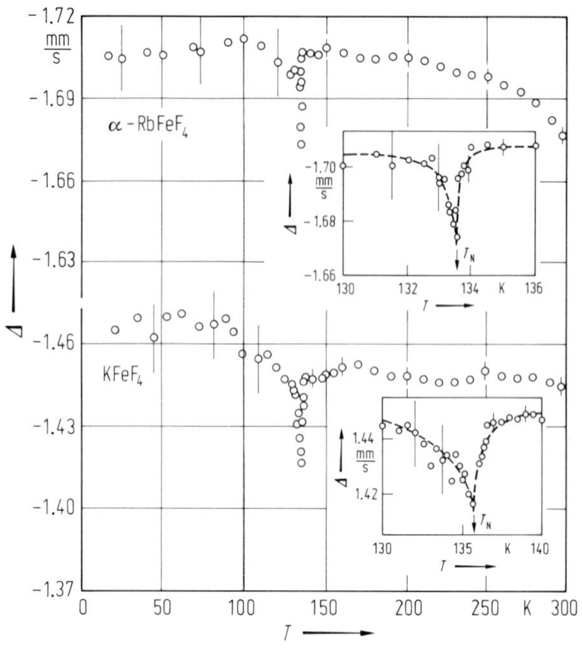

Fig. 18. $KFeF_4$, α-$RbFeF_4$, single crystals. Quadrupole splitting Δ vs. temperature T as derived from Mössbauer spectra. Inset: enlarged detail [83K].

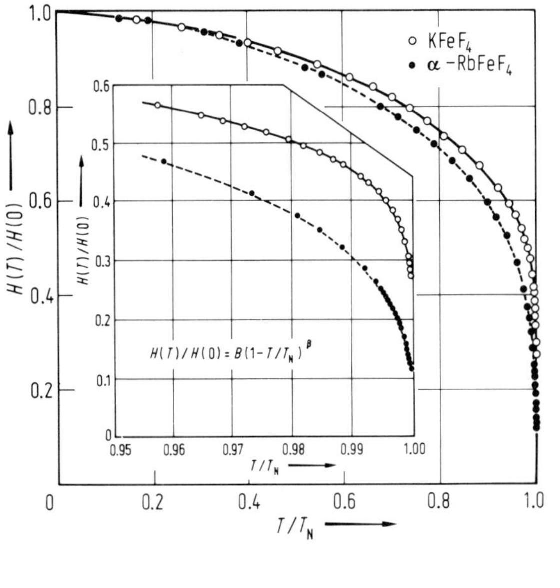

Fig. 19. $KFeF_4$, α-$RbFeF_4$, single crystals. Reduced hyperfine field $H_{hf}(T)/H_{hf}(0)$ vs. reduced temperature T/T_N. $H(0) = 540(4)$ kOe for $KFeF_4$ and 537 (4) kOe for α-$RbFeF_4$. The critical region is expanded [83K]; for $KFeF_4$, see aslo [72E1, 72H].

α-RbFeF₄

For a survey of the many crystallographic phases and for a Raman study of the corresponding phase transitions, see [90M2, 90P]. Idealised crystal structure Fig. 15. The octahedra $[FeF_6]$ are tilted by $\approx 18°$ with respect to the [001] axis, which is perpendicular to the layers [69T]. Magnetic susceptibility Fig. 20. Paramagnetic moment $p_{eff} = 5.25\mu_B$/FU. Magnetic structure similar to $CsFeF_4$, Fig. 21. Mössbauer parameters Table 6 and Fig. 22. The presence of only one single type of iron site is found. In the region just below $T_N = 133.568$ (5) K the critical exponent β is determined also by [83K], with β = 0.316 (5), indicating 3D critical behavior; see also Table 5. For ferroelastic effects, see [72A].

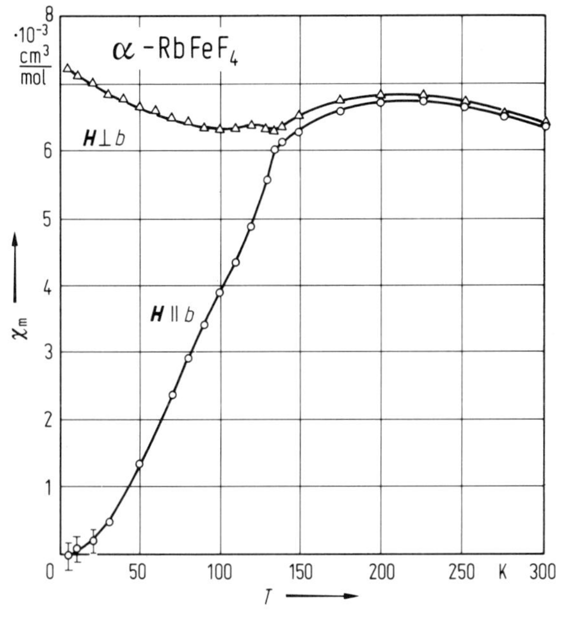

Fig. 20. α-$RbFeF_4$, single crystal. Molar magnetic susceptibility χ_m vs. temperature T for two directions of the magnetic field with respect to the crystallographic b axis. Magnetic layers perpendicular to the b axis [71E].

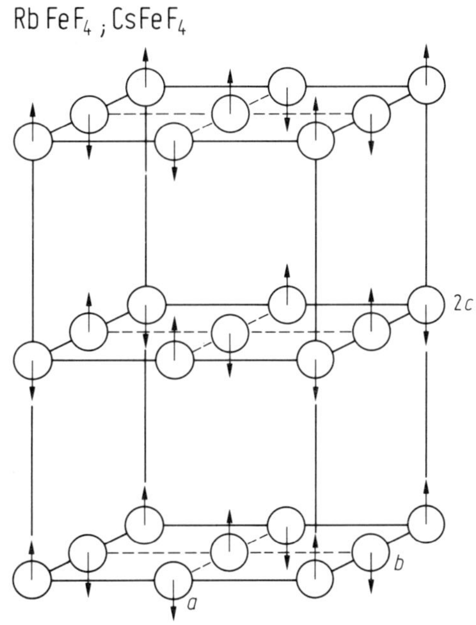

Fig. 21. $RbFeF_4$; $CsFeF_4$, powder samples. Schematic 3-dimensional antiferromagnetic structure for both α and β phases, as derived from neutron diffraction spectra obtained at 4.2K [74B].

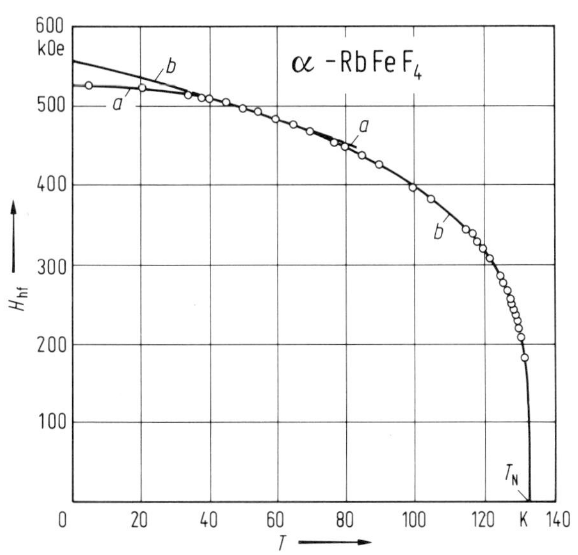

Fig. 22. α-$RbFeF_4$, powder. Magnetic hyperfine field H_{hf} vs. temperature T as derived from Mössbauer spectra. Solid curves are calculated according to (curve a) spin-wave theory for a two-dimensional quadratic-layer antiferromagnet with $J/k = -22$K and (b) critical behavior below T_N according to $H_{hf}(T) = H_{hf}(0) \cdot D(1 - \frac{T}{T_N})^\beta$ with $T_N = 133.40$ (5) K, $D = 1.18$ (1) and $\beta = 0.245$ (5) [71E].

α-$CsFeF_4$

Magnetic susceptibility Fig. 23. Curie-Weiss approximation at high T: $\Theta_p = -390$ K, $C_m = 4.36$ cm^3 Kmol^{-1}. Exchange integral calculated from $\chi_\perp$ at T=0 K: $J/k = -27.7$ (9) K, and from T_{max}: $J/k = -23.7$ (15)K [72E2]. Magnetic structure of $CsFeF_4$ Fig. 21. Mössbauer data at 4.2 K: $H_{hf} = 540$ (2)kOe, $\Delta = -1.54$ (5) mms^{-1}, η=0.3 (3) and δ=0.55 (1) mms^{-1} relative to iron metal at RT. The angle between H_{hf} and EFG principal axis is 5 (4)° [74E]. See also Table 6 and Fig. 24. Critical exponent: β = 0.278 (10) [72E2, 74E]; see also Table 5.

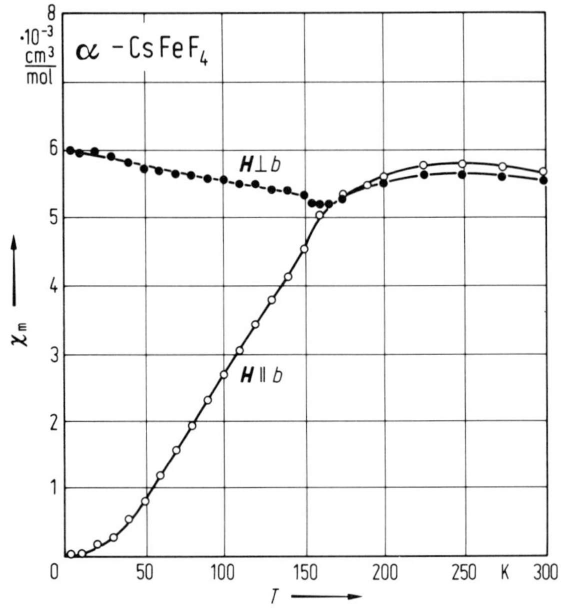

Fig. 23. α-$CsFeF_4$, single crystal. Molar magnetic susceptibility χ_m in two crystallographic directions as a function of temperature T [72E2]

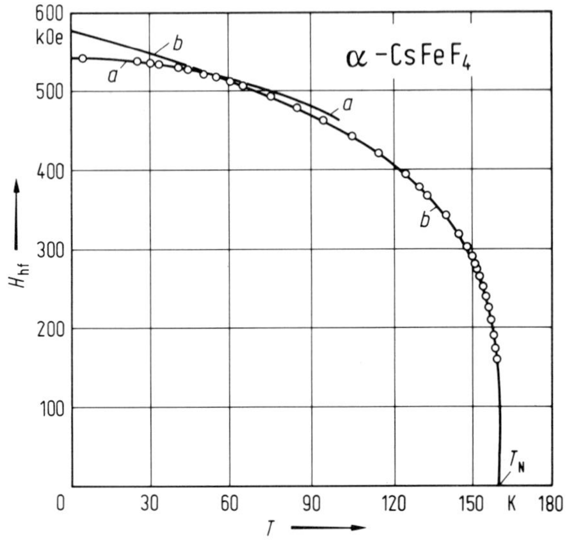

Fig. 24. α-$CsFeF_4$, powder. Magnetic hyperfine field H_{hf} vs. temperature T as derived from Mössbauer spectra. Solid curves calculated (curve *a*) with $J/k = -28.0$ (5) K and (*b*) with β = 0.278 (10); see also Fig. 22 [74E].

NH_4FeF_4

Magnetic susceptibility Fig. 25. Curie-Weiss law above 400K with $\Theta_p = -390$ K, $C_m = 4.45$ cm^3 Kmol^{-1} and $p_{eff} = 5.97$ μ_B/FU. Mössbauer data: Table 6. For the critical exponent is found β=0.26 (1) for the temperature range down to T/T_N=0.57 [77M2].

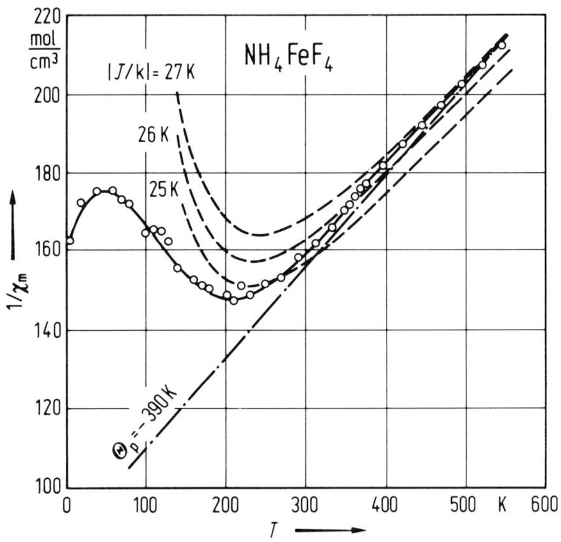

Fig. 25. NH_4FeF_4, powder. Inverse molar magnetic susceptibility χ_m^{-1} vs. temperature T. Dashed curves are calculated for three different values of the intrachain exchange constant J/k [77M2].

9.7.2 $AMF_4 \cdot H_2O$ compounds

The compounds $AMF_4 \cdot H_2O$ have a monoclinic crystal structure with parallel chains of two types of octahedra arrangements. For A = K, Rb, Tl and M = Mn the zig-zag chains consist of alternating $[MnF_6]$ and $[Mn(H_2O)_2F_4]$ octahedra linked by trans-corners. For A = Na, Cs, NH_4 the chains consist of isolated octahedral units linked by hydrogen bridges. Crystal structure data in Tables 8, 9. For exact coordinates of the various structures is referred to the original literature. Below a temperature T_N these weak ferromagnets show antiferromagnetic ordering of the magnetic moments of the M ions with a small canting angle between the magnetic sublattices. For a survey of the magnetic properties, see Table 10.

For Table 8, see next page.

Table 9. $RbMnF_4 \cdot H_2O$; $TlMnF_4 \cdot H_2O$, powders. Crystallographic unit cell parameters derived from neutron diffraction patterns [91N, see also 91P].

Compounds	T [K]	a [Å]	b [Å]	c [Å]	β [°]
$RbMnF_4 \cdot H_2O$	20	13.6775 (4)	6.5865 (2)	10.3890 (3)	103.176 (1)
	1.3	13.677 (2)	6.5841 (8)	10.386 (1)	103.12 (1)
$TlMnF_4 \cdot H_2O$	20	13.838 (7)	6.429 (5)	10.486 (1)	103.97 (1)
	4	13.820 (7)	6.420 (5)	10.469 (1)	103.97 (1)

Table 8. $AMF_4 \cdot H_2O$, $AMF_4 \cdot 2H_2O$. RT crystallographic parameters of the monoclinic unit cell as derived from X-ray diffraction patterns.

Compound	Space group; Z	a [Å]	b [Å]	c [Å]	β [°]	Ref.
$KMnF_4 \cdot H_2O$	C2/c; 8	13.891 (2)	6.212 (1)	10.484 (2)	104.74 (2)	88M
	C2/c; 8	13.907 (1)	6.2136 (2)	10.492 (1)	104.69 (1)	88P2
$RbMnF_4 \cdot H_2O$[b])	C2/c; 8	13.932 (2)	6.471 (1)	10.635 (1)	105.54 (1)	79K
$NaMnF_4 \cdot 2H_2O$	C2/m, C2 or Cm; 2	8.165 (4)	6.771 (1)	4.968 (2)	114.45 (3)	90M1
$NaMnF_4 \cdot 3H_2O$	C2/c; 8	16.381 (2)	6.676 (2)	11.303 (1)	103.78 (1)	90M1
$CsMnF_4 \cdot 2H_2O$	C2/c[a]); 4	11.907 (2)	6.597 (1)	9.316 (2)	121.77 (1)	77B2
	C2[a]); 4	11.891 (2)	6.589 (1)	10.558 (1)	131.46 (1)	77D3
$TlMnF_4 \cdot H_2O$[b])	C2/c	13.784 (1)	6.631 (1)	10.537 (1)	103.66 (1)	91N
$CsFeF_4 \cdot 2H_2O$	I2/a	9.066	6.731	10.272	104.73	91B
$CsCrF_4 \cdot 2H_2O$	I2/a	8.353	6.691	10.145	104.40	91B
$RbVF_4 \cdot 2H_2O$	B2/b[a])	11.921	9.114	6.343	124.04	76B

[a]) The diffraction patterns can also be described in the space group I2/a according to [91B].
[b]) See also Table 9.

Table 10. $AMnF_4 \cdot H_2O$. Survey of the magnetic properties.

Compounds	Figs.	J/k[a])[K] HTE	SFM	g	T_N [K]	Ref.
$KMnF_4 \cdot H_2O$		− 5.8	− 6.1			86M
	26	− 6.5		2.05		88P2
	27a, 28a				8.3	91P
$RbMnF_4 \cdot H_2O$	29	− 6.2		2.03		86P
			− 5.6			87N
	5 of chap. 9.8					
	27b, 28b				8.5	91P
	30				8.3 (3)	91N
$TlMnF_4 \cdot H_2O$	30, 31	− 3.9		2.05	8.1 (3)	91N

[a]) HTE: intrachain exchange J/k calculated from high-temperature $\chi(T)$ curve, SFM: id. calculated according to the scale Fisher model for approach of H_{hf} to the Curie temperature.

$KMnF_4 \cdot H_2O$

The crystal structure consists of infinite zig-zag chains (parallel to the *ac* plane) of alternating $[MF_6]$ and $[Mn(H_2O)_2F_4]$ octahedra sharing trans-corners. AF ordering below $T_N = 8.3$ K. Magnetic moments parallel to the *ac* plane with a canting angle of about 1.5° [88P1, 91P]. Magnetic susceptibility Fig. 26. Magnetisation curves Fig. 27a. Spontaneous magnetic moment extrapolated to 0 K: p_0 (0) = 0.11 μ_B/FU [91P]. See also Fig. 28a.

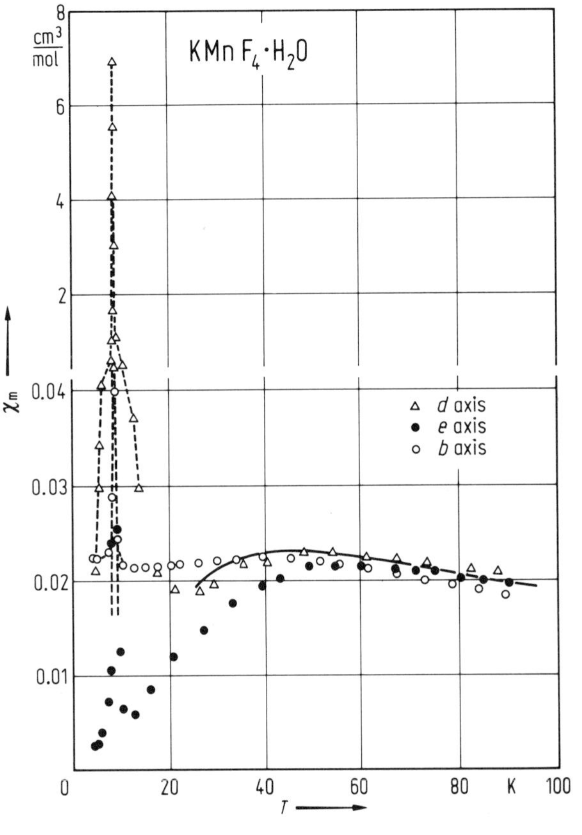

Fig. 26. $KMnF_4 \cdot H_2O$, single crystal. Molar magnetic susceptibility χ_m vs. temperature T along three directions. Directions *d* and *e* in the *ac* plane parallel and perpendicular to the chains, respectively. The *b* axis is perpendicular to the *ac* plane [88P2]. For $T <$ 10K a dashed line is used as guide to the eye. The solid line is theoretical.

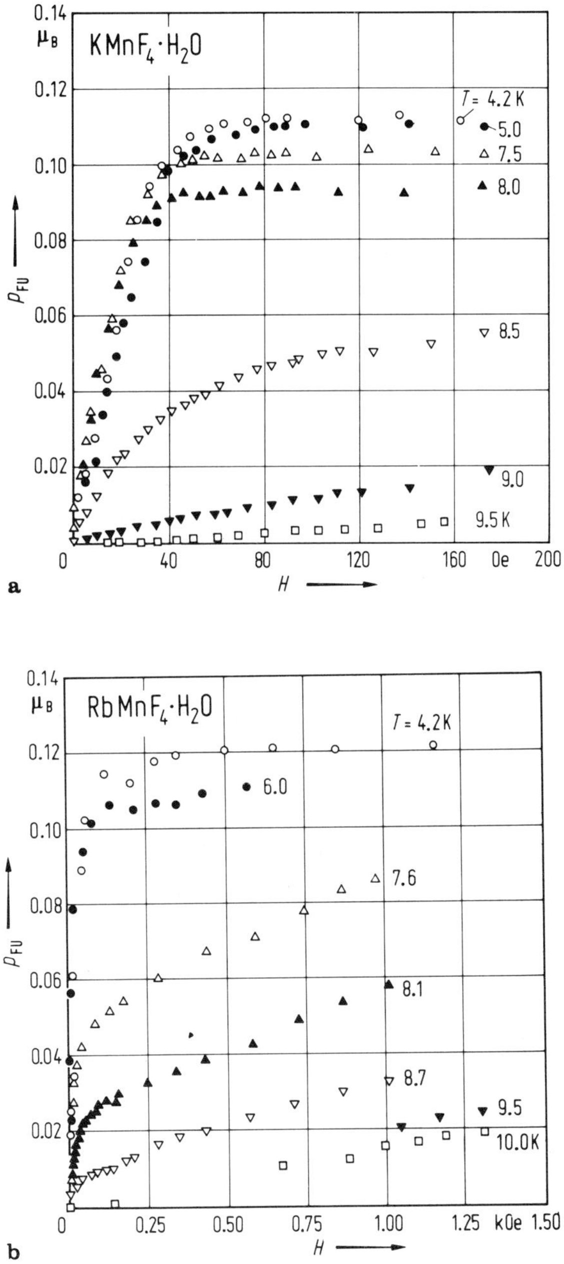

Fig. 27. $KMnF_4 \cdot H_2O$, $RbMnF_4 \cdot H_2O$, single crystals. Magnetic moment per formula unit p_{FU} vs. applied magnetic field H at various temperatures T. Magnetic field applied in the direction of the spontaneous weak ferromagnetism [91P].

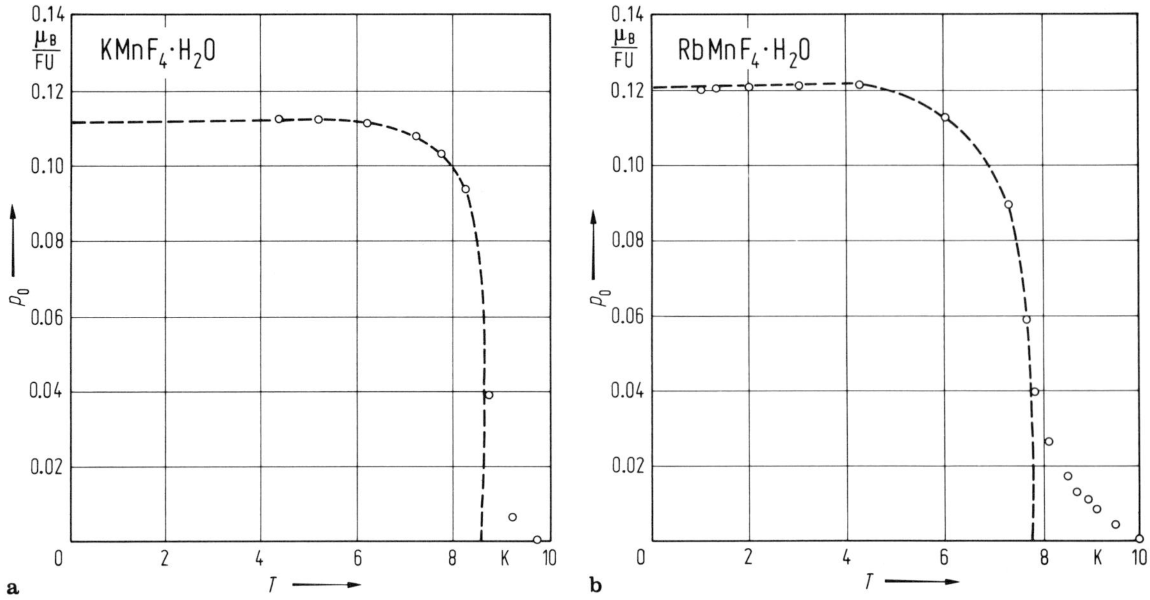

Fig. 28. $KMnF_4 \cdot H_2O$, $RbMnF_4 \cdot H_2O$, single crystals. Spontaneous magnetic moment p_0 vs. temperature T perpendicular to the *ac* plane [91P].

$RbMnF_4 \cdot H_2O$

Crystal structure similar to that of $KMnF_4 \cdot H_2O$. Magnetic susceptibility Fig. 5 in subchapter 9.8 and Fig. 29. Magnetic structure below $T_N = 8.5$K in Fig. 30a. The canting of the AF structure by a few degrees results in weak ferromagnetic behavior along the *b* axis [86P]. Magnetic moments of 2.0 (1)μ_B and 3.35 (5) μ_B along the *a* axis and *c* axis, respectively, corresponding to a total moment of 3.46 (6)μ_B per Mn ion [91N]. Magnetisation curves and spontaneous magnetic moment Fig. 27b and 28b.

For Fig. 29, see next page.

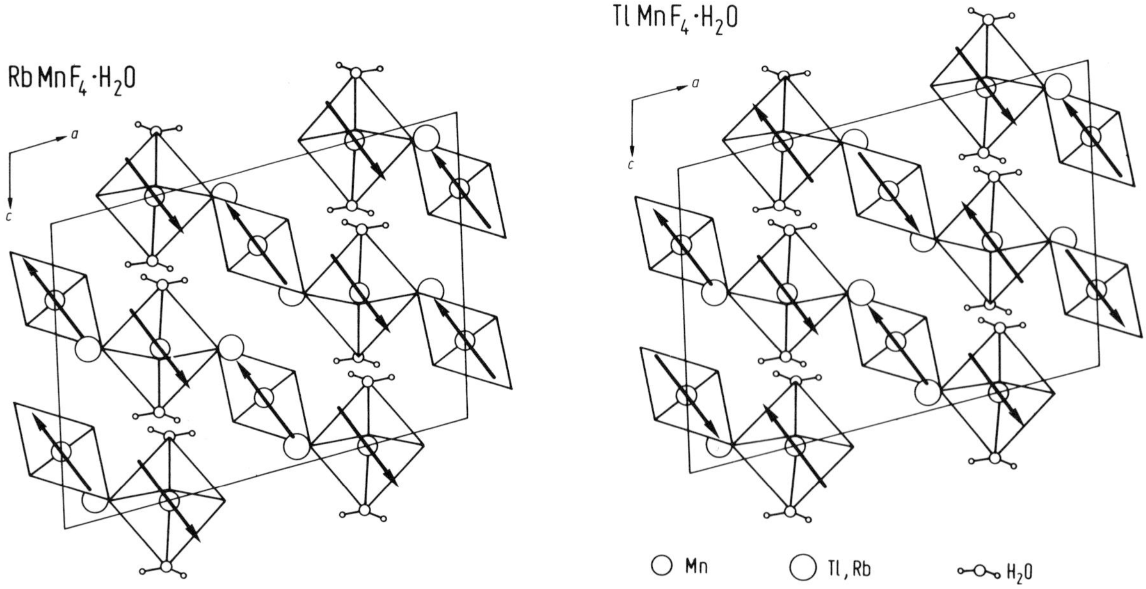

Fig. 30. $RbMnF_4 \cdot H_2O$, $TlMnF_4 \cdot H_2O$. Magnetic structures of both compounds in the AF ordered state. The main difference between the two structures is found in the interchain coupling [91N].

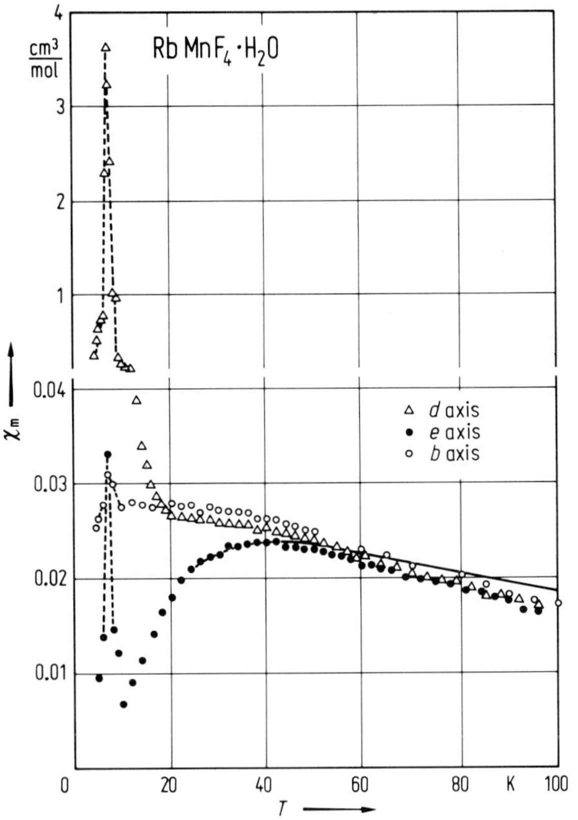

Fig. 29. $RbMnF_4 \cdot H_2O$, single crystal. Similar to Fig. 26 [86P].

$TlMnF_4 \cdot H_2O$

Crystallographic structure similar to that of $KMnF_4 \cdot H_2O$. Magnetic susceptibility Fig. 31. At high temperature a Curie-Weiss law is obeyed with $\Theta_p = -40\,(2)$ K and $C_m = 3.1\ cm^3\ Kmol^{-1}$. Below $T_N = 8.1\,(3)$ K AF ordering, see Fig. 30. Components of magnetic moments of 1.6 (1) μ_B and 3.37 (6) μ_B along the *a* and *c* axis, respectively, corresponding to 3.39 (6) μ_B for a Mn ion [91N].

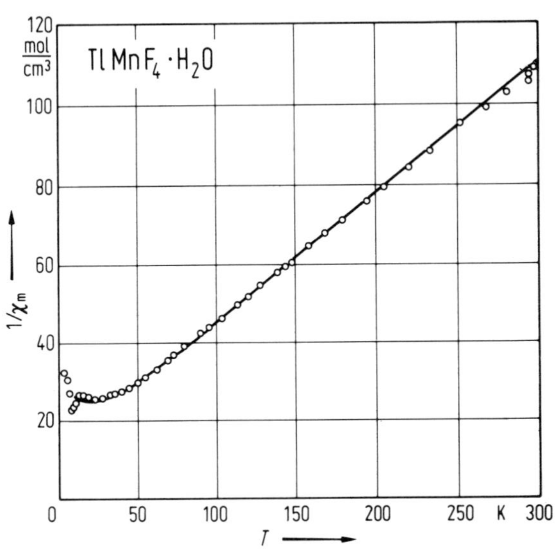

Fig. 31. $TlMnF_4 \cdot H_2O$, powder. Inverse molar magnetic susceptibility χ_m^{-1} vs. temperature *T*. Curve calculated with the parameters $J/k = -3.9$ K and $g = 2.05$ [91N].

9.7.3 References for 9.7

61H Hoppe, R., Liebe, W., Dähne, W.: Z. Anorg. Allg. Chem. **307** (1961) 276.

68B Babel, D.: Z. Naturforsch. **23**a (1968) 1417.

68K Kozak, A. de, Cousseins, J.-C.: C.R. Acad. Sci. (Paris) C**267** (1968) 74.

69T Tressaud, A., Galy, J., Portier, J.: Bull. Soc. Fr. Minéral. Cristallogr. **92** (1969) 335.

70T Tressaud, A., Portier, J., Pape, R. de, Hagenmuller, P.: J. Solid State Chem. **2** (1970) 269.

71E Eibschütz, M., Guggenheim, H.J., Holmes, L.: J. Appl. Phys. **42** (1971) 1485.

71H Heger, G., Geller, R., Babel, D.: Solid State Commun. **9** (1971) 335.

72A Abrahams, S.C., Bernstein, J.L.: Mater. Res. Bull. **7** (1972) 715.

72E1 Eibschütz, M., Davidson, G.R., Guggenheim, H.J.: Am. Inst. Phys. Conf. Proc. **5** (1972) 679.

72E2 Eibschütz, M., Guggenheim, H.J. Holmes, L., Bernstein, J.L.: Solid State Commun. **11** (1972) 457.

72H Heger, G., Geller, R.: Phys. Status Solidi b**53** (1972) 227.

72K Kündig, W., Denison, A.B., Rüegsegger, P.: Phys. Lett. **42**A (1972) 199.

73C Cretenet, J.-C.: Rev. Chim. Minérale **10** (1973) 399.

74B Babel, D., Wall, F., Heger, G.: Z. Naturforsch. **29**b (1974) 139.

74E Eibschütz, M., Davidson, G.R., Guggenheim, H.J.: Phys. Rev. B**9** (1974) 3885.

75K Kozak, A. de, Samouël, M.: J. Less-Common Met. **40** (1975) 185.

76B Buckveckij, B.V. Muradjan, L.A., Davidovic, R.L., Simonov, V.I.: Koord. Khim. **2** (1976) 1129 (in Russian).

76D Dance, J.-M., Sabatier, R., Ménil, F., Wintenberger, M., Cousseins, J.-C., Flem, G. Le, Tressaud, A.: Solid State Commun. **19** (1976) 1059.

76O Omaly, J., Batail, P., Grandjean, D., Avignant, D., Cousseins, J.-C.: Acta Crystallogr. B**32** (1976) 2106.

76R Rush, J.D., Simopoulos, A., Thomas, M.F., Wanklyn, B.M.: Solid State Commun. **18** (1976) 1039.

77B1 Bukovec, P., Šiftar, J.: Monatshefte für Chemie **108** (1977) 387.

77B2 Bukovec, P., Kaučič, V.: J. Chem. Soc. Dalton Trans. (1977) 945.

77D1 Dewan, J.C., Edwards, A.J.: J. Chem. Soc. Chem. Commun. (1977) 533.

77D2 Dance, J.M., Ménil, F., Hanžel, D., Sabatier, R., Tressaud, A., Flem, G. Le, Hagenmuller, P.: Physica **86-88**B (1977) 699.

77D3 Dubler, E., Linowsky, L., Matthieu, J.-P., Oswald, H.-R.: Helv. Chim. Acta **60** (1977) 1589.

77G Guillou, C.le, Zinn-Justin, J.: Phys. Rev. Lett. **39** (1977) 95.

77M1 Massa, W.: Inorg. Nucl. Chem. Lett. **13** (1977) 253.

77M2 Ménil, F., Tressaud, A., Sabatier, R., Flem, G. Le: Mater. Res. Bull. **12** (1977) 983.

78B Babel, D., Knoke, G.: Z. Anorg. Allg. Chem. **442** (1978) 151.

78K Köhler, P., Massa, W., Reinen, D., Hofmann, B., Hoppe, R.: Z. Anorg. Allg. Chem. **446** (1978) 131.

79K Kaučič, V., Bukovec, P.: J. Chem. Soc. Dalton Trans. (1979) 1512.

80M Massa, W., Steiner, M.: J. Solid State Chem. **32** (1980) 137.

81S Savić, I.M., Keller, H., Kündig, W., Meier, P.F.: Phys. Lett. **83**A (1981) 471.

82F Fleischer, T., Hoppe, R.: Z. Naturforsch. **37**b (1982) 1132.

83K Keller, H. Savić, I.M.: Phys. Rev. B**28** (1983) 2638.

86M Massa, W., Pebler, J.: 3rd Eur. Conf. Solid State Chem. Regensburg (1986) Abstracts 1, 175.

86P Palacio, F., Andrés, M., Horne, R., Duyneveldt, A.J. van: J. Magn. Magn. Mater. **54–57** (1986) 1487.

87N Nuñez, P., Darriet, J., Bukovec, P., Tressaud, A., Hagenmuller, P.: Mater. Res. Bull. **22** (1987) 661.

87P Pebler, J., Massa, W., Lass, H., Ziegler, B.: J. Solid State Chem. **71** (1987) 87.
87W Wandner, K.-H., Hoppe, R.: Z. Anorg. Allg. Chem. **546** (1987) 113.
88C Chadwick, J., Thomas, M.F., Johnson, C.E., Jones, D.H.: J. Phys. C: Solid State Phys. **21** (1988) 6159.
88M Massa, W., Baum, G., Drueeke, S.: Acta Crystallogr. C**44** (1988) 167.
88P1 Palacio, F., Andrés, M., Noort, D. van, Duyneveldt, A.J. van: J. Phys. (Paris) **49**-C8 (1988) 819.
88P2 Palacio, F., Andrés, M., Esteban-Calderon, C., Martinez-Ripoll M., Garcia-Blanco, S.: J. Solid State Chem. **76** (1988) 33.
90M1 Massa, W., Schmidt, R.E.: Z. Naturforsch. **45**b (1990) 593.
90M2 Morón, M.C., Bulou, A., Pique, C., Fourquet, J.L.: J. Phys.: Condens. Matter **2** (1990) 8269.
90P Pique, C., Bulou, A., Morón, M.C., Burriel, R., Fourquet, J.L., Rousseau, M.: J. Phys.: Condens. Matter **2** (1990) 8277.
91B Bentrup, U., Massa, W.: Z. Anorg. Allg. Chem. **593** (1991) 207.
91M Molinier, M., Massa, W., Khairoun, S., Tressaud, A., Soubeyroux, J.L.: Z. Naturforsch. **46**b (1991) 1669.
91N Nuñez, P., Tressaud, A., Hahn, F., Massa, W., Babel, D., Boireau, A., Soubeyroux, J.L.: Phys. Status Solidi a**127** (1991) 505.
91P Palacio, F., Andrés, M., Rodriguez-Carvajal, J., Pannetier, J.: J. Phys.: Condens. Matter **3** (1991) 2379.
92M Molinier, M., Massa, W.: Z. Naturforsch. **47**b (1992) 783.
92N Nuñez, P., Tressaud, A., Grannec, J., Hagenmuller, P., Massa, W., Babel, D., Boireau, A., Soubeyroux, J.L.: Z. Anorg. Allg. Chem. **609** (1992) 71.
93M1 Molinier, M., Frommen, C., Massa, W., Pebler, J.: Z. Naturforsch. **48**a (1993) 1054.
93M2 Morón, M.C., Palacio, F., Rodriguez-Carvajal, J.: J. Phys.: Condens. Matter **5** (1993) 4909.
93M3 Morón, M.C., Palacio, F., Rodriguez-Carvajal, J.: J. Appl. Phys. **73** (1993) 6540

9.8 A_2MX_5 and $A_2MX_5 \cdot H_2O$ compounds (A = Li, Na, K, Rb, Cs, Tl, NH_4; M = 3d element; X = F, Cl, Br)

(For symbols and abbreviations see Section 9.6.0.2.)

9.8.1 Introduction

The compounds of the groups A_2MX_5 and $A_2MX_5 \cdot H_2O$ have mostly the orthorhombic crystal symmetry. The structure consists of (MX_6) or (MX_5O) octahedra, respectively, containing the magnetic M ions in their center. These octahedra can be ordered in infinite zig-zag chains. The chains are either formed by cis-linked or by trans-linked octahedra, i.e. with shared adjacent or shared opposite vertices, respectively. The chains are separated from each other by the large A ions. In a chain, the magnetic moments of the M-ions are antiferromagnetically coupled, as expressed by an intrachain exchange constant J. The spins of the M^{3+} ions in adjacent chains are mostly only relatively weakly coupled, as expressed by an interchain exchange constant J'. The result is mostly a quasi one-dimensional magnetic structure. The ratio J'/J gives a measure for the degree of one-dimensionality of the magnetic system. Also the magnetocrystalline anisotropy has an influence on the magnetic behaviour of the system. The magnetic ordering temperature T_N will be higher in a system with higher anisotropy (e.g. Ising system) than in the case of zero anisotropy (Heisenberg system). An externally applied magnetic field can also be considered as an additional source of magnetic anisotropy with its own influence on the magnetic ordering temperature.

In the case of a Heisenberg system the intrachain exchange constant J can be derived from a series expansion of the inverse susceptibility versus temperature curve above the magnetic ordering temperature [64F, 68S, 68W]. An analysis of experimental data of the compounds with M = Mn and X = Cl seems to indicate that the absolute value of J/k for such compounds depends more on the Mn–F–Mn bridge angle than on the Mn–Mn or Mn–F distances [87P]. The ratio J'/J has been calculated for the Heisenberg model as being dependent on the ratio T_N/C_m, where T_N is Neél temperature and C_m the Curie-Weiss constant [64O].

An idea of the dimensionality of the magnetic order can also be obtained from the temperature dependence of the magnetic hyperfine field at temperatures just below the ordering temperature [52C]: $H_{hf} \propto (1 - T/T_N)^{\beta}$. The exponent $\beta = 0.125$ for pure chain magnetism and 0.33 for 3D magnetism.

9.8.2 A_2MX_5 compounds

9.8.2.1 Crystallographic properties of A_2MX_5 compounds

A survey of the crystallographic properties of the A_2MX_5 compounds is given in Table 1. The K_2FeF_5 structure, Fig. 1, presents zig-zag chains of cis corner sharing $(FeF_6)^{3-}$ octahedra running along the c axis. The chains are separated by the K ions. The same type of structure applies to α-$(NH_4)_2FeF_5$. The Rb_2FeF_5 and Rb_2CrF_5 compounds consist of $(FeF_6)^{3-}$ and $(CrF_6)^{3-}$ octahedra, respectively, which share two adjacent vertices parallel to the b axis, thus forming chains isolated by the Rb ions.

The compounds $(NH_4)_2MnF_5$ and β-$(NH_4)_2FeF_5$ consist of trans-linked zig-zag chains of octahedra parallel to the b axis and isolated by $(NH_4)^+$ ions (Fig. 2).

Table 1. A_2MF_5. RT crystallographic parameters of the orthorhombic cell as derived from X-ray diffraction data.

Compound	Space group; character of linear chains Z	Lattice parameters [Å]			Ref.
		a	b	c	
K_2CrF_5	$Pna2_1$; ‖c Z = 16	7.37	12.84	19.60	69K, 70T1
Rb_2CrF_5	Pnma; cis ‖b Z = 4	7.515(5)	5.724(4)	11.985(6)	74J^{a})
$(NH_4)_2MnF_5$^{f})	Pnma; trans ‖b Z = 4	6.20(3)	7.94(1)	10.72(1)	69S^{a})
Li_2MnF_5	C2/c; trans ‖b Z = 4	7.408(1)	4.948(1)	10.016(1)	87P^{a})
Na_2MnF_5	monoclinic^{e}); trans‖a $P2_1/c$	7.719(1)	5.236(1)	10.862(2)	86M^{a})
Rb_2MnF_5	P4/mmn or P4/m Z = 1	6.103		4.14	78G2
Cs_2MnF_5	P4/mmm; trans ‖c Z = 1	6.420(1)		4.229(1)	90H^{a})
K_2FeF_5	$Pna2_1$; cis ‖b Z = 16^{b})	7.399(5)	5.10(1)^{b})	12.84(1)	77V^{a})
	Pbcn, Z = 16	7.4059(4)	20.4282(13)	12.8771(9)	90L^{a})
Rb_2FeF_5	Pnma; cis ‖b Z = 4	7.535	5.785	11.983	80D, 81T
Cs_2FeF_5	unknown; cis ‖b Z = 4	7.84	5.95	12.567	80D
$N_2H_6FeF_5$	unknown; cis Z = 32^{c})	7.8	5.05	13.5	77H
α-$(NH_4)_2FeF_5$^{f})	Pbcn; cis ‖c Z = 8^{d})	7.6223(5)	6.7204(8)^{d})	10.4167(6)	89F
β-$(NH_4)_2FeF_5$	Pnma; trans ‖b Z = 4	6.3379(6)	7.6165(4)	11.0305(9)	89F

^{a}) Detailed data on atomic positions given in this reference.
^{b}) $4b = 20.39(1)$ Å with 16 formula units per crystallographic unit cell.
^{c}) $2a = 15.6$ Å, $4b = 21.8$ Å, with 32 formula units per crystallographic unit cell.
^{d}) $2b = 13.4408(8)$ Å, with 8 formula units per crystallographic unit cell.
^{e}) $\beta = 108.99(1)°$ [86M].
^{f}) For preliminary data of $(NH_4)_2Mn_{1-x}Fe_xF_5$ see [87M].

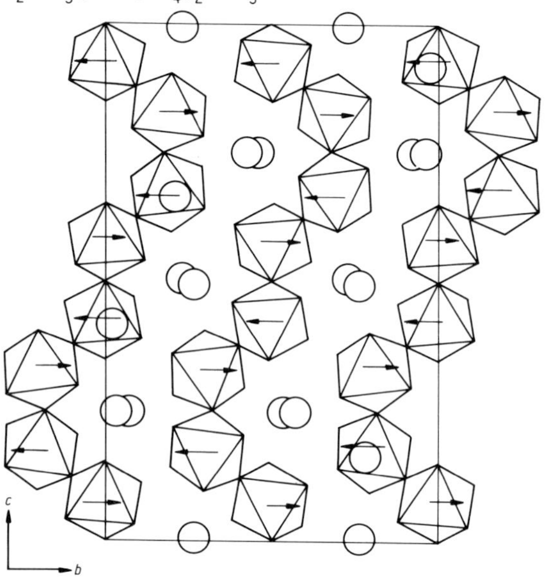

Fig. 1. K_2FeF_5; α-$(NH_4)_2FeF_5$. Cis-linked chains of $(FeF_6)^{3-}$ octahedra isolated by K (or N) ions, the latter represented by the circles. Shown is the projection on the (1 0 0) plane of the position of the ions. The arrows represent the orientation of the magnetic moments of the Fe^{3+} ions [89F].

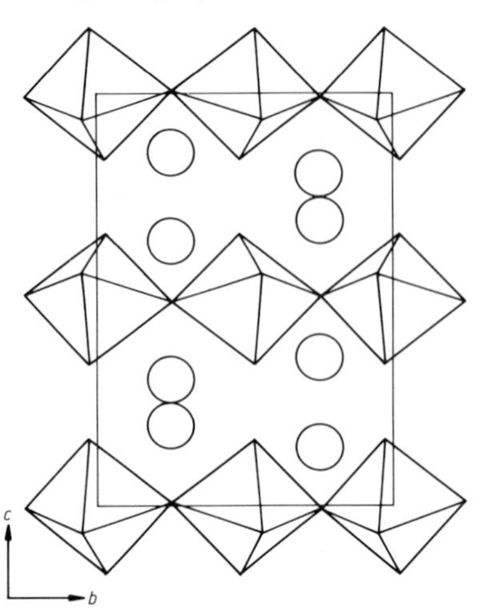

Fig. 2. $(NH_4)_2MnF_5$; β-$(NH_4)_2FeF_5$. Trans-linked chains of $(FeF_6)^{3-}$ or $(MnF_6)^{3-}$ octahedra isolated by $(NH_4)^{1+}$ ions. The circles represent the N atoms included in the unit cell. Shown is the projection of the positions of the ions on the (1 0 0) plane [89F].

9.8.2.2 Magnetic properties of A_2MF_5 compounds

For a survey see Table 2.

Table 2. A_2MF_5. Survey of the magnetic properties. Θ_p, C_m and p_{eff}: Curie–Weiss law constants. T_N: Néel point derived from Mössbauer spectra (ME) or from neutron diffraction spectra (NS). Exchange parameters J and J' derived from susceptibility versus temperature curves. Magnetic hyperfine field at $T = 0$ K, $H_{hf}(0)$, and magnetic moment of the iron ions in the ordered state $p_{Fe^{3+}}$.

Compound	Θ_p [K]	C_m [cm^3 K mol^{-1}]	p_{eff} [μ_B/FU]	T_N [K]		$-J/k$ [K]	J'/J [$\cdot 10^{-3}$]	$H_{hf}(0)$ [kOe]	$p_{Fe^{3+}}$ [μ_B] (at 0 K)	Figs; Tables	Ref.
Li_2MnF_5	− 66		4.58			6.15(15)				3	87P
Na_2MnF_5	− 91		4.01			8.25(15)				4	87P
Rb_2MnF_5	− 400	4.29				22.6				6	87N
Cs_2MnF_5	− 345	4.29				19.4				7	87N
$(NH_4)_2MnF_5$	− 153		4.00	7.5		10.6				8,9	
	[87P]		[87P]	[73K]		[87P]					
K_2FeF_5	− 125	4.39	5.93	6.95	(ME)	9.45	3.8	410(5)	3.0(1)	10⋯17;	
	[79S]	[79S]	[79S]		[86P]	[79S]	[80D]	[77G]	[79S]	Tab. 3	
				7.093	(ME)						
					[88C]						
				10	(NS)						
					[80D]						
Rb_2FeF_5	− 125	4.32		9.3(5)	(ME)	8.78	2.4	430(5)	3.5(2)	11,18⋯23;	
	[80D]	[80D]			[78G]	[80D]	[80D]	[78G]	[81T]	Tab. 4	
				8.0(5)	(NS)		⩽ 1				
					[81T]		[78G]				
Cs_2FeF_5	− 135	4.52		5	(NS)	8.01	1.1	340		11; Tab 5	80D
$N_2H_6FeF_5$	− 150	4.42	5.95	9(2)	(ME)	10.2	2			24; Tab. 6	77H
α-$(NH_4)_2FeF_5$				7.5(5)	(ME)			Fig. 25		25; Tab. 7	91C
β-$(NH_4)_2FeF_5$				13.1	(ME)			440		26; Tab. 8	91C

Li_2MnF_5, Na_2MnF_5, Rb_2MnF_5, Cs_2MnF_5

Short-range AF interaction in the Mn–F–Mn linear chains [87P]. Magnetic susceptibilities Figs. 3–7.

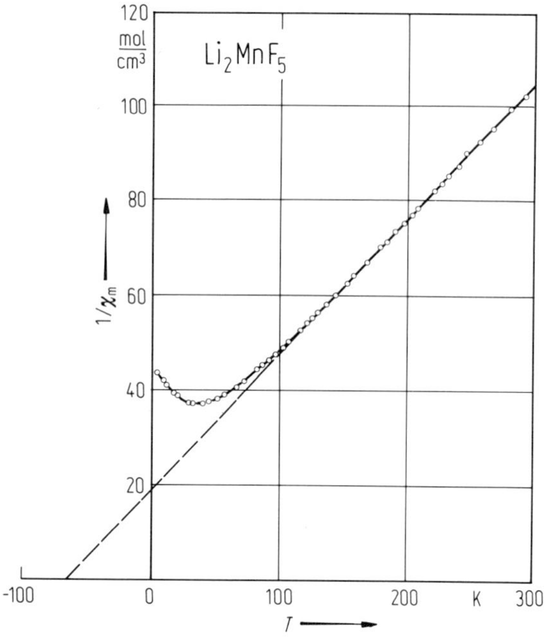

Fig. 3. Li_2MnF_5, powder. Inverse molar magnetic susceptibility χ_m^{-1} vs. temperature T. Straight line represents Curie-Weiss law for data above 150 K. Solid curve is calculated for $J/k = -6.15$ K according to [64F] and [68S]; from [87P].

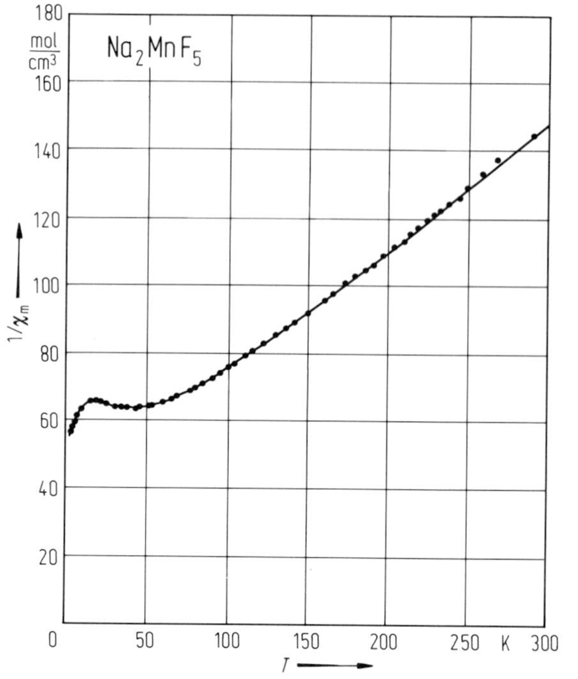

Fig. 4. Na_2MnF_5, powder. Similar to Fig. 3 with $J/k = -8.25(5)$ K [87P].

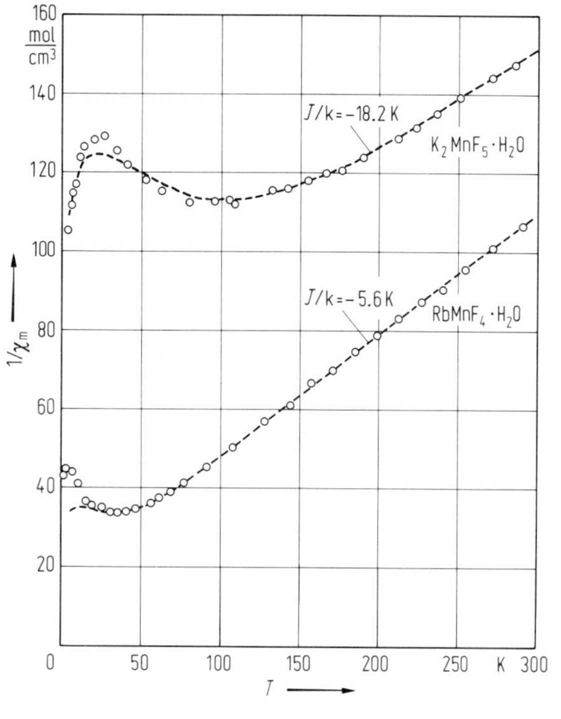

Fig. 5. $K_2MnF_5 \cdot H_2O$, $RbMnF_4 \cdot H_2O$. Similar to Fig. 3 with $J/k = -18.2$ K and -5.6 K, respectively [87N].

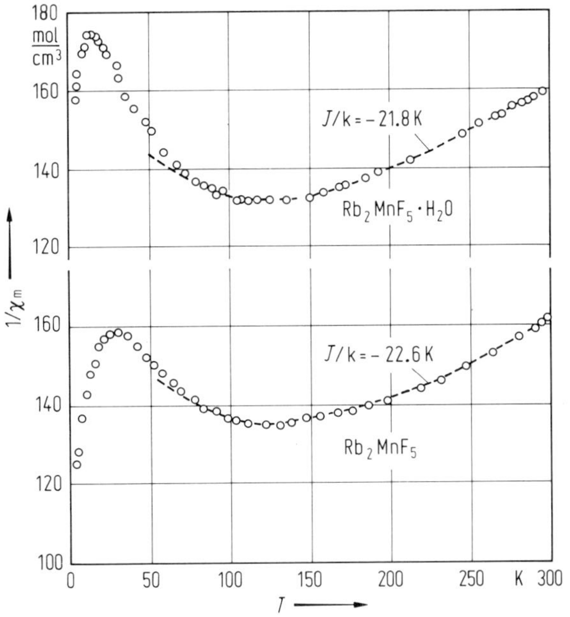

Fig. 6. Rb_2MnF_5, $Rb_2MnF_5 \cdot H_2O$. Similar to Fig. 3 with $J/k = -22.6$ K and -21.8 K, respectively [87N].

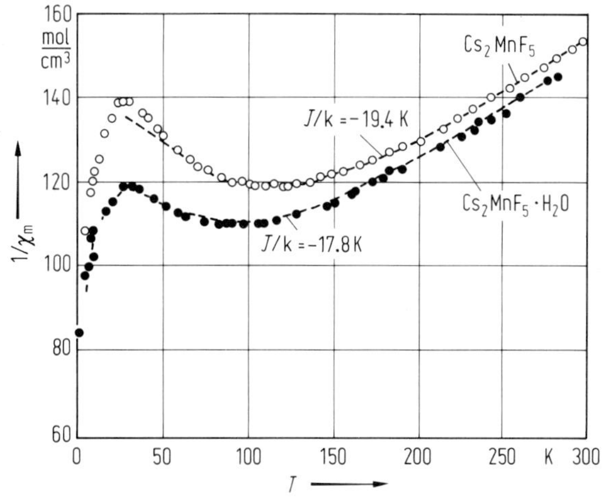

Fig. 7. Cs_2MnF_5, $Cs_2MnF_5 \cdot H_2O$. Similar to Fig. 3 with $J/k = -19.4$ K and -17.8 K, respectively [87N].

$(NH_4)_2MnF_5$

Magnetic susceptibility Fig. 8, $T_N = 7.5$ K. Magnetisation measurements at 4.2 K show hysteresis and residual magnetisation. 3D long range ordering with spins parallel to the *b* axis, with a small component of about 0.0045 in fraction along the *a* axis [73K]. Specific heat Fig. 9.

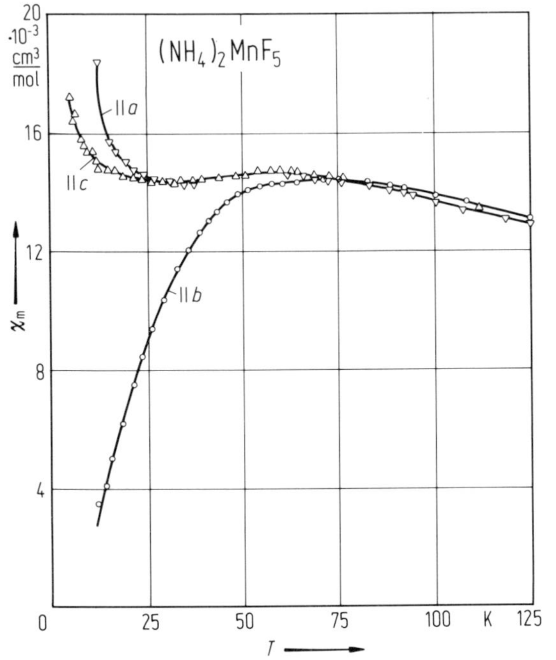

Fig. 8. $(NH_4)_2MnF_5$, single crystal. Molar magnetic susceptibility χ_m vs. temperature T for the three principal crystallographic directions [73K].

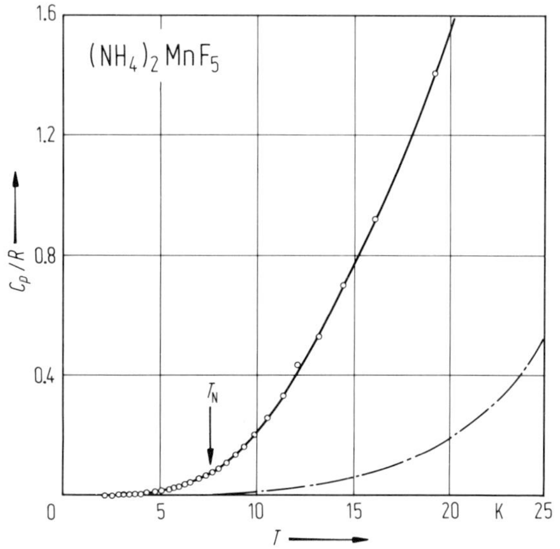

Fig. 9. $(NH_4)_2MnF_5$. Specific heat C_p expressed in units R vs. temperature T. Broken curve is the magnetic contribution calculated from an Ising linear chain model [73K].

K_2FeF_5

Paramagnetic properties Figs. 10, 11. Above 200 K a Curie–Weiss law is obeyed with $\Theta_p = -125$ K, $C_m = 4.39$ cm³ Kmol⁻¹ and $p_{eff} = 5.93\ \mu_B$/FU [79S]. At temperatures well above T_N normal isotropic paramagnetism is found. However at temperatures just above T_N the susceptibility is slightly anisotropic, and in the Mössbauer spectra considerable line broadening is found. This may be related to the correlation of the spins in one dimension (one chain) above T_N and spin relaxation effects associated with the gradual formation of three-dimensional ordered clusters when the temperature is decreased to T_N [82C]. From a Mössbauer analysis of similar critical fluctuations just below the transition temperature a value $T_N = 7.093$ K with $\beta = 0.32$ is derived [88C]. The type of the quasi one-dimensional antiferromagnetic ordering below $T_N = 10$ K as found by neutron diffraction is illustrated in Fig. 1, corresponding to the A_y mode of the Γ_1 type. The magnitude of the magnetic moments is $p_{Fe^{3+}} = 3.0(1)\ \mu_B$, instead of the free ion value of 5 μ_B [79S]. This deviation is called spin reduction. Mössbauer data: Table 3 and Fig. 12. The high-spin ferric ion has the anomalously low saturation hyperfine field of $H_{hf} = 410(5)$ kOe at 0 K, corresponding to the above mentioned spin reduction [77G]. For the influence of an externally applied magnetic field on H_{hf}, see Figs. 13–15. The critical field for spin-flop in the case of an applied field in the direction of the initial antiferromagnetism is $H_{SF} = 36.5$ kOe, Fig. 16 [86P]. The angle between the magnetic axis and the principal axis of the electric field gradient EFG is 50(5)°, independent of the applied field [78G1]. From the field H_{SF} a ratio $H_A/H_E \simeq 8.5 \cdot 10^{-4}$ can be derived. For the (H, T) phase diagram see Fig. 17. For $H = 100$ kOe, is found $T_N = 8.5$ K and 6.95 K in the case of the field applied along the easy b axis or parallel to the ac plane, respectively [82B].

Table 3. K_2FeF_5, powder. Temperature dependence of the Mössbauer parameters. δ: chemical shift relative to iron metal at RT, Δ: quadrupole splitting. The asymmetry parameter $\eta = 0.57(5)$ [77G].

T [K]	δ [mm s⁻¹]	Δ [mm s⁻¹]
300	0.42(1)	− 0.72(1)
133	0.49(1)	− 0.72(1)
77	0.52(1)	− 0.70(1)
11.2	0.49(1)	− 0.68(4)
4.2	0.53(1)	≈ 0.7[a])

[a]) Because of the value of the angle between the magnetic axis and the principal axis of the electric field gradient, the quadrupole splitting cannot be accurately determined in the magnetic region.

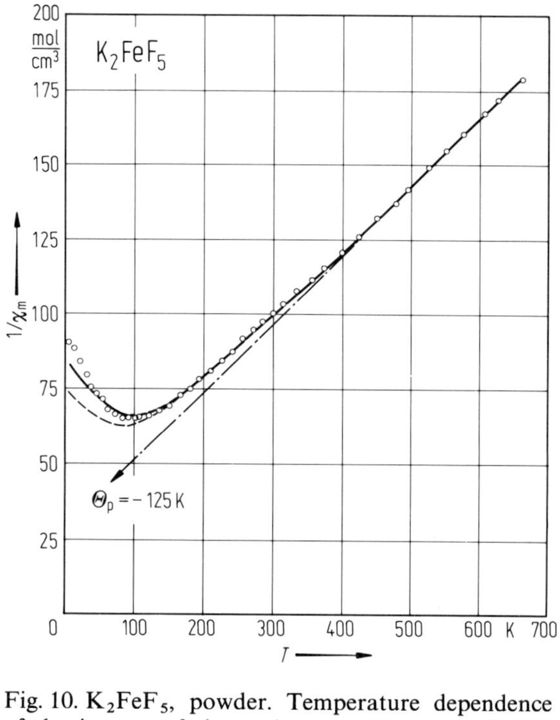

Fig. 10. K_2FeF_5, powder. Temperature dependence of the inverse of the molar magnetic susceptibility χ_m^{-1}, measured in low magnetic fields. The drawn and dashed curves are results calculated for $J/k = -9.45$ K according to [68W] and [68S], respectively. The curve pointing to $\Theta_p = -125$ K is the Curie-Weiss approximation [79S].

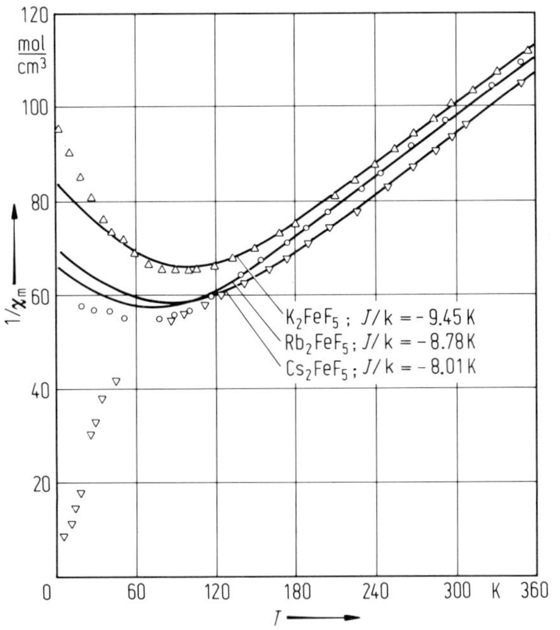

Fig. 11. K_2FeF_5, Rb_2FeF_5, Cs_2FeF_5, powder samples. Similar to Fig. 10. The curves have been fit with calculations according to [68W], [80D].

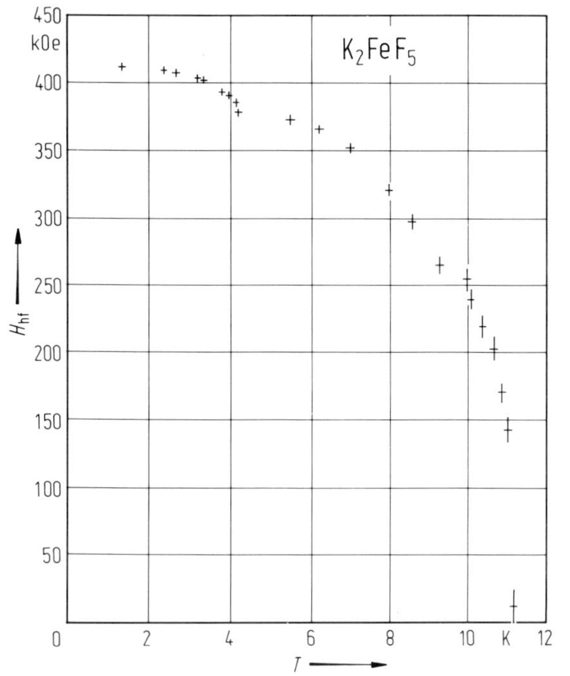

Fig. 12. K_2FeF_5, powder. The magnetic hyperfine field H_{hf} vs. temperature T [77G].

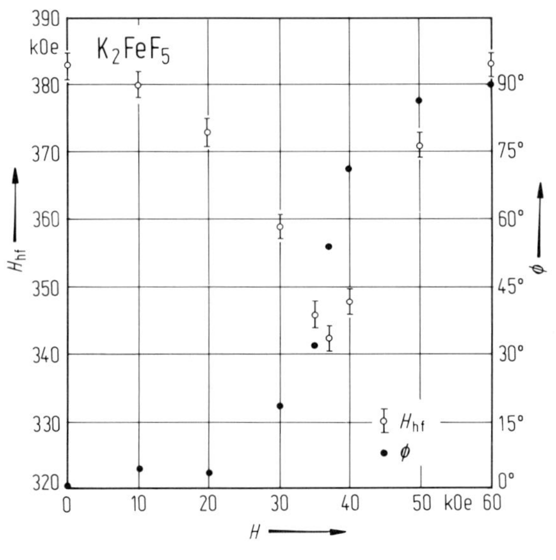

Fig. 13. K_2FeF_5, single crystal, $T = 4.2$ K. Hyperfine field H_{hf} vs. applied field H perpendicular to the *ac* plane of the sample and the angle ϕ between the magnetic axis and the direction of the applied field, showing the spin-flop transition. Data derived from Mössbauer spectra [78G1].

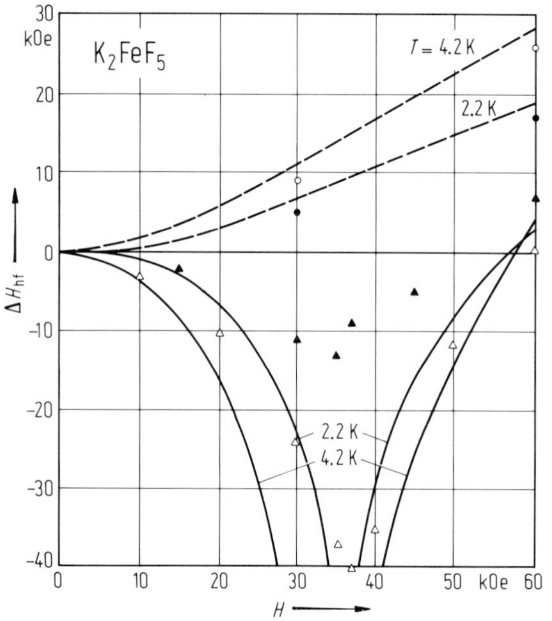

Fig. 14. K_2FeF_5, single crystal. Spin reduction presented in the form of the resulting change of the hyperfine field ΔH_{hf} for $H_{hf}(0) = 410(5)$ kOe, as derived from Mössbauer spectra at 4.2 and 2.2 K (open and closed symbols, respectively). Triangles and full curves for applied field H parallel to the initial direction of antiferromagnetism, circles and broken curve for the applied field perpendicular to this direction. The curves have been calculated assuming the hypothetical value $H_{hf}(0) = 600$ kOe, corresponding to $S = 5/2$. Error bar for experimental results is also shown [79G].

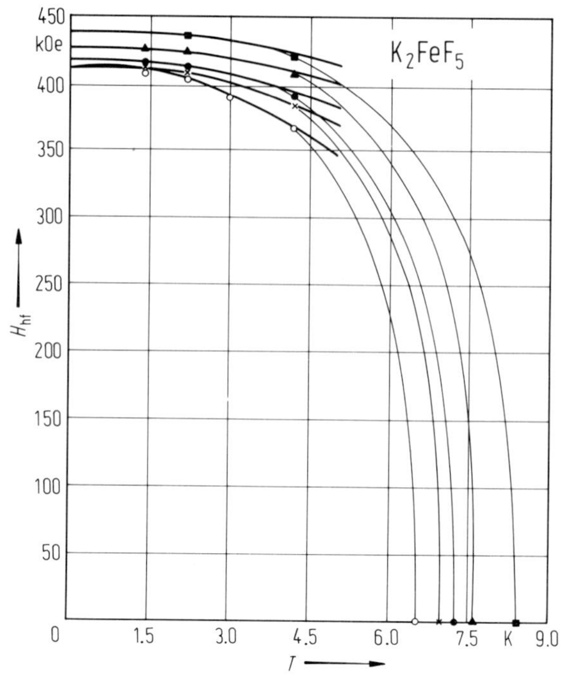

Fig. 15. K_2FeF_5, single crystal. Hyperfine field H_{hf} in the presence of an applied magnetic field H vs. temperature T. x: zero magnetic field; ● ▲ , ■: $H = 30$, 60 and 100 kOe parallel to the a axis, i.e. perpendicular to the easy axis. ○: $H = 30$ kOe parallel to the b axis, the easy axis of magnetisation. The Néel temperatures are taken from [81C2]. The heavy curves have been calculated for the various conditions, the thin curves are reasonable approximations [82C2].

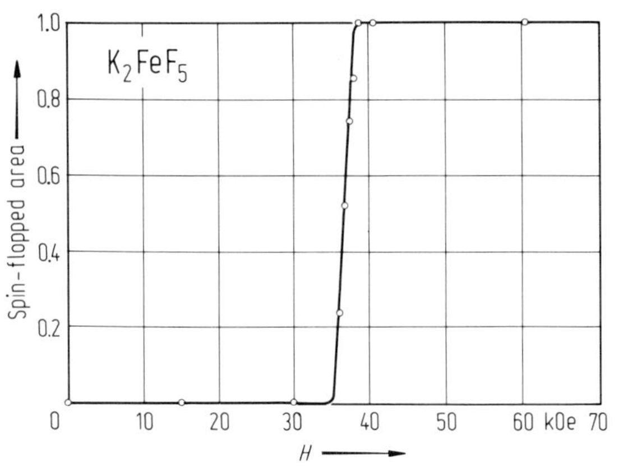

Fig. 16. K_2FeF_5, single crystal. Relative spectral area of flopped to unflopped components of the Mössbauer spectra in the case of an applied field H parallel to the antiferromagnetic axis [86P].

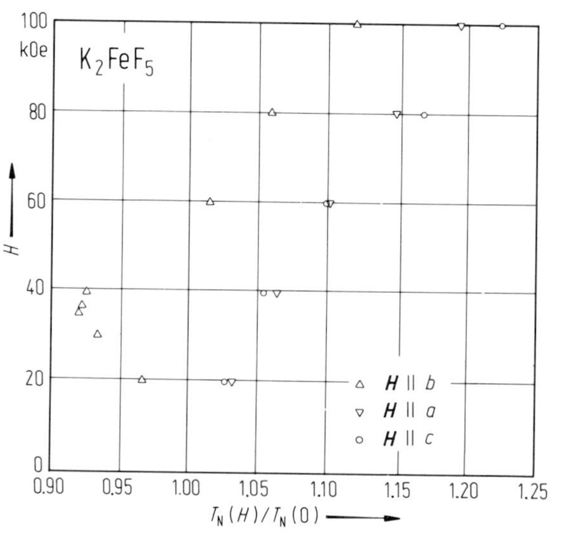

Fig. 17. K_2FeF_5, single crystal. Magnetic phase diagram for fields applied along the three principal crystallographic axes. The b axis is the direction of the magnetic moments in the absence of a magnetic field [81C2].

Rb_2FeF_5

Paramagnetic properties Figs. 11, 18. Above 220 K a Curie–Weiss law with $\Theta_p = -125$ K and $C_m = 4.32$ cm³ Kmol⁻¹ is obeyed [80D]. An analysis of the curve results in $J/k = -8.78$ K and $J'/J = 2.4 \cdot 10^{-3}$ [80D]. Mössbauer spectra indicate the presence of an ordering of the magnetic moments of the iron ions into a four sublattice antiferromagnetic state below $T_N = 9.3$ (5) K. Further Mössbauer data Table 4 and Figs. 19, 20. The saturated value of the hyperfine field is $H_{hf}(0) = 430(5)$ kOe for all sublattices. This field makes an angle of 15(5)° and − 90(5)° with respect to the EFG axes. Also a value $H_A/H_E \simeq 4 \cdot 10^{-3}$ is derived from this analysis [78G]. The magnetic structure derived from neutron diffraction data at 2 K is antiferromagnetic of the mode A_zG_x [80D]. The magnetic moments within a chain are coupled antiparallel and they are oriented parallel to the *ac* plane, with an angle of 25° with the *c* axis, with $p_{Fe^{3+}} = 3.5(2)\,\mu_B$ at 0 K. This low moment points to spin reduction, see also Fig. 21. From the magnetisation curves, Fig. 22, in the direction of the *c* axis is found a critical spin-flop field $H_{SF} = 65$ kOe, above which the moments become oriented perpendicular to the *c* axis, the antiferromagnetism in the chains being retained [81T]. See also Fig. 23.

Table 4. Rb_2FeF_5, powder. Temperature dependence of the Mössbauer parameters. Chemical shift δ, relative to iron metal at RT, and the quadrupole splitting Δ. Asymmetry parameter $\eta \simeq 0.5$.

T [K]	δ [mm s⁻¹]	Δ [mm s⁻¹]	Ref.
300	0.42(1)	− 0.72(1)	78G
150	0.50(1)	− 0.71(1)	
77	0.53(1)	− 0.70(1)	
9.3	0.54(1)	− 0.68(2)	
4.2	0.55(1)	− 0.68(2)	
300	0.41	− 0.68	80D
4.2	0.54		

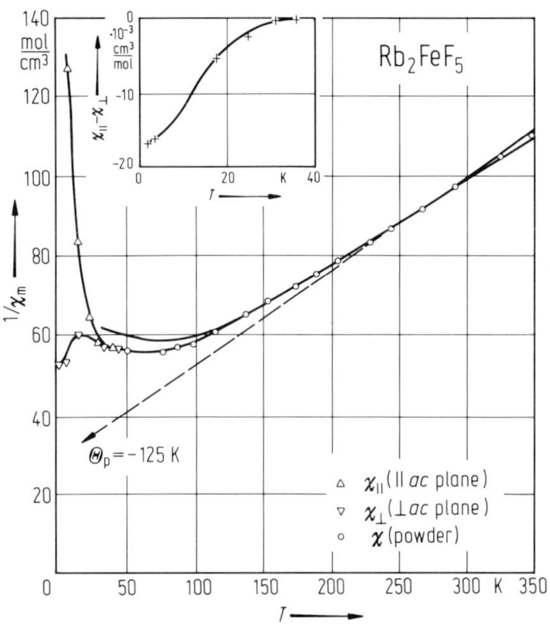

Fig. 18. Rb_2FeF_5. Temperature dependence of the inverse molar magnetic susceptibility χ_m^{-1} for an assembly of small crystals ("powder") and for a single crystal measured parallel or perpendicular to the *ac* plane in relatively low magnetic fields. The curve drawn for $T > T_N = 8.0(5)$ K is calculated for $J/k = -8.8$ K [81T].

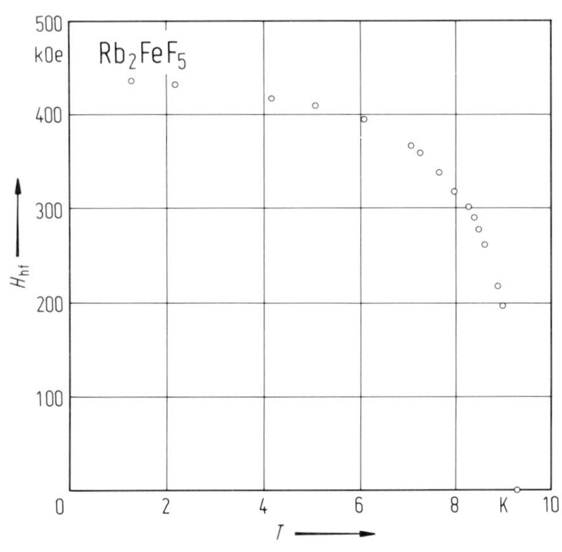

Fig. 19. Rb_2FeF_5, single crystal. The magnetic hyperfine field H_{hf} vs. temperature T as derived from Mössbauer spectra [78G].

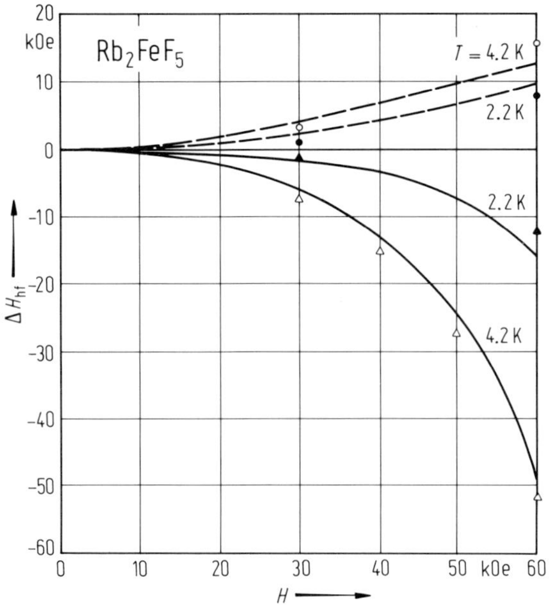

Fig. 20. Rb_2FeF_5, single crystal. Spin reduction represented similar to Fig. 14. $H_{hf}(0) = 430$ kOe [79G].

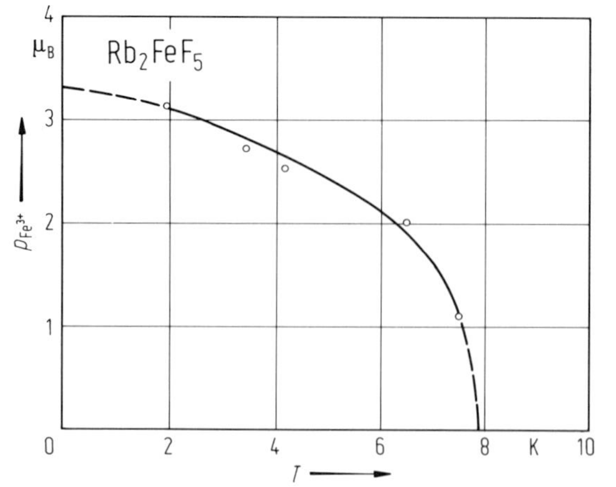

Fig. 21. Rb_2FeF_5, single crystal. Temperature dependence of the magnetic moment $p_{Fe^{3+}}$ of the iron ions as derived from neutron diffraction data [81T].

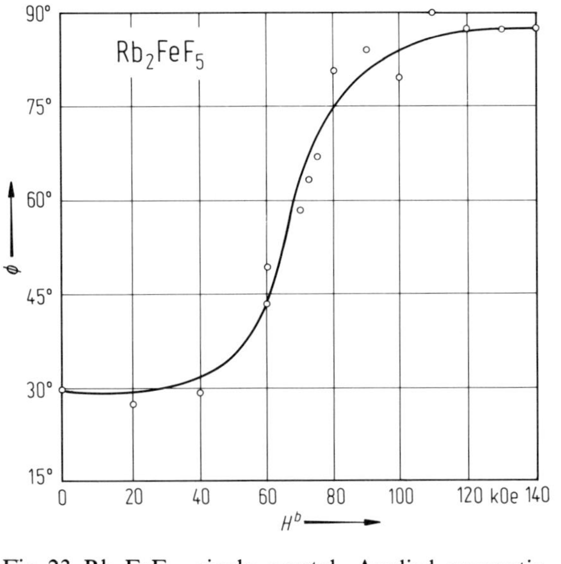

Fig. 23. Rb_2FeF_5, single crystal. Applied magnetic field H^b parallel to the b axis. Angle ϕ between spin orientation and magnetic field at 4.2 K [86P].

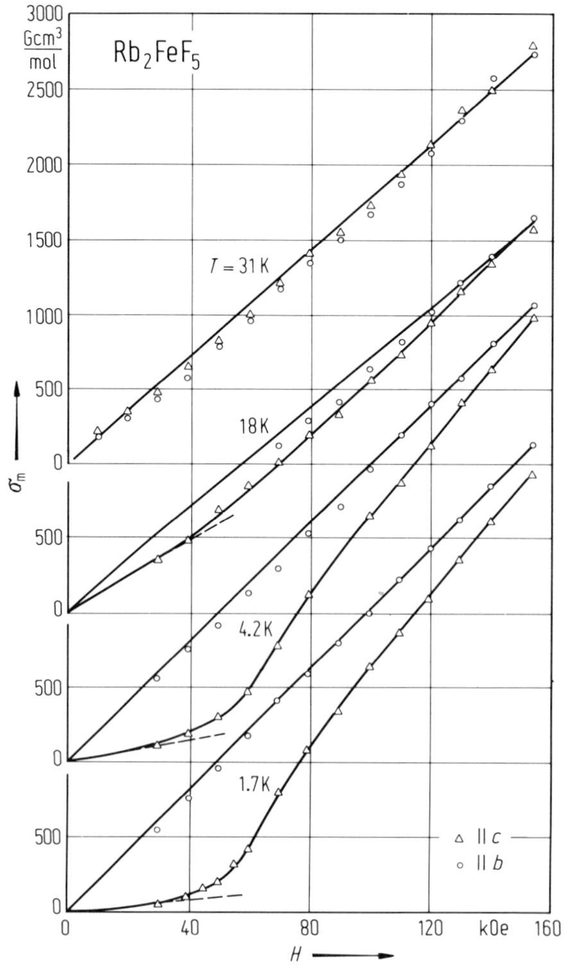

Fig. 22. Rb_2FeF_5, single crystal. The molar magnetisation σ_m vs. applied magnetic field H at various temperatures T. The direction of the field is either parallel to the crystallographic b axis (○) or c axis (△) [81T].

Cs_2FeF_5

Paramagnetic properties Fig. 11. Curie–Weiss law above 200 K with $\Theta_p = -135$ K and $C_m = 4.52$ cm^3 Kmol^{-1}. An analysis of the curve can be made with $J/k = -8.01$ K and $J'/J = 1.1 \cdot 10^{-3}$. Mössbauer data Table 5 with $H_{hf}(0) = 340$ kOe. Neutron diffraction spectra show that magnetic ordering occurs below $T_N = 5$ K. The magnetic structure is similar to that of Rb_2FeF_5 [80D].

Table 5. Cs_2FeF_5, powder. Temperature dependence of the Mössbauer parameters. Chemical shift δ, relative to iron metal at RT, and quadrupole splitting Δ [80D].

T [K]	δ [mm s^{-1}]	Δ [mm s^{-1}]
300	0.43	0.83
4.2	0.54[a])	0.76
	0.58[b])	

[a]) Value corresponding to the central doublet.
[b]) Value corresponding to the magnetic sextuplet.

$N_2H_6FeF_5$

Paramagnetic properties Fig. 24. Above 320 K a Curie–Weiss law is obeyed with $\Theta_p = -150$ K, $C_m = 4.42$ cm^3 Kmol^{-1} and $p_{eff} = 5.95\ \mu_B$/FU. On the basis of one-dimensional short range antiferromagnetic interactions within chains is calculated $J/k = -10.2$ K and $J'/J = 2 \cdot 10^{-3}$. Mössbauer data Table 6. Below $T_N = 9(2)$ K the compound shows antiferromagnetic ordering [77H].

Table 6. $N_2H_6FeF_5$, powder. Mössbauer parameters at various temperatures T. Isomer shift δ, quadrupole splitting Δ, asymmetry parameter η, hyperfine field H_{hf} and angle θ between principal axis of EFG and H_{hf}.

$T \pm 0.5$ [K]	δ[a]) [mm s^{-1}]	Δ or 2ε[b]) [mm s^{-1}]	η	H_{hf} [kOe]	θ [deg]	Ref.
4.2	0.55(2)	0.45(2)	< 0.40	409(2)	88(1)	77H
5.5	0.55(2)	0.48(2)	< 0.40	371(2)	89(1)	
12	0.53(2)	0.41(2)				
77	0.522(5)	0.412(5)				74H
136	0.496(5)	0.424(5)				
200	0.491(5)	0.417(5)				
280	0.473(5)	0.425(5)				
300	0.465(5)	0.421(5)				
400	0.459(5)	0.431(5)				

[a]) Relative to metallic iron at 293 K.
[b]) Quadrupole splitting of the external Zeeman lines.

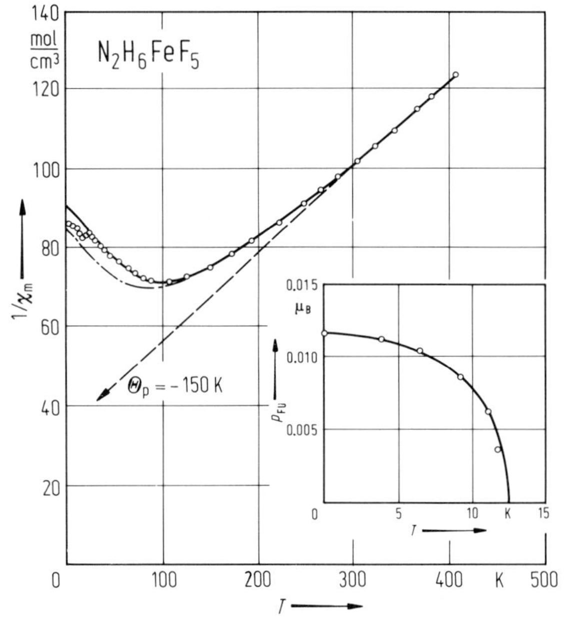

Fig. 24. $N_2H_6FeF_5$, powder. Inverse molar magnetic susceptibility χ_m^{-1} vs. temperature T. Circles: experimental values; full and dot-dashed curve: theoretical values after Weng's calculations [68W] for $S = 5/2$, $J/k = -10.2$ K and Smith, Friedberg's model [68S] for $S = 5/2$, $J/k = -10.5$ K, respectively. Inset: a ferromagnetic contribution to the magnetisation [77D, 77H].

α-$(NH_4)_2FeF_5$

Mössbauer data Table 7 and Fig. 25. $T_N = 7.5(5)$ K. The angle θ between the principal component V_{zz} of the EFG tensor and H_{hf} is either 47° or 64° as derived from the data of Table 7 [91C].

β-$(NH_4)_2FeF_5$

From differential scanning calorimetry is found that a structural transition seems to occur at 168(1) K in the heating and at 187(3) K in the cooling curve. Mössbauer data Table 8. Below $T_N = 13.1$ K the spectra show two well-resolved sextets, corresponding to two magnetically different iron ions in the unit cell, see H_{hf} in Fig. 26. For both sites an extrapolation results in $H_{hf}(0) = 440$ kOe [91C].

Table 7. α-$(NH_4)_2FeF_5$. Mössbauer data at various temperatures. Isomer shift δ and quadrupole splitting Δ or 2ε of the outermost Zeeman lines and the hyperfine field H_{hf} [91C].

T [K]	δ [a]) [mm s^{-1}]	Δ or 2ε [mm s^{-1}]	H_{hf} [kOe]
300	0.435(5)	0.645(4)	–
77	0.536(5)	0.5884)	–
7	0.55(1)	– 0.12(2)	179(2)
4.2	0.55(1)	– 0.12(2)	360(2)

[a]) Relative to metallic iron at 300 K.

Table 8. β-$(NH_4)_2FeF_5$. Mössbauer data similar to Table 7 [91C].

T [K]	δ[a]) [$mm\,s^{-1}$]	Δ or 2ε[b]) [$mm\,s^{-1}$]	H_{hf} [kOe]	%[d])
300	0.436(4)	1.157(4)	–	
77[c])	0.476(4)	1.167(4)	–	
11	0.51(1)	– 0.49(2)	246(2)	
	0.56(1)	– 0.49(2)	356(2)	
10	0.54(1)	– 0.50(2)	294(2)	
	0.55(1)	– 0.49(2)	372(2)	
9	0.54(1)	– 0.50(2)	323(2)	81
	0.54(1)	– 0.49(2)	385(2)	19
8	0.55(1)	– 0.51(2)	347(2)	81
	0.55(1)	– 0.51(2)	397(2)	19
4.2	0.54(1)	– 0.50(2)	396(2)	80
	0.54(1)	– 0.50(2)	428(2)	20

[a]) Relative to Fe-metal at 300 K.
[b]) Quadrupole shift of the sextet.
[c]) Average values.
[d]) Percentage of iron on site 1 and site 2, respectively.

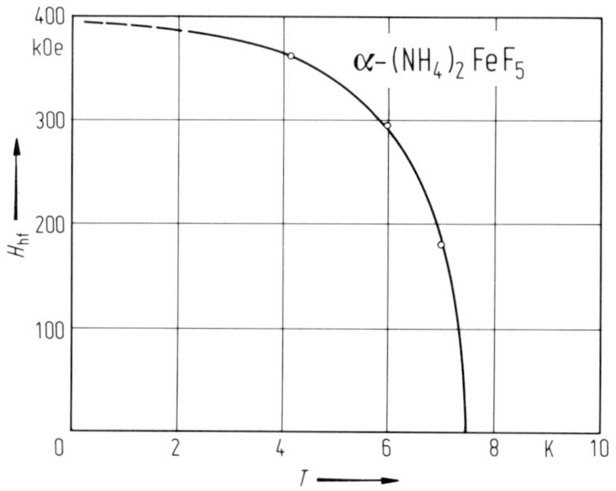

Fig. 25. α-$(NH_4)_2FeF_5$. Magnetic hyperfine field H_{hf} vs. temperature T [91C].

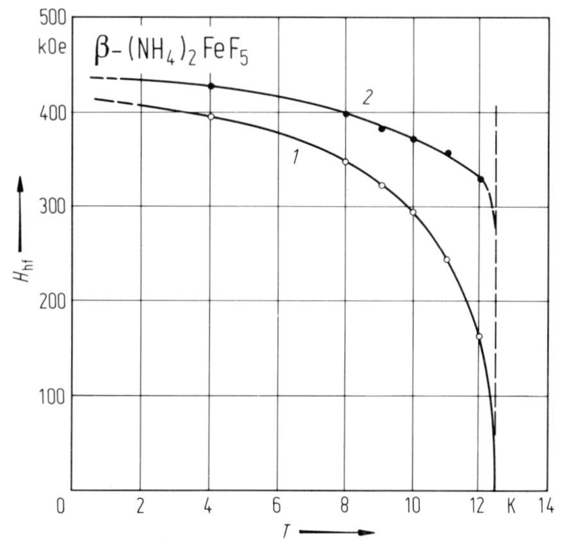

Fig. 26. β-$(NH_4)_2FeF_5$. Magnetic hyperfine field H_{hf} vs. temperature T for two distinct iron sites *1* and *2*. 80% of the iron ions are on site *1* [91C].

9.8.3 $A_2MX_5 \cdot H_2O$ compounds

9.8.3.1 Crystallographic properties of $A_2MX_5 \cdot H_2O$ compounds

The compounds $A_2MX_5 \cdot H_2O$ have generally an orthorhombic structure, see Table 9, in which some features of the structures as well as the various principal crystallographic parameters are given. For details of the atomic positions is referred to the original literature. The unit cell contains four formula units ($Z = 4$). The structure is built up of MX_6 or MX_5O octahedra forming trans-linked zig-zag chains (Fig. 27) or hydrogen-bonded zig-zag chains (Fig. 28) along a major crystallographic direction, respectively.

Table 9. $A_2MX_5 \cdot H_2O$. Crystallographic parameters at RT as derived from X-ray diffraction data.

Compound	Crystal structure, space group	Chains of octahedra		Lattice parameters [Å]			Ref.
		type	orientation	*a*	*b*	*c*	
$K_2MnF_5 \cdot H_2O$	M[a]), $P2_1/m$	MnF_6	∥*c*	6.04(1)	8.20(1)	5.94(1)	71E
$Rb_2MnF_5 \cdot H_2O$	O, Cmcm	MnF_6	∥*c*	9.383(2)	8.214(3)	8.348(2)	78B
$Cs_2MnF_5 \cdot H_2O$	O, Cmcm	MnF_6	∥*c*	9.727(8)	8.686(11)	8.508(4)	78K
	Cmm2; $Z = 2$	MnF_6	∥*c*	9.740(2)	8.674(1)	4.260(1)	90H
$Tl_2MnF_5 \cdot H_2O$	O, Cmcn	MnF_6	∥*c*	9.688(2)	8.002(1)	8.339(1)	92N
$K_2FeF_5 \cdot H_2O$	M[a]), C2/c	FeF_5O	∥*c*	9.71(1)	7.79(1)	7.95(1)	72E
α-$Rb_2FeF_5 \cdot H_2O$	unknown	–	–	–	–	–	81F, 82F1
β-$Rb_2FeF_5 \cdot H_2O$	O, Cmcm	FeF_6	∥*c*	9.691(3)	8.446(3)	7.947(3)	81F, 82F1
$Cs_2FeF_5 \cdot H_2O$	O, Cmcm	FeF_5O	∥*b*	10.361(4)	8.266(3)	8.401(3)	82B1
$(NH_4)_2FeF_5 \cdot H_2O$	O, Pbcn	FeF_5O	∥*c*	10.491(5)	8.090(3)	7.997(3)	88F
$K_2FeCl_5 \cdot H_2O$[b])	O, Pnma	$FeCl_5O$	∥*b*	13.75	9.92	6.93	78M
$Rb_2FeCl_5 \cdot H_2O$	O, Pnma	$FeCl_5O$	∥*b*	13.825(4)	9.918(3)	7.100(2)	79O
$Cs_2FeCl_5 \cdot H_2O$	O, Cmcm	$FeCl_5O$	∥*a*	7.442(3)	17.307(7)	8.077(7)	79O
	O, Cmcm	$FeCl_5O$	∥*a*	7.426(4)	17.306(7)	8.064(2)	80G
$(NH_4)_2FeCl_5 \cdot H_2O$[c])	O, Pnma	$FeCl_5O$	∥*b*	13.706(2)	9.924(1)	7.024(1)	78F
$Rb_2FeBr_5 \cdot H_2O$	O, Pnma	–	–	14.72	10.94	7.4	77C
$Cs_2FeBr_5 \cdot H_2O$	O, Pnma	–	–	14.7	10.7	7.6	77C

[a]) $\beta = 96.5°$.
[b]) See also Fig. 39.
[c]) See also Fig. 62.

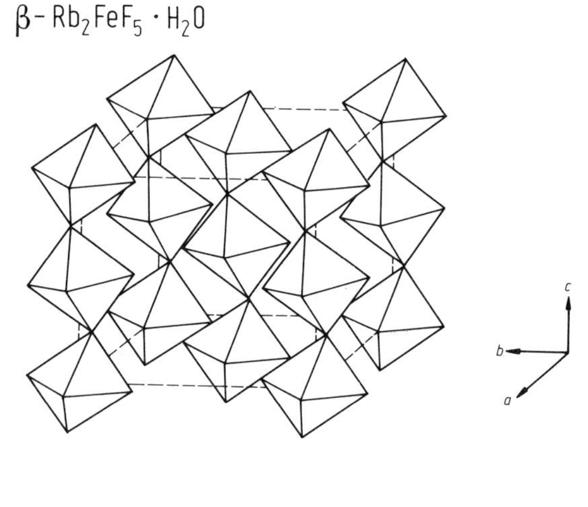

Fig. 27. β-$Rb_2FeF_5 \cdot H_2O$. The trans-linked (FeF_6) octahedra chains running along the crystallographic *c* axis [82F1].

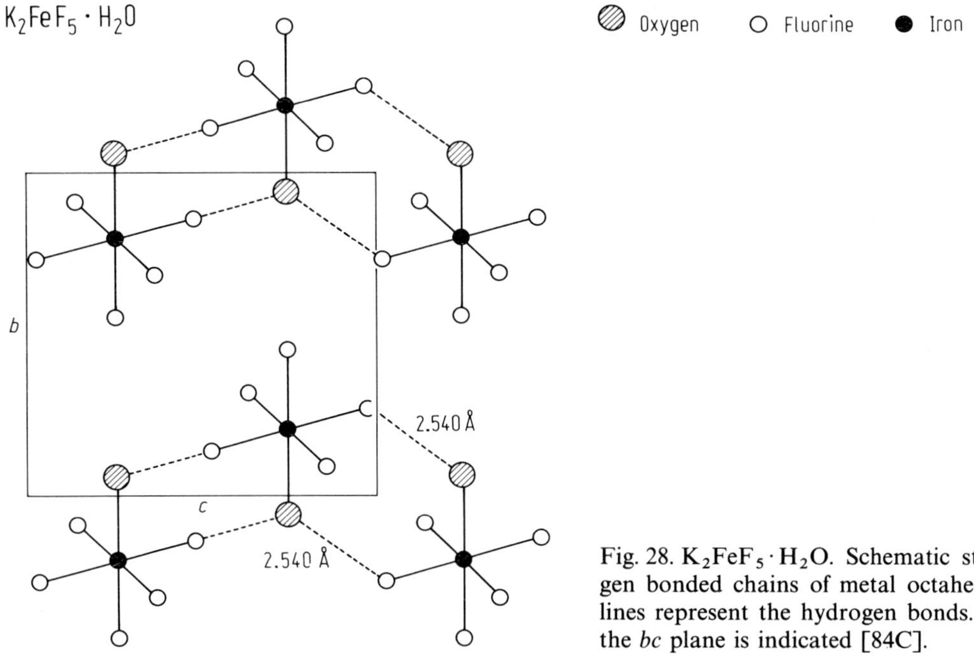

Fig. 28. $K_2FeF_5 \cdot H_2O$. Schematic structure. Hydrogen bonded chains of metal octahedra. The dashed lines represent the hydrogen bonds. The unit cell in the *bc* plane is indicated [84C].

9.8.3.2 Magnetic properties of $A_2MX_5 \cdot H_2O$ compounds

For a survey of the magnetic properties see Table 10.

$K_2MnF_5 \cdot H_2O$, $Rb_2MnF_5 \cdot H_2O$, $Cs_2MnF_5 \cdot H_2O$
Magnetic susceptibilities Figs. 5–7.

$Tl_2MnF_5 \cdot H_2O$
Crystallographic data Table 9, without a crystallographic phase transition down to 8K. Strongly anisotropic thermal expansion in the *ab* plane with $\alpha = 4.74 \cdot 10^{-5} K^{-1}$ and $5.56 \cdot 10^{-5} K^{-1}$ along the *a* and *b* axis, respectively. Along the *c* axis $\alpha = 0.91 \cdot 10^{-5} K^{-1}$.

Magnetic susceptibility Figs. 29 and 30. Below $T_N = 28(1)$K 3D-antiferromagnetic ordering. The magnetic structure at 2K is according to the Pm′cm space group, i.e. antiparallel ordering in the *ab* plane between the spins of the Mn ions located in neighboring chains, Fig. 31.

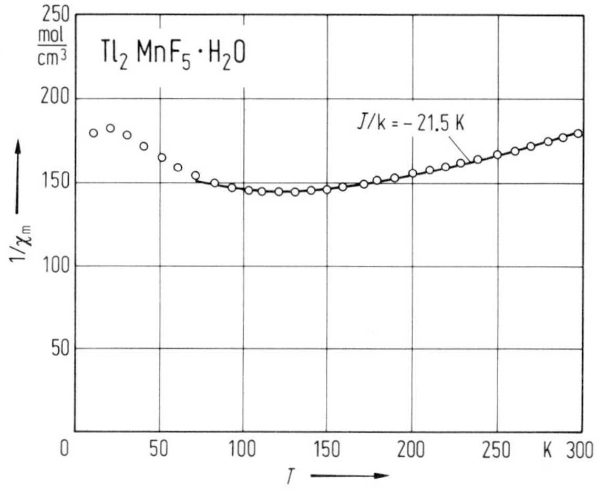

Fig. 29. $Tl_2MnF_5 \cdot H_2O$, powder. Inverse molar magnetic susceptibility χ_m^{-1} vs. temperature *T*. Solid curve is calculated with $J/k = -21.5$ K and $g = 1.99$ [92N].

Table 10. $A_2MX_5 \cdot H_2O$. Survey of the magnetic properties. Θ_p: paramagnetic Curie point; T_N: Néel point derived from Mössbauer spectra (ME), electron spin resonance (ESR), specific heat measurements (C_p) or magnetisation measurements (χ); exchange parameters J and J'; magnetic hyperfine field $H_{hf}(0)$ at $T = 0$ K.

Compound	Θ_p [K]	C_m [cm^{3}K/mol]	g	J/k [K]	J'/J [$\cdot 10^{-3}$]	T_N [K]	$H_{hf}(0)$ [kOe]	AF direction	Figs.; Tables	Ref.
$K_2MnF_5 \cdot H_2O$	−295	3.95	2.06	−18.2					5	87N
$Rb_2MnF_5 \cdot H_2O$	−428	[a])	2.02	−19.4(6)						87P
	−445	4.54	2.09	−21.8					6	87N
$Cs_2MnF_5 \cdot H_2O$	−331	[a])	2.01	−17.8(13)						87P
	−360	4.41	2.04	−17.9					7	87N
$Tl_2MnF_5 \cdot H_2O$	−470	4.30	1.99	−21.5		28(1)			29···31	92N
$K_2FeF_5 \cdot H_2O$			1.97	−0.40(2)		0.80(5)(χ)		$\perp b$	32,33	84C
						0.8···1.1(ME)	520			89T
α-$Rb_2FeF_5 \cdot H_2O$	−7									82F1
β-$Rb_2FeF_5 \cdot H_2O$	−150			−13.1	2.2	9.7(ME)	396(3)	$\perp c$	34, 35;	82F1,
									Tab. 11	82F2
$Cs_2FeF_5 \cdot H_2O$						2.4(1) (ESR)		$\parallel a$	38	83K
$(NH_4)_2FeF_5 \cdot H_2O$	−8.5								Tab. 13	88F
$K_2FeCl_5 \cdot H_2O$[b])				−1.55(15)	0.49	14.0(2)(χ)		$\parallel a$	39···47;	78M
									Tab. 14	
$Rb_2FeCl_5 \cdot H_2O$	−35.0[79O]			−1.39[77C]	0.33[77C]	10.03(5)(C_p)[77C]				
				−1.45(6)	0.40	10.00(2)(C_p)			48···52	85P2
$Cs_2FeCl_5 \cdot H_2O$	−14.0[79O]			−0.65[77C]	0.18[77C]	6.54(2)(C_p)[82P]	520[83G]	$\parallel a$	53···61	
				−0.31[78P]		6.57(5)(C_p)[77C]				
						6.43(5)(χ)[77C]				
						6.75(50)(ME)[83G]				
						6.75(10)(ME)[87J]				
						6.63(ME)[88C]				
$(NH_4)_2FeCl_5 \cdot H_2O$				−1.24(8)	0.58	6.87(1); 7.25(1)(C_p)			62···64;	78M
									Tab. 14	
$Rb_2FeBr_5 \cdot H_2O$				−2.5	0.25	22.9(1)(χ)		$\parallel a$	68	81C1
$Cs_2FeBr_5 \cdot H_2O$				−1.6	0.17	14.2(1)(χ)		$\parallel a$	69	81C1

[a]) $p_{eff} = 4.00\ \mu_B$/FU at 300 K [87P].
[b]) For $K_2Fe(Cl_{1-x}Br_x)_5 \cdot H_2O$ see Figs. 65–67.

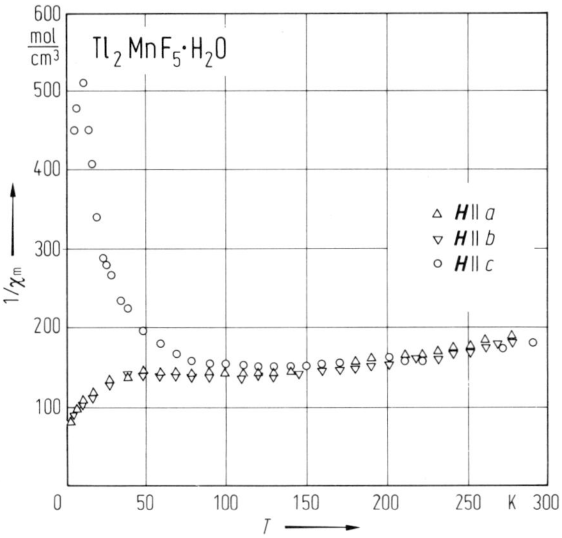

Fig. 30. $Tl_2MnF_5 \cdot H_2O$, single crystal. Inverse of the molar magnetic susceptibility, χ_m^{-1}, vs. temperature T for the three principal crystallographic directions [92N].

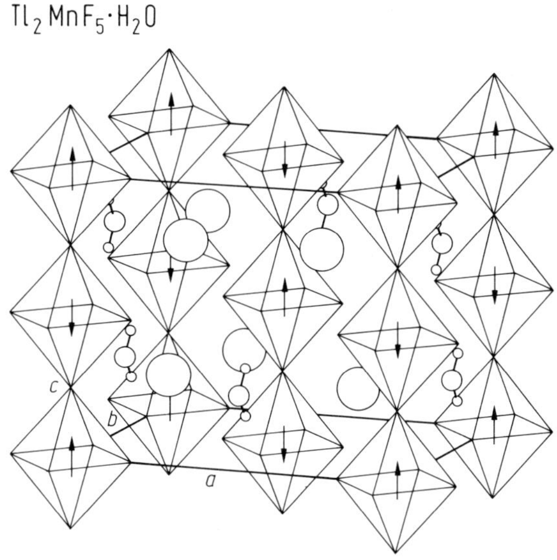

Fig. 31. $Tl_2MnF_5 \cdot H_2O$. Magnetic structure at 2K as derived from neutron diffraction patterns [92N].

$K_2FeF_5 \cdot H_2O$

Magnetic susceptibility Fig. 32. $T_N = 0.80(5)$ K below which temperature the system becomes long-range antiferromagnetic with a spin alignment in the *ac* plane, pointing at 50° from the *c* axis toward the negative direction of the *a* axis. In the range 10···280 K the paramagnetic susceptibility can be fitted with $g = 1.97(2)$ and $J/k = -40(2)$ K, with negligible interchain interaction [84C]. Mössbauer spectra show magnetic ordering and paramagnetic components between 0.8 and 1.1 K. Hyperfine field Fig. 33, and $H_{hf}(0) \simeq 520$ kOe. Spectra obtained in the presence of an applied magnetic field suggest that the *b* axis is the hard magnetic direction for the AF state, with a spin-flop transition field $H_{SF} \simeq 10^4$ kOe at low T [89T].

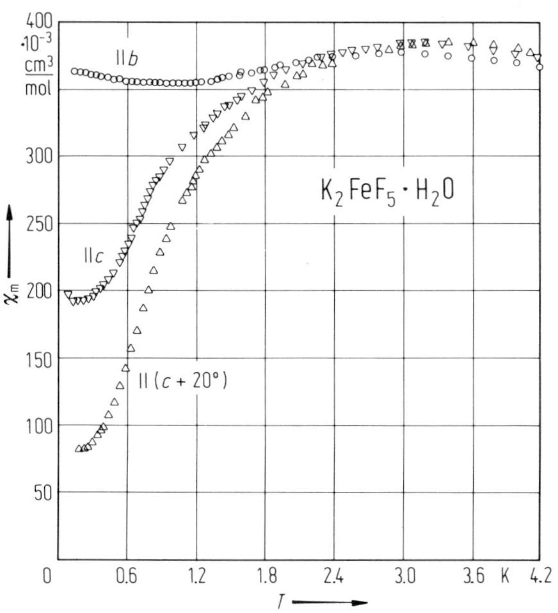

Fig. 32. $K_2FeF_5 \cdot H_2O$, single crystal. Zero-field molar magnetic susceptibility χ_m in different crystallographic directions below $T = 4.2$ K. The *b* axis is the hard axis, the *c* axis and the direction $c + 20°$ in the *ac* plane form angles of 50° and 30°, respectively, with the easy axis of magnetisation [84C].

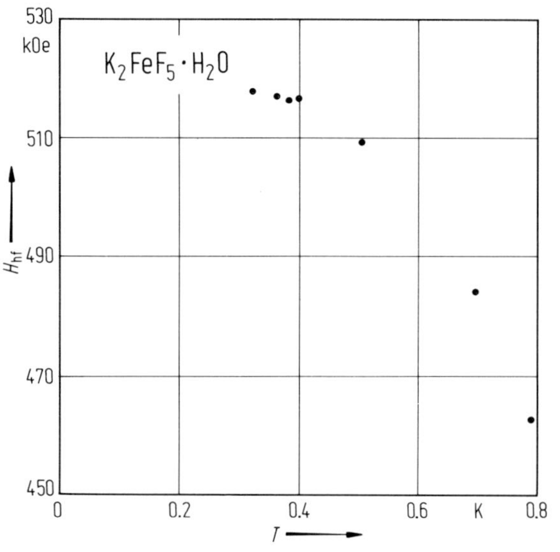

Fig. 33. $K_2FeF_5 \cdot H_2O$, powder. Temperature dependence of the hyperfine field H_{hf} in the magnetically ordered state [89T].

α-$Rb_2FeF_5 \cdot H_2O$
Paramagnetic down to 4 K, $\Theta_p = -7$ K, $C_m = 4.42$ cm^3 Kmol^{-1} [82F1].

β-$Rb_2FeF_5 \cdot H_2O$
Magnetic susceptibility Fig. 34. A fit based on linear chain AF with an $S = 5/2$ Heisenberg model shows good agreement for $J/k = -13.1$ K with $\Theta_p = -150$ K. Combined with $T_N = 9.7(1)$K (Mössbauer result) this leads to $J'/J = 2.2 \cdot 10^{-3}$, a value typical for 1D character of interactions [82F1, 82F2]. Mössbauer data Table 11 and Fig. 35. $H_{hf}(4.2\text{ K}) = 396(3)$kOe, with AF ordering perpendicular to the c axis. At 4.2 K spin-flop takes place in the ab plane [82F1].

$Rb_2Mn_{1-x}Fe_xF_5 \cdot H_2O$ (x ⩽ 0.50)
Lattice parameters Fig. 36. Magnetic susceptibility Fig. 37. The broad minimum between 110···130 K is attributed to short-range AF interaction within the linear chains. The sharper low-temperature minimum, especially for higher x values, indicates the presence of 3D antiferromagnetic order at T_N, the latter are listed in Table 12. Mössbauer spectra reveal three Fe positions corresponding to the interactions in the three configurations (1) Mn–Fe–Mn, (2) Mn–Fe–Fe and (3) Fe–Fe–Fe. For further Mössbauer data see Table 12. The spectra show relaxation effects attributed to short-range order with temperature-dependent relaxation times [89P].

Table 11. β-$Rb_2FeF_5 \cdot H_2O$, powder. Mössbauer parameters at various temperatures T. Chemical shift δ relative to metallic iron at RT, quadrupole splitting Δ and hyperfine field H_{hf} [82F1].

T [K]	δ [mm s^{-1}]	Δ [mm s^{-1}]	H_{hf} [kOe]
300	0.425(5)	1.36(1)	
77	0.516(5)	1.36(1)	
20	0.537(5)	1.34(1)	
4.2	0.52(3)	$2\varepsilon = -0.82(1)$	396(3)

Table 12. $Rb_2Mn_{1-x}Fe_xF_5 \cdot H_2O$. Mössbauer data for Fe ions in the configurations (1) Mn–Fe–Mn, (2) Mn–Fe–Fe and (3) Fe–Fe–Fe. Saturation hyperfine field $H_{hf,i}$ at 4.2 K, quadrupole splittings Δ_i at RT, relative iron contents configurations A_i derived from the spectra; 3D ordering temperature T_N from susceptibility curves of Fig. 37, as well as the intrachain exchange energy J_{FeMn}/k derived from these curves [89P].

x	$H_{hf,1}$ [kOe]	$H_{hf,2}$ [kOe]	$H_{hf,3}$ [kOe]	A_1 [%]	A_2 [%]	A_3 [%]	Δ_1 [mms^{-1}]	Δ_2 [mms^{-1}]	Δ_3 [mms^{-1}]	T_N [K]	J_{FeMn}/k [K]
0.01	511(3)			100(2)			0.841(8)			23.2(5)[a]	− 20.8(8)
0.10	510(3)	492(3)		93(2)	7(3)		0.776(8)	1.035(8)		21.0(10)	− 19.0(8)
0.20	505(3)	483(3)		90(2)	10(3)		0.788(8)	1.146(8)		19.0(10)	− 18.7(8)
0.30	502(3)	483(3)		84(2)	16(2)		0.777(8)	1.127(6)		17.1(10)	− 17.3(8)
0.40	498(3)	483(3)		70(2)	30(2)		0.773(8)	1.136(7)		15.5(5)	− 16.0(8)
0.50	493(3)	477(3)	429(6)	57(2)	34(3)	9(5)	0.786(8)	1.146(8)	1.39(10)	14.5(5)	− 15.3(8)
1.00[b]			396		100				1.360	9.7	

[a]) From neutron diffraction experiments, private communication J. L. Soubeyroux in [89P].
[b]) [82F1]

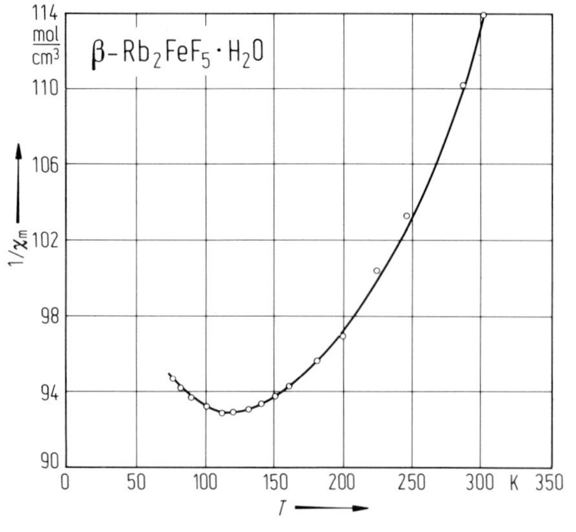

Fig. 34. β-$Rb_2FeF_5 \cdot H_2O$, powder. Temperature dependence of the inverse molar magnetic susceptibility χ_m^{-1} [82F1].

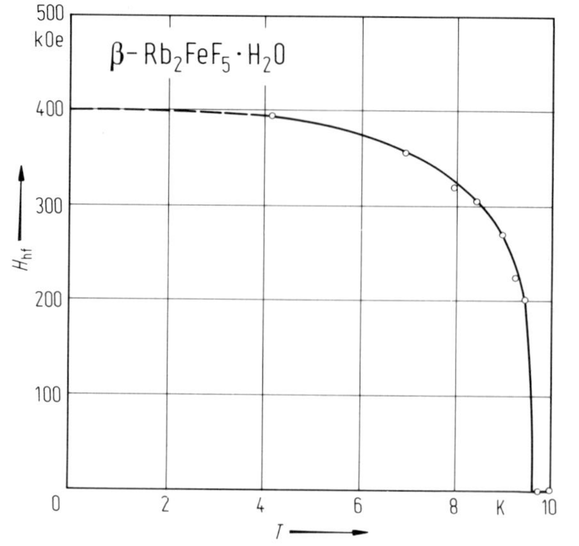

Fig. 35. β-$Rb_2FeF_5 \cdot H_2O$, powder. Temperature dependence of the magnetic hyperfine field H_{hf} [82F1].

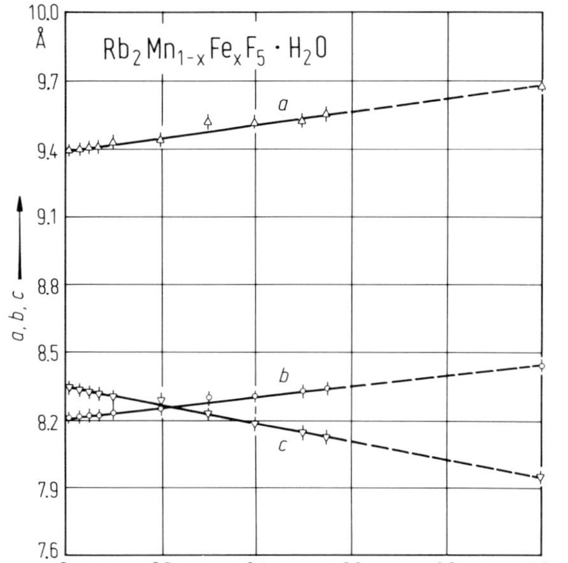

Fig. 36. $Rb_2Mn_{1-x}Fe_xF_5 \cdot H_2O$, powder. Lattice parameters *a*, *b*, and *c* of the orthorhombic structure vs. composition parameter x. The values for x = 1.0 are from [82F1]. In the range 0.55 < x < 1.0 a phase mixture of $Rb_2Mn_{1-x}Fe_xF_5 \cdot H_2O$ and $Rb_2Mn_{1-x}Fe_xF_5$ occurs [89P].

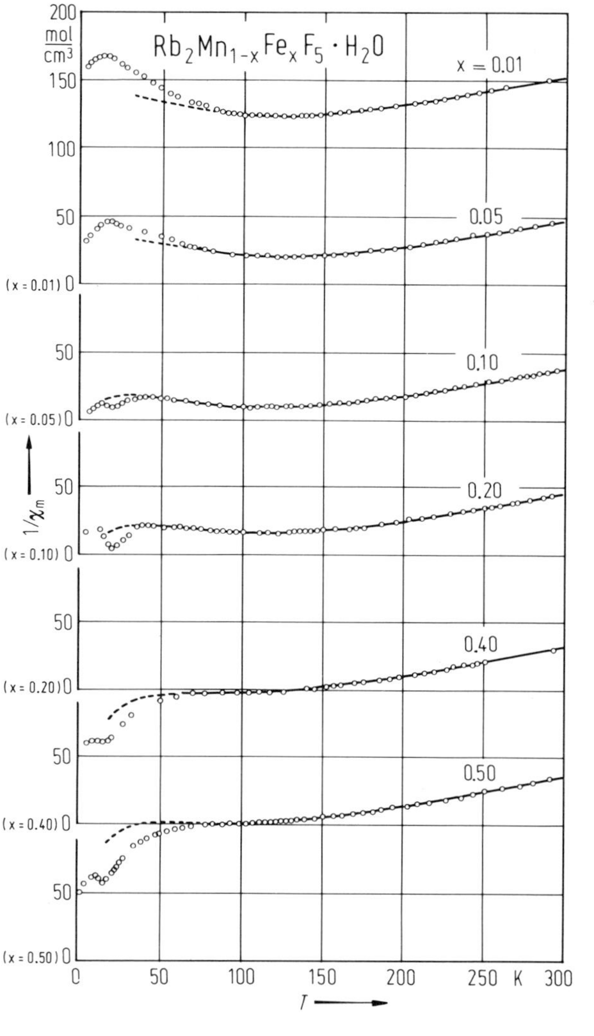

Fig. 37. $Rb_2Mn_{1-x}Fe_xF_5 \cdot H_2O$, powder. Inverse magnetic molar susceptibility χ_m^{-1} vs. temperature *T* in magnetic fields up to 20 kOe. Drawn curves theoretical values for a linear isotropic Heisenberg chain [89P].

$Cs_2FeF_5 \cdot H_2O$

Electron spin resonance indicates paramagnetism above $T_N = 2.4(1)$ K with $g = 2.0$. Below this temperature AF with easy axis $\parallel a$. At 1.8 K the susceptibility perpendicular to the a axis is $0.32(3) cm^3 mol^{-1}$ [82B1]. Antiferromagnetic resonance spectra confirm the magnetic ordering temperature T_N (Fig. 38).

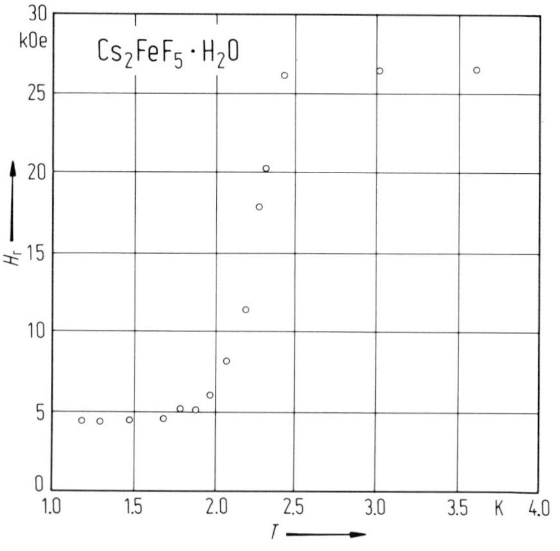

Fig. 38. $Cs_2FeF_5 \cdot H_2O$, powder. Temperature dependence of the antiferromagnetic resonance field H_r for a frequency of 74.0 GHz [83K].

$(NH_4)_2FeF_5 \cdot H_2O$

Paramagnetic down to 4.2K. Curie–Weiss law with $p_{eff} = 5.78$ μ_B/FU and $\Phi_p = -8.5$K [88F]. Mössbauer data Table 13.

Table 13. $(NH_4)_2FeF_5 \cdot H_2O$, powder. Mössbauer parameters at various temperatures T [88F].

T [K]	δ[a]) [mms^{-1}]	Δ [mms^{-1}]
300	0.417(4)	0.355(4)
77	0.526(4)	0.420(4)
4.2	0.536(4)	0.410(4)

[a]) Relative to iron metal at RT.

$K_2FeCl_5 \cdot H_2O$

Thermal expansion Table 14 and Fig. 39. Specific heat Fig. 40. Magnetic susceptibility Fig. 41, magnetic transition at $T_N = 14.0(2)$K to an AF state parallel to the *a* axis [78M]. A molecular field analysis results in $J/k = -1.55$ K and $J'/J = 0.49$. For a detailed analysis of the exchange interactions see [85P2]. Magnetic phase diagram Fig. 42. Bicritical point at $T = 13.6$ K, $H = 34.1$Oe. The extrapolated spin-flop field at 0 K is $H_{SF}(0) = 26$ kOe, see also Fig. 43. $H_A = 1.7$ kOe, $H_E = 199$ kOe, $H_A/H_E = 8.5 \cdot 10^{-3}$ [80P]. For the influence of hydrostatic pressure on the magnetisation, see Figs. 44–46. Mössbauer data Fig. 47. At 4.2 K $\varepsilon = -0.026$ mm s^{-1} [88P] and at RT: $\Delta = 0.213(4)$ mm s^{-1}, $\delta = 0.410(6)$ mm s^{-1} with respect to metallic iron [85P1].

Table 14. $K_2FeCl_5 \cdot H_2O$; $(NH_4)_2FeCl_5 \cdot H_2O$, single crystals. Thermal expansion coefficient α (in 10^{-5} K^{-1}) in the three principal crystallographic directions at RT [88P].

	α_a	α_b	α_c
$K_2FeCl_5 \cdot H_2O$	5.2(7)	4.6(9)	2.1(9)
$(NH_4)_2FeCl_5 \cdot H_2O$	8.6(7)	0.8(9)	4.1(9)

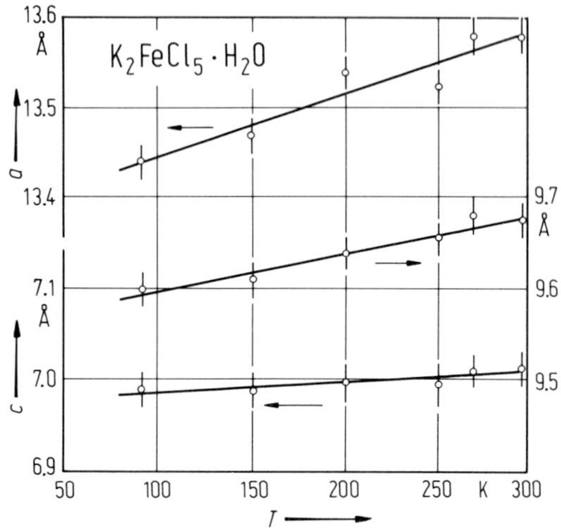

Fig. 39. $K_2FeCl_5 \cdot H_2O$. The orthorhombic lattice parameters *a*, *b* and *c* vs. temperature *T* [88P]. See also Table 9 for appreciably different data at RT.

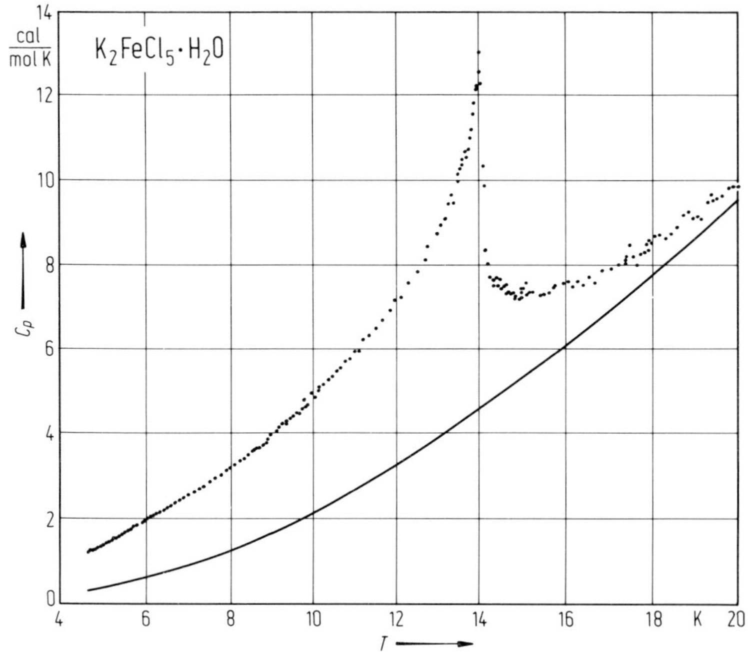

Fig. 40. $K_2FeCl_5 \cdot H_2O$. Molar specific heat C_p vs. temperature *T*. The curve represents the estimated lattice contribution [78M].

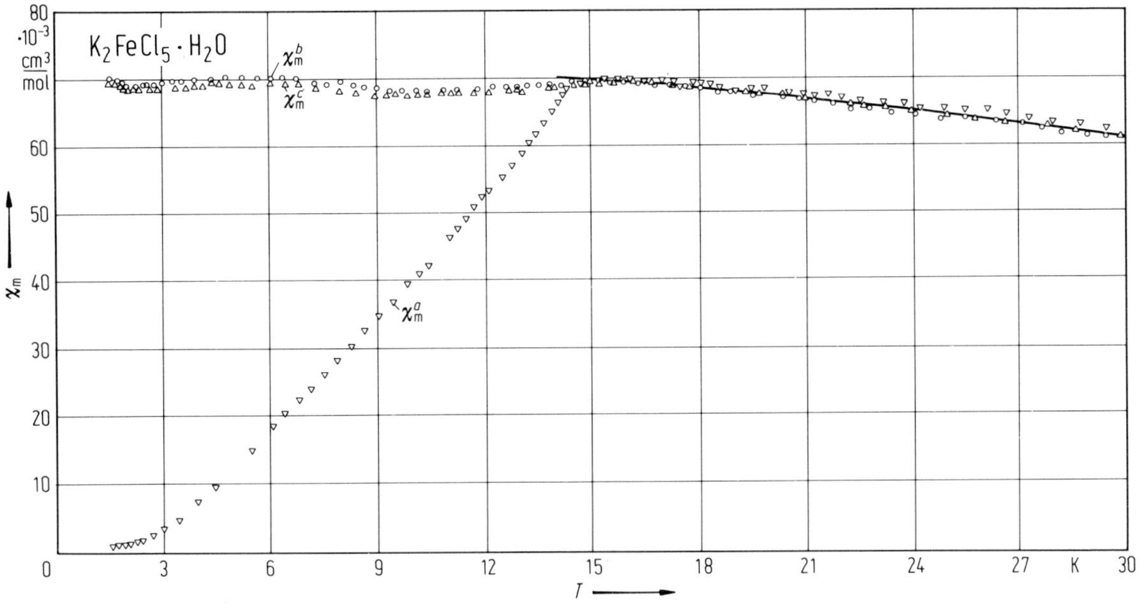

Fig. 41. $K_2FeCl_5 \cdot H_2O$, single crystal. Molar magnetic susceptibility χ_m vs. temperature T for the three principal crystallographic directions. The curve represents a calculated fit with $g = 2.00$ and $J/k = -1.55(15)$ K [78M].

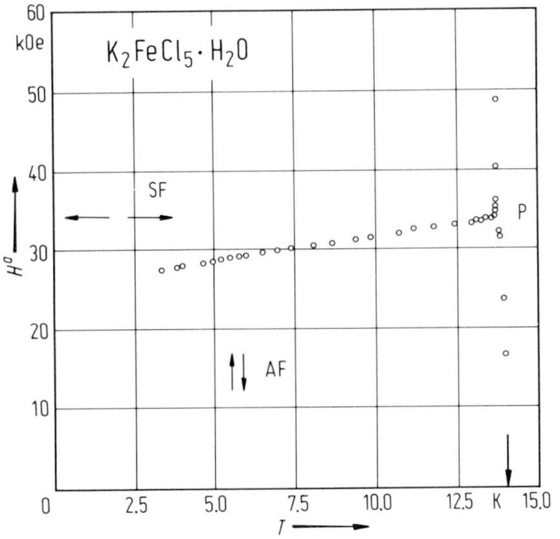

Fig. 42. $K_2FeCl_5 \cdot H_2O$, single crystal. Magnetic phase boundaries as derived from ac susceptibility measurements in applied magnetic fields H^a parallel to the easy magnetic a axis [80P].

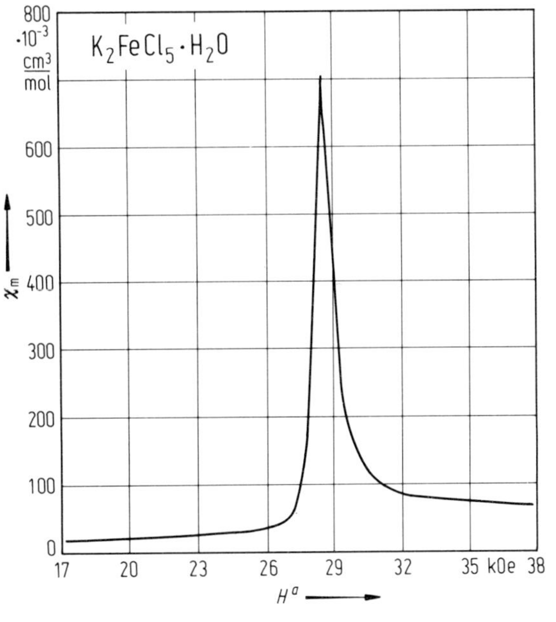

Fig. 43. $K_2FeCl_5 \cdot H_2O$, single crystal, $T = 4.2$ K. Low-field molar magnetic susceptibility χ_m in the presence of a magnetic field H^a applied parallel to the easy magnetic axis. The peak corresponds to a first-order AF-SF transition [91P].

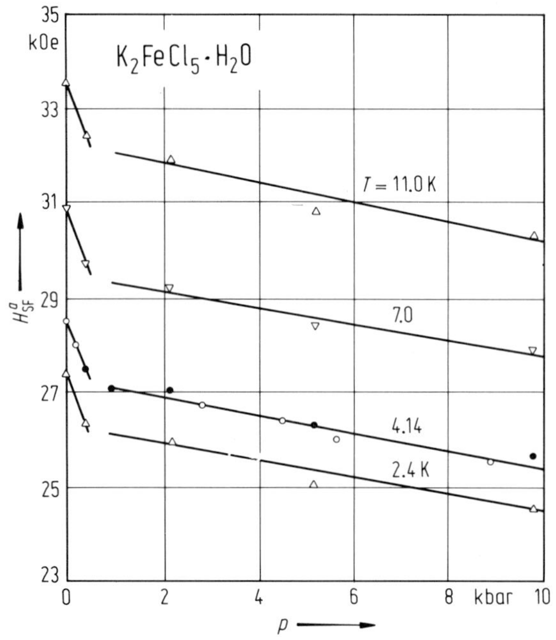

Fig. 44. $K_2FeCl_5 \cdot H_2O$, single crystal. Influence of hydrostatic pressure p on the spin-flop field H^a_{SF} applied in the easy magnetic direction, measured at various temperatures T and derived from magnetisation measurements [81O].

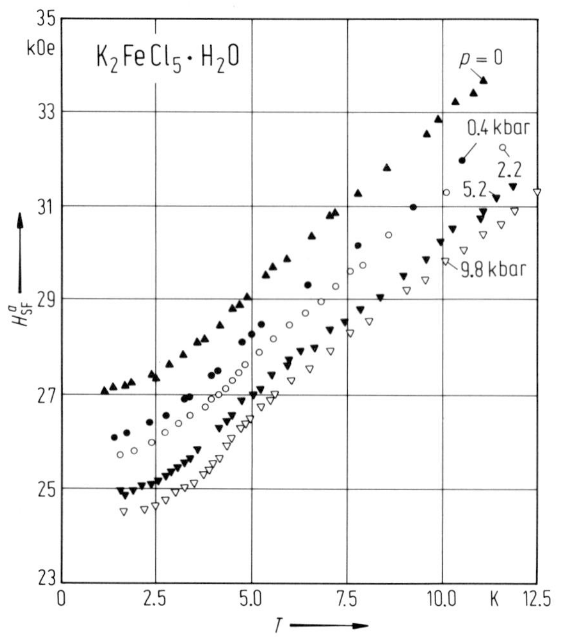

Fig. 45. $K_2FeCl_5 \cdot H_2O$, single crystal. Similar to Fig. 44 but now with the pressure p as a parameter [81O].

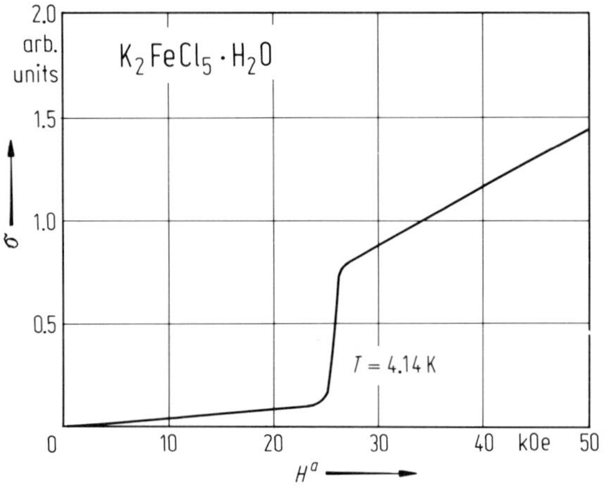

Fig. 46. $K_2FeCl_5 \cdot H_2O$, single crystal under a hydrostatic pressure of 5.2 kbar. Magnetisation σ in arbitrary units as a function of an applied field H^a in the crystallographic a direction [81O].

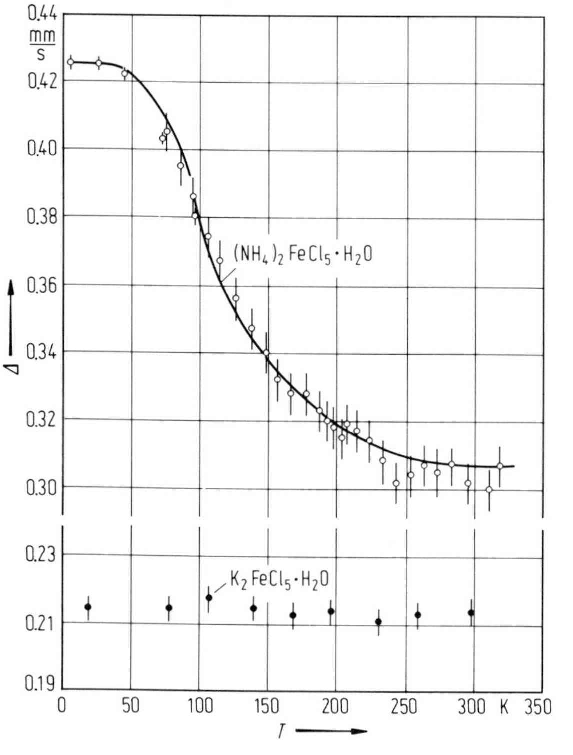

Fig. 47. $K_2FeCl_5 \cdot H_2O$; $(NH_4)_2FeCl_5 \cdot H_2O$. Quadrupole splitting Δ vs. temperature T in the paramagnetic state [88P, 85P1]. Solid line: theoretical fit.

$Rb_2FeCl_5 \cdot H_2O$

Specific heat Fig. 48. Anomaly at $T_N = 10.00(2)$ K [85P2], to be compared to 10.03(5) in [77C]. Magnetic susceptibility Figs. 49–51. High-temperature analysis gives $\Theta_p = -35.0$ K, $J/k = -1.39$ K and $J'/J = 0.33$ [77C]. Details [85P2]. Magnetic phase diagram Fig. 52 with the triple point $T = 9.75$ K and $H = 17.8$ kOe [79O] or $T = 10.0(1)$ K and $H = 18.0(2)$ kOe, with $H_{SF}(0) = 14$ kOe, $H_A = 0.6$ kOe, $H_E = 172$ kOe, $H_A/H_E = 3.4 \cdot 10^{-3}$ [80P]. Mössbauer data at RT: $\Delta = 0.3(1)$ mm s^{-1} and $\delta = 0.20(1)$ mm s^{-1} with respect to metallic iron [85P1].

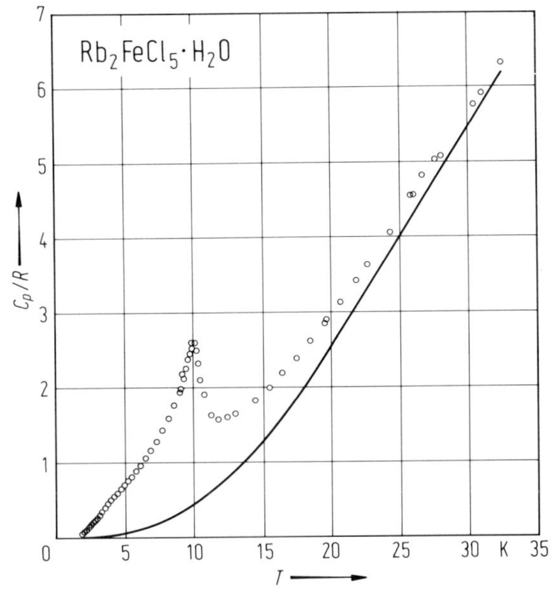

Fig. 48. $Rb_2FeCl_5 \cdot H_2O$, powdered crystals. Specific heat C_p in units of the gas constant R vs. temperature T. The curve represents a calculated lattice contribution [85P2].

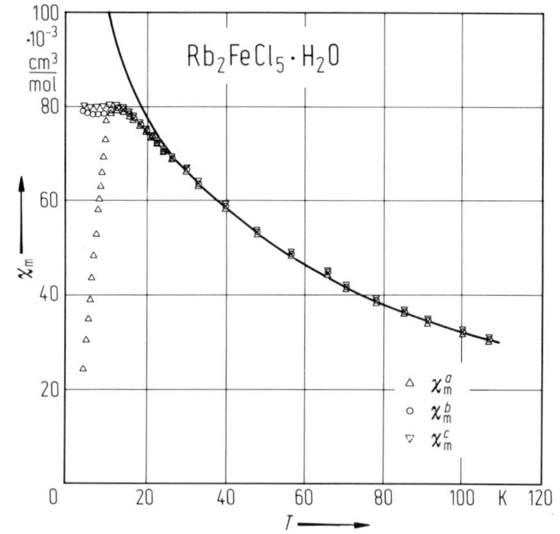

Fig. 49. $Rb_2FeCl_5 \cdot H_2O$, single crystal. Temperature dependence of the low-field molar magnetic susceptibility χ_m along the principal crystallographic axes. The solid curve represents a Curie-Weiss law [79O].

For Fig. 50, see next page.

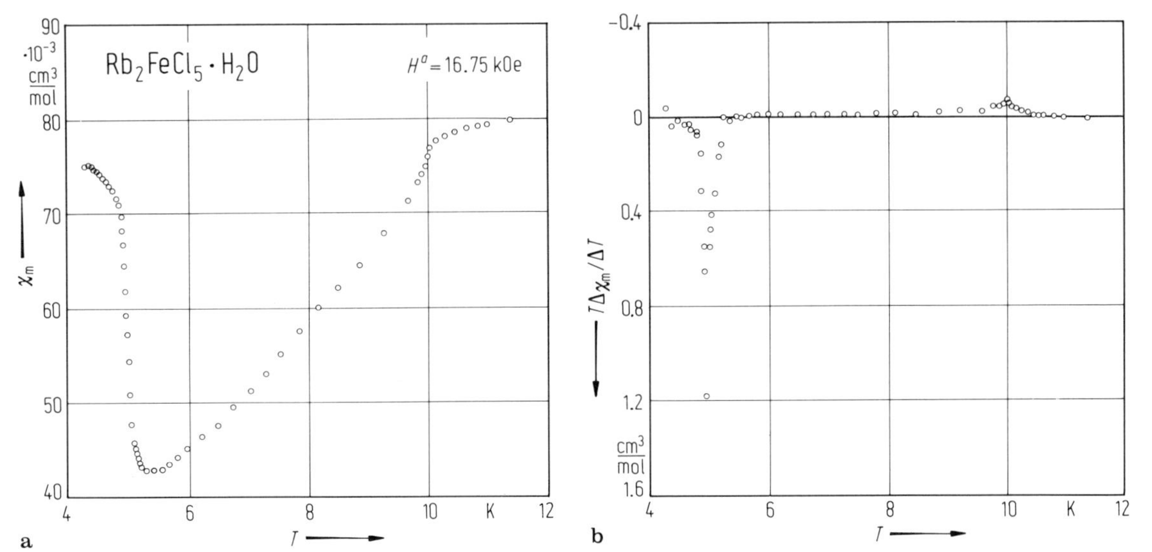

Fig. 51. $Rb_2FeCl_5 \cdot H_2O$. (a) similar to Fig. 50 but with expanded scale, (b) first derivative of the curve in (a): $T \cdot \Delta\chi_m/\Delta T$ [79O].

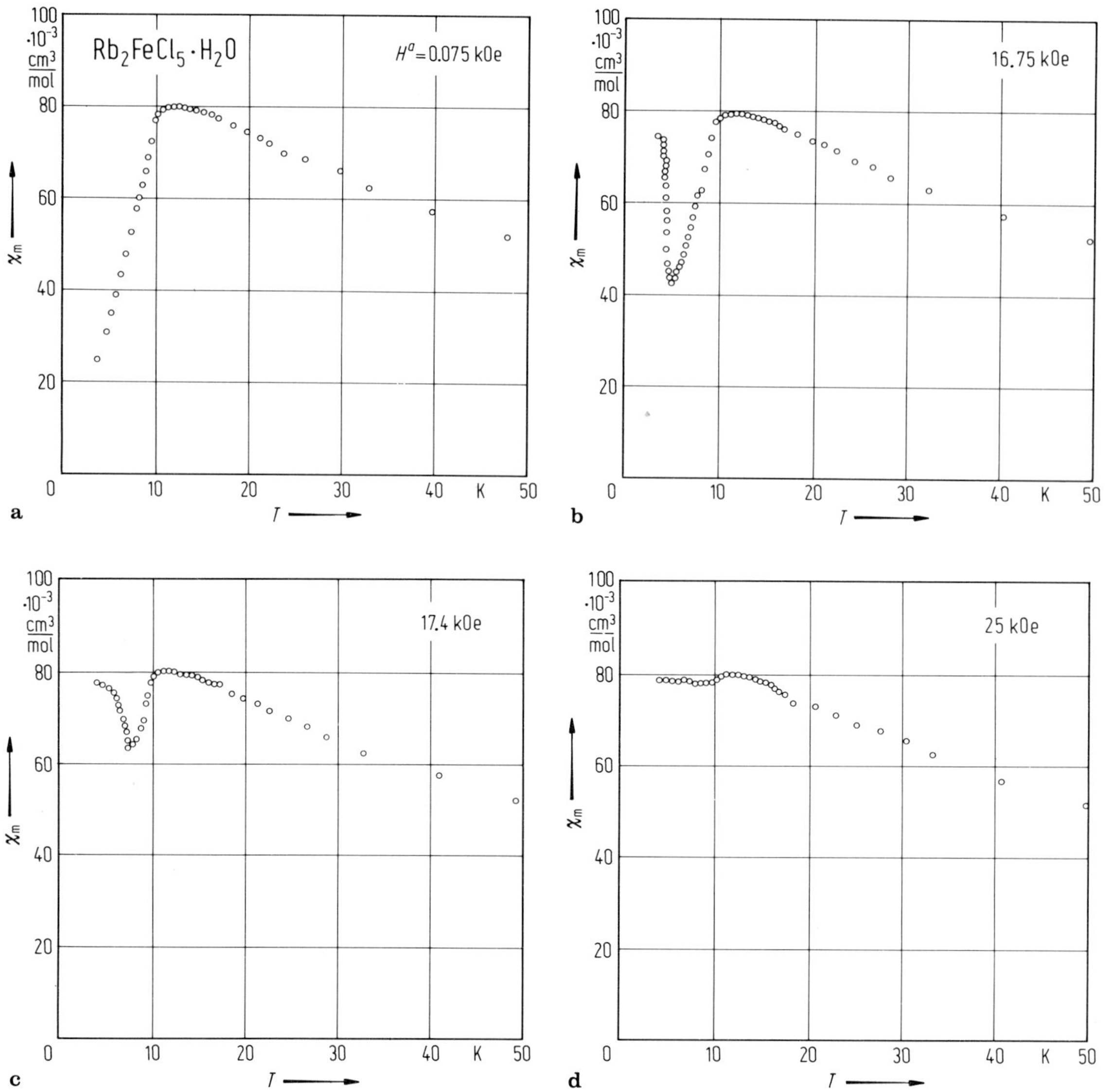

Fig. 50. $Rb_2FeCl_5 \cdot H_2O$, single crystal. Molar magnetic susceptibility χ_m vs. temperature T measured in magnetic fields of various strengths H^a applied parallel to the easy magnetic a axis. Uncorrected for demagnetising effects [79O].

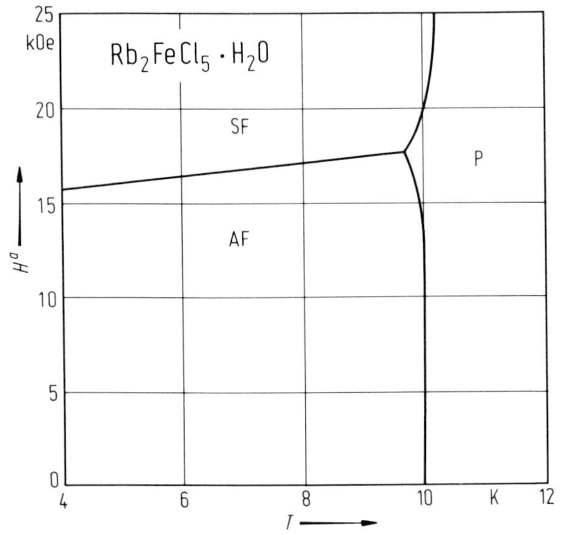

Fig. 52. $Rb_2FeCl_5 \cdot H_2O$, single crystal. (H, T) phase diagram for an applied field along the magnetically preferential crystallographic a axis [79O].

$Cs_2FeCl_5 \cdot H_2O$

Specific heat Figs. 53, 54. $T_N = 6.54(2)$ K [82P]. Magnetic susceptibility Fig. 55, $\Theta_p = -14.0$ K, $g = 1.97$ [79O]. High-temperature analysis: simple cubic 3D Heisenberg AF system parallel to the a axis, though some uncertainty remains [82P]. $J/k = -0.310$ K(5) [77C, 78P]. From spin wave theory applied to the Mössbauer spectra is derived $H_E = 65$ kOe, $H_A = 2$ kOe with $H_A/H_E = 0.03$ [87J]. Magnetic phase diagrams Figs. 56, 57. Bicritical point $T = 6.25$ K, $H = 14.5$ kOe, $H_A/H_E = 1.2 \cdot 10^{-2}$ and $H_{SF}(0) = 11.5$ kOe [79O, 78P].

Mössbauer data: Néel temperature $T_N = 6.75(50)$ K [83G] and 6.75(10) [87J].

- At 7 K, just above T_N, the spectra consist of a doublet with $\Delta = 0.30$ mm s^{-1}, $\delta = 0.53$ or 0.41 mm s^{-1} with respect to the ^{57}Co/Rh source or metallic iron at RT, respectively [87J].
- At 77 K the principal component V_{zz} of the EFG and the angles between V_{zz} and the crystal axes are given by one of the combinations in the following table:
  $V_{zz} = -0.30$ mm s^{-1}, $\theta_a = 52°$, $\theta_b = 49°$, $\theta_c = 63°$ or
  $V_{zz} = +0.30$ mm s^{-1}, 58°, 61°, 47° [87J].
- At 151.5(5) K a crystallographic phase transition takes place, the EFG becoming symmetric in the high-temperature phase, in the low-temperature phase its saturation value $\Delta = 0.295(5)$ mm s^{-1}, see Fig. 58 [87C].

$H_{hf}(0) = 520.0(5)$ kOe [83G] or 520(5) kOe [87J], see Fig. 59. For the influence of an externally applied magnetic field on the effective hyperfine field see Figs. 60, 61. Critical fluctuations of the ordering near $T_N = 6.63$ K which depend on the critical parameter $\beta = 0.35$ have been measured by [88C].

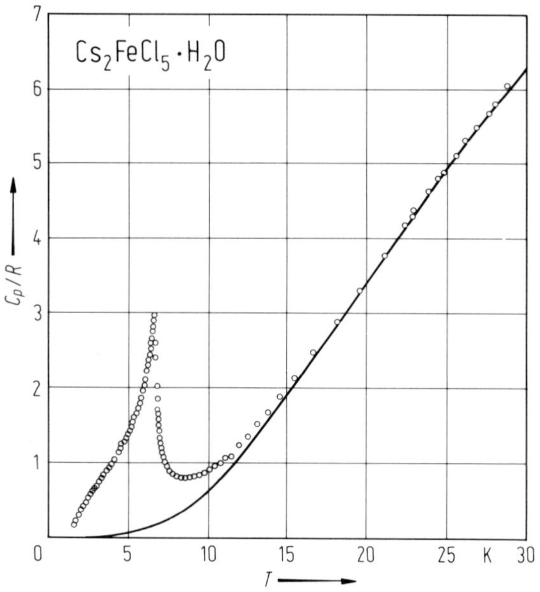

Fig. 53. $Cs_2FeCl_5 \cdot H_2O$. Temperature dependence of the heat capacity C_p in units of the gas constant R. Solid curve is calculated lattice contribution [82P].

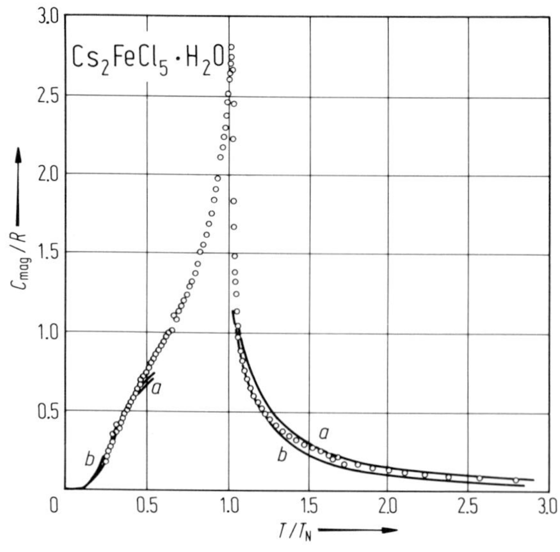

Fig. 54. $Cs_2FeCl_5 \cdot H_2O$. Temperature dependence of the magnetic contribution to the heat capacity C_{mag}, measured in units of the gas constant R. $T_N = 6.54(2)$ K. Curves calculated from spin-wave theory for the Heisenberg $S = 5/2$ antiferromagnet for an sc (a) and a bcc (b) lattice [82P].

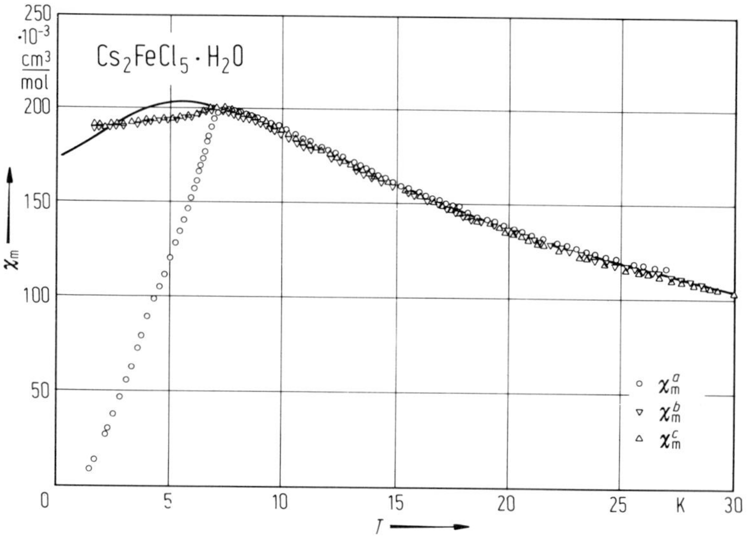

Fig. 55. $Cs_2FeCl_5 \cdot H_2O$, single crystal. Low-field molar magnetic susceptibility χ_m vs. temperature T measured in the direction of the three principal crystallographic directions. The drawn curve represents data fitted to a Curie-Weiss law [77C, see also 79O].

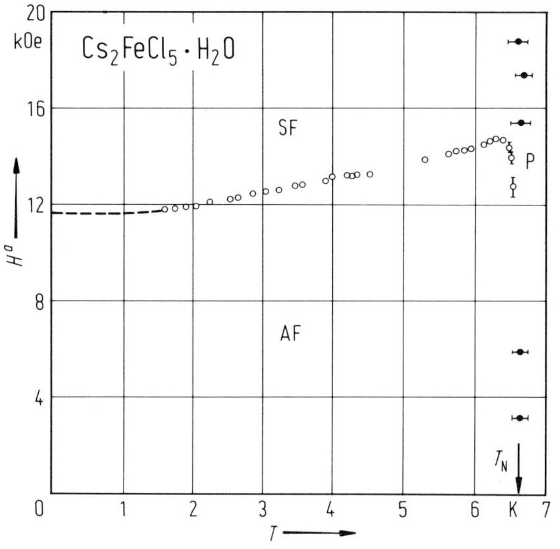

Fig. 56. $Cs_2FeCl_5 \cdot H_2O$, single crystal. Magnetic phase diagram for fields applied along the magnetically preferred *a* axis, as derived from magnetic susceptibility curves. ○ measured at constant *T*, ● measured at constant *H* [78P].

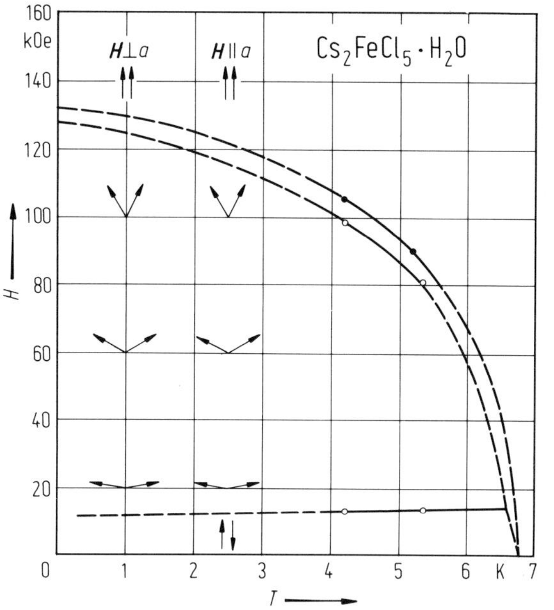

Fig. 57. $Cs_2FeCl_5 \cdot H_2O$, single crystal. Sketch of magnetic phase diagram as derived from Mössbauer spectra measured in the presence of a magnetic field *H*, directed parallel or normal to the magnetic easy *a* axis [87J].

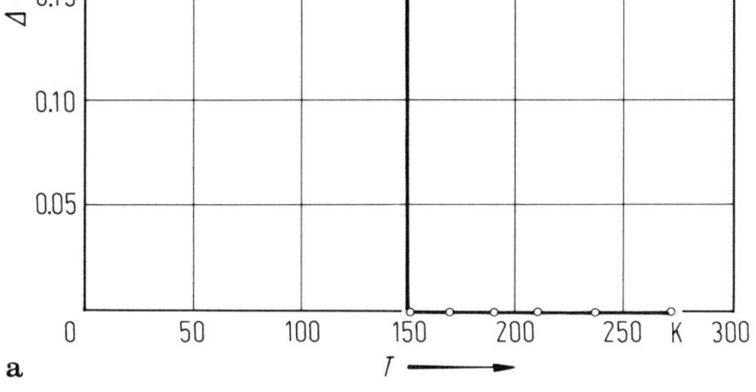

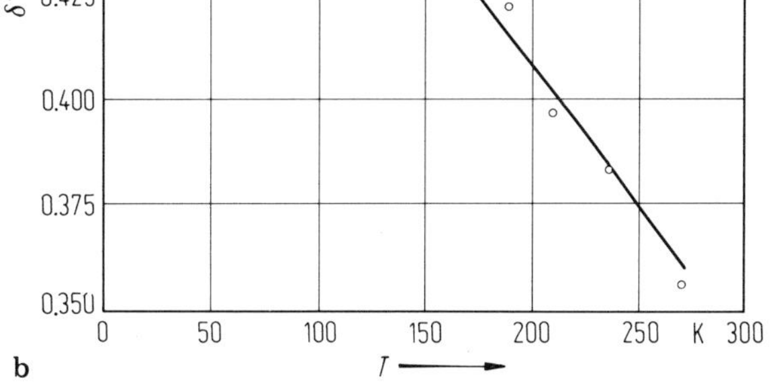

Fig. 58. $Cs_2FeCl_5 \cdot H_2O$, polycrystalline sample. Temperature dependence of (a) the quadrupole splitting Δ and (b) the Mössbauer spectrum centre shift δ'. The curves are fits to a critical exponent model near T_c and to a theoretical expression incorporating the Debye model with $\Theta_D = 301(10)$ K, respectively [87C].

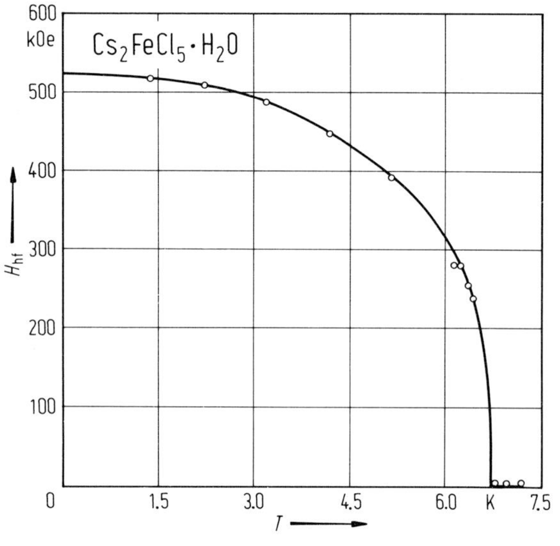

Fig. 59. $Cs_2FeCl_5 \cdot H_2O$, polycrystalline sample. The hyperfine field H_{hf} vs. temperature T [87J].

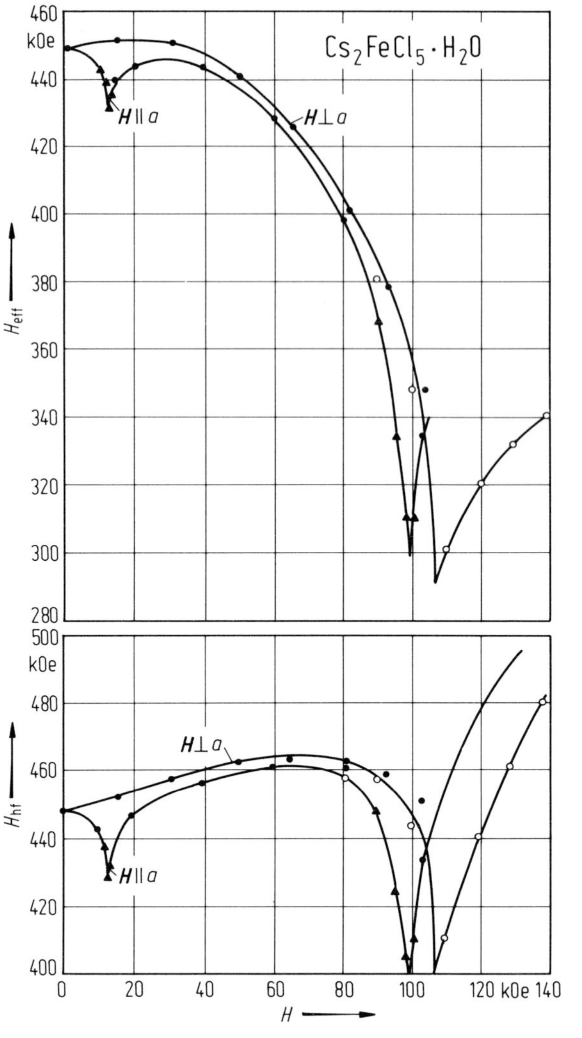

Fig. 61. $Cs_2FeCl_5 \cdot H_2O$, single crystal. The effective internal field H_{eff} and the hyperfine field H_{hf} vs. applied field H as derived from Mössbauer spectra for two orientations of the magnetic field with respect to the crystallographic a axis [87J]. Different symbols correspond to different samples.

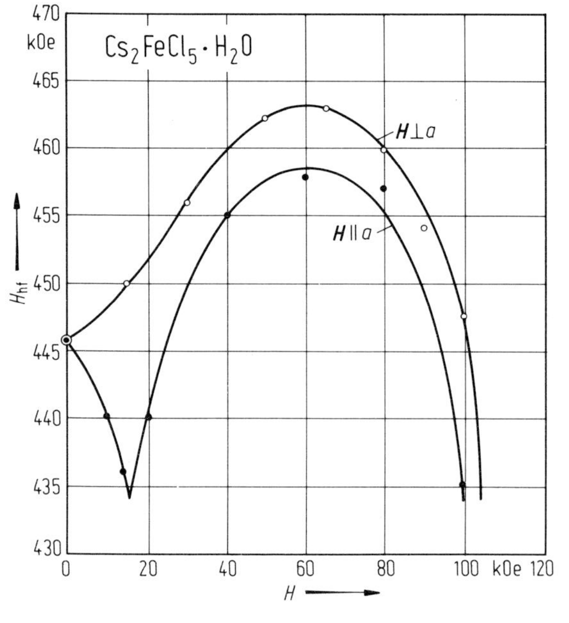

Fig. 60. $Cs_2FeCl_5 \cdot H_2O$, single crystal. $T = 4.2$ K. Magnetic hyperfine field H_{hf} vs. applied magnetic field H. The latter is either parallel or perpendicular to the preferential magnetic axis [83G].

$(NH_4)_2FeCl_5 \cdot H_2O$

Thermal expansion Table 14, Fig. 62. Specific heat Fig. 63. Magnetic susceptibility Fig. 64. Two magnetic transitions at 6.87(1) and 7.25(1) K, $J/k = -1.24(8)$ K, $J'/J = 0.58$ [78M], for a detailed analysis see [85P2]. Mössbauer data Table 15 and Fig. 47. Two phase transitions observed between 7···9 K [88P, 90C].

Table 15. $(NH_4)_2FeCl_5 \cdot H_2O$, Mössbauer parameters [90C].

T [K]	δ [a]) [mm s^{-1}]	Δ or 2ε [mm s^{-1}]	H_{hf} [kOe]
300 [a])	0.42 [c])	0.30 [d])	–
200	0.48	0.32	–
77	0.54	0.41	–
20	0.58	0.42	–
9.5	0.58	0.42	–
9	0.58	0.41	–
4.2 [b])	0.58; 0.55	– 0.47; + 0.12	{442; 442 418; 418

[a]) Isomer shift relative to Fe-metal at room temperature.
[b]) Fitted with two distinct Fe^{3+} ions, each showing two hyperfine fields.
[c]) 0.415(7) acc. to [85P1].
[d]) 0.302(6) acc. to [85P1].

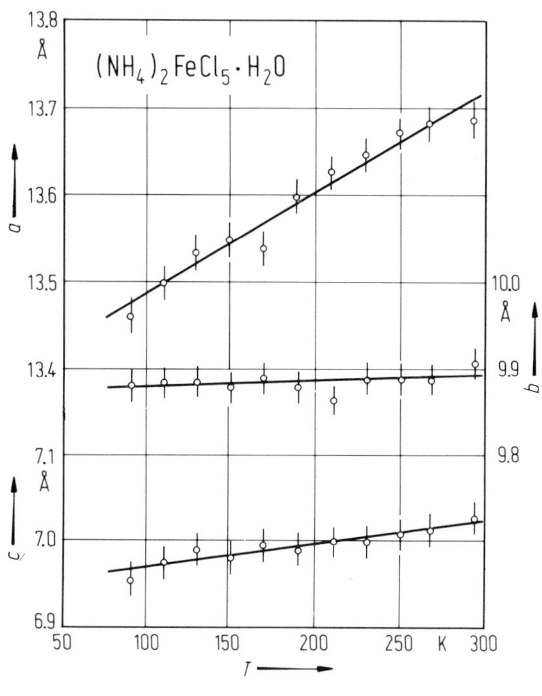

Fig. 62. $(NH_4)_2FeCl_5 \cdot H_2O$. The orthorhombic lattice parameters a, b and c vs. temperature T [88P].

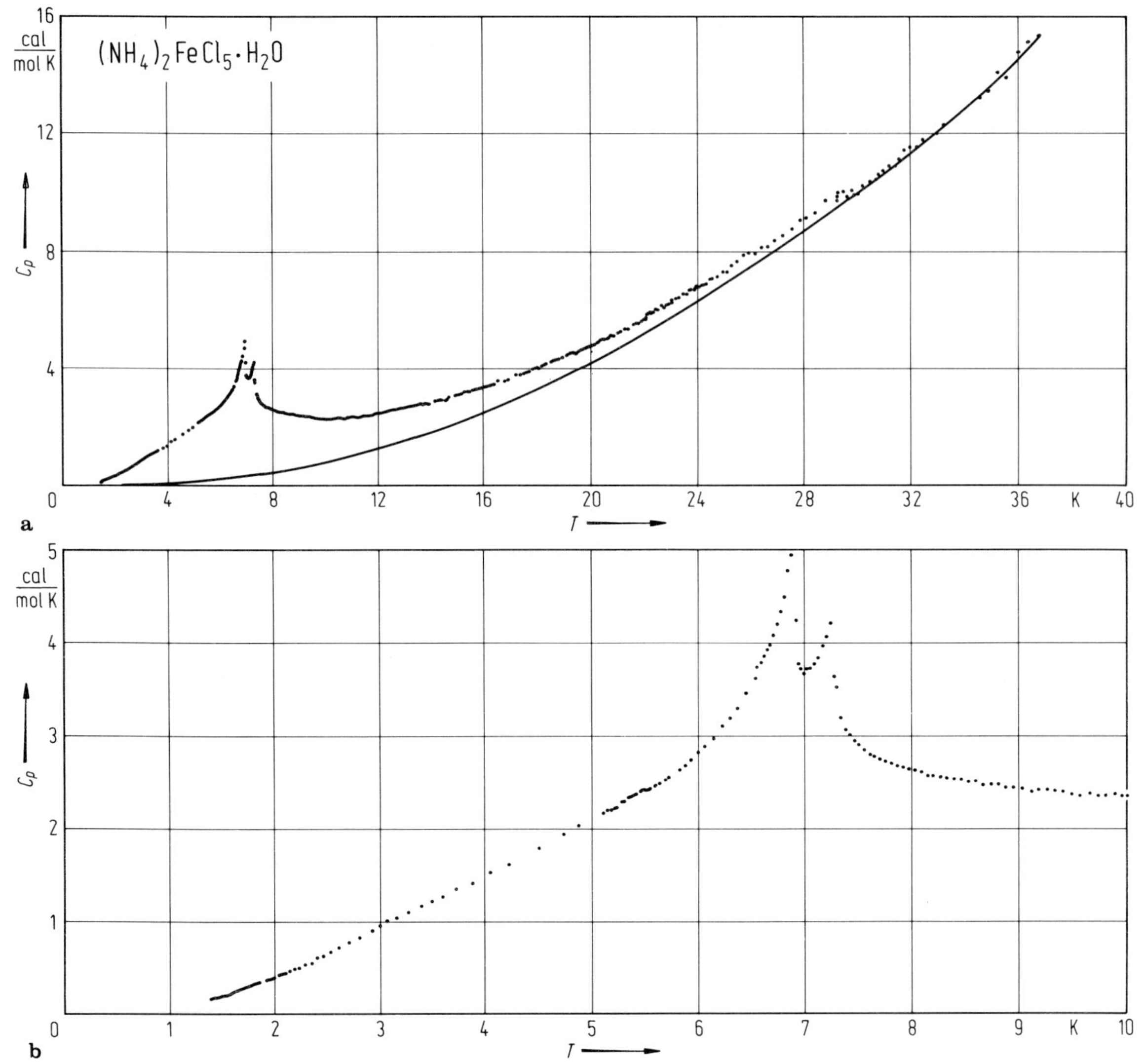

Fig. 63. $(NH_4)_2FeCl_5 \cdot H_2O$. Molar specific heat C_p vs. temperature T. The solid curve in (a) represents the estimated lattice contribution. Fig.(b) shows C_p vs. T on an enlarged scale in the low T-range [78M].

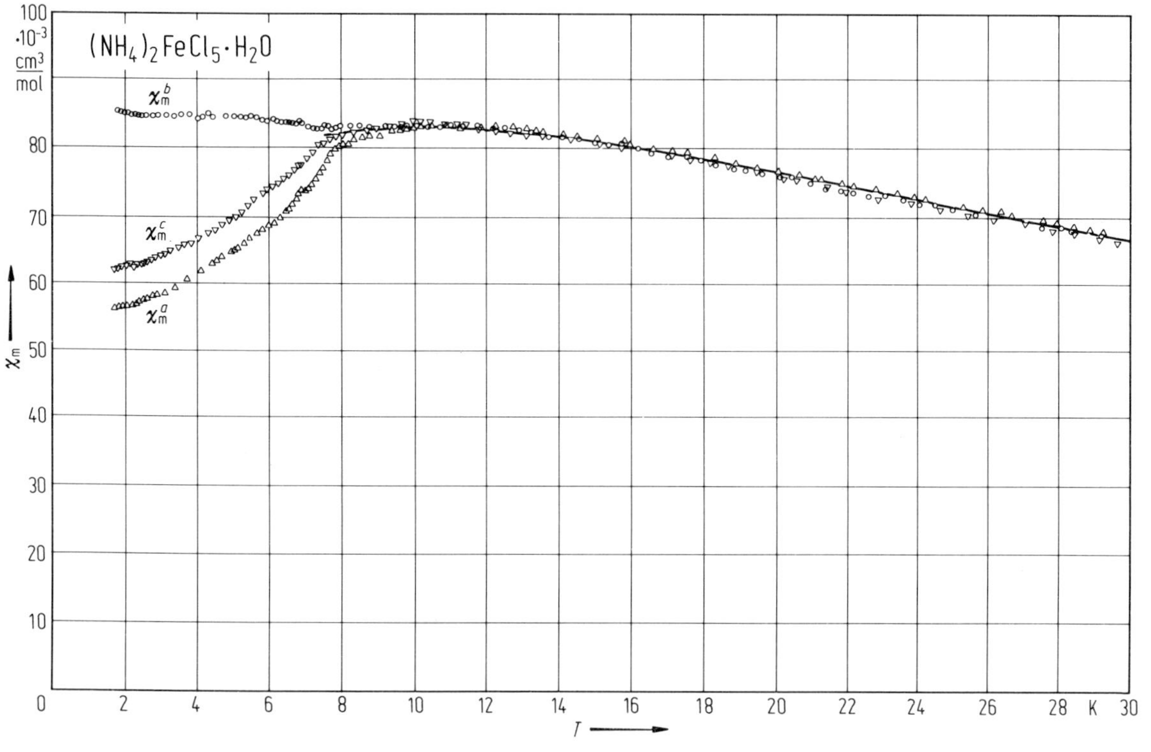

Fig. 64. $(NH_4)_2FeCl_5 \cdot H_2O$, single crystal. Molar magnetic susceptibility χ_m vs. temperature T for the three principal crystallographic directions. The solid curve represents a calculated fit with $g = 2.00$ and $J/k = -1.24(8)$ K [78M].

$K_2Fe(Cl_{1-x}Br_x)_5 \cdot H_2O$

Magnetic susceptibility Fig. 65, $T_N = 15.2$ K for x = 0.3 [91P]. The low-field susceptibility shows a complicated dependence both on the amplitude of the applied measuring field and on the simultaneous presence of a strong magnetic bias field in the direction of easy magnetisation, Figs. 66, 67. These phenomena have been related to hysteresis effects in the spin-flop region in the case that bromium ions are present [91P, 92P].

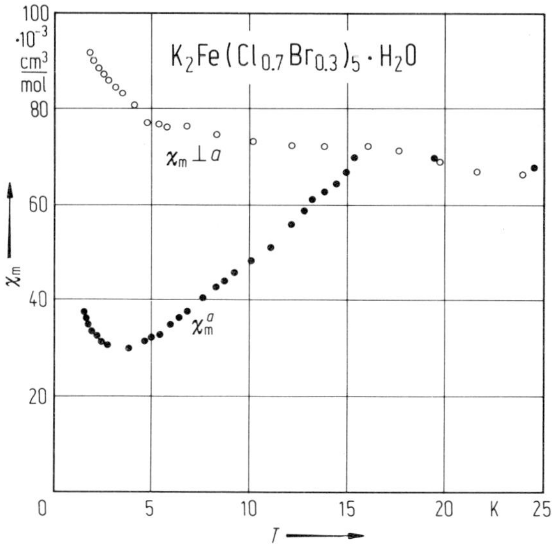

Fig. 65. $K_2Fe(Cl_{0.7}Br_{0.3})_5 \cdot H_2O$, single crystal. Molar magnetic susceptibility, extrapolated to zero applied field χ_m vs. temperature T measured parallel (●) or perpendicular (○) to the easy magnetic a axis. $T_N = 15.2$ K [91P].

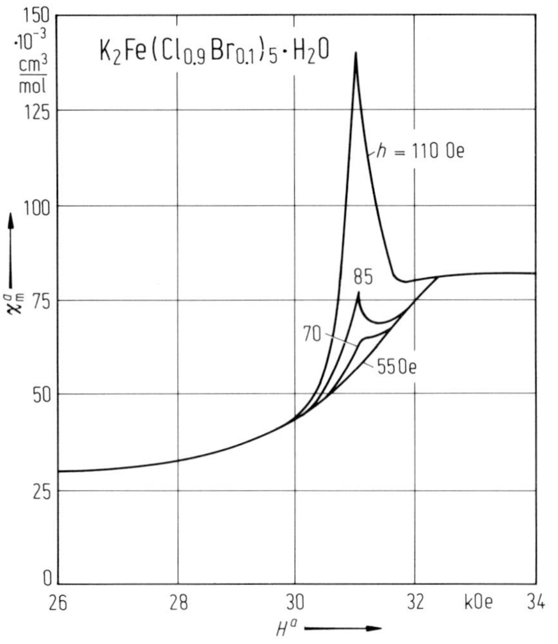

Fig. 66. $K_2Fe(Cl_{0.9}Br_{0.1})_5 \cdot H_2O$, single crystal. $T = 4.2$ K. Dependence of the ac low-field ($h < 160$ Oe) molar magnetic susceptibility χ_m^a in the direction of the a axis, on the presence of an applied magnetic field H^a parallel to the easy magnetic a axis. The peaks correspond to AF↔SF transitions and their height depends on the peak to peak values h of the measuring field [91P].

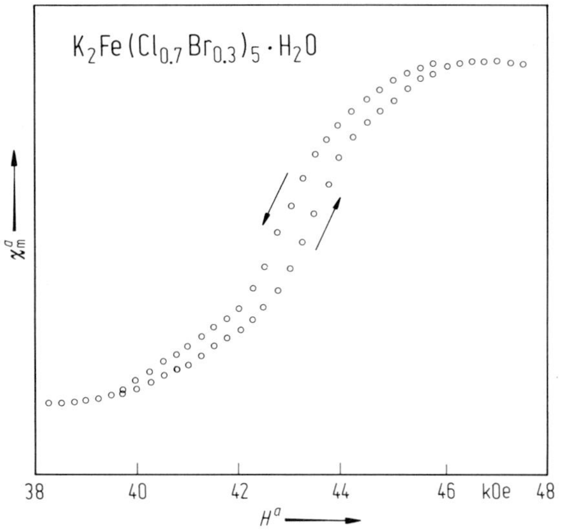

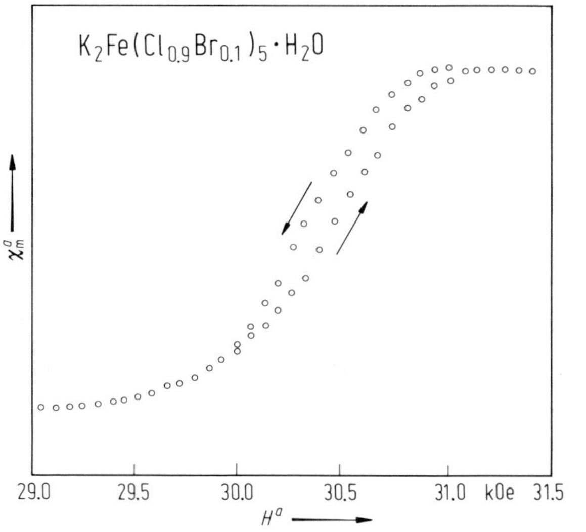

Fig. 67. $K_2Fe(Cl_{0.9}Br_{0.1})_5 \cdot H_2O$, $K_2Fe(Cl_{0.7}Br_{0.3})_5 \cdot H_2O$; $T = 2.0$ K. Measurement as in Fig. 66 with $h = 1$ Oe for increasing and decreasing H^a values. For a bromium-free compound no hysteresis is observed [91P].

$Rb_2FeBr_5 \cdot H_2O$

Magnetic susceptibility Fig. 68, $T_N = 22.9(1)$ K [81C1]. Up to $H = 55$ kOe no spin-flop transition is found [80P].

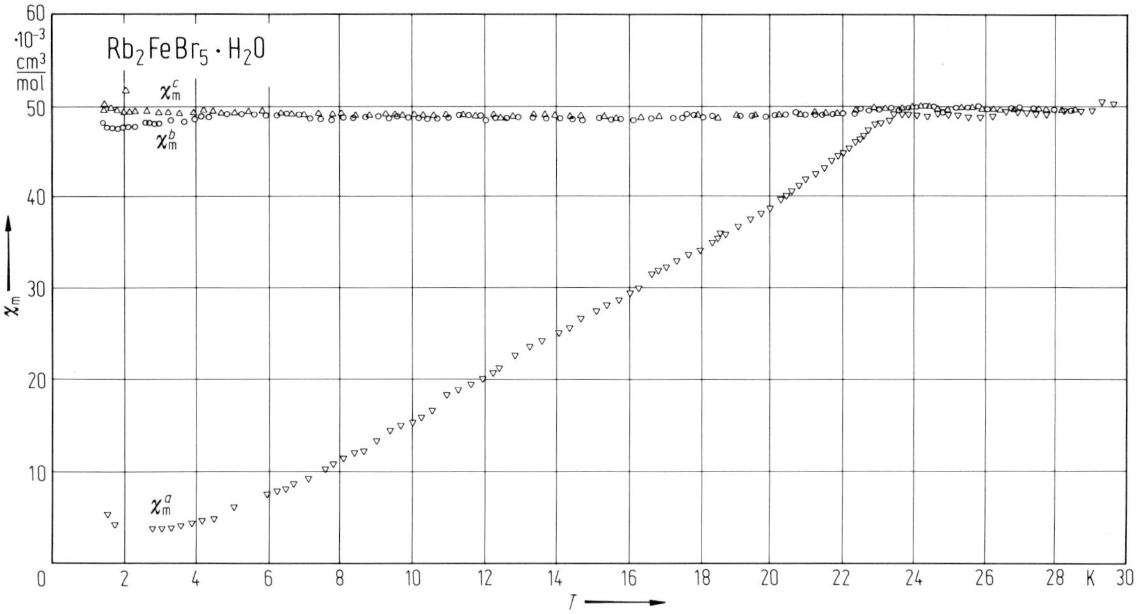

Fig. 68. $Rb_2FeBr_5 \cdot H_2O$, single crystal. Temperature dependence of the molar magnetic susceptibility χ_m in the three principal crystallographic directions [81C1].

$Cs_2FeBr_5 \cdot H_2O$

Magnetic susceptibility Fig. 69, $T_N = 14.2(1)$ K [81C1]. Up to $H = 55$kOe no spin-flop transition is found [80P].

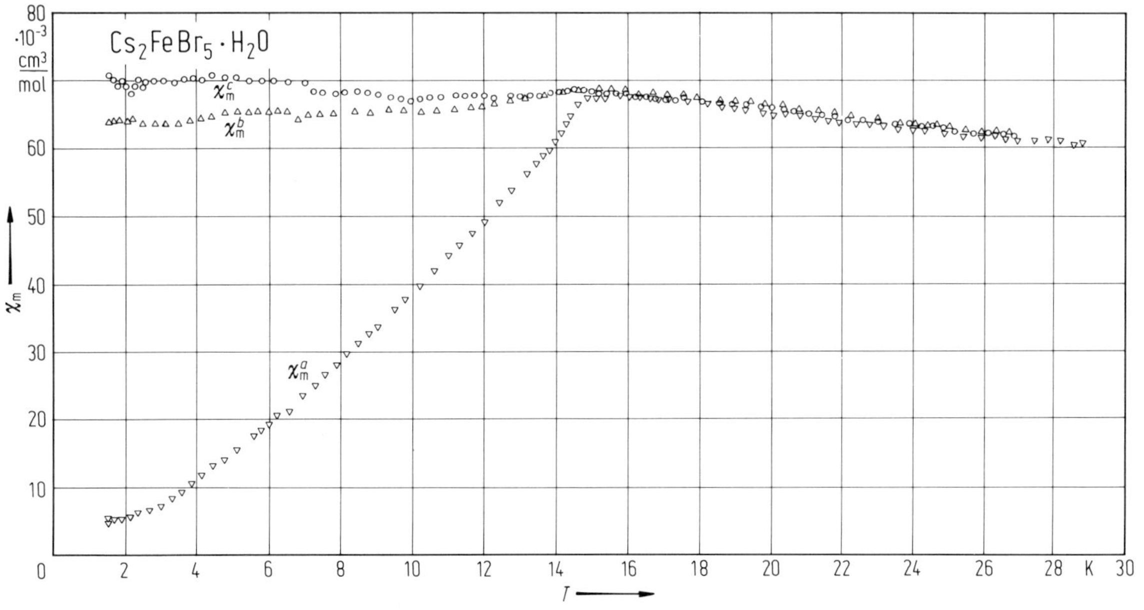

Fig. 69. $Cs_2FeBr_5 \cdot H_2O$, single crystal. Temperature dependence of the molar magnetic susceptibility χ_m in the three principal crystallographic directions [81C1].

9.8.4 References for 9.8

47L Lindqvist, I.: Ark. Kemi, Mineral. Geol. **24 A** (1947) 1.

48B Bellanca, A.: Periodico di Mineralogia **17** (1948) 59.

52C Chang, C. H.: Phys. Rev. **88** (1952) 1422.

64F Fisher, M.: Am. J. Phys. **32** (1964) 343.

64O Oguchi, T.: Phys. Rev. **133 A** (1964) 1098.

68S Smith, T., Friedberg, S. A.: Phys. Rev. **176** (1968) 660.

68W Weng, Chi.Yuan: PhD Thesis, Carnegie Mellon University, Pittsburgh, USA (1968).

69K de Kozak, A.: C. R. Acad. Sci. (Paris) **268 C** (1969) 416.

69S Sears, D. R., Hoard, J. L.: J. Chem. Phys. **50** (1969) 1066.

70T1 Tressaud, A.; Portier, J., de Pape, R., Hagenmuller, P.: J. Solid State Chem. **2** (1970) 269.

71E Edwards, A. J.: J. Chem. Soc. **A** (1971) 2653.

72E Edwards, A. J.: J. Chem. Soc. Dalton Trans. (1972) 816.

73K Kida, J., Watanabe, T.: J. Phys. Soc. Jpn. **34** (1973) 952.

74H Hanžel, D., Moljk, A., Slivnik, J.: Inorg. Nucl. Chem. Lett. **10** (1974) 1.

74J Jacoboni, C., de Pape, R, Poulain, M., leMarquille, J. Y., Grandjean, D.: Acta Crystallogr. **B 30** (1974) 2688.

77C Carlin, R. L., Bhatia, S. N., O'Connor, C. J.: J. Am. Chem. Soc. **99** (1977) 7728.

77D Dance, J. M., Menil, F., Hanžel, D., Sabatier, R., Tressaud, A., le Flem, G., Hagenmuller, P.: Physica **86-88B** (1977) 699.

77G Gupta, G. P., Dickson, D. P. E., Johnson, C. E., Wanklyn, B. M.: J. Phys. C: Solid State Phys. **10** (1977) L459.

77H Hanžel, D., Tressaud, A., Dance, J. M., Hagenmuller, P.: Solid State Commun. **22** (1977) 215.

77V Vlasse, M., Matejka, G., Tressaud, A., Wanklyn, B. M.: Acta Crystallogr. **B33** (1977) 3377.

78B Bukovec, P., Kaučič, V.: Acta Crystallogr. **B34** (1978) 3339.

78F Figgis, B. N., Raston, C. L., Sharma, R. P., White, A. H.: Aust. J. Chem. **31** (1978) 2717.

78G Gupta, G. P., Dickson, D. P. E., Johnson, C. E., Wanklyn, B. M.: J. Phys. C: Solid State Phys. **11** (1978) 3889.

78G1 Gupta, G. P., Dickson, D. P. E., Johnson, C. E.: J. Phys. C: Solid State Phys. **11** (1978) 215.

78G2 Günter, J. R., Matthieu, J. P., Oswald, H. R.: Helv. Chim. Acta **61** (1978) 328.

78K Kaučič, V., Bukovec, P.: Acta Crystallogr. **B 34** (1978) 3337.

78M McElearney, J. N., Merchant, S.: Inorg. Chem. **17** (1978) 1207.

78P Paduan-Filho, A., Palacio, F., Carlin, R. L.: J. Phys. (Paris) **39** (1978) L-279.

79G Gupta, G. P., Dickson, D. P. E., Johnson, C. E.: J. Phys. C: Solid State Phys. **12** (1979) 2411.

79N Negita, K., Nakamura, N., Chihara, H.: Chem. Phys. Lett. **63** (1979) 187.

79O O'Connor, C. J., Deaver, Jr., B. S., Sinn, E.: J. Chem. Phys. **70** (1979) 5161.

79S Sabatier, R., Soubeyroux, J. L., Dance, J. M., Tressaud, A., Wintenberger, M., Fruchart, D.: Solid State Commun. **29** (1979) 383.

80D Dance, J. M., Soubeyroux, J. L., Sabatier, R., Fournes, L., Tressaud, A., Hagenmuller, P.: J. Magn. Magn. Mater. **15–18** (1980) 534.

80G Greedan, J. E., Hewitt, D. C., Faggiani, R., Brown, I. D.: Acta Crystallogr. **B36** (1980) 1927.

80O Ortiz, W. A., Paduan-Filho, A., Missell, F. P.: Phys. Lett. **77 A** (1980) 183.

80P Palacio, F., Paduan-Filho, A., Carlin, R. L.: Phys. Rev. **B 21** (1980) 296.

81C1 Carlin, R. L., O'Connor, C. J.: Chem. Phys. Lett. **78** (1981) 528.

81C2 Cooper, D. M., Dickson, D. P. E., Johnson, C. E.: Hyperfine Interact. **10** (1981) 783.

81F Fourquet, J. L., Plet, F., de Pape, R.: Rev. Chim. Minér. **18** (1981) 19.

81O Ortiz, W. A., Paduan-Filho, A., Missell, F. P.: J. Magn. Magn. Mater. **24** (1981) 67.

81T Tressaud, A., Soubeyroux, J. L., Dance, J. M., Sabatier, R., Hagenmuller, P.: Solid State Commun. **37** (1981) 479.

82B Boersma, F., Cooper, D. M., de Jonge, W. J. M., Dickson, D. P. E., Johnson, C. E., Tinus, A. M. C.: J. Phys. C: Solid State Phys. **15** (1982) 4141.

82B1 Belov, N. V., Golovastikov, N. I., Ivashchenko, A. N., Kotyuzhanskii, B. Ya., Mel'nikov, O. K., Filippov, V. I.: Kristallografiya **27** (1982) 511; Sov. Phys. Crystallogr. (English Transl.) **27** (1982) 309.

82C1 Cooper, D. M., Dickson, D. P. E., Johnson, C. E.: J. Phys. C: Solid State Phys. **15** (1982) 1025.

82C2 Cooper, D. M., Gupta, G. P., Dickson, D. P. E., Johnson, C. E.: J. Phys. C: Solid State Phys. **15** (1982) 3391.

82F1 Fourquet, J. L., de Pape, R., Teillet, J., Varret, F., Papaefthymiou, G. C.: J. Magn. Magn. Mater. **27** (1982) 209.

82F2 Forquet, J. L., de Pape, R., Teillet, J.: Solid State Chem. 1982 Proceedings of the Second European Conference, Veldhoven, The Netherlands 7–9 June 1982. 1983 Elsevier Scientific Publishing Company, Amsterdam, p. 597.

82P Puértolas, J. A., Navarro, R., Palacio, F., Bartolomé, J., González, D., Carlin, R. L.: Phys. Rev. **B 26** (1982) 395.

83C Carlin, R. L., Burriel, R.: Phys. Rev. **B 27** (1983) 3012.

83G Gupta, G. P., Baines, J. A., Dickson, D. P. E., Johnson, C. E., Thomas, M. F.: J. Magn. Magn. Mater. **31–34** (1983) 763.

83K Kotyuzhanskii, B. Ya.· Fiz. Tverd. Tela (USSR) **25** (1983) 1898; Sov. Phys. Solid State (English Transl.) **25** (1983) 1097.

84C Carlin, R. L., Burriel, R., Rojo, J. A., Palacio, F.: Inorg. Chem. **23** (1984) 2213.

85P1 Partiti, C. S. M., Piccini, A., Rechenberg, H. R.: Solid State Commun. **56** (1985) 687.

85P2 Puértolas, J. A., Navarro, R., Palacio, F., Bartolomé, J., Gonzáles, D., Carlin, R. L.: Phys. Rev. **B31** (1985) 516.

86M Massa, W.: Acta Crystallogr. **C42** (1986) 644.

86P Pankhurst, Q. A., Johnson, C. E., Thomas, M. F.: Hyperfine Interact. **29** (1986) 1361.

87C Chadwick, J., Thomas, M. F.: J. Phys. C: Solid State Phys. **20** (1987) 3979.

87J Johnson, J. A., Johnson, C. E., Thomas, M. F.: J. Phys. C: Solid State Phys. **20** (1987) 91.

87M Massa, W., Pebler, J., Hahn, F., Babel, D.: NATO ASI Series B **168** (1987) 429.

87N Nuñez, P., Darriet, J., Bukovec, P., Tressaud, A., Hagenmuller, P.: Mater Res. Bull. **22** (1987) 661.

87P Pebler, J., Massa, W., Lass, H., Ziegler, B.: J. Solid State Chem. **71** (1987) 87.

88C Chadwick, J., Thomas, M. F., Johnson, C. E., Jones, D. H.: J. Phys. C: Solid State Phys. **21** (1988) 6159.

88F Fourquet, J. L., Plet, F., Calage, Y.: J. Solid State Chem. **74** (1988) 34.

88P Partiti, C. S. M. Rechenberg, H. R., Sanchez, J. P.: J. Phys. C: Solid State Phys. **21** (1988) 5825.

89F Fourquet, J. L., le Bail, A., Duroy, H., Moron, M. C.: Eur. J. Solid State Inorg. Chem. **26** (1989) 435.

89P Pebler, J.: Inorg. Chem. **28** (1989) 1038.

89T Takács, L., Reiff, W. M.: J. Phys. Chem. Solids **50** (1989) 33.

90C Calage, Y., Dormann, J. L., Moron, M. C., Palacio, F.: Hyperfine Interact. **54** (1990) 483.

90H Hahn, F., Massa, W.: Z. Naturforsch. **45b** (1990) 1341.

90L LeBail, A., Desert, A., Fourquet, J. L.: J. Solid State Chem. **84** (1990) 408.

91C Calage, Y., Moron, M. C., Fourquet, J. L., Palacio, F.: J. Magn. Magn. Mater. **98** (1991) 79.

91P Paduan-Filho, A., Becerra, C. C., Palacio, F.: Phys. Rev. **B 43** (1991) 11107.

92N Nuñez, P., Tressaud, A., Darriet, J., Hagenmuller, P., Hahn, G., Frenzen, G., Massa, W., Babel, D.: Inorg. Chem. **31** (1992) 770.

92P Paduan-Filho, A., Becerra, C. C., Westphal, C., Gabás, M., Palacio, F.: J. Magn. Magn. Mater. **104–107** (1992) 269.

9.9 $A_2MM'X_6$ compounds (A, M, M' = metal, X = halogen element)

9.9.0 Introduction

The table and figures of this section present data on magnetic and related properties of ionic quaternary and pseudoquaternary perovskite-type compounds of the formula $A_2MM'X_6$ which contain besides the halogen elements F, Cl, Br and I the monovalent elements of group 1A i.e. Na, K, Rb, Cs represented by the symbol A and/or M as well as the trivalent elements of the 4f transition series Ce, Pr, Nd, Eu, Gd, Tb, Dy, Ho, Er, Tm, Yb, Lu, the 5f transition series U, Np, Pu and the elements Al, Ga, Sc, Y or La, all represented by the symbol M'.

The compounds listed in the table and figures are designated by their chemical formula. The arrangement in the table is according to the position of their elements in the periodic system.

Thus the halogen elements (X) are given in the following order

F, Cl Br, I.

Within one halide system the sequence of the group 1A elements (A, M) is

Na, K, Rb, Cs.

The 4f, 5f transition elements and Al, Ga, Sc, Y, La (M') are listed in the following order:

Al, Ga, Sc, Y, La
Ce, Pr, Nd, Eu, Gd, Tb, Dy, Ho, Er, Tm, Yb, Lu (4f elements)
U, Np, Pu, Am (5f elements)
(V, Cu).

For remarks concerning the information given in the table see the Introduction of 9.6.
For symbols and abbreviations see Section 9.6.02.
For references of this section see Section 9.6.3.

9.9.1 Table

Compound	Θ [K]	T_C [K]	T_N [K]	χ	p_A [μ_B]	p_{eff} [μ_B]	Remarks	Ref.
K_2NaGaF_6: Cr^{3+}							magneto-optical study	80D1
$Rb_2NaY_{0.9}Er_{0.1}F_6$			< 0.45					74B1
Rb_2NaEuF_6				6.38			$\chi(0) \cdot 10^{-3}$ cm^3/mol, Van Vleck susceptibility	74B1
Rb_2NaGdF_6			< 0.45					74B1
Rb_2NaTbF_6				0.312 Fig. 1			$\chi(0)$ cm^3/mol	74B1
Rb_2NaHoF_6				1.04 Figs. 2, 3			$\chi(0)$ cm^3/mol, Van Vleck paramagnetism, below 1 K	74B1
					10.5		p_{Ho} from NMR at low T, $p_{Ho}(H)$: Fig. 4, second order structural phase transition to the tetragonal phase at 172 K from X-ray and at 170 K from NMR	82V1
Rb_2NaErF_6			< 0.45					74B1
Rb_2NaTmF_6				0.0569 Fig. 1			$\chi(0)$ cm^3/mol	74B1
Rb_2NaYbF_6			< 0.45					74B1
Rb_3PrF_6				Fig. 5			not fcc structure	74B1
Cs_2NaAlF_6: Cr^{3+}							ESR position of lines depending on H	90F1
Cs_2NaHoF_6			0.62	Fig. 2			magnetic transition of pure exchange origin or at least partly hyperfine induced suggested	74B1
				Fig. 6				80U1
					Figs. 7, 9		$p_{Ho}(H)$: Fig. 7, ^{19}F NMR frequency shift: Fig. 8, Van Vleck paramagnet with non-magnetic Γ_3-doublet ground state and an enhanced magnetic moment, degeneracy of ground state lifted at phase transition at 395 mK: Fig. 10	83V2

Cs_2KPrF_6		0.0044		$\chi(0)$ cm^3/mol	74B1
		Fig. 5			
Cs_2KHoF_6		Fig. 6			80U1
Cs_2RbHoF_6		Fig. 6			80U1
$K_2Cu_2Cl_6$		Fig. 305 of 9.6			67M1
$Cs_2NaScCl_6$: Sb^{3+}				magnetic field effect on the luminescence of Sb^{3+} ion, for explanation see $Cs_2NaLaCl_6$	88M2
Cs_2NaYCl_6				magneto-optical investigations	90S1
Cs_2NaYCl_6: Sb^{3+}				magnetic field effect of the luminescence of Sb^{3+} ion, for explanation see $Cs_2NaLaCl_6$	88M2
Cs_2NaYCl_6: Cf^{3+}	2.8	Fig. 11	7.36	for $2.2 < T < 14$ K, $g = 6.723$, CEF parameters $W = 1.9$ K, $x = -0.8$	76K3
	-13.5		10.0	for $20 < T < 100$ K	
$Cs_2NaLaCl_6$: Sb^{3+}				magnetic field effects on the luminescence of Sb^{3+}, luminescence spectra and the decay curves of Sb^{3+} at $0 < \mu_0 H < 6$T and at 1.9 K. For $\boldsymbol{H} \parallel$ [100] the decay curves are bi-exponential with a field-dependent fast component and a component which equals the decay at $H = 0$	88M2
$Cs_2NaCeCl_6$	-1.45		1.28	for $3.2 < T < 27$ K, ground level Γ_7	74H1
	-2.20		1.54	for $27 < T < 42$ K	
		Fig. 12		CEF parameters $W = 334$ K, $x = 1$, CEF splitting: Fig. 15, crystal structure: Fig. 13	75D1
				CEF parameters $B_2 = 0$, $B_4 = 438$ K, $B_6 = 24.5$ K	79N1
$Cs_2NaPrCl_6$		8.12		$\chi_m \cdot 10^3$ cm^3/mol, constant from 2.5 to 90 K Γ_1 ground state level suggested	74H1

(continued)

Compound	Θ [K]	T_C [K]	T_N [K]	χ	p_A [μ_B]	p_{eff} [μ_B]	Remarks	Ref.
$Cs_2NaNdCl_6$	0.76					2.76	for $2.4 < T < 15$ K	74H1
	−10.4					3.42	for $15 < T < 70$ K	
				Fig. 14			CEF splitting: Fig. 15, crystal structure: Fig. 13, CEF parameters $W = -19$ K, $x = 0.76$, ground state level Γ_8	75D1
							CEF parameters $B_4 = 322$ K, $B_6 = 25.8$ K	78A1
							CEF parameter $B_2 = 10$ K, $p_{Nd}(H)$: Fig. 16	79N1
$Cs_2NaGdCl_6$						7.64	for $2.5 < T < 30$ K	74H1
$Cs_2NaTbCl_6$	−7.51					9.65	for $25 < T < 75$ K	74H1
				Fig. 17			CEF splitting: Fig. 15, crystal structure: Fig. 13, CEF parameters $W = 1.80$ K, $x = -0.8$	75D1
$Cs_2NaDyCl_6$	−4.07					10.77	$C_m = 14.49$ cm^3 K/mol, for $11 < T < 90$ K	74H1
				Fig. 18			crystal splitting: Fig. 15, crystal structure: Fig. 13, CEF parameters $W = 0.92$ K, $x = -0.63$, ground state level Γ_6	75D1
$Cs_2NaHoCl_6$	−1.66					10.38	for $15 < T < 80$ K	74H1
			0.62	Fig. 19			CEF splitting: Fig. 15, structure: Fig. 13, CEF parameters $W = -1.12$ K, $x = 0.72$, ground state Γ_3	75D1
							CEF parameters $B_2 = 0$, $B_4 = 299$ K, $B_6 = 15.7$ K	79N1
				Figs. 20 ⋯ 23				82S1
			1.5 mK Fig. 25	Fig. 24			heat capacity vs. T and H: Figs. 25 ⋯ 29 T_N calculated 0.48 mK	86S1
			4.8 mK	Fig. 30		0.395		88S1

$Cs_2NaErCl_6$	− 1.63	Fig. 31	8.95	for $2.5 < T < 75$ K	74H1
$Cs_2NaTmCl_6$	− 16.20		7.94	for $42 < T < 65$ K	74H1
		Fig. 32		CEF splitting: Fig. 15, crystal structure: Fig. 13, CEF parameters $W = -2.50$ K, $x = -0.90$	75D1
$Cs_2NaYbCl_6$	0.95		2.46	for $2.5 < T < 20$ K, from ESR $g = 2.584$ for Yb^{3+},	71 K1
	70		4.47	for $45 < T < 100$ K, ground state level Γ_6	
		Fig. 33		CEF splitting: Fig. 15, CEF parameters $W = 20$ K, $x = -0.85$	75D1
$Cs_2NaLuCl_6$: Cm^{3+}	3.87	Fig. 34	7.90	for $7.5 < T < 25$ K, $C_m = 7.81$ cm³ K/mol,	74H2
	1.15		7.48	p_{eff}/C_m, for $25 < T < 45$ K, $C_m = 6.99$ cm³ K/mol	
$Cs_2NaLuCl_6$: Bk^{3+}		Fig. 35			74H2
Cs_2NaUCl_6	0.53	Fig. 36	2.49	p_{eff} below 20 K, $C_m = 0.778$ cm³ K/mol	74H2
$Cs_2NaNpCl_6$	0.47	Fig. 37	1.92	$C_m = 0.46$ cm³ K/mol	74H2
$Cs_2NaPuCl_6$	1.3	Fig. 38	0.97	p_{eff} below 21 K, $C_m = 0.117$ cm³ K/mol	74H2
$Cs_2NaAmCl_6$		Fig. 39			74H2
		Fig. 40 650		at low temperature temperature-independent paramagnetism, χ in 10^{-6}cm³/mol	86S1
$Cs_3VCl_6 \cdot 4H_2O$				$g_{\parallel} = 1.93$, $g_{\perp} = 1.74$, $D/k = 11.3$ K, $p_V(H)$: Figs. 41, 42	79S2

9.9.2 Figures

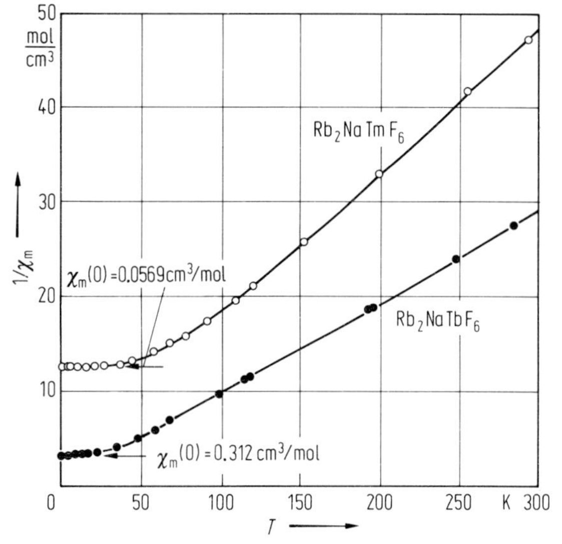

Fig. 1. Rb_2NaTbF_6, Rb_2NaTmF_6. Temperature dependence of χ_m^{-1} [74B1].

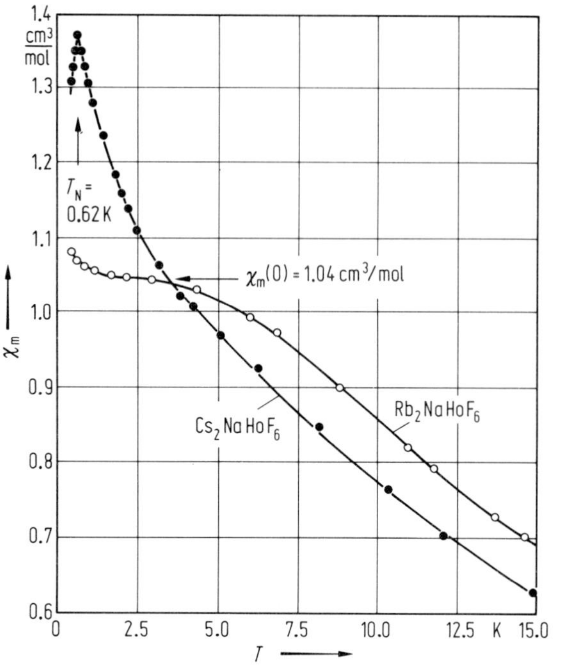

Fig. 2. Rb_2NaHoF_6, Cs_2NaHoF_6. Temperature dependence of χ_m [74B1].

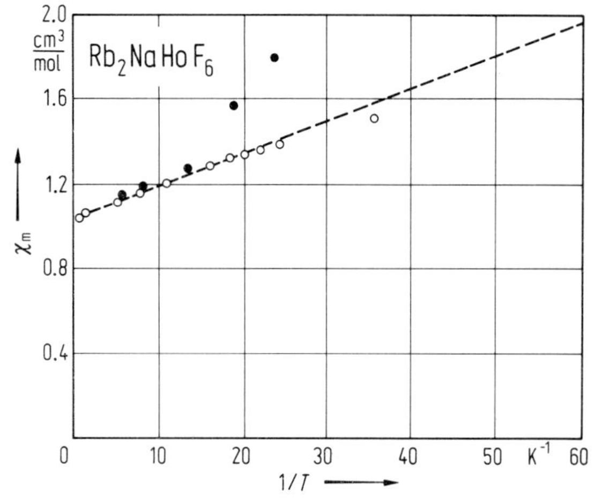

Fig. 3. Rb_2NaHoF_6. χ_m at very low temperatures. Open circles: cooling in 200 Oe, full circles: warming in 200 Oe after demagnetization. Dashed line theoretical [74B1].

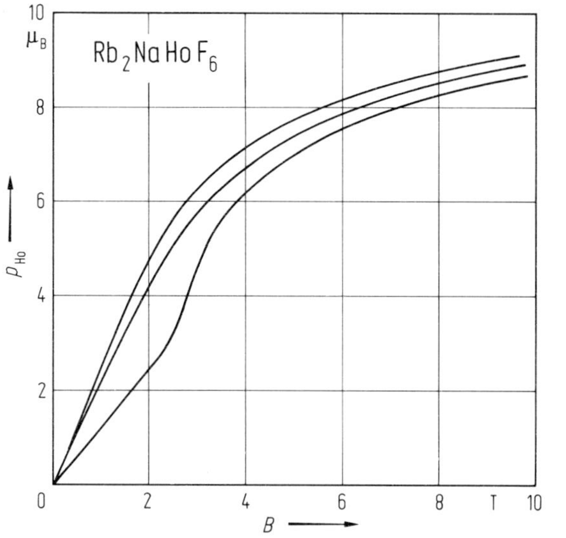

Fig. 4. Rb_2NaHoF_6. Magnetic field dependence of p_{Ho} at 4.2 K along "cubic" axes. The lower line belongs to the [001] axis [82V1].

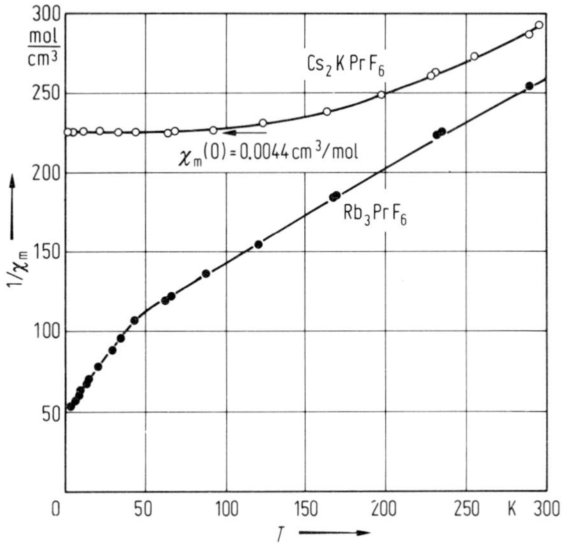

Fig. 5. Rb_3PrF_6, Cs_2KPrF_6. Temperature dependence of χ_m^{-1} [74B1].

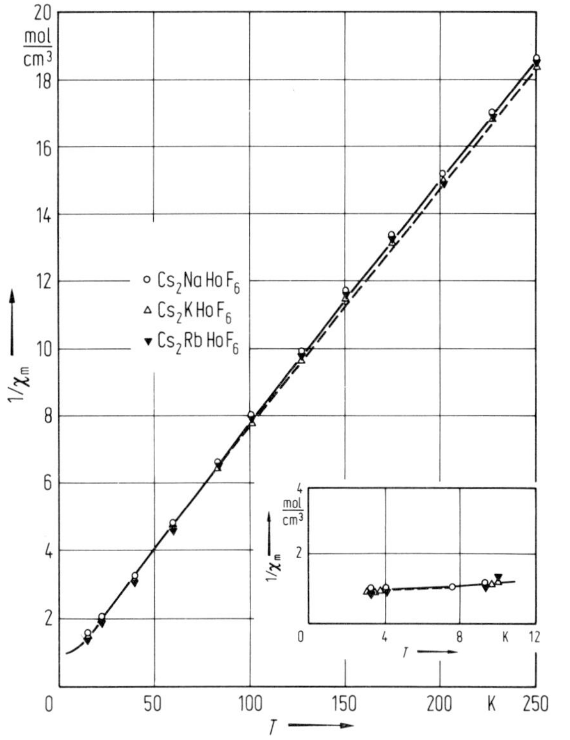

Fig. 6. Cs_2NaHoF_6, Cs_2KHoF_6, Cs_2RbHoF_6. Temperature dependence of χ_m^{-1} [80U1].

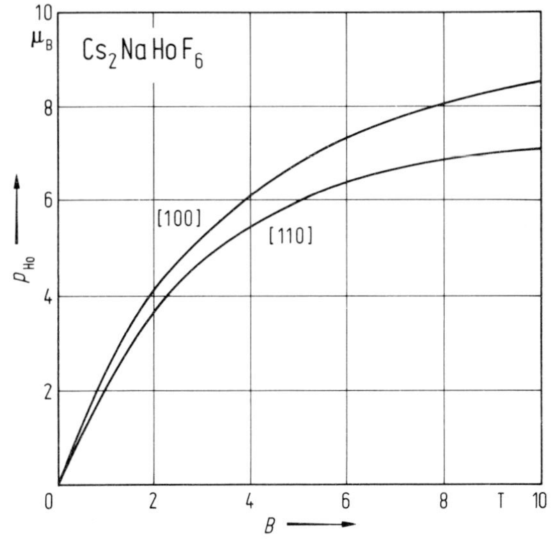

Fig. 7. Cs_2NaHoF_6. Magnetic field dependence of p_{Ho} at 1.2 K for $\boldsymbol{H}$‖ to the [100] and [110] crystal axes [83V2].

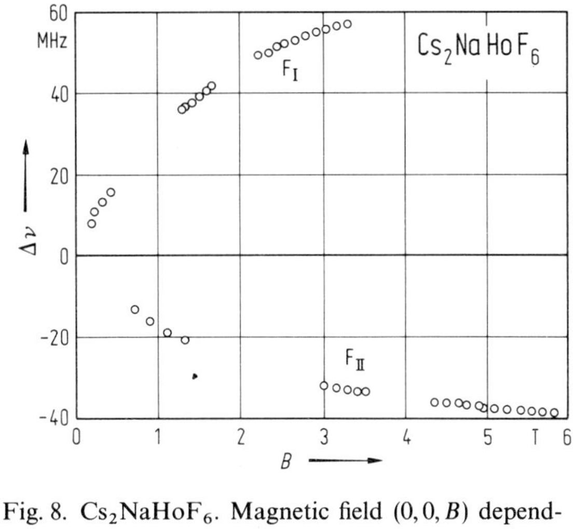

Fig. 8. Cs_2NaHoF_6. Magnetic field $(0, 0, B)$ dependence of the paramagnetic frequency shift, $\Delta\nu$, of the ^{19}F resonance. FI and FII denote frequency shift of fluorines on the [001] axis and the (001) plane, respectively [83V2].

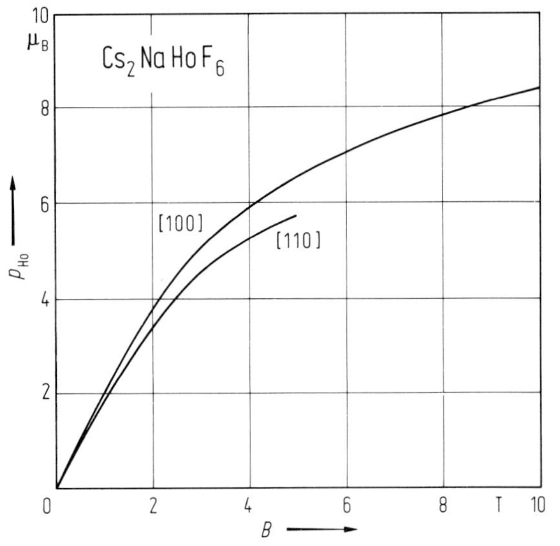

Fig. 9. Cs_2NaHoF_6. Magnetic field dependence of p_{Ho} at 4.2 K for $\boldsymbol{H}\|$ to the [100] and [110] crystal axes [83V2].

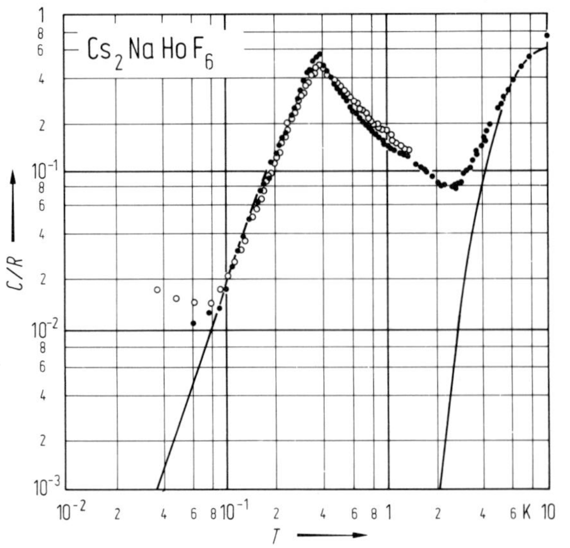

Fig. 10. Cs_2NaHoF_6. Temperature dependence of heat capacity. The maximum corresponding to a phase transition occurs at $T_t = 393(5)$ mK. The drawn line indicates the calculated heat capacity contribution of the excited Γ_4-state at an energy $W_4/k = 25.6$ K. Full circles: single crystal, open circles: fragments ($d \approx 1$ mm) [83V2].

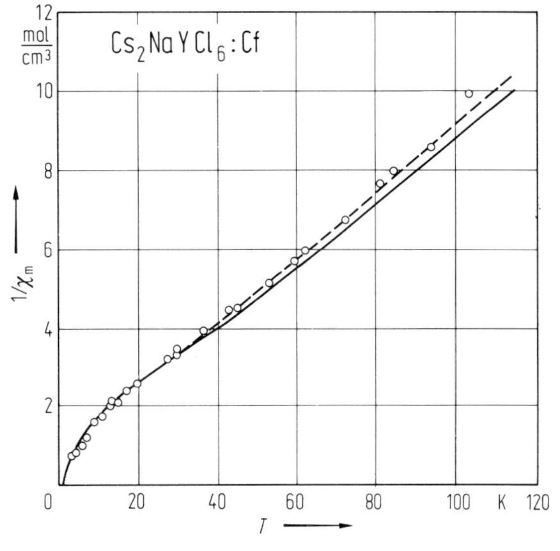

Fig. 11. Cs_2NaYCl_6:Cf. Temperature dependence of χ_m^{-1}. Solid line: calculated for $x = -0.9$, $g_J = 1.254$, dashed line: calculated for $x = -0.9$, $g_J = 1.240$ [76K3]. g_J: free ion Landé factor.

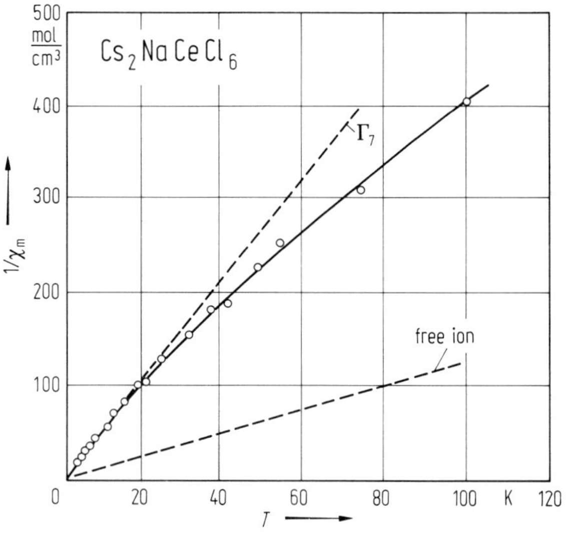

Fig. 12. $Cs_2NaCeCl_6$. Temperature dependence of χ_m^{-1}. Solid line is a fit using CEF parameters $W = 334$ K, $x = 1$. Susceptibilities appropriate to the isolated ground state Γ_7 and the free Ce ion are also shown [75D1]. See also [74H1].

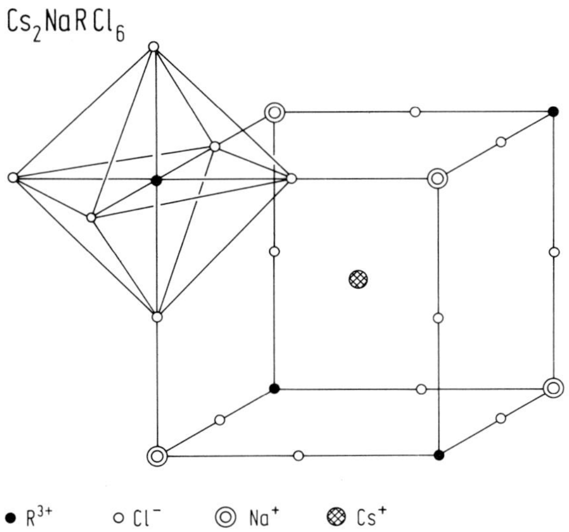

Fig. 13. Cs_2NaRCl_6. Crystal structure [75D1]. R: rare earth ion.

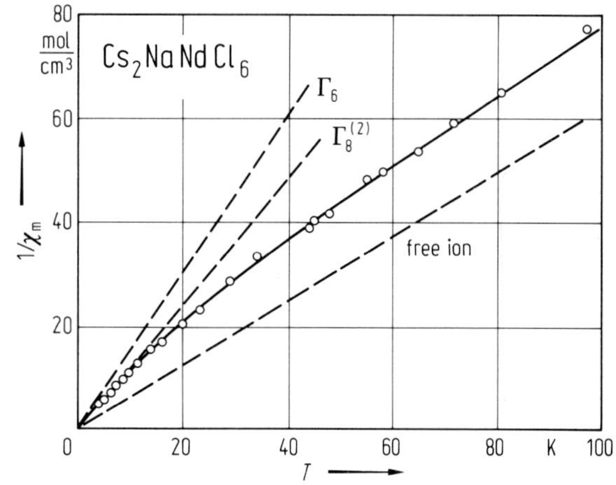

Fig. 14. $Cs_2NaNdCl_6$. Temperature dependence of χ_m^{-1}. Solid line is a fit using CEF parameters $W = -19.0$ K, $x = 0.76$. Susceptibilities for Γ_6, $\Gamma_8^{(2)}$ states of Nd and the free Nd ion are also shown [75D1]. See also [74H1].

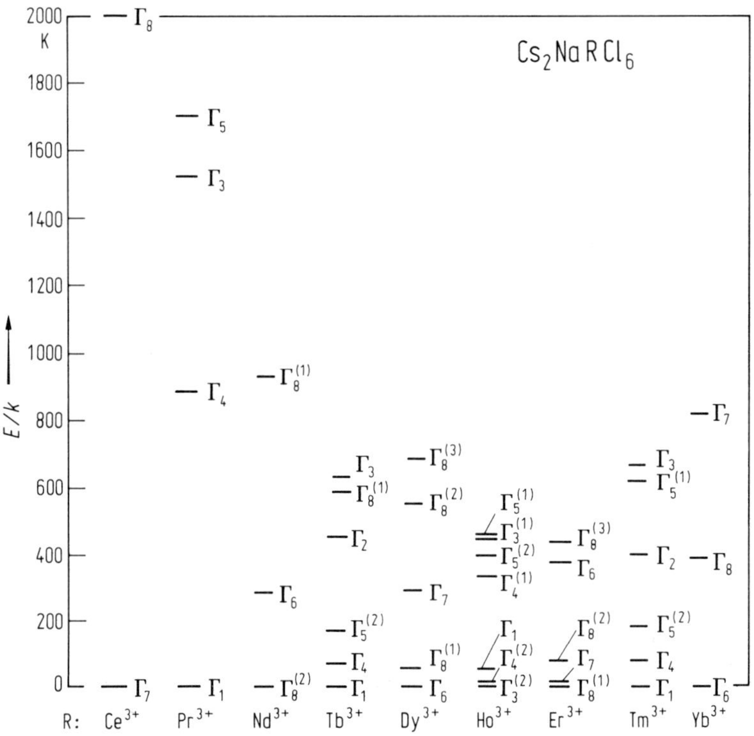

Fig. 15. Cs_2NaRCl_6, R = Ce, Pr, Nd, Tb, Dy, Ho, Er, Tm, Yb. CEF splitting using hte CEF parameters W and x [75D1].

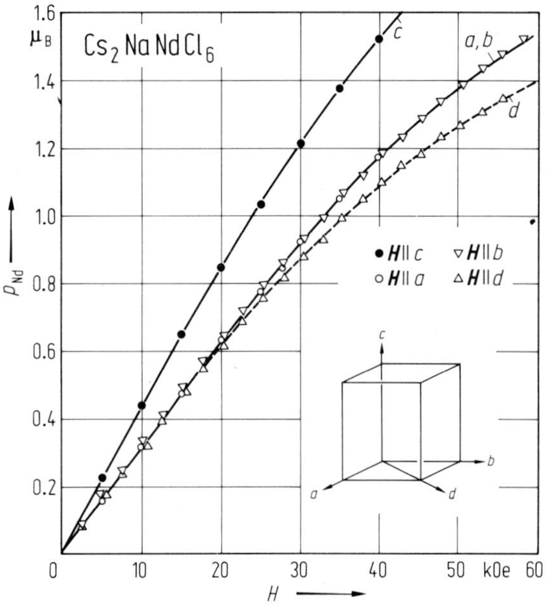

Fig. 16. $Cs_2NaNdCl_6$. Magnetic moment for various directions of applied field. Equal slopes for low fields in the basal plane and a steeper slope along *c* direction indicates the tetragonal symmetry [79N1].

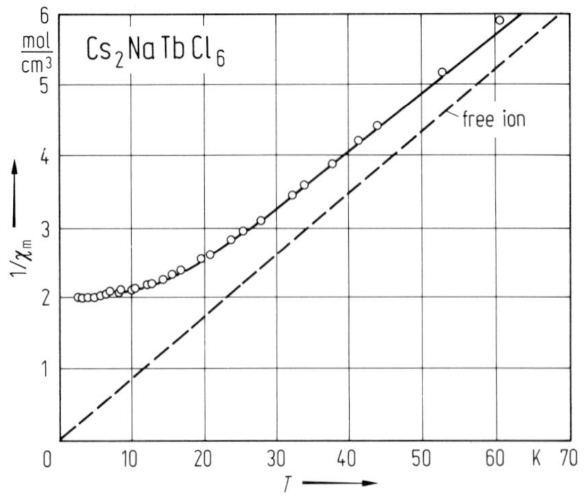

Fig. 17. $Cs_2NaTbCl_6$. Temperature dependence of χ_m^{-1}. Solid line is a fit using CEF parameters $W = -1.84$ K, $x = -0.83$ [75D1]. See also [74H1].

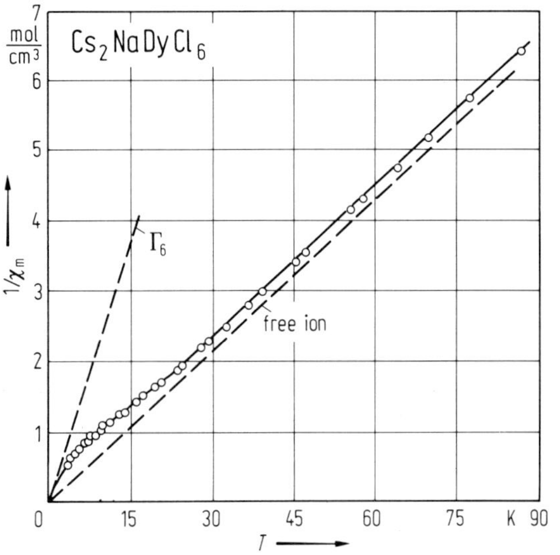

Fig. 18. $Cs_2NaDyCl_6$. Temperature dependence of χ_m^{-1}. Solid line is a fit using CEF parameters $W = 0.92$ K, $x = -0.63$. Susceptibilities for Γ_6 state and free Dy ion are also shown [75D1]. See also [74H1].

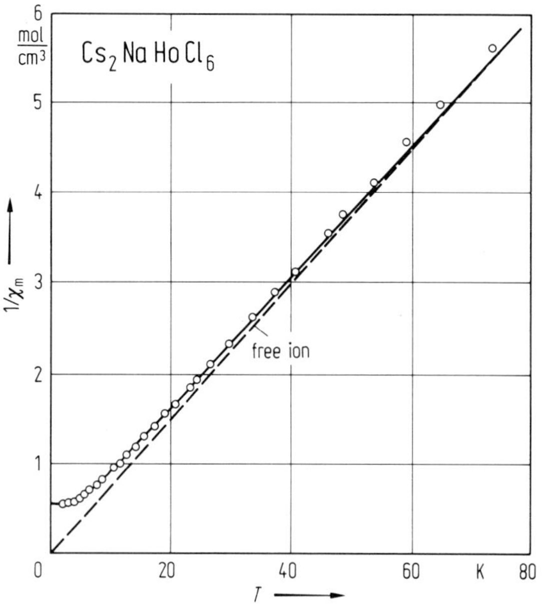

Fig. 19. $Cs_2NaHoCl_6$. Temperature dependence of χ_m^{-1}. Solid line is a fit using CEF parameters $W = -1.12$ K, $x = 0.72$ [75D1]. See also [74H1].

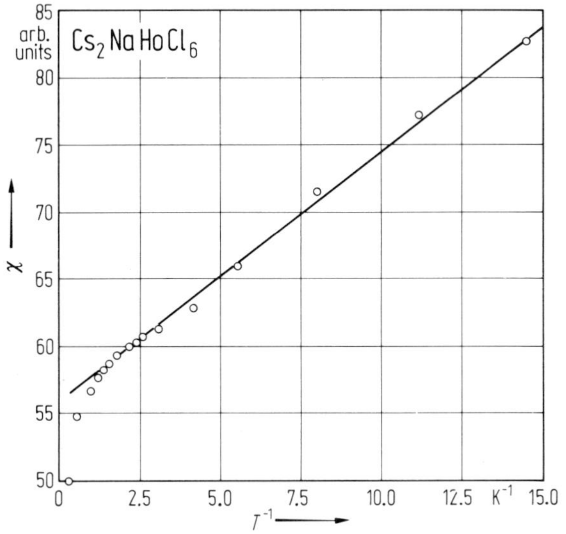

Fig. 20. $Cs_2NaHoCl_6$. Temperature dependence of ac χ along the [532] direction [82S1].

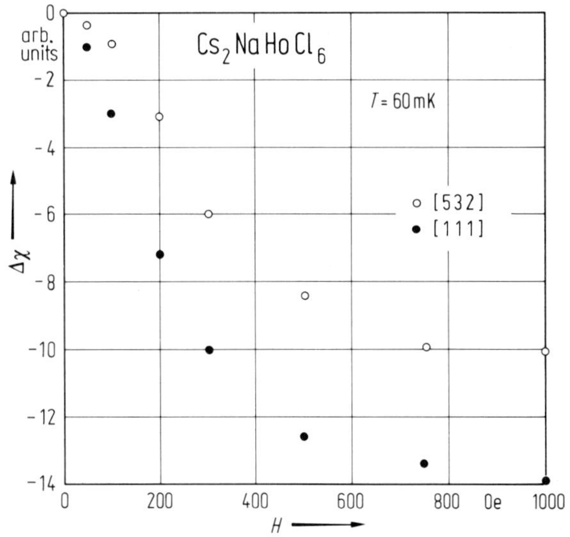

Fig. 21. $Cs_2NaHoCl_6$. Magnetic field dependence of $\Delta\chi$ along [532] and [111] direction at 60 mK. The vertical axis represents the deviation of the magnetic susceptibility [82S1]. See also caption of Fig. 22.

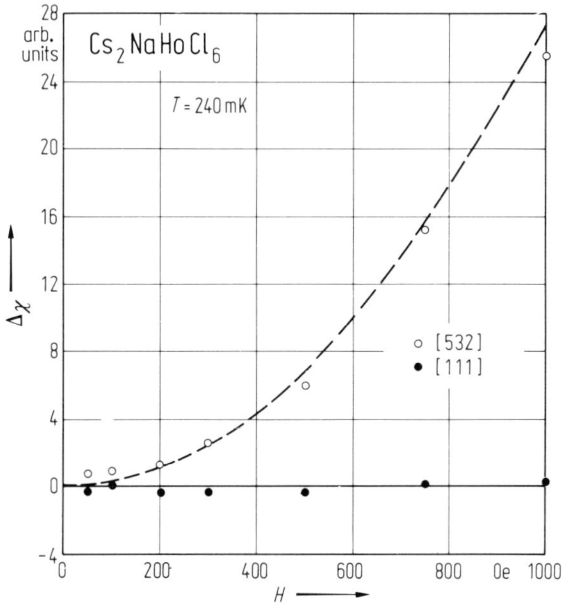

Fig. 22. $Cs_2NaHoCl_6$. Magnetic field dependence of the deviation $\Delta\chi$ of the ac ($\nu = 80$ Hz) magnetic susceptibility in the magnetic field from the value under zero applied field. The dashed line is calculated and normalized with respect to the value at 300 Oe of equation $\chi = 2\,(g_J\mu_B)^2\,[2a + Ka\tanh(\delta/kT)]$, where $\delta = Ka(g_J\mu_B B)^2$. First term is the ordinary Van Vleck temperature independent susceptibility, second term arises from the splitting of the Γ_3 doublet levels [82S1].

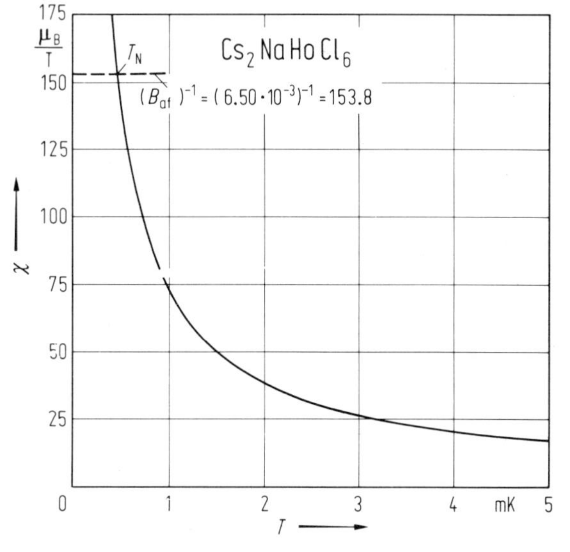

Fig. 23. $Cs_2NaHoCl_6$. Temperature dependence of χ in mK region, in μ_B per Tesla. $\chi = \dfrac{\chi_0}{1 - \chi_0 B_{a.f}}$, χ_0: magnetic susceptibility of the enhanced nuclear spins and of electrons. $B_{a.f}$: coefficient of local magnetic field in AF lattice, χ: magnetic susceptibility of the sublattice moment. The antiferromagnetic state occurs at the condition $\chi_0 = B_{a.f}^{-1}$ [82S1].

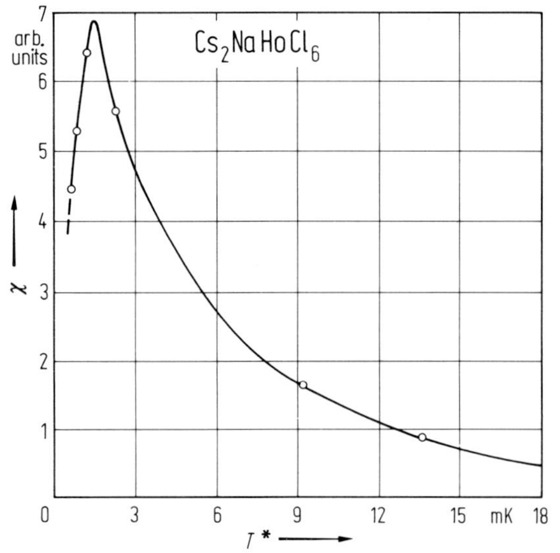

Fig. 24. $Cs_2NaHoCl_6$. Temperature dependence of χ, following magnetic cooling from various initial fields. The abscissa represents the final temperature $T^* = T_i b/(H_i^2 + b^2)^{1/2}$, where $b = 217$ Oe, T_i: initial temperature (≈ 25 mK), H_i: initial field [86S1].

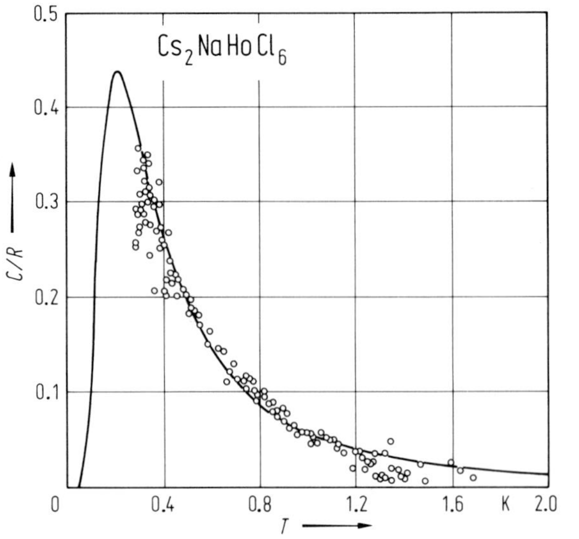

Fig. 25. $Cs_2NaHoCl_6$. Temperature dependence of the heat capacity. Solid line calculated of the Schottky-type heat capacity assuming two levels separated by $\Delta = 0.5$ K [86S1].

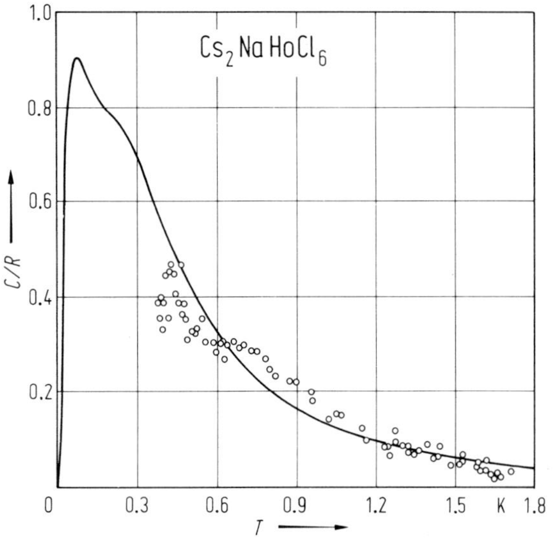

Fig. 26. $Cs_2NaHoCl_6$. Temperature dependence of the heat capacity at $\mu_0 H = 0.5$ T. Solid line calculated of the Schottky-type heat capacity assuming two levels separated by $\Delta = 0.9$ K [86S1].

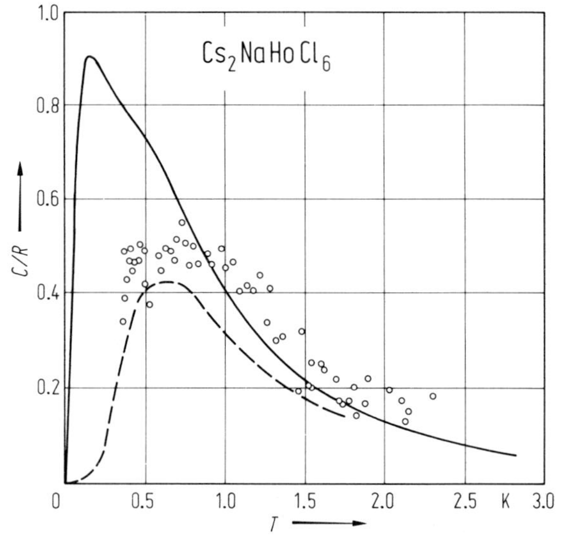

Fig. 27. $Cs_2NaHoCl_6$. Temperature dependence of heat capacity at $\mu_0 H = 1$ T. Solid line calculated of the Schottky-type heat capacity assuming two levels separated by $\Delta = 1.65$ K. Dashed line is the electronic Schottky-type heat capacity [86S1].

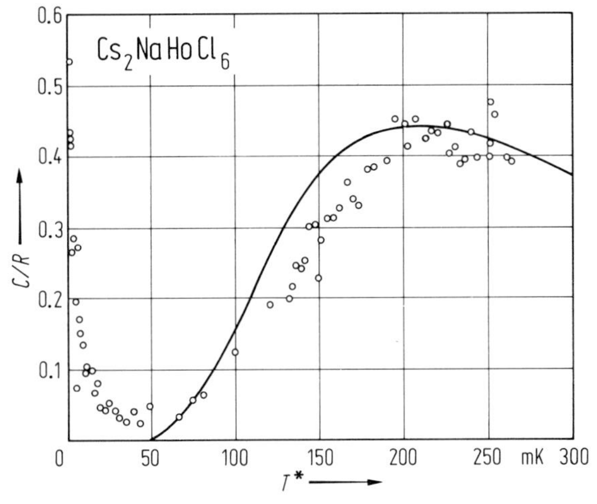

Fig. 28. $Cs_2NaHoCl_6$. Temperature dependence of the heat capacity measured by rf heating method for T^* (see Fig. 24). Solid line is normalized to the maximum value around 210 mK of the calculated Schottky heat capacity shown in Fig. 27 [86S1].

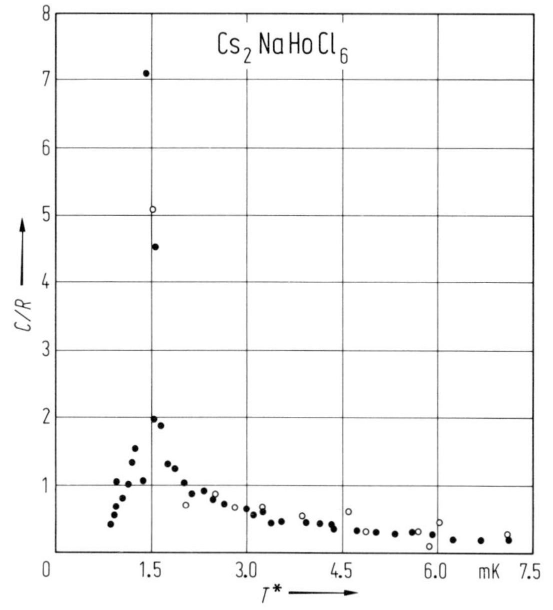

Fig. 29. $Cs_2NaHoCl_6$. Temperature dependence of heat capacity around T_N measured by rf heating method. Open circles are the results of second run made half a year later [86S1]. For T^* see Fig. 24.

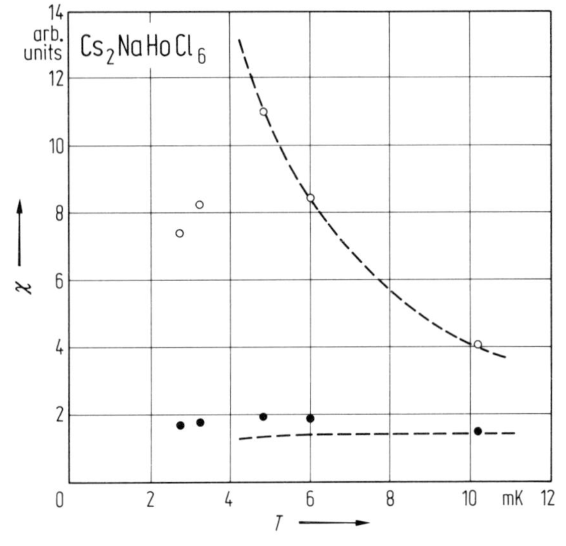

Fig. 30. $Cs_2NaHoCl_6$. Temperature dependence of χ. Open circles: results for the [110] direction, full circles: results for the [001] direction, dashed lines: calculated assuming quadrupole splittings. The measured χ are normalized to the calculated value at 11 mK [88S1].

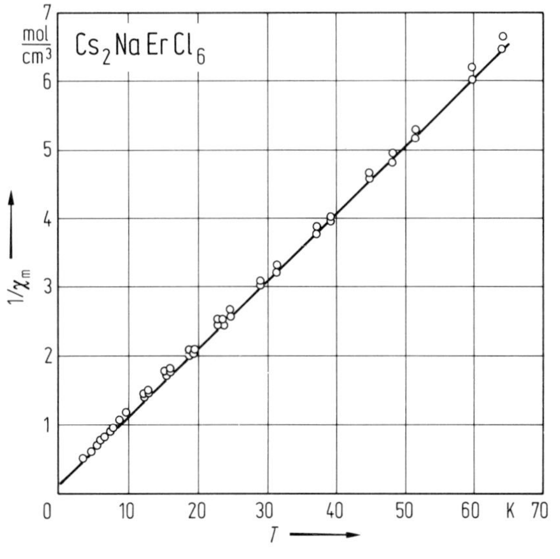

Fig. 31. $Cs_2NaErCl_6$. Temperature dependence of χ_m^{-1} [74H1].

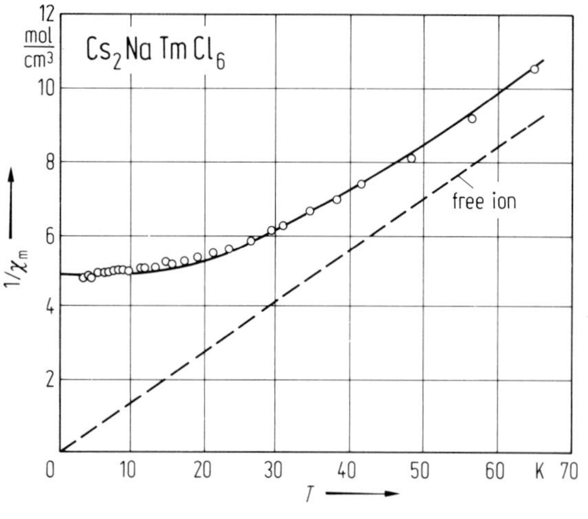

Fig. 32. $Cs_2NaTmCl_6$. Temperature dependence of χ_m^{-1}. Solid line is a fit using CEF parameters $W = -2.55$ K, $x = -0.90$. Susceptibility for free ion is also shown [75D1]. See also [74H1].

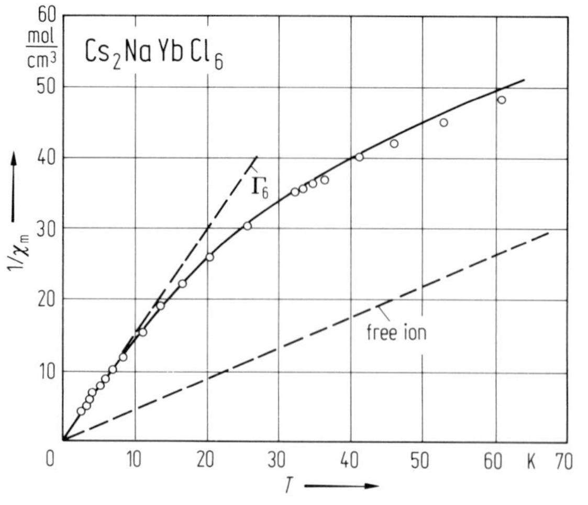

Fig. 33. $Cs_2NaYbCl_6$. Temperature dependence of χ_m^{-1}. Solid line is a fit using CEF parameters $W = 27.1$ K, $x = -0.9$. Susceptibilities for the free ion and the isolated Γ_6 level of Yb are also shown [75D1]. See also [71K1].

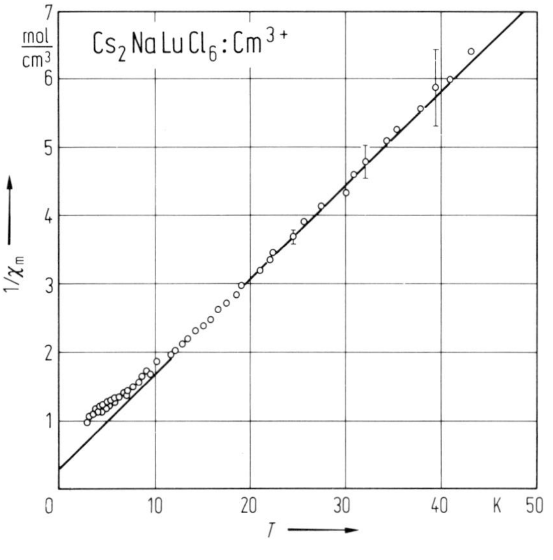

Fig. 34. $Cs_2NaLuCl_6$: Cm^{3+}. Temperature dependence of χ_m^{-1}. Deviation from free-ion moment at lower temperatures attributed to CEF splitting in Cm^{3+} [74H2].

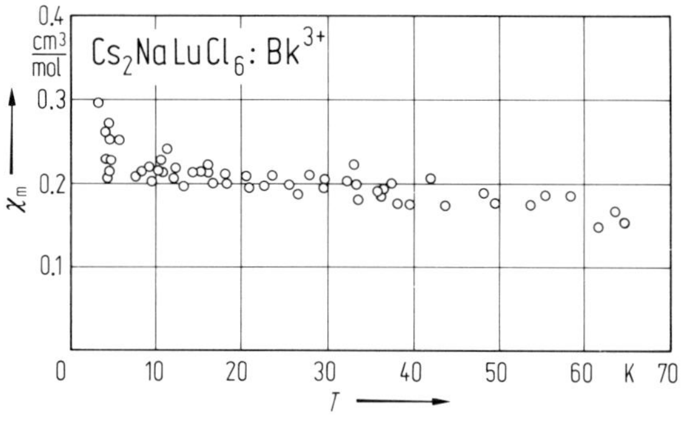

Fig. 35. $Cs_2NaLuCl_6$: Bk^{3+}. Temperature dependence of χ_m. Correction for Cf^{3+} impurity is negligible [74H2].

Fig. 36. Cs_2NaUCl_6. Temperature dependence of χ_m^{-1}. The solid line below 20 K gives $p_{eff} = 2.49\mu_B$ and above 25 K, $p_{eff} = 2.92\mu_B$ [74H2]. ▶

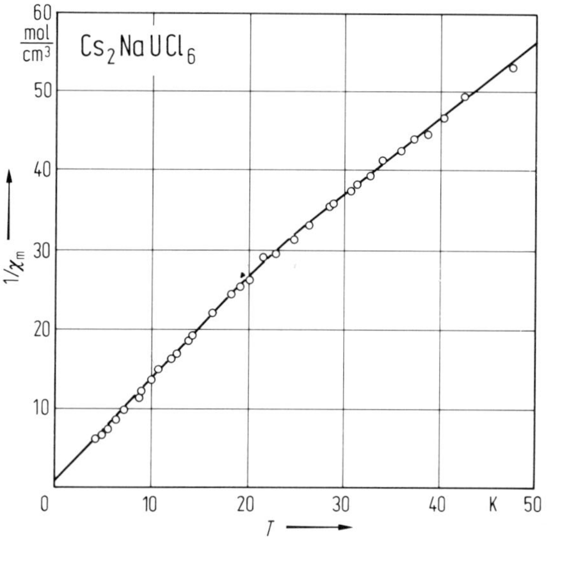

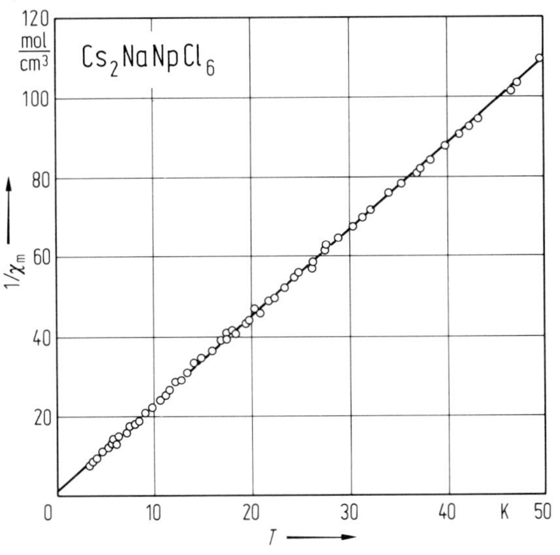

Fig. 37. $Cs_2NaNpCl_6$. Temperature dependence of χ_m^{-1} [74H2].

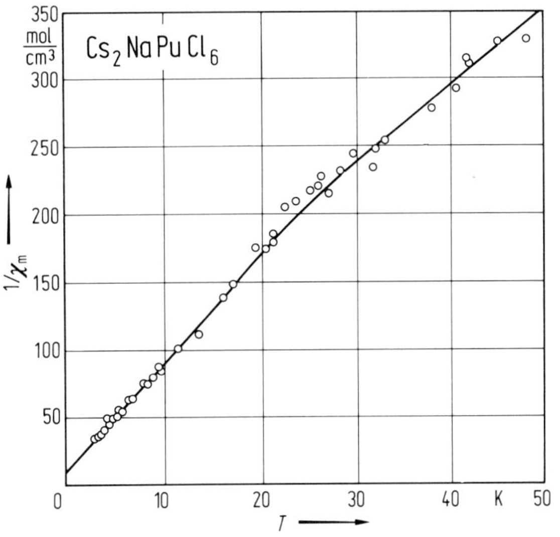

Fig. 38. $Cs_2NaPuCl_6$. Temperature dependence of χ_m^{-1}. The solid line below 21 K gives $p_{eff} = 0.97\mu_B$, and above 25 K, $p_{eff} = 1.16(8)\mu_B$ [74H2].

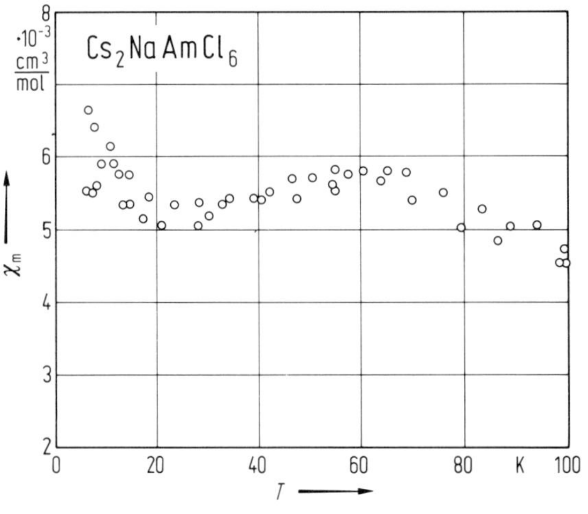

Fig. 39. $Cs_2NaAmCl_6$. Temperature dependence of χ_m corrected for Cm^{3+} impurity [74H2].

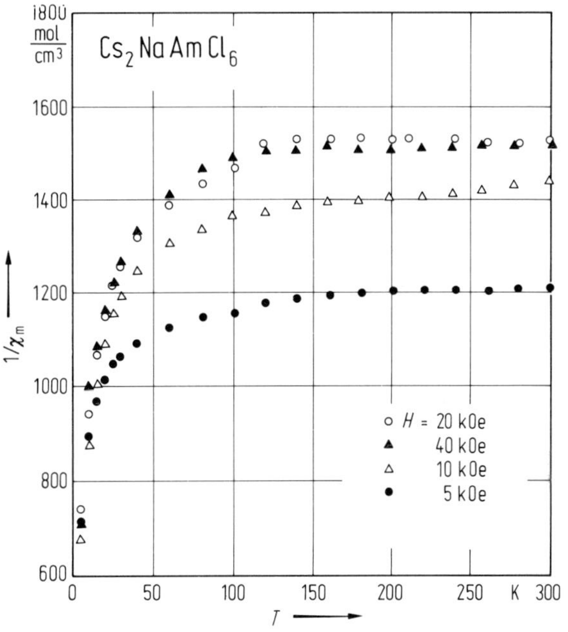

Fig. 40. $Cs_2NaAmCl_6$. Temperature dependence of χ_m^{-1} at 5, 10, 20 and 40 kOe. A correction for sample diamagnetism ($-2.53 \cdot 10^{-4}$ cm^3/mol) has been applied to the data [86S1].

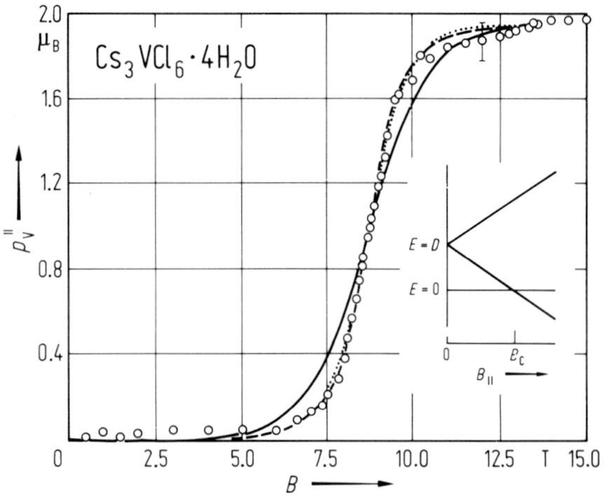

Fig. 41. $Cs_3VCl_6 \cdot 4H_2O$. Magnetic field dependence of parallel p_v measured with a starting temperature of 1.15 K. Theoretical isothermal curves are calculated with $g_{\parallel} = 1.94$ and $D/k = 11.5$ K (with $\chi_{\parallel}^{TIP} = 0$ and $zJ/k = 0$) at $T = 1.15$ K (solid line) and at $T = 0.80$ K (dotted line). The dashed line is calculation with $g_{\parallel} = 1.94$ and $D/k = 12.3$ K and $zJ/k = 1$ K at $T = 1.15$ K ($\chi_{\parallel}^{TIP} = 0$) [79S2]. Insert shows energy of the levels vs. magnetic field B.

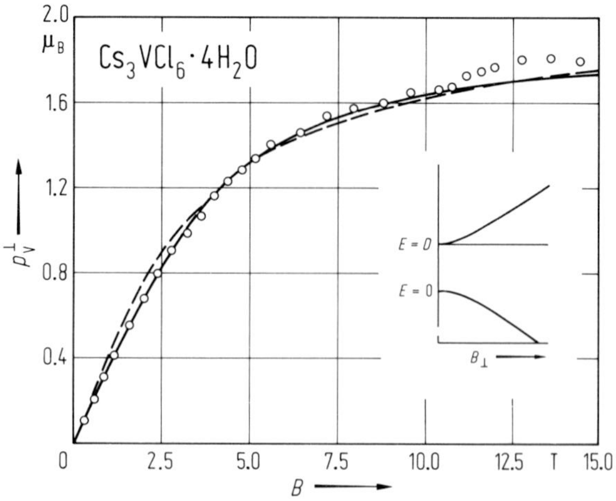

Fig. 42. $Cs_3VCl_6 \cdot 4H_2O$. Magnetic field dependence of perpendicular p_v at a starting temperature of 1.15 K. Calculated isothermal curves are for $D/k = 11.3$ K, $g_{\perp} = 1.80$, $zJ/k = 0$, $\chi_{\perp}^{TIP} = 0$ (solid line) and for $D/k = 12.3$ K, $g_{\perp} = 1.63$, $zJ/k = 1$ K, $\chi_{\perp}^{TIP} = 7.5 \cdot 10^{-2}$ cm^3/mol (dashed line) [79S2].

9.10 $A_2M^{2+}M^{3+}F_7$ compounds (A = Na, Ag; M = 3d element, Mg, Al, or In)

(For symbols and abbreviations see Section 9.6.0.2)

9.10.1 Introduction

All the $A_2M^{2+}M^{3+}F_7$ compounds, for which the magnetic properties are compiled in this chapter, have a weberite-related crystal structure. From a magnetic point of view the major crystallographic character is the occurrence of $(M^{2+}F_6)^{4-}$ and $(M^{3+}F_6)^{3-}$ octahedra with a diamagnetic or a paramagnetic ion M in the centre. Each type of the octahedra is ordered in chains. At low temperature three types of ordered magnetism can result, dependent on the chemical composition. For diamagnetic M^{3+} and paramagnetic M^{2+} ions, an antiferromagnetic ordering in the $(M^{2+}F_6)^{4-}$ chains is found and mutatis mutandis for paramagnetic M^{3+} and diamagnetic M^{2+} antiferromagnetism in the $(M^{3+}F_6)^{3-}$ chains. When both types of M ions are paramagnetic, the result can be ferrimagnetism with parallel magnetic ordering within the chains and antiparallel ordering between the chains.

9.10.2 Crystallographic properties of $A_2M^{2+}M^{3+}F_7$ compounds

The compounds $A_2M^{2+}M^{3+}F_7$ generally crystallize in the structure of the material weberite Na_2MgAlF_7 with an orthorhombic symmetry, see e.g. Fig. 1. Phase containing Mn^{2+} are trigonal and copper-based weberites are monoclinic. For crystallographic data see Table 1. The space group of the material has been reclassified as Imma [89L] with a new refinement with $Z = 4$ formula units in the unit cell, though some ambiguity remains. The various weberite-related structures are shown in Fig. 2.

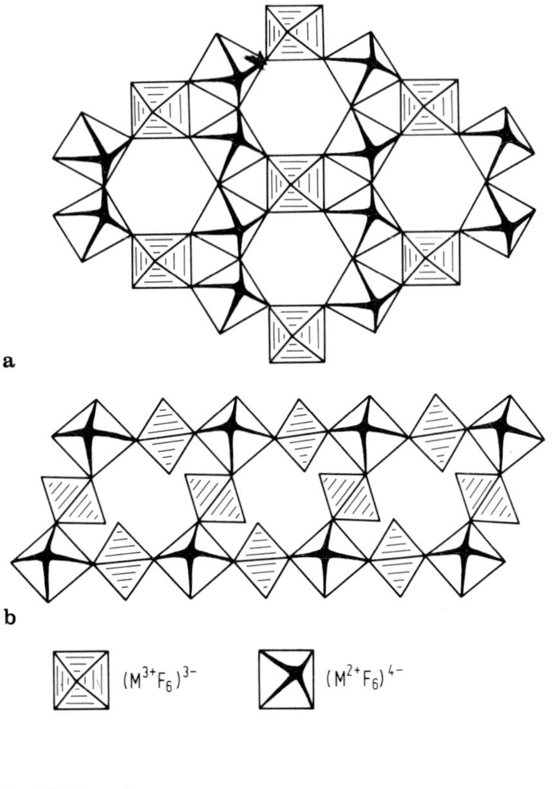

Fig. 1. (a) Idealized projection of a crystal layer of the compounds $A_2M^{2+}M^{3+}F_7$ with weberite structure on the (011) plane, showing the chains of $(M^{3+}F_6)^{3-}$ and $(M^{2+}F_6)^{4-}$ octahedra and the tunnels with hexagonal or triangular cross sections between them, giving room for the A^{1+} ions. (b) The layers of Fig. 1a are interconnected by $(M^{3+}F_6)^{3-}$ octahedra [70C].

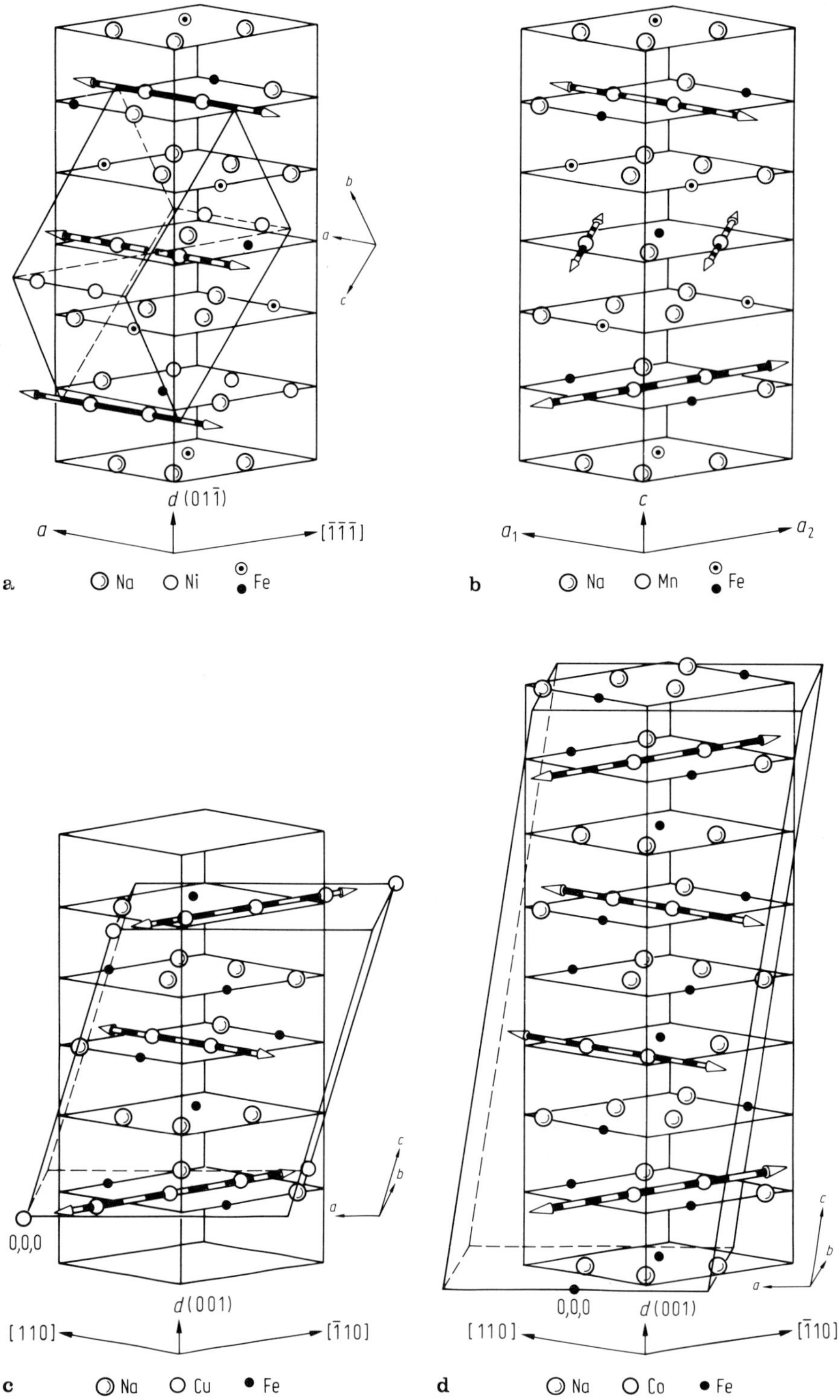

Fig. 2. Various weberite-related structures. (a) orthorhombic type e.g. Na_2NiFeF_7, (b) trigonal type e.g. Na_2MnFeF_7, (c) monoclinic type ($Z = 8$) e.g. Na_2CuFeF_7 and (d) id. ($Z = 16$) e.g. Na_2CoFeF_7. The arrows indicate the directions of the $(M^{2+}F_6)^{4-}$ chains of corner-sharing octahedra [93B].

Table 1. $A_2M^{2+}M^{3+}F_7$ compounds[a]). Various weberite-related structures with the principal crystallographic parameters (in Å) at RT. Number of FU per unit cell *Z*. Unit cell O: orthorhombic, M: monoclinic, T: trigonal.

Compound	Unit cell; space group	Fig.	*Z*	*a*	*b*	*c*	*β*	*V* [Å³]	Ref.
Na_2MgAlF_7	O; [b])		4	7.051(1)	9.968(2)	7.285(1)		512.1(3)	82K1 [c])
Na_2MgVF_7	O; Imm2		4	7.24	10.30	7.45		555.5	69C
Na_2MgCrF_7	O; Imm2		4	7.15	10.20	7.39		538.4	69C
Na_2MgFeF_7	O; Imm2		4	7.25	10.26	7.49		557.3	69C
Na_2FeAlF_7	O; Imm2		4	7.21(2)	10.20(2)	7.23(2)		–	74T
Na_2CoAlF_7	M; C2/c	2d	16	12.378(4)	7.210(3)	24.019(9)	99.67(2)°	2113.1	92G [c])
Na_2NiAlF_7	O; Imma	2a	4	7.074(2)	10.038(2)	7.315(1)		–	92S [c])
Na_2MnGaF_7	T; $P3_121$	2b	6	7.401(1)	–	18.091(2)		858.2	88C [c])
Na_2CuGaF_7	M; C2/c		8	12.325(5)	7.318(1)	12.780(5)	109.29(2)°	1087.8	92R [c])
Na_2NiInF_7	O; Pmnb		4	7.356(1)	10.334(2)	7.523(1)		571.9	92F [c])
Na_2CuInF_7	O; Pmnb		4	7.318(1)	10.602(2)	7.712(1)		598.3	92R [c])
Ag_2NiAlF_7	O; Imm2		4	7.210(6)	10.139(15)	7.564(6)		–	74D
Ag_2NiInF_7	O; Imm2		4	7.499(6)	10.622(7)	7.822(6)		–	74D
Na_2MnVF_7	T; $P3_121$	2b	6	7.492(2)	–	18.261(7)		–	78V [c])
Na_2MnCrF_7	T; $P3_121$	2b	6	7.421(32)	–	18.166(6)		866.4	88C [c])
Na_2CoCrF_7	M; C2/c	2d	16	12.591(7)	7.326(5)	24.449(10)	99.67(4)°	2223(3)	93B [c])
Na_2NiCrF_7	O; Imma	2a	4	7.183(1)	10.224(1)	7.414(1)		–	87L [c,d])
Na_2CuCrF_7	O; Pmnb		4	7.100	10.338	7.518		551.8	88K [c])
Na_2ZnCrF_7	O;		4	7.315	10.336	7.299		–	70H
Na_2MnFeF_7	T; $P3_121$	2b	6	7.488(2)	–	18.257(6)		–	78V [c,e])
Na_2FeFeF_7	M; C2/c		16	12.676(3)	7.422(1)	24.710(5)	99.97(3)°		93Y
Na_2CoFeF_7	M; C2/c	2d	16	12.622(10)	7.360(4)	24.516(20)	99.71(5)°	2244.9(29)	92W [c])
Na_2NiFeF_7	O; Imma	2a	4	7.2338(3)	10.3050(3)	7.4529(3)		555.57	89L [c,e])
	O; Imm2		4	7.245(1)	10.320(1)	7.458(1)		–	78H [c])
Na_2CuFeF_7-I	M; C2/c	2c	8	12.46(1)	7.363(8)	12.93(1)	109.36(7)°	–	90Y [c])
-II	M; C2/c		16	7.343(1)	24.672(5)	12.444(2)	99.27(3)°	2225.0(7)	92W [c])
Na_2NiCoF_7	O; Imm2		4	7.20(2)	10.24(2)	7.40(2)		–	70C
$Na_2Cu_3F_7$	M; C2/c		4	12.124(7)	7.344(5)	6.924(5)	120.59(4)°	530.7	88R [c])
Ag_2NiCrF_7	O; Imm2		4	7.305(6)	10.285(9)	7.673(6)		–	74D
Ag_2NiFeF_7	O; Imm2		4	7.345(6)	10.345(7)	7.692(6)		–	74D

[a]) For more compounds see [82K2, 85K].
[b]) Mineral "weberite". Space groups Imma, Imm2 and $I2_12_12_1$ indistinguishable from X-ray data [82K1].
[c]) In this reference refined structure analysis and detailed ion positions and relative distance are given.
[d]) Neutron powder diffraction results.
[e]) Result of both X-ray and neutron diffraction spectra.

9.10.3 Magnetic properties of $A_2M^{2+}M^{3+}F_7$ compounds

A survey of the paramagnetic properties and the properties in the magnetically ordered state are given in Table 2 and Figs. 3–10. Further data are included under the special heading for each of the compounds following this survey.

Table. 2. $A_2M^{2+}M^{3+}F_7$ compounds. Survey of the magnetic properties. Curie–Weiss constants Θ_p, C_m (experimental, and calculated for spin-only moments) and the effective magnetic moment p_{eff} in the paramagnetic temperature range. Néel point T_N and extrapolated spontaneous magnetic moment p_0 at 0 K as derived from magnetisation measurements in the magnetically ordered state.

Compound	Paramagnetic state				Magnetically ordered state			Ref.
	Θ_p [K]	C_m[cm^3 K mol^{-1}] exp.	theor.	p_{eff} [μ_B/FU]	T_N [K]	$p_0(0)$ [μ_B/FU]	Fig.; Tables	
Na_2MgFeF_7	− 2.1				⩽ 2			91P
	0(2)	4.23(5)	4.38		[a]			74T
Na_2FeAlF_7	− 20(2)	3.58(5)	3		19.1	0.32(1)	3	74T
Na_2CoAlF_7	− 107(5)	6.13(5)	–		2.9(3)		7	93B
Na_2NiAlF_7	− 24(2)	1.27(5)	1		18(1)	0.083(5)	3	74T
	− 25	–	–	3.1	11.0(5)		4	73H
					12(2)	0.15	10	92F
Na_2NiInF_7	− 35(2)	1.20			9	0.027	10	92F
Ag_2NiAlF_7	− 29(2)	1.46(5)	1		18(1)	0.083(5)	5	74D
Ag_2NiInF_7	− 39(2)	1.19(5)	1		18(1)	0.028(5)	5	74D
Na_2CoCrF_7	− 40(5)	6.39(5)	–		5.5(5)	0.054	7	93B
Na_2NiCrF_7	− 35	2.90	2.87		4	–		77T
Na_2MnFeF_7	− 134(5)	8.19(5)	8.76		97(2)		3	74T
					94(1) [b]			77P
Na_2FeFeF_7	− 104(5)	7.42(5)	7.38		84(2)	0.75(1)	6	74T
Na_2CoFeF_7	− 100(5)	7.35(5)	6.26		80(2)	1.2(1)	6,7	93B, 74T
Na_2NiFeF_7	− 50(5)	4.60(5)	5.38		90		Tab. 4	70C
	− 50	–	–	4.4	88(1)		8	72H, 73H
Na_2NiCoF_7	− 88(5)	4.27(5)	4		126(2)	0.9(1)	6	74T
					≈ 130	–		70C
Ag_2NiCrF_7	− 28(2)	3.08(5)	2.87		10(2)	–	9	74D
Ag_2NiFeF_7	− 61(5)	4.51(5)	5.38		103(2)	2.1(1)	9	74D

[a]) Paramagnetic.
[b]) From ME data.

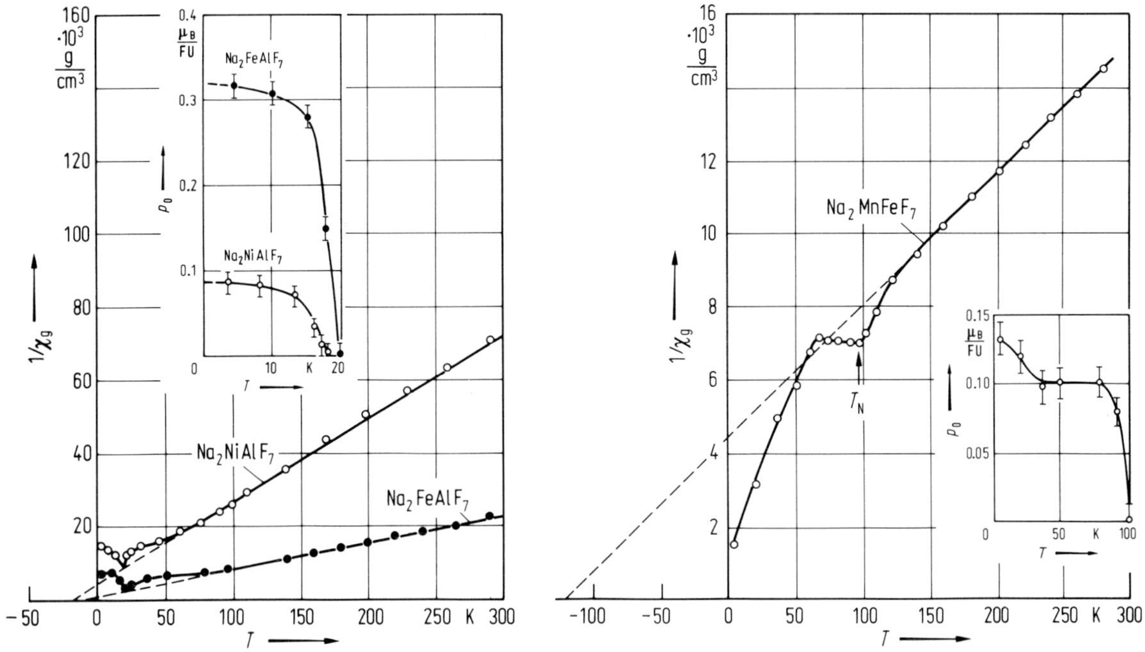

Fig. 3. Na_2FeAlF_7; Na_2NiAlF_7; Na_2MnFeF_7. Inverse magnetic mass susceptibility χ_g^{-1} vs. temperature T. Inset: Spontaneous magnetic moment p_0 vs. temperature T [74T].

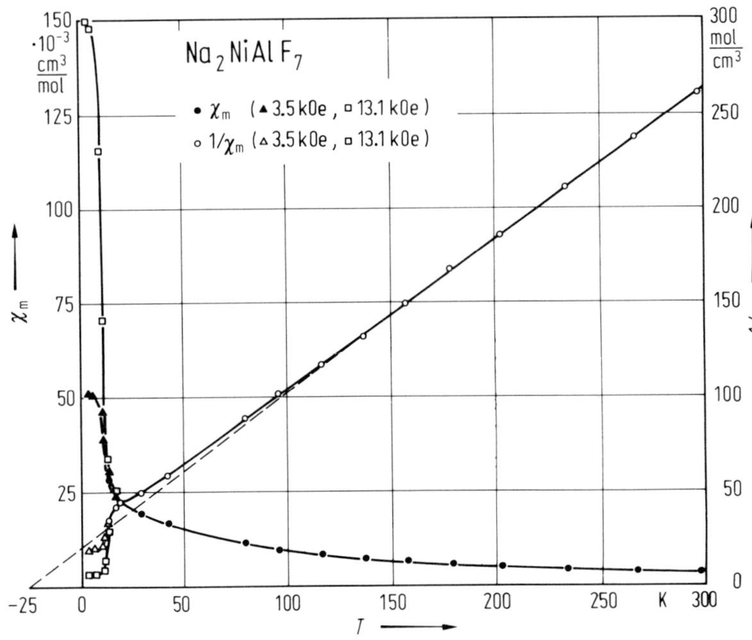

Fig. 4. Na_2NiAlF_7, powder. Temperature dependence of the molar magnetic susceptibility χ_m and its inverse value χ_m^{-1} as measured in magnetic fields of 3.5 (triangles) and 13.1 kOe (squares). Above $T = 13.5$ K no field dependence is found [73H].

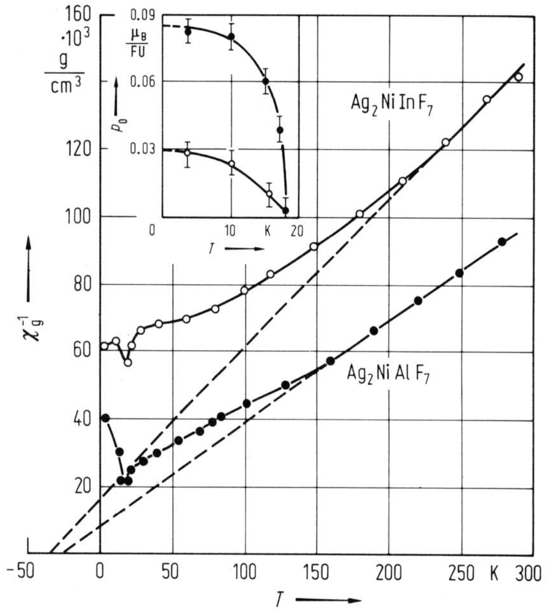

Fig. 5. Ag_2NiAlF_7; Ag_2NiInF_7. Inverse magnetic mass susceptibility χ_g^{-1} vs. temperature T. Inset: Spontaneous magnetic moment p_0 vs. T [74D].

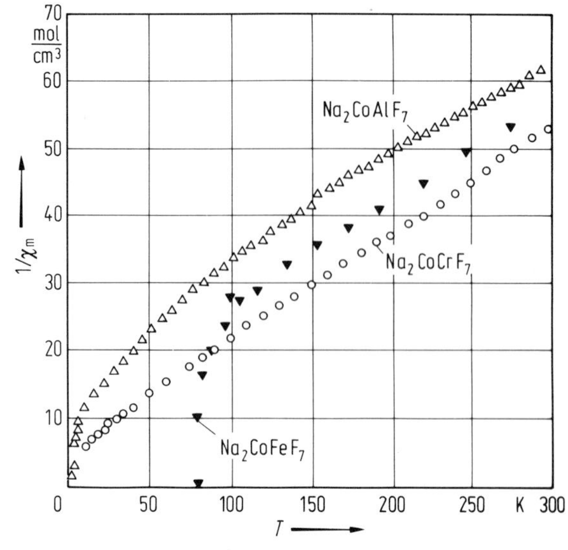

Fig. 7. Na_2CoAlF_7; Na_2CoCrF_7; Na_2CoFeF_7. Inverse molar magnetic susceptibility χ_m^{-1} vs. temperature T [93B].

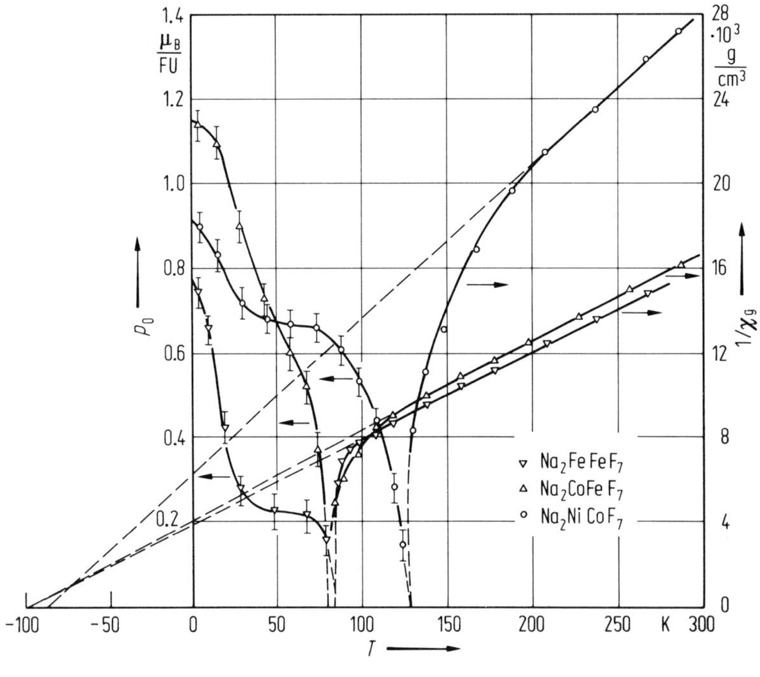

Fig. 6. Na_2FeFeF_7; Na_2CoFeF_7; Na_2NiCoF_7. Temperature dependence of the inverse magnetic mass susceptibility χ_g^{-1} and the spontaneous magnetic moment p_0 [74T].

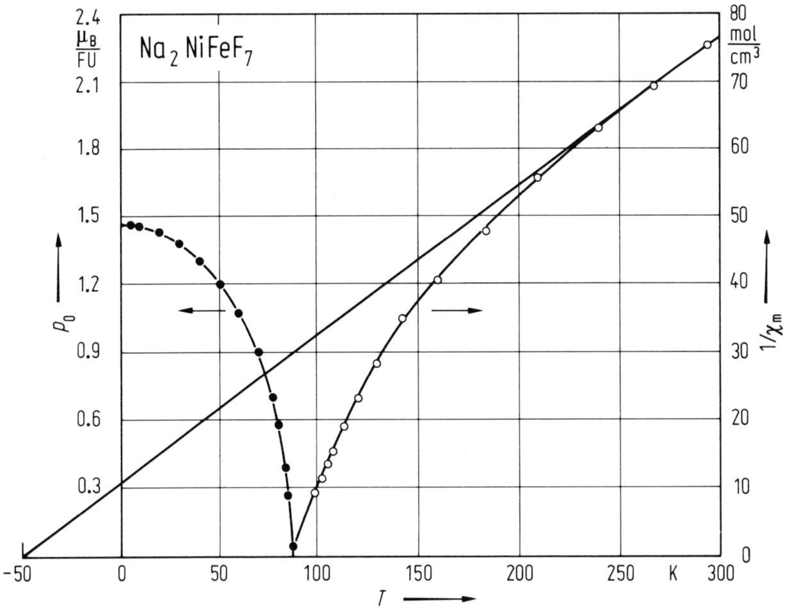

Fig. 8. Na_2NiFeF_7, polycrystalline. Inverse magnetic molar susceptibility χ_m^{-1} and spontaneous magnetic moment p_0 vs. temperature T [72H, 73H, see also 70C].

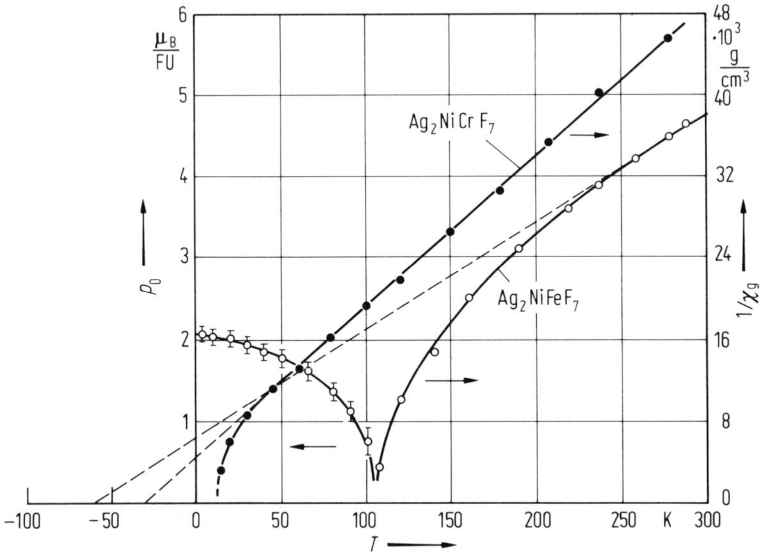

Fig. 9. Ag_2NiCrF_7; Ag_2NiFeF_7. Temperature dependence of the inverse magnetic mass susceptibility χ_g^{-1} and the spontaneous magnetic moment p_0 [74D].

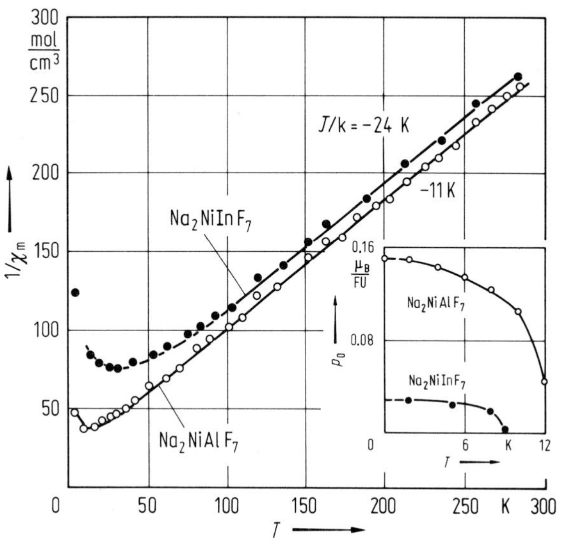

Fig. 10. Na_2NiAlF_7; Na_2NiInF_7. Inverse molar magnetic susceptibility χ_m^{-1} vs. temperature T. Inset: spontaneous magnetic moment p_0 vs. T. The curves have been calculated using $J/k = -24$ K for the In and $J/k = -11$ K for the Al compound [92F].

Na_2MgFeF_7

Diamagnetic Mg^{2+} ions form linear chains isolating the magnetic Fe^{3+} ions. Mössbauer spectra obtained in the presence of magnetic fields at 4.2 K can be understood by accepting AF exchange between the Fe^{3+} ions, Fig. 11, the spins are aligned along the a axis. The ME data at RT and 4 K are $\delta = 0.43$ mm s^{-1} and 0.55 mm s^{-1}, respectively with respect to α-Fe at RT. The quadrupole splitting Δ is 0.49 mm s^{-1} at both temperatures [91P].

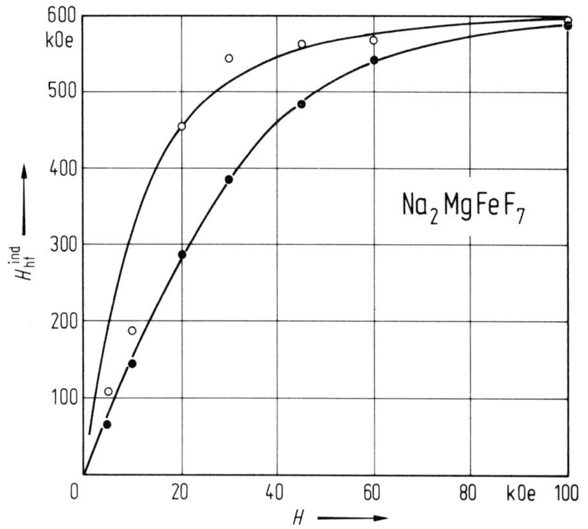

Fig. 11. Na_2MgFeF_7, powder. Induced hyperfine field H_{hf}^{ind} antiparallel to the applied field at 4.2 K. Solid circles represent the major phase in the sample, open circles an unidentified impurity. Curves are fits to a Curie–Weiss model [91P].

Na_2CoAlF_7

Neutron diffraction data are explained as the result of a collinear AF spin arrangement, see Fig. 12 with magnetic moments $p_{FU} = 1.9(1)\mu_B$ in the *ac* plane with components $p_x = 1.4(2)\mu_B$/FU and $p_z = 1.5(2)\mu_B$/FU. The magnetic space group C2′/c′ is suggested [93B].

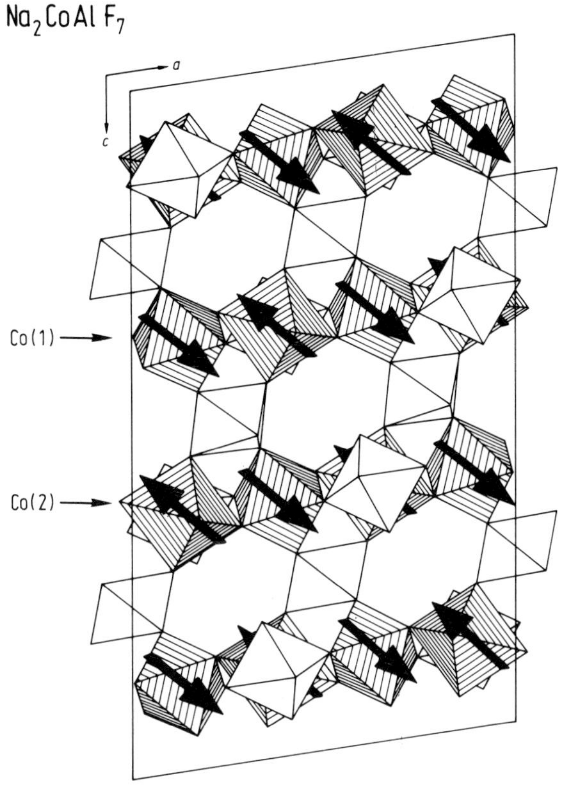

Fig. 12. Na_2CoAlF_7. Magnetic structure at 1.6 K, as derived from neutron diffraction data. The magnetic moments between Co(1) and Co(2) are AF coupled within a chain and ferromagnetically coupled along the [101] direction with respect to two neighboring chains [93B].

Na_2NiAlF_7

The high-temperature susceptibility data can be fitted with an intrachain exchange constant $J/k = -11(2)$ K. Neutron diffraction shows that the magnetic ordering at low temperature is one-dimensional canted AF with a small spontaneous magnetic moment parallel to the *a* axis, Fig. 13. At 4.2 K the ordered magnetic moments of the Ni^{2+} ions have a value $p_{Ni^{2+}} = 1.1\ \mu_B$. The spins lie in the *ac* plane with an angle of about 40° against the direction the *c* axis [73H].

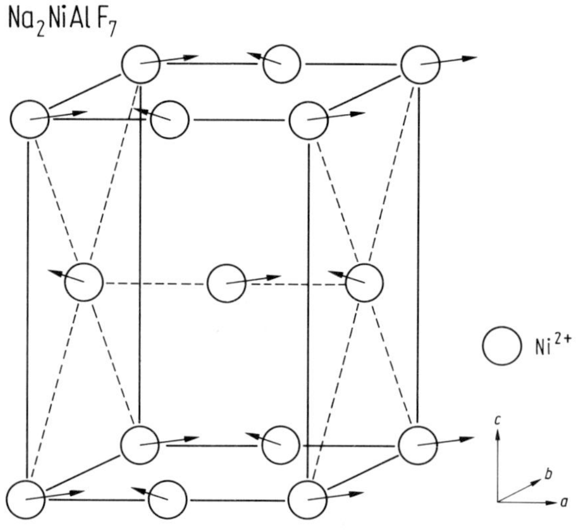

Fig. 13. Na_2NiAlF_7. Canted antiferromagnetic structure [73H].

Na_2NiInF_7

High-temperature susceptibility fitted by an intrachain constant $J/k = -24$ K. At low temperature ordering similar to that in Na_2NiAlF_7 [92F].

Ag_2NiAlF_7; Ag_2NiInF_7

At low temperature AF phases with a small ferromagnetic component [74D].

Na_2CoCrF_7

Paramagnetic with weak ferromagnetism at $T = 4$ K, pointing to AF with some spin canting or to another ferromagnetic contribution [93B].

Na_2NiCrF_7

The low-temperature magnetic structure derived from neutron powder diffraction spectra is shown in Fig. 14. The magnetic mode is Γ_4 with $(+G_x, +F_z)$ for Ni^{2+} and $(-G_y, -F_z)$ for Cr^{3+} components. The frustrated magnetic moments of the magnetic sublattices of Ni^{2+} and Cr^{3+} are quasi orthogonal, the angle between the magnetic moments of each sublattice is 103.9° at 1.5 K and 97.3° at 2.13 K. The magnetic moments in the Ni^{2+} sublattice are mainly coupled AF along [100] with a canting along [001], causing a ferromagnetic component in this direction. Canting angle 7° and 24° at 2.13 K and 1.5 K, respectively [87L]. For the magnitude of the magnetic moments, see Fig. 15.

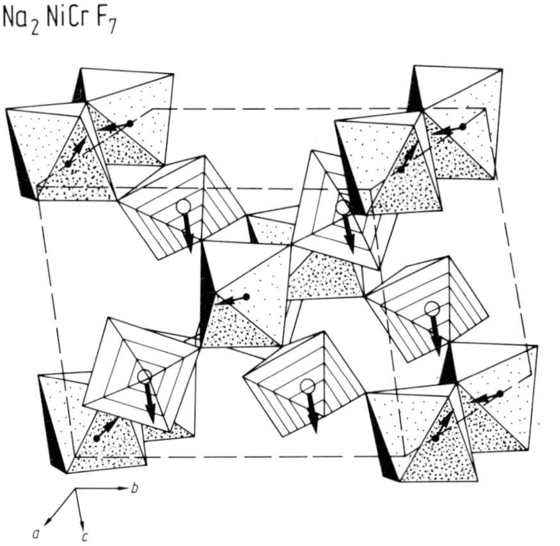

Fig. 14. Na_2NiCrF_7. Perspective view of the magnetic ordering at low temperature. $(Ni^{2+}F_6)^{4-}$ octahedra dot-shaded, the $(Cr^{3+}F_6)^{3-}$ octahedra hatched. The arrows represent the magnetic moments [87L].

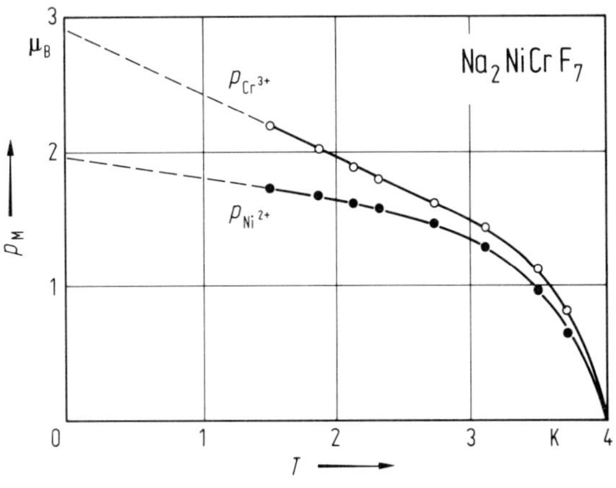

Fig. 15. Na_2NiCrF_7. Temperature dependence of the magnitude of the magnetic moments of the Ni^{2+} and Cr^{3+} ions as derived from neutron diffraction spectra [87L].

Na_2MnFeF_7

Mössbauer data see Table 3.

Table 3. Na_2MnFeF_7. Mössbauer data for the Fe-ions on two different sites (1) and (2) as a function of temperature. Hyperfine field H_{hf}, quadrupole splitting of the external Zeeman lines Δ or 2ε and the isomer shift δ [77P].

T [K]	$H_{hf,1}$ [kOe]	$H_{hf,2}$ [kOe]	δ_1 a)	[mms^{-1}]	δ_2 a)	Δ_1 or $2\varepsilon_1$ b)	[mms^{-1}]	Δ_2 or $2\varepsilon_2$ b)
4.2	554	544	0.444		0.449	0.952		0.376
74	387	374	0.405		0.396	0.707		0.672
83	312	297	0.397		0.387	0.710		0.762
90	255	245	0.385		0.367	0.766		0.861
100				0.362			0.537	
110				0.359			0.536	
158				0.343			0.527	
297				0.263			0.500	

a) Relative to ^{57}Co–Pd.
b) Quadrupole splitting of the external Zeeman lines.

Na_2NiFeF_7

Magnetisation curves Fig. 16, 17, susceptibility Figs. 8, 18. Magnetisation vs. temperature Fig. 19. Ferrimagnetic below about 90 K with the magnetic modes $\Gamma_3(+F_x, +G_z)$ for the Ni^{2+} and $\Gamma_3(-F_x)$ for the Fe^{3+} sublattices [89L], different from earlier results [72H]. The magnitude of the components of the sublattices: $p_{Fe^{3+}} = 4.93(11)\mu_B$. $p_{Ni^{2+}} = 1.36(21)\mu_B$ at 4.2 K and $p_{Fe^{3+}} = 4.34(12)\mu_B$, $p_{Ni^{2+}} = 0.97(22)\mu_B$ at 55 K. At 50 K a significant change in spin canting is observed without a modification of the model [89L]. For Mössbauer data see Table 4 and Figs. 20, 21. The critical law $H_{hf}(T)/H_{hf}(0) = D(1 - T/T_N)^\beta$ is fitted to the data of Fig. 20 with $\beta = 0.30(1)$ and $D = 1.02(3)$ [89L].

Table 4. Na_2NiFeF_7. Mössbauer data. Isomer shift δ, quadrupole splitting of the external Zeeman lines Δ or 2ε and the hyperfine field H_{hf} at various temperatures T [89L].

T [K]	δ [a]) [mm s^{-1}]	Δ or 2ε [b]) [mm s^{-1}]	H_{hf} [kOe]
300	0.444(4)	0.494(4)	–
100	0.529(4)	0.525(4)	–
77	0.54(1)	– 0.26(2)	290
65	0.56(1)	– 0.26(2)	438(2)
55	0.54(1)	– 0.27(2)	483(2)
50	0.55(1)	– 0.27(2)	504(2)
45	0.51(1)	– 0.29(2)	527(2)
40	0.54(1)	– 0.27(2)	544(2)
20	0.55(1)	– 0.26(2)	578(2)
4.2	0.54(1)	– 0.27(2)	584(2)

a) Isomer shift relative to metallic iron at 300 K.
b) Quadrupolar splitting of the external lines of the Zeeman spectrum.

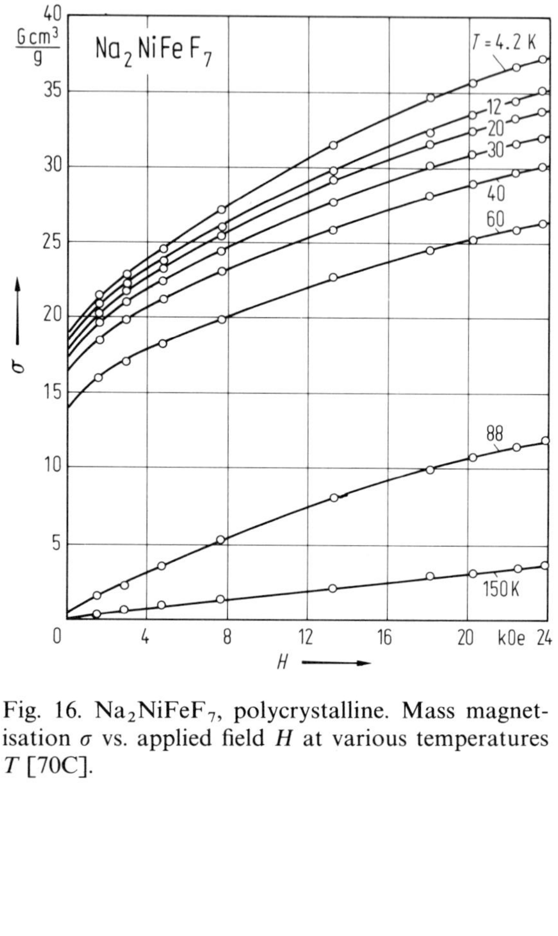

Fig. 16. Na_2NiFeF_7, polycrystalline. Mass magnetisation σ vs. applied field H at various temperatures T [70C].

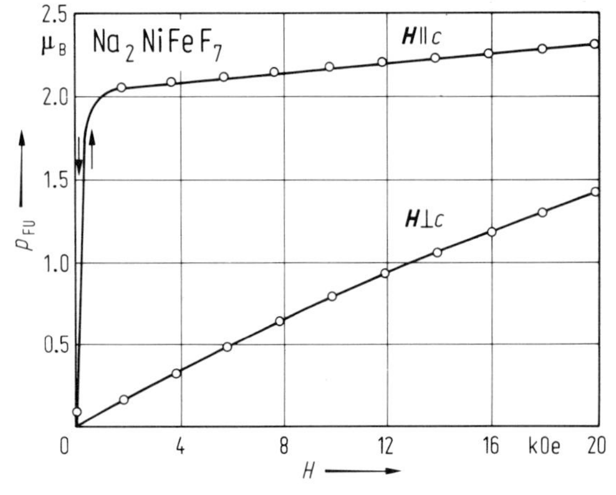

Fig. 17. Na_2NiFeF_7, single crystal. Magnetisation p_{FU} per formula unit vs. applied magnetic field H at 4.2 K parallel or perpendicular to the c axis [76T].

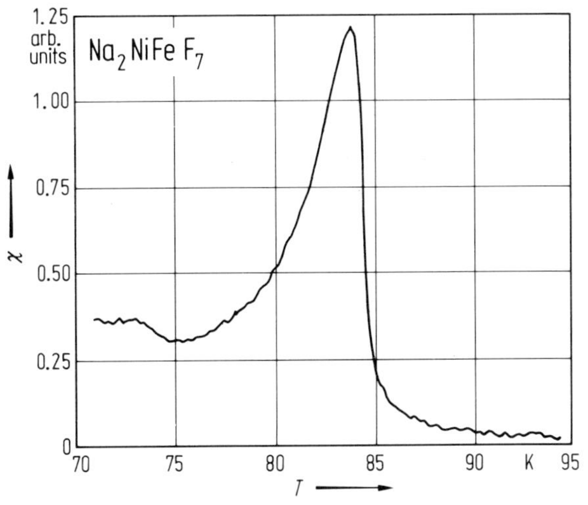

Fig. 18. Na_2NiFeF_7. Temperature dependence of the magnetic susceptibility χ (in arbitrary units) measured in a magnetic field of 10^{-2} Oe [92T].

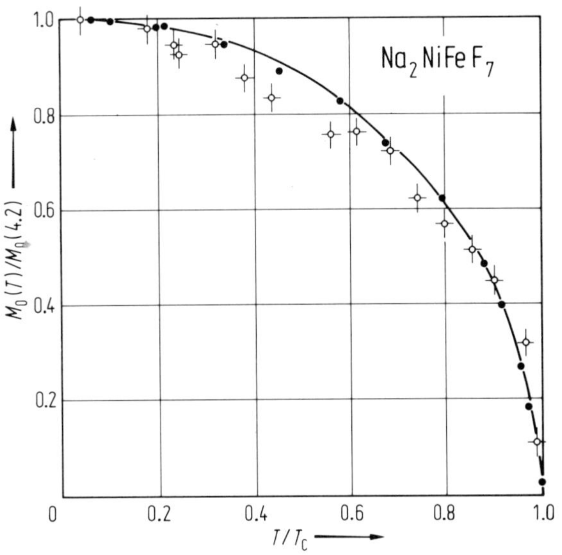

Fig. 19. Na_2NiFeF_7. Reduced spontaneous magnetisation $M_0(T)/M_0(4.2\ K)$ vs. reduced temperature T/T_C ($T_C = 88$ K). Full circles derived from zero field magnetisation measurements, open circles from sublattice magnetisations derived from magnetic contributions to the (110, 002) peak in the neutron powder diffraction pattern. Solid curve: Brillouin function for $S = 3/2$ [72H, 73H].

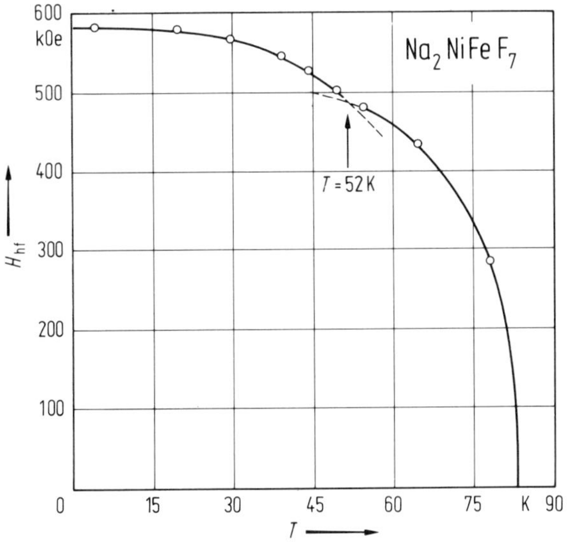

Fig. 20. Na_2NiFeF_7. Magnetic hyperfine field H_{hf} vs. temperature T. The dashed curves are extrapolations of the lower- and higher-temperature parts of the solid curves in the neighbourhood of the spin canting temperature $T = 52$ K [89L].

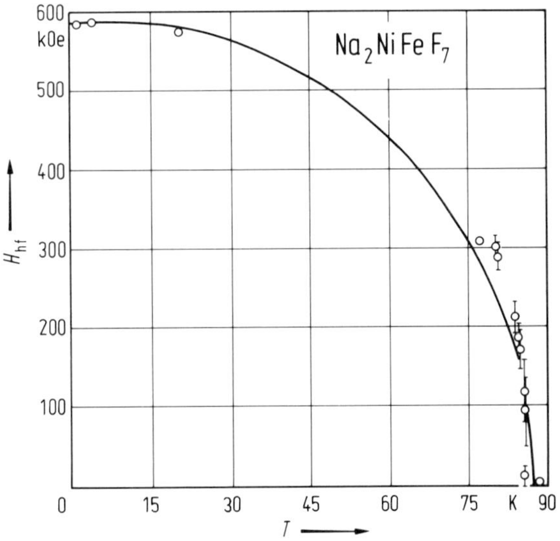

Fig. 21. Na_2NiFeF_7. Similar to Fig. 20. The solid line represents the Brillouin function for $S = 5/2$ [92T].

Ag_2NiCrF_7; Ag_2NiFeF_7

Ferrimagnetic at low temperature [74D].

9.10.4 References for 9.10

69C Chassaing, J.: C. R. Acad. Sci. (Paris) **C268** (1969) 2188.
70H Hänsler, R., Rüdorff, W.: Z. Naturforsch. **25b** (1970) 1306.
72H Heger, G., Viebahn-Hänsler, R.: Solid State Commun. **11** (1972) 1119.
73H Heger, G.: Int. J. Magn. **5** (1973) 119.
74D Dance, J.-M., Grannec, J., Jacobini, Ch., Tressaud, A.: C. R. Acad. Sci. (Paris)' **C279** (1974) 601.
74T Tressaud, A., Dance, J.-M., Portier, J., Hagenmuller, P.: Mater. Res. Bull. **9** (1974) 1219.
76T Tressaud, A. Dance, J.-M., Vlasse, M., Portier, J.: C. R. Acad. Sci. (Paris) **C 282** (1976) 1105
77P Pebler, J., Schmidt, K., Babel, D., Verscharen, W.: Z. Naturforsch. **32**b (1977) 369.
77T Tressaud, A., Dance, J.-M.: Adv. Inorg. Chem. Radiochem., H.J. Emeléns, A. G. Sharpe (eds.) **20** (1977) 133.
78H Haegele, R., Verscharen, W., Babel, D., Dance, J.-M., Tressaud, A.: J. Solid State Chem. **24** (1978) 77.
78V Verscharen, W., Babel, D.: J. Solid State Chem. **24** (1978) 405.
82K1 Knop, O., Cameron, T. S., Jochem, K.: J. Solid State Chem. **43** (1982) 213.
82K2 Koch, J., Hebecker, C., John, H.: Z. Naturforsch. **37b** (1982) 1659.
85K Koch, J., Hebecker, C.: Naturwissenschaften **72** (1985) 431.
87L Laligant, Y., Ferey, G., Heger, G., Pannetier, J.: Z. Anorg. Allgem. Chem. **553** (1987) 163.
88C Courbion, G., Férey, G., Holler, H., Babel, D.: Eur. J. Solid State Inorg Chem. **25** (1988) 435.
88K Kummer, S., Massa, W., Babel, D.: Z. Naturforsch. **43b** (1988) 694.
88R Renaudin, J., Leblanc, M., Ferey, G., de Kozak, A., Samouël, M.: J. Solid State Chem. **73** (1988) 603.
89L Laligant, Y., Calage, Y., Heger, G. Pannetier, J., Ferey, G.: J. Solid State Chem. **78** (1989) 66.
90Y Yakubovich, O. V., Urusov, V. S., Frenzen, G., Massa, W., Babel, D.: Eur. J. Solid State Inorg. Chem. **27** (1990) 467.
91P Pankhjurst, Q. A., Johnson, C. E., Wanklyn, B. M.: J. Magn. Magn. Mater. **97** (1991) 126.
92F Frenzen, G., Massa, W., Babel, D., Ruchaud, N., Grannec, J., Tressaud, A., Hagenmuller, P.: J. Solid State Chem. **98** (1992) 121.
92G Gravereau, P., Boireau, A., Dance, J.-M., Trut, L., Tressaud, A.: Acta Crystallogr. **C 48** (1992) 2108.
92R Ruchaud, N., Grannec, J., Gravereau, P., Nunez, P., Tressaud, A., Massa, W., Frenzen, G., Babel, D.: Z. Anorg. Allgem. Chem. **610** (1992) 67.
92S Schmidt, R. E., Massa, W., Babel, D.: Z. Anorg. Allgem. Chem. **615** (1992) 11.
92T Thompson, G. R., Pankhurst, Q. A., Johnson, C. E.: J. Magn. Magn. Mater. **104–107** (1992) 893.
92W Welsch, M., Babel, D.: Z. Naturforsch. **47**b (1992) 685.
93B Boireau, A., Gravereau, P., Dance, J.-M., Tressaud, A., Hagenmuller, P., Soubeyroux, J. L., Welsch, M., Babel, D.: Mater. Res. Bull. **28** (1993) 27.
93Y Yakubovich, O., Urusov, V., Massa, W., Frenzen, G., Babel, D.: Z. Anorg. Allgem. Chem. **619** (1993) 1903.

9.11 $A_5M_3F_{14}$ compounds (A = Na, K or Ag; M = 3d element or Al)

(For symbols and abbreviations see Section 9.6.0.2)

9.11.1 Crystallographic properties of $A_5M_3F_{14}$ compounds

The compounds $A_5M_3F_{14}$ generally crystallize in the structure of the mineral chiolite $Na_5Al_3F_{14}$ [38B] with tetragonal crystal structure, see Fig. 1. The structure essentially consists of parallel layers of $(MF_6)^{3-}$ octahedra (with M in the centre). The octahedra are connected by their vertices and they can be very much deformed leading to crystals with orthorhombic or monoclinic structure. The M^{3+} ions occupy two different crystallographic sites, 2(a) and 4(d), respectively in the relation 1:2. The layers are perpendicular to the *c* axis and they can be considered as forming chains of octahedra parallel to the *a* and *b* directions. The A^+ atoms are either filling the "holes" in these layers, or are intercalated between the layers. For details on the crystal structures of the various compounds see Table 1.

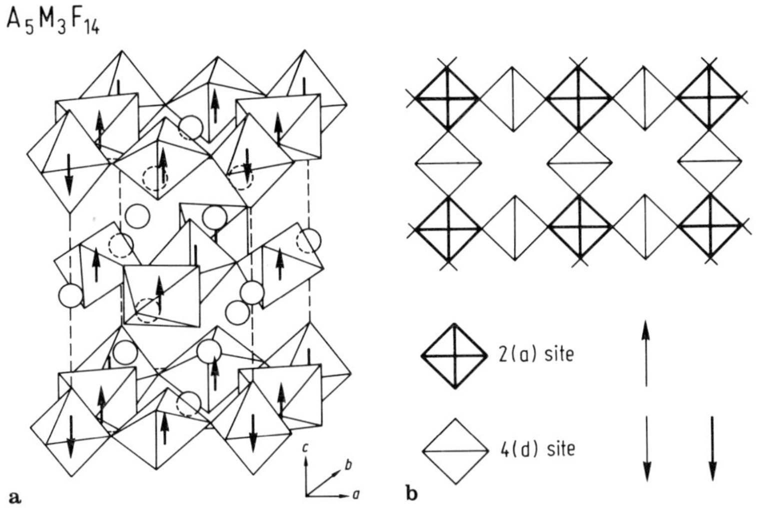

a b

Fig. 1. $A_5M_3F_{14}$. (a) The chiolite structure. The $(MF_6)^{3-}$ octahedra have been drawn schematically. For the case that M^{3+} is a magnetic ion, the relative orientation of the magnetic moments of these ions at the centre of the octahedra is indicated by an arrow. The circles represent the A^+ ions. (b) Idealised representation of the $(M_3F_{14})_n^{5n-}$ layer of the chiolite structure showing the two differently coordinated fluor octahedra. The relative direction of the magnetic moments of their central magnetic M^{3+} ions is also shown [75W, 79D, 85D].

Table 1. $A_5M_3F_{14}$ (A = Na, K, Ag, M = Al, 3d element), chiolite-type compounds with tetragonal crystal structure, unless otherwise stated. Crystallographic data at RT derived from X-ray spectra of a powder (P) or a crystal (C).

Compound	Space group; *Z*	Lattice parameters *a* [Å]	*c* [Å]	P/C	Ref.
$Na_5Al_3F_{14}$[a])	P4/mnc; 2	7.00(1)	10.39(1)	P	38B
	id.	7.0138(8)	10.402(2)	C	81J
$Na_5Ti_3F_{14}$	$P4_2/n$; 2	7.497	10.287	C	76A
$Na_5V_3F_{14}$	—	7.33(1)	10.36(1)	P	72M
$K_5V_3F_{14}$	P4/mnc or P4nc	7.71	11.57	C	76V
α-$Na_5Cr_3F_{14}$[c])	orthorhombic	a = 10.49(1)	10.21(1)	C	75M
	$C222_1$; 4	b = 10.19(1)			
β-$Na_5Cr_3F_{14}$[d])	$P4_22_12$	7.32(2)	10.24(2)	C	75M
$Na_5Mn_3F_{14}$	orthorhombic	a = 12.4455(3)	12.021(2)	C	87W
	$Pna2_1$; 4	b = 7.482(1)			
α-$Na_5Fe_3F_{14}$[c])	monoclinic	a = 12.72(3)	73.23(10)[b])	C	58K
	C2/m, Cm or C2; 4	b = 7.46(2)			
	id.	a = 12.81	73.06[b])	C	76V
		b = 7.37			
β-$Na_5Fe_3F_{14}$[e])	P2/m, Pm or P2; 2	a = 7.48	10.47[b])	C	76V
		b = 7.45			
γ-$Na_5Fe_3F_{14}$[d,f])	$P4_22_12$; 2	7.34	10.38(2)	C	58K
	id.	7.345(7)	10.400(7)	C	76V
$Na_5Fe_2CoF_{14}$		7.33(1)	10.34(1)	P	72M
$Na_5Co_3F_{14}$		7.30(1)	10.21(1)	P	72M
$Ag_5V_3F_{14}$	P4/mnc[e])	7.429	10.934	P	88G
$Ag_5Cr_3F_{14}$	P4/mnc[e])	7.281	11.151	P	88G
$Ag_5Fe_3F_{14}$	P4/mnc[e])	7.408	11.050	P	88G

[a]) The mineral chiolite [38B].
[b]) Monoclinic angle $\beta = 90.0\ (5)°$.
[c]) Slowly cooled from 750···350 °C, low-temperature phase, stable below 622(5) °C.
[d]) Quenched from above 600 °C, high-temperature phase, stable above 640(5) °C.
[e]) Variety with allotropic reversible transformations to the other crystal forms at 622(5) and 640(5) °C.
[f]) Neutron diffraction spectra confirm the magnetic group $P4_22_12$ [75W].

9.11.2 Magnetic properties of $A_5M_3F_{14}$ compounds

The compounds show a simple ferrimagnetic ordering below a Curie temperature T_C. The resulting magnetic structure is shown in Fig. 1. A survey of the magnetic properties is given in Table 2 in which is also referred to the Figs. 2–13. Hyperfine parameters derived from the Mössbauer resonance spectra of γ-$Na_5Fe_3F_{14}$ [76V] and $Ag_5Fe_3F_{14}$ [88G] at RT are characteristic of high spin Fe^{3+} ions in a distorted octahedral environment. Only below the magnetic ordering temperature a difference for the two different iron lattice sites is found, see Table 3. For ferrimagnetic resonance properties and optical properties of γ-$Na_5Fe_3F_{14}$ see Fig. 14 and Figs. 15 and 16, respectively.

Table 2. $A_5M_3F_{14}$ (A = Na, K, Ag; M = 3d element) compounds. Survey of the magnetic properties. Curie-Weiss constants Θ_p, C_m and the effective magnetic moment p_{eff} per M atom in the paramagnetic temperature range. Curie point T_C and extrapolated spontaneous magnetic moment $p_o(0)$ at 0K as derived from magnetisation measurements in the magnetically ordered state.

Compound	Paramagnetic range Θ_p [K]	p_{eff} [μ_B]	C_m [$cm^3\,Kmol^{-1}$]	Ferrimagnetic range T_C [K]	$p_o(0)$ [μ_B/FU]	Fig.	Ref.
$Na_5V_3F_{14}$	−48(5)	2.55(3)	2.44(3)	21.2	1.94(2)	2, 3	72M
$K_5V_3F_{14}$				18(1)	2.15	8–11	77C
60 ··· 190K	−45	2.77	2.86				
190 ··· 500K	−77	2.93	3.22				
α-$Na_5Cr_3F_{14}$	−32(5)	3.81(5)	5.44(5)	18(2)	2.96(5)	4–6	75M
α-$Na_5Fe_3F_{14}$				76(2)	2.7	7	76T
γ-$Na_5Fe_3F_{14}$				80			58K
	−95(5)	5.55(3)	11.55(3)	90(2)	4.98(2)[a]	2, 3	72M
$Na_5Fe_2CoF_{14}$	−100(5)		10.53(3)	94(2)	4.33(2)	2, 3	72M
$Na_5Co_3F_{14}$	−110(5)	4.76(3)	8.50(3)	108(2)	3.10(2)	2, 3	72M
$Ag_5V_3F_{14}$	−34(5)	2.66(3)	2.66(3)	12(2)	1.98(2)	12, 13	88G
$Ag_5Cr_3F_{14}$	−33(5)	3.81(3)	5.45(3)	14(2)	2.96(2)	12, 13	88G
$Ag_5Fe_3F_{14}$	−172(5)	5.56(3)	11.60(3)	80(2)	4.98(2)	12, 13	88G

[a]) From neutron diffraction spectra can be derived a magnetic moment $p_{Fe^{3+}} = 5.2(5)\,\mu_B$ [75W].

Table 3. γ-$Na_5Fe_3F_{14}$; $Ag_5Fe_3F_{14}$. Mössbauer data at various temperatures. Quadrupole interaction Δ, isomer shift δ relative to α-Fe at 300 K, magnetic hyperfine field H_{hf} and linewidth Γ at half height (below T_C the values correspond to the outer lines of the sextuplet).

Compound	T [K]	H_{hf} [kOe]	Δ [mms^{-1}]	δ [mms^{-1}]	Γ [mms^{-1}]	Ref.
γ-$Na_5Fe_3F_{14}$						76V
	293		0.987(5)	0.424(5)	0.31(1)	
4(d) site[a])	4.2	565(5)		0.57(2)	0.33(2)	
2(a) site[a])	4.2	537(5)		0.54(2)	0.30(2)	
$Ag_5Fe_3F_{14}$						88G
	300		0.817(3)	0.451(5)	0.310(5)	
	80		0.857(3)	0.551(5)	0.427(5)	
4(d) site[b])	4.2	584(1)		0.579(5)	0.45(4)	
2(a) site[c])	4.2	547(1)		0.557(5)	0.37(4)	

[a]) Intensity spectrum of 4(d) sites close to twice that of the 2(a) sites.
[b]) 64% of the intensity of the spectrum.
[c]) 36% of the intensity of the spectrum.

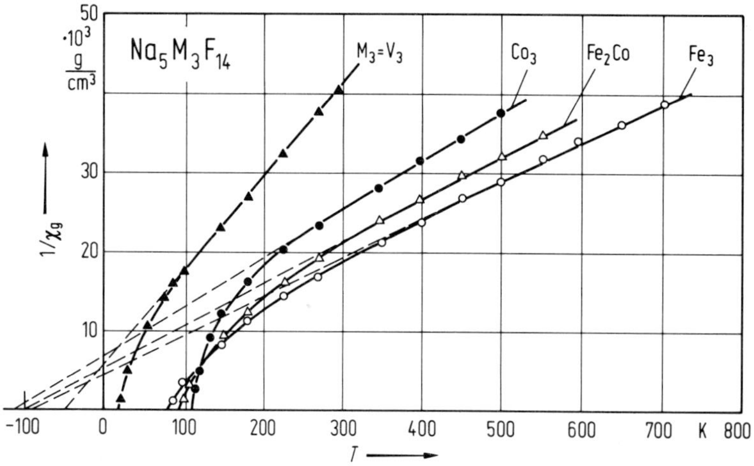

Fig. 2. $Na_5M_3F_{14}$, powder. Inverse magnetic mass susceptibility χ_g^{-1} vs. temperature T [72M].

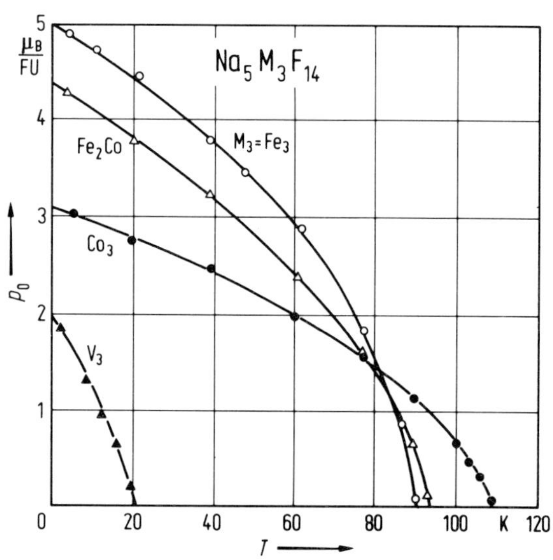

Fig. 3. $Na_5M_3F_{14}$, powder. Spontaneous magnetisation expressed as magnetic moment p_o per formula unit vs. temperature T measured in a magnetic field of 20 kOe [72M, see also 63S].

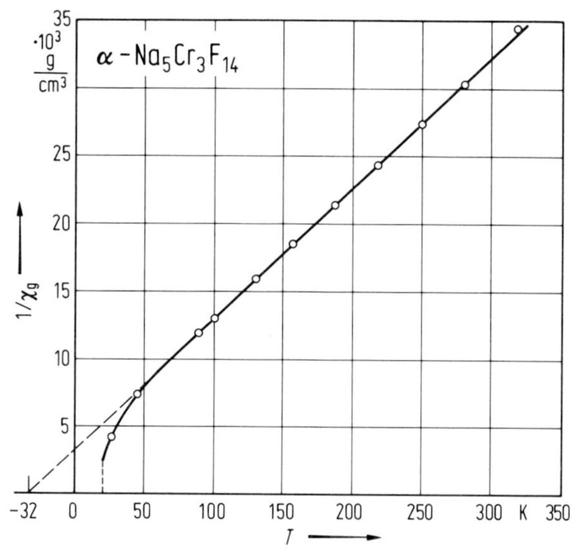

Fig. 4. α-$Na_5Cr_3F_{14}$, powder. Inverse magnetic mass susceptibility χ_g^{-1} vs. temperature T [75M].

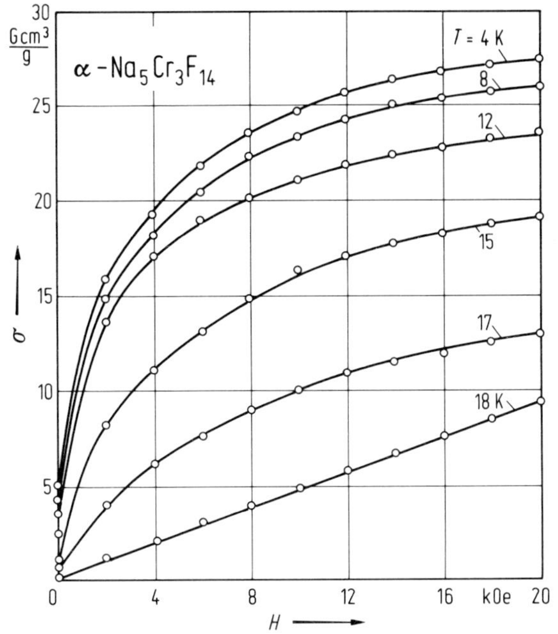

Fig. 5. α-$Na_5Cr_3F_{14}$, powder. Mass magnetisation σ vs. applied field H at various temperatures T. [75M].

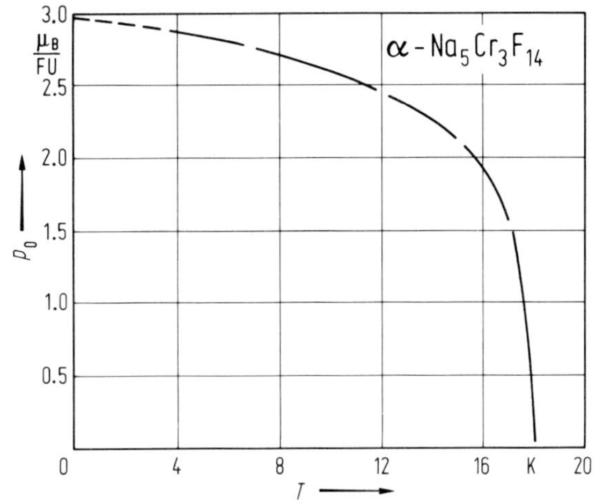

Fig. 6. α-$Na_5Cr_3F_{14}$, powder. Spontaneous magnetisation expressed as magnetic moment per formula unit p_o vs. temperature T as obtained by extrapolation of the curves of Fig. 5 according to σ vs. $1/H$ [75M].

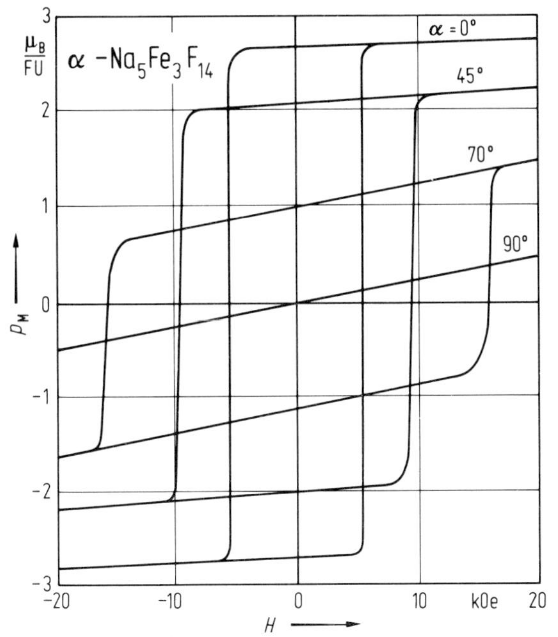

Fig. 7. α-$Na_5Fe_3F_{14}$, single crystal. Hysteresis cycles at 4.2K. The measuring field makes an angle α with the magnetically preferred direction in the crystal, which itself is perpendicular to the b axis [76T].

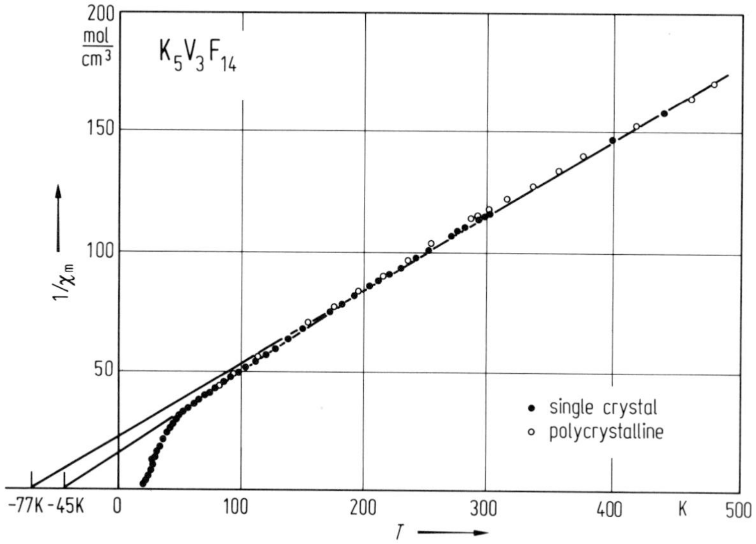

Fig. 8. $K_5V_3F_{14}$. Inverse molar magnetic susceptibility χ_m^{-1} vs. temperature T. Extrapolations for high and low temperature range, see Table 2 [77C].

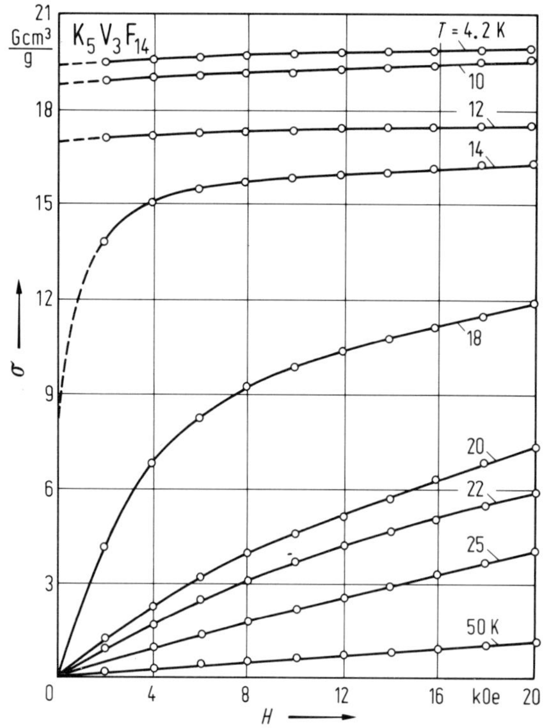

Fig. 9. $K_5V_3F_{14}$, single crystal. Mass magnetisation σ vs. applied magnetic field H at various temperatures. The applied field is oriented in the easy direction of magnetisation (i.e. the c axis) of the crystal [77C].

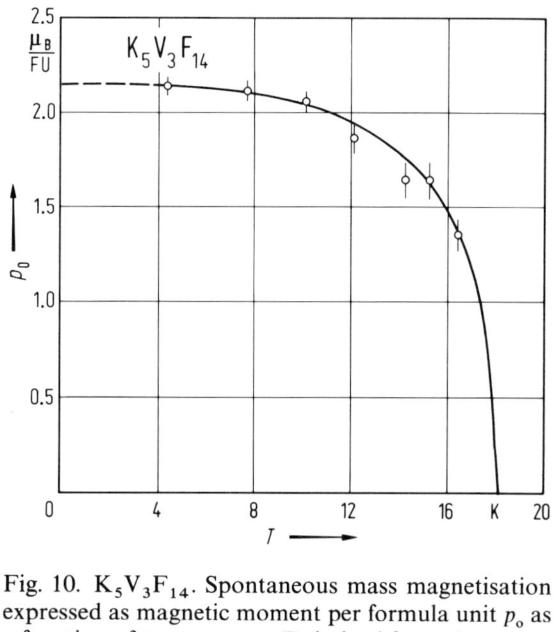

Fig. 10. $K_5V_3F_{14}$. Spontaneous mass magnetisation expressed as magnetic moment per formula unit p_o as a function of temperature, T, derived from the curves of Fig. 9. $p_o(0) = 2.15\mu_B/V^{3+}$ and $T_C = 18(1)$K [77C].

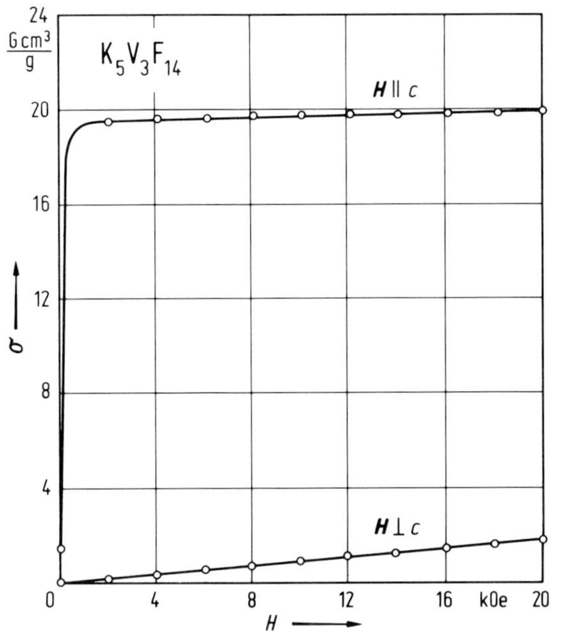

Fig. 11. $K_5V_3F_{14}$, single crystal. Mass magnetisation σ vs. applied fields parallel or perpendicular to the c axis of the crystal at 4.2 K [77C].

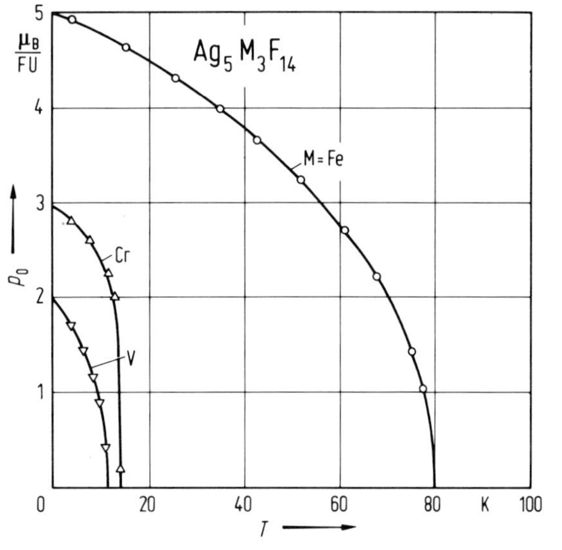

Fig. 13. $Ag_5M_3F_{14}$, M = V, Cr or Fe, annealed powders. Spontaneous mass magnetisation expressed as magnetic moment per formula unit p_o vs. temperature T [88G].

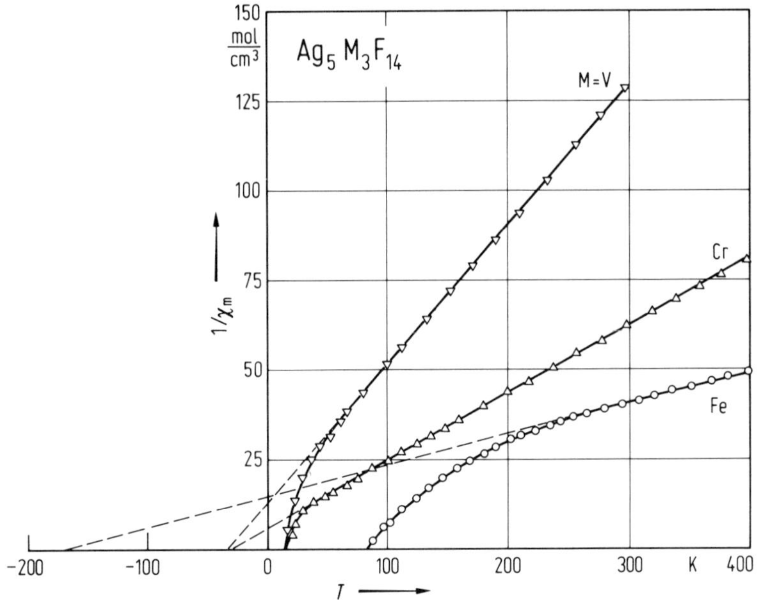

Fig. 12. $Ag_5M_3F_{14}$, M = V, Cr or Fe, annealed powders. Inverse molar magnetic susceptibility χ_m^{-1} vs. temperature T [88G].

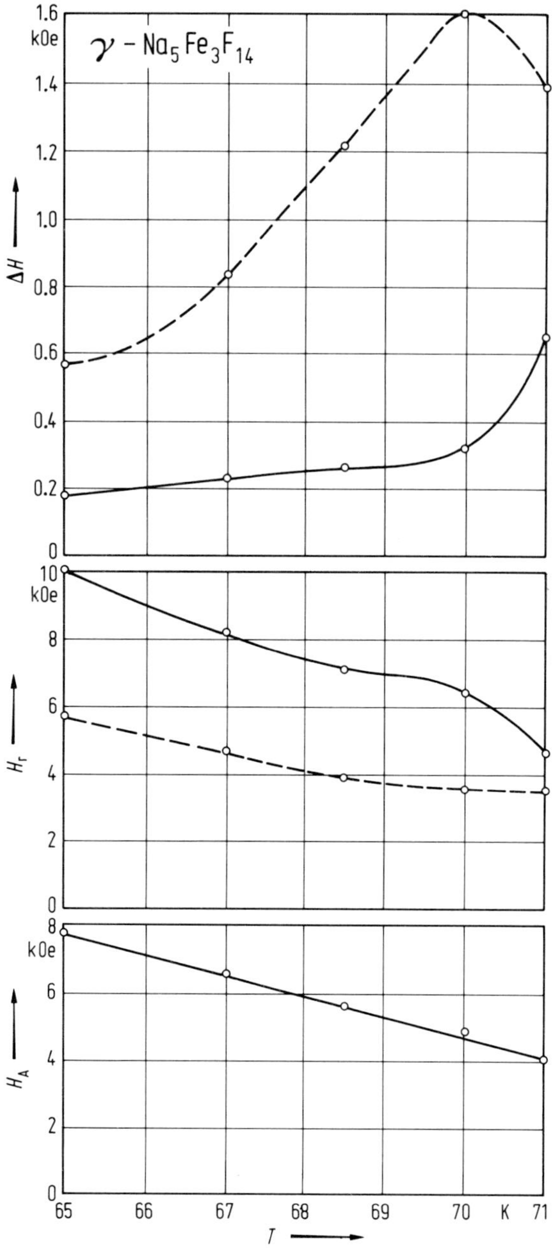

Fig. 14. γ-$Na_5Fe_3F_{14}$, single crystal. Ferrimagnetic resonance absorption at various temperatures with the dc field along a hard magnetic axis. ΔH: linewidth, H_r: resonance field strength for the two resonances occurring at each temperature, H_A: magnetic anisotropy field derived from the two values of H_r at each temperature [63S].

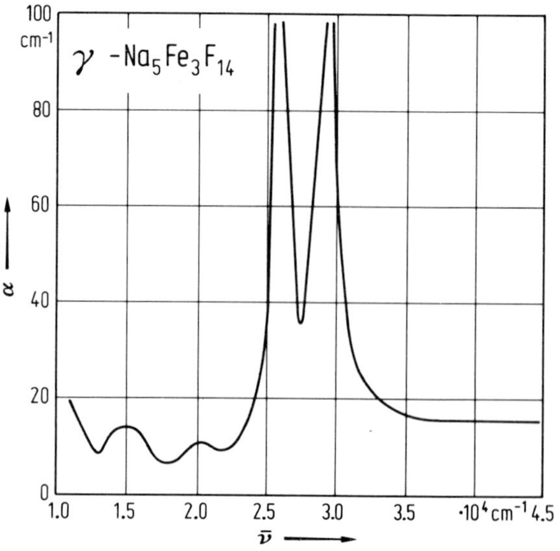

Fig. 15. γ-$Na_5Fe_3F_{14}$, single crystal plate, thickness 0.25 mm. Optical absorption α at RT as dependent on wave number $\bar{\nu}$. Light propagation along a twofold axis of symmetry, i.e. a hard magnetic axis [63S].

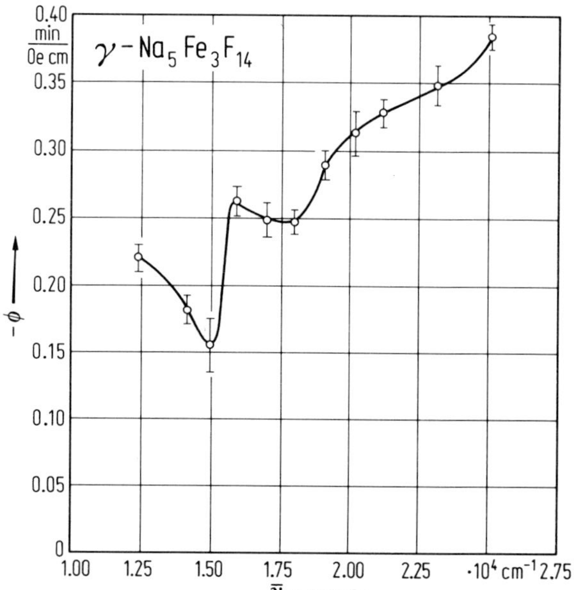

Fig. 16. γ-$Na_5Fe_3F_{14}$, single crystal plate. Verdet constant ϕ vs. wavenumber $\bar{\nu}$ at 300 K. Experimental arrangement as in Fig. 15 [63S].

9.11.3 References for 9.11

38B Brosset, C.: Z. Anorg. Allgem. Chem. **238** (1938) 201.

58K Knox, K., Geller, S.: Phys. Rev. **110** (1958) 771.

63S Spencer, E.G., Berger, S.B., Linares, R.C., Lenzo, P.V.: Phys. Rev. Lett. **10** (1963) 236.

72M McKinzie, H., Dance, J.-M., Tressaud, A., Portier, J., Hagenmuller, P.: Mater. Res. Bull. **7** (1972) 673.

75M Miranday, J.-P., Ferey, G., Jacoboni, C., Dance, J.-M., Tressaud, A., de Pape, R.: Rev. Chim. Minér. **12** (1975) 187.

75W Wintenberger, M., Dance, J.-M., Tressaud, A.: Solid State Commun. **17** (1975) 1355.

76A Avignant, D., Cousson, A., Cousseins, J.-C., Vedrine, A.: Bull. Soc. Chim. Fr. **11–12** (1976) 1662.

76T Tressaud, A., Dance, J.-M., Vlasse, M., Portier, J.: C.R. Acad. Sci. (Paris) **C282** (1976) 1105.

76V Vlasse, M., Menil, F., Moriliere, C., Dance, J.-M., Tressaud, A., Portier, J.: J. Solid State Chem. **17** (1976) 291.

77C Cros, C., Dance, J.-M., Grenier, J.-C., Wanklyn, B.M., Garrard, B.J.: Mater. Res. Bull. **12** (1977) 415.

79D Dance, J.-M., Ravez, J.: C.R. Acad. Sci. (Paris) **C289** (1979) 247.

81J Jacoboni, C., Lable, A., Rousseau, J.J.: J. Solid State Chem. **36** (1981) 297.

85D Dance, J.-M., Tressaud, A.: "Inorganic Solid Fluorides"; Hagenmuller, P. (ed.) London: Academic Press, Inc. 1985, p. 371.

87W Wandner, K.-H., Hoppe, R.: Z. Anorg. Allgem. Chem. **551** (1987) 123.

88G Grannec, J., Yacoubi, A., Tressaud, A., Fournes, L., Hagenmuller, P.: Mater. Res. Bull. **23** (1988) 1611.